Introduction to Graphene-Based Nanomaterials

Beginning with an introduction to carbon-based nanomaterials, their electronic properties, and general concepts in quantum transport, this detailed primer describes the most effective theoretical and computational methods and tools for simulating the electronic structure and transport properties of graphene-based systems.

Transport concepts are clearly presented through simple models, enabling comparison with analytical treatments, and multiscale quantum transport methodologies are introduced and developed in a straightforward way, demonstrating a range of methods for tackling the modeling of defects and impurities in more complex graphene-based materials. The authors also discuss the practical applications of this revolutionary nanomaterial, contemporary challenges in theory and simulation, and long-term perspectives.

Containing numerous problems for solution, real-life examples of current research, and accompanied online by further exercises, solutions, and computational codes, this is the perfect introductory resource for graduate students and researchers in nanoscience and nanotechnology, condensed matter physics, materials science, and nanoelectronics.

Luis E. F. Foa Torres is a Researcher at the Argentine National Council for Science and Technology (CONICET) and an Adjoint Professor at the National University of Córdoba, Argentina, specializing in quantum transport with emphasis on inelastic effects and driven systems.

Stephan Roche is an ICREA Research Professor at the Catalan Institute of Nanoscience and Nanotechnology (ICN2), where he is Head of the Theoretical and Computational Nanoscience Group, focusing on quantum transport phenomena in materials such as graphene.

Jean-Christophe Charlier is a Professor of Physics at the University of Louvain, Belgium, whose interests include condensed matter physics and nanosciences. His main scientific expertise focuses on first-principles computer modeling for investigating carbon-based nanomaterials.

Torres, Roche, and Charlier have written a very attractive book on graphene-based materials that takes a reader or student with no prior exposure to this topic to a level where he or she can carry out research at a high level and work in this area professionally, assuming a standard background of a condensed matter physics graduate student. The material is nicely organized into chapters which can be subdivided into daily learning segments and with problem sets that could be helpful for either formal course presentation or self study. Four appendices with more detailed presentations allow readers to develop the skills needed for using and extending present knowledge to advancing the science of few-layered materials or for developing applications based on these materials. All in all I would expect this to become a popular text for present and future researchers who will be active in the present decade, advancing science and launching technological innovation.

Mildred Dresselhaus, *Massachusetts Institute of Technology*

This book covers the fundamental aspects of graphene, starting from the very beginning. By reading this book, most basic subjects on graphene and some special theoretical methods can be understood at a high level. Starting with the current status of graphene science, the authors proceed to a self-contained description of the electronic structure of graphene, then an especially detailed description of the transport properties of graphene, such as back scattering, Klein tunnelling, quantum dots, Landau levels etc., based on the authors' work. Methods introduced to investigate these subjects range from the tight binding method to *ab initio* calculations, so that the readers can select their preferred method, and the appendices contain useful mathematical explanations for these methods. Thus, without reading the other textbooks, the reader can understand the text. The book should be useful not only for theoretical researchers but also for graduate students and experimental researchers, who will quickly understand the theorists' perspective. This book will be an important basic textbook on the physics of graphene.

R. Saito, *Tohoku University*

Introduction to Graphene-Based Nanomaterials

From Electronic Structure to Quantum Transport

LUIS E. F. FOA TORRES

Argentine National Council for Science and Technology (CONICET) and National University of Córdoba, Argentina

STEPHAN ROCHE

ICREA and Catalan Institute of Nanoscience and Nanotechnology

JEAN-CHRISTOPHE CHARLIER

University of Louvain, Belgium

CAMBRIDGE
UNIVERSITY PRESS

CAMBRIDGE
UNIVERSITY PRESS

University Printing House, Cambridge CB2 8BS, United Kingdom

Published in the United States of America by Cambridge University Press, New York

Cambridge University Press is part of the University of Cambridge.

It furthers the University's mission by disseminating knowledge in the pursuit of education, learning and research at the highest international levels of excellence.

www.cambridge.org
Information on this title: www.cambridge.org/9781107030831

First published 2014

Printed in the United Kingdom by TJ International Ltd. Padstow Cornwall

A catalog record for this publication is available from the British Library

Library of Congress Cataloging-in-Publication Data

Foa Torres, Luis E. F.
Introduction to graphene-based nanomaterials: from electronic structure to quantum transport /
Luis E. F. Foa Torres, Stephan Roche, Jean-Christophe Charlier.
pages   cm
ISBN 978-1-107-03083-1 (hardback)
1. Nanostructured materials.   2. Graphene.   3. Quantum theory.
I. Roche, Stephan.   II. Charlier, Jean-Christophe.   III. Title.
TA418.9.N35F63 2013
620.1'15–dc23   2013022966

Additional resources for this publication at www.cambridge.org/foatorres

Contents

Preface

Once deemed impossible to exist in nature, graphene, *the first truly two-dimensional nanomaterial ever discovered*, has rocketed to stardom since being first isolated in 2004 by Nobel Laureates Konstantin Novoselov and Andre K. Geim of the University of Manchester. Graphene is a single layer of carbon atoms arranged in a flat honeycomb lattice. Researchers in high energy physics, condensed matter physics, chemistry, biology, and engineering, together with funding agencies, and companies from diverse industrial sectors, have all been captivated by graphene and related carbon-based materials such as carbon nanotubes and graphene nanoribbons, owing to their fascinating physical properties, potential applications and market perspectives.

But what makes graphene so interesting? Basically, graphene has redefined the limits of what a material can do: it boasts record thermal conductivity and the highest current density at room temperature ever measured (a million times that of copper!); it is the strongest material known (a hundred times stronger than steel!) yet is highly mechanically flexible; it is the least permeable material known (not even helium atoms can pass through it!); the best transparent conductive film; the thinnest material known; and the list goes on ...

A sheet of graphene can be quickly obtained by exfoliating graphite (the material that the tip of your pencil is made of) using sticky tape. Graphene can readily be observed and characterized using standard laboratory methods, and can be mass-produced either by chemical vapor deposition (CVD) or by epitaxy on silicon carbide substrates. Driven by these intriguing properties, graphene research is blossoming at an unprecedented pace and marks the point of convergence of many fields. However, given this rapid development, there is a scarcity of tutorial material to explain the basics of graphene while describing the state-of-the-art in the field. Such materials are needed to consolidate the graphene research community and foster further progress.

The dearth of up-to-date textbooks on the electronic and transport properties of graphene is especially dramatic: the last major work of reference in this area, written by Riichiro Saito, Gene Dresselhaus, and Mildred Dresselhaus, was published in 1998. Seeking to answer the prayers of many colleagues, who have had to struggle in a nascent field characterized by a huge body of research papers but very little introductory material, we decided to write this book. It is the fruit of our collective research experience, dating from the early days of research on graphene and related materials, up through the past decade, when each of us developed different computational tools and theoretical approaches to understand the complex electronic and transport properties in realistic models of these materials.

We have written *Introduction to Graphene-Based Nanomaterials: From Electronic Structure to Quantum Transport* for everyone doing (or wishing to do) research on the electronic structure and transport properties of graphene-related systems. Assuming basic knowledge of solid state physics, this book offers a detailed introduction to some of the most useful methods for simulating these properties. Furthermore, we have made additional resources (computational codes, a forum, etc.) available to our readers at cambridge.org/foatorres, and at the book website (*introductiontographene.org*), where additional exercises as well as corrections to the book text (which will surely appear) will be posted.

Graphene and related materials pertain to a larger family that encompasses all kinds of two-dimensional materials, from boron nitride lattices, to transition-metal dichalcogenides (MoS_2, WS_2), to the silicon analogue of graphene, silicene, a recently discovered zero-gap semiconductor. Researchers are beginning to explore the third dimension by shuffling two-dimensional materials and by fabricating three-dimensional heterostructures (BN/graphene, BN/MoS_2/graphene, etc.) with unprecedented properties.

Interestingly, low-energy excitations in two-dimensional graphene (and in one-dimensional metallic carbon nanotubes), known as massless Dirac fermions, also develop at the surface of topological insulators (such as $BiSe_2$, Bi_2Te_3, etc.), which are bulk insulators. Topological insulators thus share commonalities with graphene, such as Berry's phase-driven quantum phenomena (Klein tunneling, weak antilocalization,. . .), and exhibit other features such as spin-momentum locking that offer different and ground-breaking perspectives for spintronics. Therefore, we believe that our presentation of the fundamentals of electronic and transport properties in graphene and related materials should prove useful to a growing community of scientists, as they touch on advanced concepts in condensed matter physics, materials science, and nanoscience and nanotechnology.

The book starts with an introduction to the electronic structures and basic concepts in transport in low-dimensional materials, and then proceeds to describe the specific transport phenomena unique to graphene-related materials. Transport concepts are then presented through simple disorder models, which in some cases enable comparison with analytical treatments. Additionally, the development of multiscale quantum transport methodologies (either within the Landauer–Büttiker or Kubo–Greenwood formalisms) is introduced in a straightforward way, showing the various options for tackling defects and impurities in graphene materials with more structural and chemical complexity: from combined *ab initio* with tight-binding models, to transport calculations fully based on first principles. To facilitate reading, the essential technical aspects concerning the formalism of Green functions, as well as transport implementation and order-N transport schemes are described in dedicated appendices.

This book encompasses years of scientific research that has enabled us to establish certain foundations in the field, a work made possible by the efforts of collaborators, including many postdoctoral and doctoral students. We are particularly indebted to Hakim Amara, Rémi Avriller, Blanca Biel, Andrés Botello-Méndez, Victoria Bracamonte, Hernán Calvo,

Jessica Campos-Delgado, Damien Connétable, Alessandro Cresti, Eduardo Cruz-Silva, Aron Cummings, Virginia Dal Lago, Xavier Declerck, Simon Dubois, Nicolas-Guillermo Ferrer Sanchez, Lucas Ingaramo, Gabriela Lacconi, Sylvain Latil, Nicolas Leconte, Aurélien Lherbier, Alejandro Lopez-Bezanilla, Thibaud Louvet, Yann-Michel Niquet, Daijiro Nozaki, César Núñez, Hanako Okuno, Frank Ortmann, Andreï Palnichenko, Pablo Pérez-Piskunow, Juan Pablo Ramos, Claudia Rocha, Xavier Rocquefelte, Luis Rosales, Haldun Sevincli, Eric Suárez Morell, Florent Tournus, François Triozon, Silvia Urreta, Gregory Van Lier, Dinh Van Tuan, François Varchon, Wu Li, Zeila Zanolli, and Bing Zheng.

We would also like to express our sincere gratitude to the following inspiring individuals with whom we have worked over the past decade: Pulickel Ajayan, Tsuneya Ando, Marcelo Apel, Adrian Bachtold, Carlos Balseiro, Florian Banhart, Robert Baptist, Christophe Bichara, Xavier Blase, Roberto Car, Antonio Castro-Neto, Mairbek Chshiev, Gianaurelio Cuniberti, Silvano De Franceschi, Hongjie Dai, Alessandro De Vita, Millie and Gene Dresselhaus, François Ducastelle, Reinhold Egger, Peter Eklund, Morinobu Endo, Walter Escoffier, Chris Ewels, Andrea Ferrari, Albert Fert, Takeo Fujiwara, Xavier Gonze, Andrea Latgé, Caio Lewenkopf, Annick Loiseau, Jose-Maria Gómez Rodriguez, Nicole Grobert, Paco Guinea, Luc Henrard, Eduardo Hernández, Jean-Paul Issi, Ado Jorio, Philip Kim, Jani Kotakoski, Vladimir Kravtsov, Philippe Lambin, Sergio Makler, Ernesto Medina, Vincent Meunier, Natalio Mingo, Costas Moulopoulos, Joel Moser, Yann-Michel Niquet, Kentaro Nomura, Kostya Novoselov, Pablo Ordejón, Pedro Orellana, Mónica Pacheco, Horacio Pastawski, Marcos Pimenta, Bertrand Raquet, Gian-Marco Rignanese, Angel Rubio, Riichiro Saito, Bobby Sumpter, Mauricio and Humberto Terrones, Gonzalo Usaj, and Sergio Valenzuela.

We thank our home institutions for supporting our research, as well as the Alexander von Humboldt Foundation (SR and LEFFT) and the Abdus Salam International Centre for Theoretical Physics (LEFFT). Finally, we are indebted to our respective wives (Sandra Rieger, Encarni Carrasco Perea and Mireille Toth-Budai) and our children (Hector and Gabriel Roche, and Ilona, Elise, and Mathilde Charlier) for their warm enthusiasm and continuous support during all these years of time-consuming work.

We hope that you find this book to be a useful companion for starting in this field and perhaps even for your day-to-day research. We recommend that you start by reading Chapter 1 and then follow the advice in the *Guide to the book* (Section 1.3).

And we wish you an exciting journey in Flatland!...

Luis E. F. Foà Torres, Stephan Roche, and Jean-Christophe Charlier.

1 Introduction to carbon-based nanostructures

Carbon is a truly unique chemical element. It can form a broad variety of architectures in all dimensions, both at the macroscopic and nanoscopic scales. During the last 20+ years, brave new forms of carbon have been unveiled. The family of carbon-based materials now extends from C_{60} to carbon nanotubes, and from old diamond and graphite to graphene. The properties of the new members of this carbon family are so impressive that they may even redefine our era. This chapter provides a brief overview of these carbon structures.

1.1 Carbon structures and hybridizations

Carbon is one of the most versatile elements in the periodic table in terms of the number of compounds it may create, mainly due to the types of bonds it may form (single, double, and triple bonds) and the number of different atoms it can join in bonding. When we look at its ground state (lowest energy) electronic configuration, $1s^2 2s^2 2p^2$, carbon is found to possess two core electrons ($1s$) that are not available for chemical bonding and four valence electrons ($2s$ and $2p$) that can participate in bond formation (Fig. 1.1(a)). Since two unpaired $2p$ electrons are present, carbon should normally form only two bonds from its ground state.

However, carbon should maximize the number of bonds formed, since chemical bond formation will induce a decrease of the system energy. Consequently, carbon will rearrange the configuration of these valence electrons. Such a rearrangement process is called *hybridization*, where only $2s$ and $2p$ electrons are affected. Indeed, one $2s$ electron will be promoted into an empty $2p$ orbital, thus forming an excited state (Fig. 1.1(b)). Carbon will thus hybridize from this excited state, being able to form at most four bonds.

One possible hybridization scheme consists in mixing the four atomic orbitals (one $2s$ orbital + three $2p$ orbitals), leading to the formation of four sp^3 hybrid orbitals, each filled with only one electron (Fig. 1.1(c)). In order to minimize repulsion, these four hybrid orbitals optimize their position in space, leading to a tetrahedral geometry where four σ bonds are formed with the carbon neighbors, each at an angle of 109.5° to each other. Methane (CH_4) is the typical molecule that satisfies this specific bonding arrangement. Diamond is the three-dimensional carbon allotropic form where the atoms are arranged in a variation of the face-centered cubic crystal structure, called a diamond lattice (Fig. 1.2(a)). In diamond, all carbon atoms are in the sp^3 hybridization and are

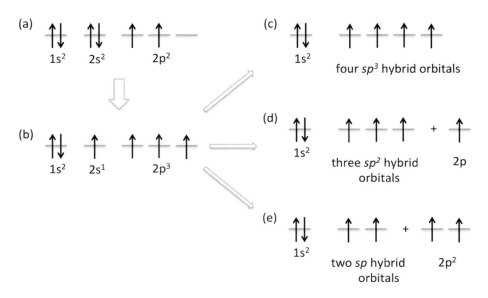

Figure 1.1 Electronic configurations of carbon: (a) ground state; (b) excited state; (c) sp^3 hybridization; (d) sp^2 hybridization; and (e) sp hybridization.

connected by σ bonds (due to the overlapping between two hybrid orbitals, each containing one electron) to four nearest neighbors with a bond length of 1.56Å. Diamond (from the ancient Greek $\alpha\delta\alpha\mu\alpha\sigma$ – *adámas* "unbreakable") is renowned as a material with extreme mechanical properties originating from the strong sp^3 covalent bonding between its atoms. In particular, diamond exhibits one of the highest hardness values, has an extremely high thermal conductivity, is an electrical insulator with a band gap of ~5.5 eV, and is transparent to visible light (Hemstreet, Fong & Cohen, 1970).

Another possible hybridization scheme consists in mixing three atomic orbitals among the four (one $2s$ orbital + two $2p$ orbitals), leading to the formation of three sp^2 hybrid orbitals, each filled with only one electron (Fig. 1.1(d)). Again, the three sp^2 hybrid orbitals will arrange themselves in order to be as far apart as possible, leading to a trigonal planar geometry where the angle between each orbital is 120°. The remaining p-type orbital will not mix and will be perpendicular to this plane. In such a configuration, the three sp^2 hybrid orbitals will form σ bonds with the three nearest neighbors and the side-by-side overlap of the unmixed pure p orbitals will form π bonds between the carbon atoms, accounting for the carbon–carbon double bond. Ethylene (C_2H_4) and aromatic molecules like benzene (C_6H_6) are typical examples of sp^2 hybridization.

Graphite is a three-dimensional crystal made of stacked layers consisting of sp^2 hybridized carbon atoms (Fig. 1.2(b)); each carbon atom is connected to another three making an angle of 120° with a bond length of 1.42Å. This anisotropic structure clearly illustrates the presence of strong σ covalent bonds between carbon atoms in the plane, while the π bonds provide the weak interaction between adjacent layers in the graphitic structure. Graphite (from the ancient Greek $\gamma\rho\alpha\phi\omega$ – *graphó* "to write") is well known

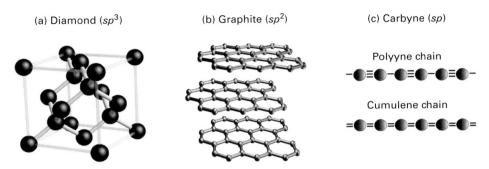

Figure 1.2 Carbon structures exhibiting different hybridizations: (a) diamond (sp^3); (b) graphite (sp^2); and (c) carbyne (sp).

for its use in pencils because of its ability to mark surfaces as a writing material, due to nearly perfect cleavage between basal planes related to the anisotropy of bonding. Under standard conditions (ordinary temperatures and pressures), the stable form of carbon is *graphite*. Unlike diamond, graphite is a famous lubricant, an electrical (semi-metal) and thermal conductor, and reflects visible light. *Natural* graphite occurs in two crystal structures: Bernal (hexagonal) (Bernal, 1924) and rhombohedral (Lipson & Stokes, 1942) structures that are characterized by different stackings of the basal planes, ..*ABABAB*.. and ..*ABCABC*.. respectively. The hexagonal and rhombohedal structures belong to the $P6_3/mmc$ (D_{6h}^4) and $R\bar{3}m$ (D_{3d}^5) space groups, respectively. Samples usually contain no more than 5–15% rhombohedral structure intermixed with Bernal form and sometimes disordered graphite (Lipson & Stokes, 1942). These disordered graphitic forms, such as pregraphitic carbon or turbostratic graphite, are mainly composed of randomly oriented basal carbon sheets. However, pure graphite crystals can be found naturally, and can also be artificially synthesized by thermolytic processes, such as the production of highly oriented pyrolytic graphite (HOPG) (Moore, 1974).

The last possible hybridization consists in mixing two atomic orbitals among the four (one $2s$ orbital + one $2p$ orbital), leading to the formation of two sp hybrid orbitals, each filled with only one electron (Fig. 1.1(e)). The geometry which results is linear with an angle between the sp orbitals of 180°. The two remaining p-type orbitals which are not mixed are perpendicular to each other. In such a configuration, the two sp hybrid orbitals will form σ bonds with the two nearest neighbors and the side-by-side overlap of the two unmixed pure p orbitals will form π bonds between the carbon atoms, accounting for the carbon–carbon triple bond (which is thus composed of one σ bond and two π bonds). Acetylene ($H - C \equiv C - H$) is the typical linear molecule that satisfies this specific bonding arrangement. Carbon also has the ability to form one-dimensional chains, called carbynes (Fig. 1.2(c)), that are traditionally classified as *cumulene* (monoatomic chains with double bonds, .. $= C = C = $ ..) or *polyyne* (dimerized chains with alternating single and triple bonds, .. $- C \equiv C - $ ..). While sp^2 and sp^3 carbon-based structures have been widely characterized, the synthesis of carbynes has been a challenge for decades due to the high reactivity of chain ends and to a strong tendency to interchain crosslinking (Heimann, Evsyukov & Kavan, 1999).

Linear carbon chains consisting of a few tens of atoms were first synthesized via chemical methods (Cataldo, 2005) by stabilizing the chain ends with nonreactive terminal groups (Kavan & Kastner, 1994, Lagow *et al.*, 1995). However, these systems consist of a mixture of carbon and other chemical elements, and the synthesis of carbynes in a pure carbon environment has only recently been achieved via supersonic cluster beam deposition (Ravagnan *et al.*, 2002, Ravagnan *et al.*, 2007) and via electronic irradiation of a single graphite basal plane (graphene) inside a transmission electron microscope (Meyer *et al.*, 2008a, Jin *et al.*, 2009, Chuvilin *et al.*, 2009).

1.2   Carbon nanostructures

Carbon nanomaterials also reveal a rich polymorphism of various allotropes exhibiting each possible dimensionality: fullerene molecule (0D), nanotubes (1D), graphite platelets and graphene ribbons (2D), and nano-diamond (3D) are selected examples (Terrones *et al.*, 2010). Because of this extraordinary versatility of nanomaterials exhibiting different physical and chemical properties, carbon nanostructures are playing an important role in nanoscience and nanotechnology.

Carbon nanoscience started with the discovery of C_{60} Buckminsterfullerene (Kroto *et al.*, 1985). This cage-like molecule of 7Å in diameter contains 60 carbons atoms laid out on a sphere (Fig. 1.3(a)). The structure of the C_{60} Buckminsterfullerene consists in a truncated icosahedron with 60 vertices and 32 faces (20 hexagons and 12 pentagons where no pentagons share a vertex) with a carbon atom at the vertices of each polygon and a bond along each polygon edge (Fig. 1.3(a)). Each carbon atom in the structure is bonded covalently with three others ($sp^{2+\delta}$ hybridization; δ is due to the curvature) with an average bond length of 1.46Å within the five-member rings (single bond) and 1.4Å for the bond connecting five-member rings (the bond fusing six-member rings). The number of carbon atoms in each fullerene cage can vary. Indeed, fullerene molecules are generally represented by the formula C_n, where n denotes the number of carbon atoms present in the cage. Anyway, the C_{60} nano-soccer ball (or *buckyball*) is the most stable and well characterized member of the fullerene family. The name of these C_n molecules was derived from the name of the noted inventor and architect Buckminster Fuller, since C_n resembles his trademark geodesic domes. The C_{60} molecule is still dominating fullerene research and stimulating the creativity and imagination of scientists, and has paved the way for a whole new chemistry and physics of nanocarbons (Dresselhaus, Dresselhaus & Eklund, 1996).

Soon after, in 1988, graphitic onions (of which the first electron microscope images were reported by Sumio Iijima in 1980 (Harris, 1999)) were suggested to be nested icosahedral fullerenes (C60@C240@C540@C960...) (Kroto & McKay, 1988) containing only pentagonal and hexagonal carbon rings (Fig. 1.3(b)). In 1992, the reconstruction of polyhedral graphitic particles into almost spherical carbon onions (nested giant fullerenes) was demonstrated by Daniel Ugarte (Ugarte, 1992) using high-energy electron irradiation inside a high-resolution transmission electron microscope (HRTEM). By analogy, the formation of C_{60} has also been very recently observed *in situ* by creating

0-D	1-D	2-D	3-D

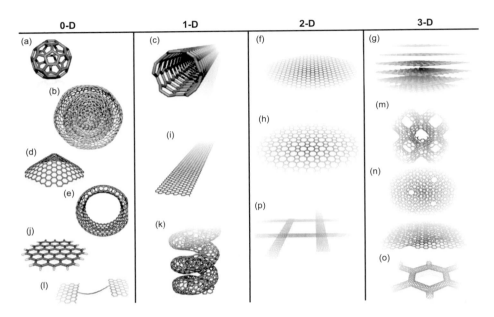

Figure 1.3 Atomistic models of various sp^2-like hybridized carbon nanostructures exhibiting different dimensionalities, 0D, 1D, 2D and 3D: (a) C$_{60}$: Buckminsterfullerene; (b) nested giant fullerenes or graphitic onions; (c) carbon nanotube; (d) nanocones or nanohorns; (e) nanotoroids; (f) graphene surface; (g) 3D graphite crystal; (h) Haeckelite surface; (i) graphene nanoribbons; (j) graphene clusters; (k) helicoidal carbon nanotube; (l) short carbon chains; (m) 3D Schwarzite crystals; (n) carbon nanofoams (interconnected graphene surfaces with channels); (o) 3D nanotube networks, and (p) nanoribbon 2D networks. Reproduced from (Terrones *et al.*, 2010).

local defects in graphene upon electron irradiation in a HRTEM (Chuvilin *et al.*, 2010). These carbon onions are quasi-spherical nanoparticles consisting of fullerene-like carbon layers enclosed by concentric graphitic shells, thus exhibiting electronic and mechanical properties different from any other carbon nanostructures due to their highly symmetric structure.

In 1976, the "ultimate" carbon fibers (later known as a multiwall carbon nanotube), produced by a modified chemical vapour deposition (CVD) method usually used to produce conventional carbon fibers, were observed using TEM (Oberlin, Endo & Koyama, 1976). However, the emergence of carbon nanotubes (CNTs) really came in 1991 after the C$_{60}$ discovery. Indeed, "graphite microtubules," multiwall nanotubes (MWNTs) produced via an arc discharge between two graphite electrodes in an inert atmosphere (same method for producing fullerenes), were first characterized using HRTEM (Iijima, 1991), thus confirming that their atomic structures consisted of nested graphene nanotubes terminated by fullerene-like caps. A couple of years later, in 1993, single-wall carbon nanotubes (SWNTs, Fig. 1.3(c)) were synthesized using the same carbon arc technique in conjunction with metal catalysts (Iijima & Ichihashi, 1993, Bethune *et al.*, 1993). CNTs are allotropic forms of carbon characterized by a long and hollow cylindrical-shaped nanostructure with a length-to-diameter ratio that may reach 10^8

(Zheng *et al.*, 2004) which is significantly larger than any other one-dimensional material. Carbon nanotubes are frequently considered as members of the fullerene family, since their ends may be capped with a buckyball hemisphere. The cylinder walls are formed by one-atom-thick sheets of carbon rolled up at specific and discrete *chiral* angles. Both the nanotube diameter and this rolling angle lead to specific properties; for example, a SWNT may behave as a metal or a semiconductor depending on its geometry (Saito, Dresselhaus & Dresselhaus, 1998), as described in detail in the next chapter. Due to long-range weak interactions (van der Waals and π-stacking), individual nanotubes naturally align into *ropes* or *bundles* (Thess *et al.*, 1996). These carbon nanotubes exhibit unusual properties, which are extremely valuable for nanotechnology, electronics, mechanics, optics and other fields of materials science. In particular, owing to their extraordinary mechanical properties, electrical and thermal conductivity, carbon nanotubes find applications as additives (primarily carbon fiber) in composite materials, as for instance in baseball bats, golf clubs, or car parts (Dresselhaus, Dresselhaus & Avouris, 2001).

After these consecutive discoveries of the fullerenes and carbon nanotubes, other graphitic-like nanostructures were successfully produced, observed and accurately characterized using various experimental techniques. The topologies associated with these new carbon nanostructures include nanocones (Fig. 1.3(d)) (Krishnan *et al.*, 1997), nanopeapods (Smith, Monthioux & Luzzi, 1998), nanohorns (Fig. 1.3(d)) (Iijima *et al.*, 1999), and carbon rings or toroids (Fig. 1.3(e)) (Liu *et al.*, 1997).

The fundamental building block in all these carbon nanostructures (except for sp^3 nanodiamond) relies on the theoretical concept of the two-dimensional crystalline allotrope of carbon, called *graphene* (Fig. 1.3(f)). Indeed, graphene is the name given to the ideally-perfect infinite one-atom-thick planar sheet of sp^2-bonded carbon atoms, densely packed in a honeycomb crystal lattice (Boehm *et al.*, 1962*a* & *b*). This ideal two-dimensional solid has thus been widely employed as a useful theoretical concept to describe the properties of many carbon-based materials, including graphite (where a large number of graphene sheets are stacked, see Fig. 1.3(g)) (Wallace, 1947), nanotubes (where graphene sheets are rolled up into nanometer-sized cylinders, see Fig. 1.3(c)), large fullerenes (where graphene sheets, according to Euler's theorem, contain at least 12 pentagons displaying a spherical shape, see Fig. 1.3(a–b)), and ribbons (where graphene is cut into strips, see Fig. 1.3(i)) (Li *et al.*, 2008). Actually, planar graphene itself was presumed not to exist in the free state, being unstable with respect to the formation of curved structures, such as soot, fullerenes, and nanotubes. However, in 2004, graphene samples were synthesized either by mechanical exfoliation (repeated peeling or micromechanical cleavage, known as the "scotch tape method") of bulk graphite (highly oriented pyrolytic graphite) (Novoselov *et al.*, 2004, Novoselov *et al.*, 2005*a*) or by epitaxial growth through thermal decomposition of SiC (Berger *et al.*, 2006). The relatively easy production of graphene using the scotch tape method and the transfer facility of a single atomic layer of carbon from the *c*-face of graphite to a substrate suitable for the measurement of its electrical properties have led to a renewed interest in what was considered to be a prototypical, yet theoretical, two-dimensional system. Graphene displays, indeed, unusual electronic properties arising from confinement of electrons in two dimensions and peculiar geometrical symmetries. Indeed, old theoretical studies of graphene (Wallace, 1947) reveal that the specific linear electronic band dispersion near the Brillouin zone corners (Dirac point) gives rise to electrons and holes that propagate as if they were massless Dirac fermions, with a velocity of the order of a few hundredths of the velocity of light. Charge excitations close to the Fermi level can thus be formally described as massless relativistic particles obeying a Dirac equation, whereas a new degree of freedom reflecting inherent symmetries (sublattice degeneracy) appears in the electronic states: the *pseudospin*. Because of the resulting pseudospin symmetry, electronic states turn out to be particularly insensitive to external sources of elastic disorder (topological and electrostatic defects) and, as a result, charge mobilities in graphene layers as large as 10^5 cm^{2}V^{-1}s^{-1} have been reported close to the Dirac point (Novoselov *et al.*, 2004). In addition, in suspended graphene, the minimum conductivity at the Dirac point approaches a universal (geometry independent) value of $4e^2/h$ at low temperature (Du *et al.*, 2008). Low temperature electron mobility approaching 2×10^5 cm^{2}V^{-1}s^{-1} has been measured for carrier density below 5×10^9 cm^{-2}. Such values cannot be attained in conventional semiconductors such as silicon or germanium. In addition, graphene has been demonstrated to exhibit anomalous quantum transport properties such as an integer quantum Hall effect (Novoselov *et al.*, 2005b, Zhang *et al.*, 2005), and also one of the most exotic and counterintuitive consequences of quantum electrodynamics: the unimpeded penetration of relativistic particles through high and wide potential barriers, known as the Klein paradox (Katsnelson, Novoselov & Geim, 2006). These discoveries have stirred a lot of interest in the scientific community as well as in the international media. The excitement behind this discovery has two main driving forces: basic science and technological implications (Geim & Novoselov, 2007). Because of its high electronic mobility, structural flexibility, and capability of being tuned from *p*-type to *n*-type doping by the application of a gate voltage, graphene is considered a potential breakthrough in terms of carbon-based nanoelectronics.

All these outstanding properties of the graphene sheet have heavily stimulated the discovery of new closely-related planar carbon-based nanostructures with sp^2 hybridization, such as bilayer graphene, trilayer graphene, few-layer graphene, and graphene nanoribbons; these that have subsequently emerged, each with novel and unusual properties that are different from both graphene and graphite. Whenever these structures exhibit ..*ABABAB*.. or ..*ABCABC*.. stackings they are considered as graphitic stacks. In fact, this distinction is made because it has been demonstrated that the properties of graphene can be recovered in systems with several sp^2-hybridized carbon layers when stacking disorder is introduced.

Theoretical works have also suggested the possibility of stable flat sp^2-hybridized carbon sheets containing pentagons, heptagons and hexagons, termed pentaheptite (2D sheets containing heptagons and pentagons only) (Crespi *et al.*, 1996) or Haeckelites (2D crystals containing pentagons, heptagons and/or hexagons, see Fig. 1.3(h)) (Terrones *et al.*, 2000). These planar structures are intrinsically metallic and could exist in damaged or irradiated graphene. However, further experiments are needed in order to both produce and identify them successfully.

When infinite perfect graphene crystals become finite, borders and boundaries appear, implying the presence of carbon atoms that exhibit a coordination below three at the edges. Among these graphene-based nanostructures are nanoribbons (Fig. 1.3(i)) and nanoclusters (Fig. 1.3(j)). In general, a graphene nanoribbon (GNR) is defined as a 1D sp^2-hybridized carbon crystal with boundaries which possesses a large aspect ratio (Fig. 1.3(i)). Edge terminations could be armchair, or zigzag, or even a combination of both. The graphene cluster concept arises when the dimensionality is lost and no periodicity is present (Fig. 1.3(j)). Finally, long carbon chains with alternating single–triple or double bonds (Fig. 1.2(c)) are also considered as a 1D nanosystem as already briefly described in the previous section.

Finally, Schwarzites are hypothetical graphitic (sp^2 hybridization) three-dimensional crystals obtained by embedding non-hexagonal carbon rings (Fig. 1.3(m)), thus spanning two different space groups, in which the most symmetrical cases belong to cubic Bravais lattices (Terrones & Terrones, 2003). These 3D carbon-based nanostructures can be visualized as nanoporous carbon (Fig. 1.3(n)), exhibiting nanochannels. From a theoretical point of view, these nanoporous carbon materials have been suggested to have outstanding performance in the storage of hydrogen due to their large surface area (Kowalczyk *et al.*, 2007). Another type of 3D array of nanocarbons consists of nanotube networks (Fig. 1.3(o)), which have been predicted to exhibit outstanding mechanical and electronic properties, besides having a large surface area (e.g. 3600 m^2/g) (Romo-Herrera *et al.*, 2006). Interestingly, these types of random 3D nanotube networks have been produced using CVD approaches (Lepro *et al.*, 2007) and further theoretical and experimental studies are still required in order to achieve crystalline 3D networks.

The series of events described above and dedicated to the most important discoveries in carbon nanoscience clearly demonstrate that carbon is a fascinating element and is able to form various morphologies at the nanoscale, possessing different physico-chemical properties, some of them yet unknown. In the following sections, we will concentrate on novel one- and two-dimensional sp^2-like carbon nanostructures. But before starting your trip we recommend that you read the *Guide to the book* below.

1.3   Guide to the book

This book deals with the electronic and transport properties of some of the most promising new forms of carbon introduced before. Chapter 2 starts by introducing the electronic properties of both pristine and defected carbon nanostructures, and also overviews the salient electronic features under magnetic fields (Aharonov–Bohm phenomenon and Landau levels). The emphasis is on tight-binding models, though widely used effective low-energy models are also introduced. When possible, the results are commented on in the light of *ab initio* simulations.

The rest of the book is mostly dedicated to the electronic transport properties of graphene-related materials. Chapter 3 offers a general overview of the tools used later on, namely, Landauer–Büttiker and Kubo–Greenwood formalisms, together with the commonly used semiclassical Boltzmann transport equation, which presents severe limitations for the exploration of the quantum transport at the Dirac point. Most of the technical details (or tricks!) concerning the numerical implementations of such transport methods are given in dedicated appendices. The first illustrations of transport properties in disordered graphene materials are given in Chapter 4, with a starting discussion concerning the limits of ballistic transport and the peculiar Klein tunneling mechanism. The role of disorder is further discussed broadly in Chapter 5, through the use of the legendary Anderson disorder model, which is the first approach for studying the main transport length scales and conduction regimes. Weak and strong localization phenomena (including weak antilocalization) are presented and related to the nature of disorder (short versus long range potential). Various forms of structural disorders are then studied including monovacancies, and polycrystalline and amorphous graphene, showing how irregularities affect mean free paths and localization lengths, eventually turning the materials to a strong Anderson insulator.

Chapter 6 gives a brief overview of quantum transport beyond DC conditions, Floquet theory for time-periodic Hamiltonians, and a succinct review of the literature on AC transport in carbon-based nanostructures. This is a currently very active field, which connects to graphene photonics and plasmonics. Many developments are expected within the next years, and theoretical study certainly needs to be further extended. The material provided here will be very useful for those researchers interested in the field.

Ab initio and multiscale transport methodologies are discussed in Chapter 7. To achieve accurate transport calculations on very-large-size disordered systems, a combination of *ab initio* approach- and order-N transport algorithms is crucial. The presentation will provide a simple description of various possible hybrid methodologies for investigating the complex transport fingerprints of chemically or structurally disordered carbon nanotubes of graphene-based materials. Some targeted functionalities such as chemical sensing will then be discussed in detail for chemically functionalized nanotubes. Sensing is often viewed as a major application for such low-dimensional carbon-based materials. The possibilities but also limitations are illustrated in this chapter.

Chapter 8 is devoted to some short presentations of several applications such as flexible electronics, graphene nanoresonators, photonics or spintronics. This chapter presents some debate and open issues, as perceived by the authors, and these issues should generate a great amount of research in the next decade. In particular, graphene spintronics offers fascinating possibilities of revolutionary information processing using the spin degree of freedom. Progress towards spin gating and spin manipulation, however, demands attention and effort in revisiting the way spin diffusion and spin relaxation mechanisms are described in graphene, as these are likely to genuinely differ from conventional relaxation effects described in metals and small-gap semiconductors.

Some of these chapters are essentially tutorial (Chapters 1, 2, 3, 4, and 6 and Appendices A–D), offering enough material for introductory lectures at the master degree level. Others are intended to give an overview of most foundational literature in the respective fields, or to shine some light on leading-edge research (Chapters 5, 7, and 8). The choice of topics, presentation, and illustrations are unavoidably biased towards the authors' own experiences, and despite the attempts to properly acknowledge the foundational papers, many citations are certainly missing. We will try to amend this in later editions.

All chapters contain a very short list of suggested material for further reading. The core tutorial chapters of this book contain lists of problems with varying levels of difficulty. Many of them are computational exercises where a *learning by doing* spirit is encouraged. Along this line, many solutions, additional exercises and miscellaneous material, as well as computational codes, are made available online at the website:

www.introductiontographene.org

www.cambridge.org/foatorres

The symbol on the right will indicate that additional material is available online. The authors intend that an updated list of typos and errors will also be available there. This will be the authors' contact point with their ultimate inspiration for this enterprise: you and your fellow readers.

Finally, the interdisciplinarity of the potential readers of this book makes it impossible (and probably pointless) to develop a book that all readers can read linearly from beginning to end. This is why the authors suggest tailoring it to your own experience and objectives. Before starting, you are recommended to get your own *Table of instructions* from the authors' website.

1.4   Further reading

- The reader may enjoy reading the personal accounts given in Dresselhaus (2011) and Geim (2011) where a flavor of the story behind the development of these materials is given.

2 Electronic properties of carbon-based nanostructures

2.1 Introduction

As described in Chapter 1, the sp^2 carbon-based family exhibits a great variety of allotropes, from the low-dimensional fullerenes, nanotubes and graphene ribbons, to two-dimensional monolayer graphene, or stacked graphene multilayers. Two-dimensional monolayer graphene stands as the building block, since all the other forms can be derived from it. Graphene nanoribbons can be seen as quasi-one-dimensional structures, with one lateral dimension short enough to trigger quantum confinement effects. Carbon nanotubes can be geometrically constructed by folding graphene nanoribbons into cylinders, and graphite results from the stacking of a very large number of weakly bonded graphene monolayers.

The isolation of a single graphene monolayer by mechanical exfoliation (repeated peeling or micromechanical cleavage) starting from bulk graphite has been actually quite a surprise, since it was previously believed to be thermodynamically unstable (Novoselov *et al.*, 2004, Novoselov *et al.*, 2005*b*). At the same time, the route for controlling the growth of graphene multilayers on top of silicon carbide by thermal decomposition was reported, and eventually led to fabrication of single graphene monolayers of varying quality depending on the surface termination (silicon or carbon termination) (Berger *et al.*, 2006). Basic electronic properties of graphene were actually well-known since the seminal work by Wallace in the late forties (Wallace, 1947), such as the electron–hole symmetry of the band structure and the specific linear electronic band dispersion near the Brillouin zone corners (Dirac point), but it was after the discovery of carbon nanotubes by Iijima from NEC (Iijima, 1991) that the exploration of electronic properties of graphene-based materials was revisited (for a review see Charlier, Blase & Roche 2007).

This chapter introduces the main electronic features of monolayer graphene and few-layer graphene together with its low-dimensional versions (carbon nanotubes and graphene nanoribbons). The first section starts with an overview of the electronic properties of graphene described using a simple nearest-neighbor *tight-binding* model, together with an extended derivation of the effective description of low-energy excitations as massless Dirac fermions. The description beyond the linear approximation is then discussed by introducing trigonal warping deformation, or extending the tight-binding model to third-nearest-neighbor coupling.

This is followed by the improved description provided by *first-principles* calculations within density functional theory (DFT) and beyond using *many-body perturbation theory within the GW approximation*, which leads to a renormalized Fermi velocity close to the Dirac point. The next part focuses on the specificities of graphene nanoribbons (GNRs), with a description of the formation of confinement-induced energy gaps which increase linearly with reducing the lateral size. These GNR structures are shown to share some commonalities with their folded versions, since carbon nanotubes (CNTs) are often pictured as the geometrical result of rolling up a graphene ribbon. Carbon nanotubes are found to be either metallic or semiconducting depending on their helical symmetry. Metallic (armchair) nanotubes are actually the best existing one-dimensional ballistic conductors, almost insensitive to the Peierls dimerization mechanism, and exhibiting quantized conductance when appropriately connected to metals such as palladium. The energy gaps in semiconducting tubes downscale linearly with the tube diameter, and eventually close for the limit of very large diameter (in accordance with the zero-gap limit of a graphene monolayer). Finally, note that there is currently great interest in analyzing the effects of chemical doping and structural defects in graphene-based materials, given the possibility to tailor the electronic properties and add novel functionalities to the related devices, to improve or complement the silicon-based CMOS technologies.

2.2   Electronic properties of graphene

2.2.1   Tight-binding description of graphene

In two-dimensional graphene, carbon atoms are periodically arranged in an infinite honeycomb lattice (Fig. 2.1(a)). Such an atomic structure is defined by two types of bonds within the sp^2 hybridization, as described in Chapter 1. From the four valence orbitals of the carbon atom (the $2s$, $2p_x$, $2p_y$, and $2p_z$ orbitals, where z is the direction perpendicular to the sheet), the (s, p_x, p_y) orbitals combine to form the inplane σ (bonding or occupied) and σ^* (antibonding or unoccupied) orbitals. Such orbitals are even with respect to the planar symmetry. The σ bonds are strongly covalent bonds determining the energetic stability and the elastic properties of graphene (Fig. 2.1(a)). The remaining p_z orbital, pointing out of the graphene sheet as shown in Fig. 2.1(a), is odd with respect to the planar symmetry and decoupled from the σ states. From the lateral interaction with neighboring p_z orbitals (called the $pp\pi$ interaction), localized π (bonding) and π^* (antibonding) orbitals are formed (Wallace, 1947). Graphite consists of a stack of many graphene layers. The unit cell in graphite can be primarily defined using two graphene layers translated from each other by a C-C distance ($a_{cc} = 1.42$ Å). The three-dimensional structure of graphite is maintained by the weak interlayer van der Waals interaction between π bonds of adjacent layers, which generate a weak but finite out-of-plane delocalization (Charlier, Gonze & Michenaud, 1994*b*).

The bonding and antibonding σ bands are actually strongly separated in energy (> 12 eV at Γ), and therefore their contribution to electronic properties is commonly

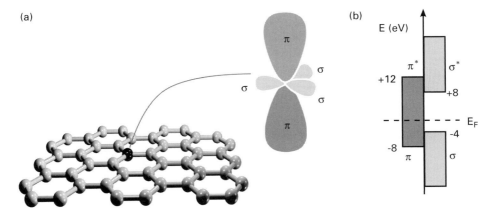

Figure 2.1 The carbon valence orbitals. (a) The three σ orbitals in graphene and the π orbital perpendicular to the sheet. The σ bonds in the carbon hexagonal network strongly connect the carbon atoms and are responsible for the binding energy and the structural properties of the graphene sheet. The π bonds are perpendicular to the surface of the sheet. The corresponding bonding and antibonding σ bands are separated by a large energy gap of ~12 eV; while (b) the bonding and antibonding π states lie in the vicinity of the Fermi level (E_F). Consequently, the σ bonds are frequently neglected for prediction of the electronic properties of graphene around the Fermi energy.

disregarded (Fig. 2.1(b)). The two remaining π bands completely describe the low-energy electronic excitations in both graphene (Wallace, 1947) and graphite (Charlier *et al.*, 1991). The bonding π and antibonding π^* orbitals produce valence and conduction bands (Fig. 2.1(b)) which cross at the charge neutrality point (Fermi level of undoped graphene) at vertices of the hexagonal Brillouin zone.

Carbon atoms in a graphene plane are located at the vertices of a hexagonal lattice. This graphene network can be regarded as a triangular Bravais lattice with two atoms per unit cell (A and B) and basis vectors ($\mathbf{a}_1, \mathbf{a}_2$):

$$\mathbf{a}_1 = a\left(\frac{\sqrt{3}}{2}, \frac{1}{2}\right), \quad \mathbf{a}_2 = a\left(\frac{\sqrt{3}}{2}, -\frac{1}{2}\right). \tag{2.1}$$

Note that $a = \sqrt{3}a_{cc}$, where $a_{cc} = 1.42$ Å is the carbon–carbon distance in graphene. In Fig. 2.2(a) A-type and B-type atoms are represented by full and empty dots respectively. From this figure we see that each A- or B-type atom is surrounded by three atoms of the opposite type.

By using the condition $\mathbf{a}_i \cdot \mathbf{b}_j = 2\pi \delta_{ij}$, the reciprocal lattice vectors ($\mathbf{b}_1, \mathbf{b}_2$) can be obtained,

$$\mathbf{b}_1 = b\left(\frac{1}{2}, \frac{\sqrt{3}}{2}\right), \quad \mathbf{b}_2 = b\left(\frac{1}{2}, -\frac{\sqrt{3}}{2}\right), \tag{2.2}$$

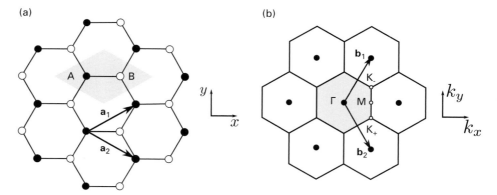

Figure 2.2 (a) Showing the basis vectors $\mathbf{a}_1$ and $\mathbf{a}_2$ in the hexagonal network of graphene. This network is a triangular Bravais lattice with a two atom-basis: A (full dots) and B (empty dots). (b) The reciprocal lattice points corresponding to the triangular Bravais lattice (full dots) as well as the associated basis vectors $\mathbf{b}_1$ and $\mathbf{b}_2$. The unit cell/Brillouin zone are shown shaded gray in a and b respectively. Highly symmetric points labeled with Γ (zone center), K_+, K_-, and M are also indicated in b.

with $b = 4\pi/(3a_{cc}) = 4\pi/a\sqrt{3}$. These vectors are shown in Fig. 2.2(b) together with the first Brillouin zone (shaded gray). This hexagonal-shaped Brillouin zone[1] is built as the Wigner–Seitz cell of the reciprocal lattice. Out of its six corners, two of them are inequivalent (the others can be written as one of these two plus a reciprocal lattice vector). These two special points are denoted with K_+ and K_-. Another high symmetry point is the one labeled with M in Fig. 2.2(b). They can be chosen as:

$$\mathbf{K}_+ = \frac{4\pi}{3a}\left(\frac{\sqrt{3}}{2}, -\frac{1}{2}\right), \quad \mathbf{K}_- = \frac{4\pi}{3a}\left(\frac{\sqrt{3}}{2}, \frac{1}{2}\right), \quad \mathbf{M} = \frac{2\pi}{\sqrt{3}a}(1,0). \quad (2.3)$$

When the carbon atoms are placed onto the graphene hexagonal network (Fig. 2.2(a)), the electronic wavefunctions from different atoms overlap. However, because of symmetry the overlap between the p_z orbitals and the s or the p_x and p_y electrons is strictly zero. Therefore, the p_z electrons which form the π bonds in graphene can be treated independently from the other valence electrons. Within this π-band approximation, the A atom (or B atom) is uniquely defined by one orbital per atom site $p_z(\mathbf{r} - \mathbf{r}_A)$ (or $p_z(\mathbf{r} - \mathbf{r}_B)$).

To derive the electronic spectrum of the total Hamiltonian, the corresponding Schrödinger equation has to be solved. According to Bloch's theorem, the eigenfunctions evaluated at two given Bravais lattice points $\mathbf{R}_i$ and $\mathbf{R}_j$ differ from each other in just a phase factor, $\exp(i\mathbf{k} \cdot (\mathbf{R}_i - \mathbf{R}_j))$. Because of the two-atom basis, the Bloch *ansatz*

---

[1] Note that the hexagonal shape of the Brillouin zone is a consequence of the triangular Bravais lattice. It is by no means connected with the two-atom basis which does not enter into the definition of the Brillouin zone.

for the eigenfunctions is a linear combination of Bloch sums[2] on each sublattice:

$$\Psi(\mathbf{k}, \mathbf{r}) = c_A(\mathbf{k})\tilde{p}_z^A(\mathbf{k}, \mathbf{r}) + c_B(\mathbf{k})\tilde{p}_z^B(\mathbf{k}, \mathbf{r}), \tag{2.4}$$

where

$$\tilde{p}_z^A(\mathbf{k}, \mathbf{r}) = \frac{1}{\sqrt{N_{\text{cells}}}} \sum_j e^{i\mathbf{k}.\mathbf{R}_j} p_z(\mathbf{r} - \mathbf{r}_A - \mathbf{R}_j), \tag{2.5}$$

$$\tilde{p}_z^B(\mathbf{k}, \mathbf{r}) = \frac{1}{\sqrt{N_{\text{cells}}}} \sum_j e^{i\mathbf{k}.\mathbf{R}_j} p_z(\mathbf{r} - \mathbf{r}_B - \mathbf{R}_j), \tag{2.6}$$

where $\mathbf{k}$ is the electron wavevector, N_{cells} the number of unit cells in the graphene sheet, and $\mathbf{R}_j$ is a Bravais lattice point. In the following we will neglect the overlap $s = \langle p_z^A | p_z^B \rangle$ between neighboring p_z orbitals. Then, the Bloch sums form an orthonormal set:

$$\langle \tilde{p}_z^\alpha(\mathbf{k}) \mid \tilde{p}_z^\beta(\mathbf{k}') \rangle = \delta_{\mathbf{k},\mathbf{k}'} \delta_{\alpha,\beta}, \tag{2.7}$$

where $\alpha, \beta = A, B$. Using these orthogonality relations in the Schrödinger equation, $\mathcal{H}\Psi(\mathbf{k}, \mathbf{r}) = E\Psi(\mathbf{k}, \mathbf{r})$, one obtains a 2×2 eigenvalue problem,

$$\begin{pmatrix} \mathcal{H}_{AA}(\mathbf{k}) & \mathcal{H}_{AB}(\mathbf{k}) \\ \mathcal{H}_{BA}(\mathbf{k}) & \mathcal{H}_{BB}(\mathbf{k}) \end{pmatrix} \begin{pmatrix} c_A(\mathbf{k}) \\ c_B(\mathbf{k}) \end{pmatrix} = E(\mathbf{k}) \begin{pmatrix} c_A(\mathbf{k}) \\ c_B(\mathbf{k}) \end{pmatrix}. \tag{2.8}$$

The matrix elements of the Hamiltonian are given by:

$$\mathcal{H}_{AA}(\mathbf{k}) = \frac{1}{N_{\text{cells}}} \sum_{i,j} e^{i\mathbf{k}.(\mathbf{R}_j - \mathbf{R}_i)} \langle p_z^{A,\mathbf{R}_i} \mid \mathcal{H} \mid p_z^{A,\mathbf{R}_j} \rangle, \tag{2.9}$$

$$\mathcal{H}_{AB}(\mathbf{k}) = \frac{1}{N_{\text{cells}}} \sum_{i,j} e^{i\mathbf{k}.(\mathbf{R}_j - \mathbf{R}_i)} \langle p_z^{A,\mathbf{R}_i} \mid \mathcal{H} \mid p_z^{B,\mathbf{R}_j} \rangle, \tag{2.10}$$

with $\mathcal{H}_{AA} = \mathcal{H}_{BB}$ and $\mathcal{H}_{AB} = \mathcal{H}_{BA}^*$, and introducing the notation $p_z^{A,\tau} = p_z(\mathbf{r} - \mathbf{r}_A - \tau)$ and $p_z^{B,\tau} = p_z(\mathbf{r} - \mathbf{r}_B - \tau)$. After simple manipulations, and by restricting the interactions to first-nearest-neighbors only, one gets:

$$\mathcal{H}_{AB}(\mathbf{k}) = \langle p_z^{A,0} | \mathcal{H} | p_z^{B,0} \rangle + e^{-i\mathbf{k}.\mathbf{a}_1} \langle p_z^{A,0} | \mathcal{H} | p_z^{B,-\mathbf{a}_1} \rangle + e^{-i\mathbf{k}.\mathbf{a}_2} \langle p_z^{A,0} | \mathcal{H} | p_z^{B,-\mathbf{a}_2} \rangle$$
$$= -\gamma_0 \alpha(\mathbf{k}), \tag{2.11}$$

where γ_0 stands for the transfer integral between first neighbor π orbitals (typical values for γ_0 are 2.9–3.1 eV (Charlier *et al.*, 1991, Dresselhaus *et al.*, 2000)), and the function $\alpha(\mathbf{k})$ is given by:

$$\alpha(\mathbf{k}) = (1 + e^{-i\mathbf{k}.\mathbf{a}_1} + e^{-i\mathbf{k}.\mathbf{a}_2}). \tag{2.12}$$

Taking $\langle p_z^{A,0} | \mathcal{H} | p_z^{A,0} \rangle = \langle p_z^{B,0} | \mathcal{H} | p_z^{B,0} \rangle = 0$ as the energy reference, we can write $\mathcal{H}(\mathbf{k})$ as:

$$\mathcal{H}(\mathbf{k}) = \begin{pmatrix} 0 & -\gamma_0 \alpha(\mathbf{k}) \\ -\gamma_0 \alpha(\mathbf{k})^* & 0 \end{pmatrix}. \tag{2.13}$$

---

[2] Alternatively, one may proceed by writing the Hamiltonian and the eigenfunctions in matrix form, as shown in the supplementary material on the authors' website.

This 2×2 Hamiltonian is very appealing and may also be written in terms of Pauli matrices as in (Haldane, 1988), thereby emphasizing the analogy with a spin Hamiltonian.[3] Section 2.2.2 derives in detail the consequences of the A/B bipartite lattice structure on the (pseudo)-spinor symmetry of (four-component) electronic eigenstates. The energy dispersion relations are easily obtained from the diagonalization of $\mathcal{H}(\mathbf{k})$ given by Eq. (2.13):

$$E_{\pm}(\mathbf{k}) = \pm \gamma_0 |\alpha(\mathbf{k})| \tag{2.14}$$

$$= \pm \gamma_0 \sqrt{3 + 2\cos(\mathbf{k}.\mathbf{a}_1) + 2\cos(\mathbf{k}.\mathbf{a}_2) + 2\cos(\mathbf{k}.(\mathbf{a}_2 - \mathbf{a}_1))}, \tag{2.15}$$

which can be further expanded as

$$E_{\pm}(k_x, k_y) = \pm \gamma_0 \sqrt{1 + 4\cos\frac{\sqrt{3}k_x a}{2}\cos\frac{k_y a}{2} + 4\cos^2\frac{k_y a}{2}}. \tag{2.16}$$

The wavevectors $\mathbf{k} = (k_x, k_y)$ are chosen within the first hexagonal Brillouin zone (BZ). Clearly, the zeros of $\alpha(\mathbf{k})$ correspond to the crossing of the bands with the $+$ and $-$ signs. One can verify that $\alpha(\mathbf{k} = \mathbf{K}_+) = \alpha(\mathbf{k} = \mathbf{K}_-) = 0$ and therefore the crossings occur at the points $\mathbf{K}_+$ and $\mathbf{K}_-$. Furthermore, with a single p_z electron per atom in the π-π^* model (the three other s, p_x, p_y electrons fill the low-lying σ band), the $(-)$ band (negative energy branch) in Eq. (2.16) is fully occupied, while the $(+)$ branch is empty, at least for electrically neutral graphene. Thus, the Fermi level E_F (or charge neutrality point) is the zero-energy reference in Fig. 2.3 and the Fermi surface is composed of the set of K_+ and K_- points. Graphene displays a metallic (zero-gap) character. However, as the Fermi surface is of zero dimension (since it is reduced to a discrete and finite set of points), the term semi-metal or zero-gap semiconductor is usually employed. Expanding Eq. (2.16) for $\mathbf{k}$ in the vicinity of $\mathbf{K}_+$ (or $\mathbf{K}_-$), $\mathbf{k} = \mathbf{K}_+ + \delta\mathbf{k}$ ($\mathbf{k} = \mathbf{K}_- + \delta\mathbf{k}$), yields a linear dispersion for the π and π^* bands near these six corners of the 2D hexagonal Brillouin zone,

$$E_{\pm}(\delta\mathbf{k}) = \pm \hbar v_F |\delta\mathbf{k}|, \tag{2.17}$$

where

$$v_F = \frac{\sqrt{3}\gamma_0 a}{2\hbar} \tag{2.18}$$

is the electronic group velocity. Graphene is thus highly peculiar for this linear energy–momentum relation and electron–hole symmetry. The electronic properties in the vicinity of these corners of the 2D Brillouin zone mimic those of *massless* Dirac fermions (developed in Section 2.2.2) forming "Dirac cones" as illustrated in Fig. 2.3. The six points where the Dirac cones touch are referred to as the Dirac points. The electronic

[3] Writing the Hamiltonian in terms of Pauli matrices allows us also to classify the terms according to their symmetries. A particularly important one is electron–hole symmetry. The Hamiltonian is said to have electron–hole symmetry if there is a transformation $\mathcal{P}$, such that $\mathcal{P}^{\dagger}\mathcal{H}\mathcal{P} = -\mathcal{H}$. This guarantees that if Ψ is an eigenstate of $\mathcal{H}$ with a positive energy E (electron function), then $\mathcal{P}\Psi$ is also an eigenstate with energy $-E$ (hole function) and the spectrum is symmetric with respect to $E = 0$. For a Hamiltonian as the one here, a term proportional to σ_z (such as a staggering potential which breaks A-B symmetry) opens a gap but preserves electron–hole symmetry.

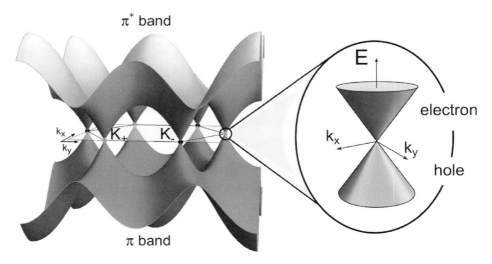

Figure 2.3 Graphene π and π^* electronic bands. In this simple approach, the π and π^* bands are symmetric with respect to the valence and conduction bands. The linear dispersion relation close to the K_+ (light grey dots) and K_- (black dots) points of the first 2D Brillouin zone gives rise to the "Dirac cones" as shown on the right. Note that close to these cones k_x and k_y are used to denote the shift from the corresponding K point.

group velocities close to those points are quite high at $\sim 8.5 \times 10^5$ m/s, and within the massless Dirac fermions analogy represent an effective "speed of light."

This simple orthogonal *tight-binding* model (Wallace, 1947) yields π and π^* zone-center Γ energies which are symmetric ($\pm\gamma_0$) with respect to E_F. In fact, the anti-bonding (unoccupied) π^* bands are located at a higher energy if the overlap integral S is not set to zero (as illustrated in Fig. 2.1(b)). A better (but more complicated) $\pi - \pi^*$ parameterization could lead to analogous results (Reich *et al.*, 2002), as well as more accurate first-principles calculations. In the following, after a presentation of the effective massless Dirac fermion model, we comment on the effects beyond nearest neighbor interactions and the so-called trigonal warping correction.

2.2.2 Effective description close to the Dirac point and massless Dirac fermions

By expanding Eq. (2.13) for the Hamiltonian around K_+ and K_- (the two inequivalent corners of the Brillouin zone) we get an approximation close to those points. To keep a compact notation in what follows, $\mathbf{k}$ measures the deviations from those points. A linear expansion then gives

$$\mathcal{H}_{K_+} = \hbar v_F \begin{pmatrix} 0 & k_x - ik_y \\ k_x + ik_y & 0 \end{pmatrix} = v_F(p_x\sigma_x + p_y\sigma_y), \tag{2.19}$$

where $p_{x(y)} = \hbar k_{x(y)}$ and the Pauli matrices are defined as usual:

$$\sigma_x = \begin{pmatrix} 0 & 1 \\ 1 & 0 \end{pmatrix}, \quad \sigma_y = \begin{pmatrix} 0 & -i \\ i & 0 \end{pmatrix}, \quad \sigma_z = \begin{pmatrix} 1 & 0 \\ 0 & -1 \end{pmatrix}. \tag{2.20}$$

The effective Hamiltonian can also be written in the more compact form:

$$\mathcal{H}_{K_+} = v_F \hat{\sigma} \cdot \mathbf{p}, \tag{2.21}$$

where $\hat{\sigma} = (\sigma_x, \sigma_y, \sigma_z)$. For the inequivalent K point one has the transposed Hamiltonian

$$\mathcal{H}_{K_-} = \mathcal{H}_{K_+}^t. \tag{2.22}$$

Substituting $\mathbf{p}$ by the corresponding operator $\hat{\mathbf{p}} = -i\hbar\hat{\nabla}$ in Eq. (2.21) (this is equivalent to the $\mathbf{k}.\mathbf{p}$ or effective mass approximation Ajiki 1993, DiVicenzo and Mele, 1984), a form equivalent to the Dirac–Weyl Hamiltonian in two dimensions is obtained, which in quantum electrodynamics follows from the Dirac equation by setting the rest mass of the particle to zero. Therefore, the low-energy excitations mimic those of massless Dirac particles of spin 1/2 (such as a massless neutrino), with velocity of light c, and inherent chirality as explained below. However, in contrast to relativistic Dirac particles, low-energy excitations of graphene have a Fermi velocity v_F about 300 times smaller than the light velocity, whereas the Pauli matrices appearing in the low-energy effective description operate on the sublattice degrees of freedom instead of spin, hence the term *pseudospin*. The low-energy quasiparticles in graphene are often referred to as massless Dirac fermions.

One of the most interesting properties of the Dirac–Weyl equation is its helical or chiral nature[4] which is a direct consequence of the Hamiltonian being proportional to the helicity operator, which here for the case of the Hamiltonian in Eq. (2.21) is defined as:

$$\hat{h} = \hat{\sigma} \cdot \frac{\mathbf{p}}{|\mathbf{p}|}. \tag{2.23}$$

The quantity $\hat{h}$ is essentially the projection of the sublattice pseudospin operator $\hat{\sigma}$ on the momentum direction. Interestingly, since $\hat{h}$ commutes with the Hamiltonian, the projection of the pseudospin is a well-defined conserved quantity which can be either positive or negative, corresponding to pseudospin and momentum being *parallel* or *antiparallel* to each other (see Fig. 2.4). At the K_- point, the Hamiltonian is proportional to $\hat{\sigma}^t.\mathbf{p}$ and involves the *left-handed* Pauli matrices $\hat{\sigma}^t$ (in contrast to the right-handed matrices $\hat{\sigma}$). Therefore, one says that chirality is inverted when passing from K_+ to K_- as represented in Fig. 2.4.

To explore this in more detail, let us rewrite once more the Hamiltonian as:

$$\mathcal{H}_\xi(\mathbf{p}) = v_F|\mathbf{p}| \begin{pmatrix} 0 & e^{-i\xi\theta_p} \\ e^{+i\xi\theta_p} & 0 \end{pmatrix}, \tag{2.24}$$

where $p_x + ip_y = \sqrt{p_x^2 + p_y^2}\,e^{i\theta_p}$, $\theta_p = \arctan(p_y/p_x)$ and ξ can take the values $\xi = +1$ which corresponds to K_+ and $\xi = -1$ to K_-. Then, one can verify that this Hamiltonian is diagonalized by the unitary operator

$$\mathcal{U}_\xi = \frac{1}{\sqrt{2}} \begin{pmatrix} -e^{-i\xi\theta_p} & e^{-i\xi\theta_p} \\ 1 & 1 \end{pmatrix}. \tag{2.25}$$

---

[4] For massless particles the two are identical and the terms are used interchangeably.

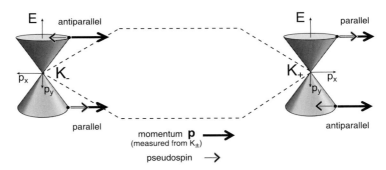

Figure 2.4 The two inequivalent Dirac cones at K_+ and K_- points of the first Brillouin zone, together with direction of the pseudospin parallel or antiparallel to the momentum **p** of selected energies in conduction and valence bands.

Indeed,

$$\mathcal{U}_\xi^\dagger(\mathbf{p})\mathcal{H}_\xi(\mathbf{p})\mathcal{U}_\xi(\mathbf{p}) = v_F \begin{pmatrix} -|\mathbf{p}| & 0 \\ 0 & |\mathbf{p}| \end{pmatrix} = -v_F|\mathbf{p}|\sigma_z, \qquad (2.26)$$

which makes explicit the linear energy dispersion $E_\pm(\mathbf{p}) = \pm v_F|\mathbf{p}|$ and the electron–hole symmetry.[5] On the other hand, the eigenstates of Eq. (2.24) can be written as:

$$|\Psi_{\xi,s}\rangle = \frac{1}{\sqrt{2}} \begin{pmatrix} 1 \\ se^{+i\xi\theta_p} \end{pmatrix}. \qquad (2.27)$$

The index $s = \pm 1$ is the band index ($s = +1$ for the conduction band and $s = -1$ for the valence band) and ξ the valley index as stated before ($\xi = +1$ (K_+), $\xi = -1$ (K_-)). Using this explicit form for the eigenstates we can directly verify that they are also eigenstates of the appropriate helicity operator (also called chirality operator) with eigenvalues ± 1.

Around K_+ ($\xi = +1$), the pseudospin of eigenstates in the conduction band is parallel to the momentum and antiparallel for eigenstates in the valence band. The chirality in this case is simply the band index. The property around K_- ($\xi = -1$) is reversed as illustrated in Fig. 2.4. This peculiarity has a strong influence in many of the most intriguing properties of graphene. For example, for an electron to backscatter (i.e. changing **p** to $-\mathbf{p}$) it needs to reverse its pseudospin. But as the pseudospin direction is locked to that of momentum, backscattering is not possible if the Hamiltonian is not perturbed by a term which flips the pseudospin (this is also termed *absence of backscattering* (Ando, Nakanishi & Saito, 1998)).

Although we are dealing all the time with both valleys separately, it is important to keep in mind that the full structure of the eigenstates is described by a four-component spinor wavefunction, $(|\Psi_{K_+,A}\rangle, |\Psi_{K_+,B}\rangle, |\Psi_{K_-,A}\rangle, |\Psi_{K_-,B}\rangle)^t$. The full Hamiltonian of

---

[5] Also, by comparison with the relativistic expression, $E(p) = \pm\sqrt{p^2 v_F^2 + m^{*2}c^4}$ enforces a zero effective mass.

ideal graphene is given by,

$$\hat{\mathcal{H}} = v_F \begin{pmatrix} 0 & \pi^\dagger & 0 & 0 \\ \pi & 0 & 0 & 0 \\ 0 & 0 & 0 & \pi \\ 0 & 0 & \pi^\dagger & 0 \end{pmatrix}, \tag{2.28}$$

with $\pi = p_x + ip_y$ and $\pi^\dagger = p_x - ip_y$. Although for this ideal case the states at both k points are decoupled, one should be aware that any perturbation which is not smooth at the atomic scale (e.g. impurities) will couple them.

Phase ambiguity and Berry phase

The existence of an inherent phase ambiguity of the quantum wavefunction is well illustrated through the Bloch theorem, which states that the eigenstates of a given Hamiltonian $\mathcal{H}$ (defining the energetics of the atomic unit cell with periodic boundary conditions) can generally be written as $|\Psi_k\rangle = e^{i\mathbf{k}.\mathbf{r}}|\psi_k\rangle$, with $|\psi_k\rangle$ defined inside the unit cell (invariant under any transformation such as $|\psi_k\rangle \rightarrow e^{\varphi_k}|\psi_k\rangle$, with e^{φ_k} an arbitrary phase function in k-space). To leave the phase ambiguity and capture the phase interferences in physical observables, one has to define the so-called Berry connection (equivalent to a vector potential) as $\mathbf{A} = i\langle\psi_k|\hat{\nabla}_k|\psi_k\rangle$. All physical quantities will be invariant under any gauge transformation $\mathcal{A} \rightarrow \mathcal{A} + \hat{\nabla}_k\varphi_k$, while the Berry phase defined as a gauge-invariant quantity

$$\gamma_c = \oint \mathbf{A}.d\mathbf{k} \tag{2.29}$$

measures the total phase accumulated upon a transformation (rotation) of the wavefunction in k-space along a closed loop. The Berry curvature $\mathbf{F} = \hat{\nabla}_k.\mathbf{A}$ is analogous to the magnetic field, while $\gamma_c = \iint \mathbf{F}.d^2k$ gives the Berry flux. The existence of a nontrivial Berry phase has been demonstrated to have many profound consequences in quantum physics (Thouless, 1998, Xiao, Chang & Niu, 2010), and in graphene and carbon nanotubes it conveys phenomena such as absence of backscattering in nanotubes, Klein tunneling, weak antilocalization, zero-energy Landau level, and an anomalous quantum Hall effect, as described in the following chapters.

Under 2π rotation, the eigenstates of the Dirac excitations get a π phase factor. Using the rotation operator $\mathcal{R}(\theta) = e^{-i\theta.\mathbf{S}/\hbar}$, with $\mathbf{S} = \hbar/2\hat{\sigma}_z$ for spin-1/2 particles, it is indeed readily shown that $\mathcal{R}(\theta = 2\pi)|\Psi_{K_\pm}(s = \pm1)\rangle = e^{i\pi\hat{\sigma}_z}|\Psi_{\xi,s}\rangle = -|\Psi_{\xi,s}\rangle$ (using $e^{-i\theta(\hat{n}.\hat{\sigma})/\hbar} = \cos\theta + i(\hat{n}.\hat{\sigma})\sin\theta$).

One can also directly compute the Berry phase from the general definition as

$$\mathbf{A} = -i\langle\psi_k|\hat{\nabla}_k|\psi_k\rangle = \frac{-i}{2}(1, e^{-i\theta}).\begin{pmatrix} 0 \\ i\nabla_k\theta e^{i\theta} \end{pmatrix} = \frac{\mathbf{e}_\theta}{2|\mathbf{k}|}, \tag{2.30}$$

($\mathbf{e}_\theta$ is a unit vector perpendicular to $\mathbf{p}$) while

$$\gamma_c = \oint \mathbf{A}.d\mathbf{k} = \int_0^{2\pi} d\mathbf{k}.\frac{\mathbf{e}_\theta}{2|\mathbf{k}|} = \pi. \tag{2.31}$$

The observation of a Dirac cone and the existence of pseudospin-related quantum phases has been confirmed, in particular through polarization-dependent angle-resolved photoemission spectroscopy (ARPES) (Hwang *et al.*, 2011, Liu *et al.*, 2011). Figure 2.5 shows the experimental photoelectron intensity maps at the Fermi level E_F versus the two-dimensional wave vector k for single-layer graphene for the two polarization geometries. The main feature in the intensity maps of both geometries is an almost circular Fermi surface centered at the K point. Additionally, the angular intensity distribution is seen to be polarization-dependent in the sense that the minimum intensity position is in the first Brillouin zone for X-polarization, while the maximum intensity position is in the first Brillouin zone for Y-polarization geometry, suggesting a π rotation of the maximum intensity in the $k_x - k_y$ plane around the K point upon rotating the light polarization by $\pi/2$, from X to Y. This result can be demonstrated by computing the wave-vector-dependent photoelectron intensity $I_k = |\langle \mathbf{k} + \mathbf{Q} | \mathcal{H}_{elm}(\mathbf{k}, \mathbf{Q}) | \Psi_{sk} \rangle|^2$, introducing the field-induced transition matrix element between graphene eigenstates and $|f_{\mathbf{k}+\mathbf{Q}}\rangle = 1/\sqrt{2}(1, 1)$, the plane-wave final state projected onto the p_z orbitals of graphene (note that all states are expressed using the basis set of Bloch sums of localized p_z orbitals at sublattices A and B). The interaction Hamiltonian coupling to electromagnetic waves of wave vector $\mathbf{Q}$ is obtained by using the velocity operator and the external vector potential as $-\frac{e}{c}\hat{\mathbf{A}}.\hat{\mathbf{v}} \ (A(\mathbf{r}, t) = A_{\mathbf{Q}} e^{i(\mathbf{Q}.\mathbf{r} - \omega t)})$, which can be approximated close to the Dirac point by $\mathcal{H}_{elm}(\mathbf{q} + \mathbf{K}) \sim \frac{e v_F}{\hbar c}(A_x^0 \sigma_x + A_y^0 \sigma_y)$. Finally, one obtains for both polarizations $I_k^{X-pol} \sim \sin(\theta_q/2)$, whereas $I_k^{Y-pol} \sim \cos(\theta_q/2)$, from which it is clear that the photoemission intensity map is rotated by π when the light polarization is rotated by $\pi/2$ (Hwang *et al.*, 2011).

The electronic properties of the 2D graphene can thus be described by an effectively massless Dirac fermion model in the vicinity of the charge neutrality point, with linearly dispersing bands and electron–hole symmetry. These properties derived close to the

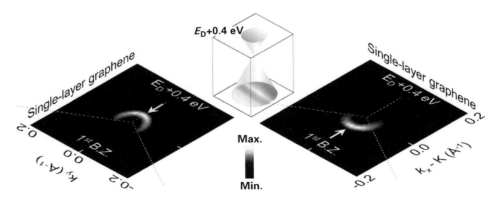

Figure 2.5 Measured intensity maps of single-layer graphene at energy $E = E_F$ with X- and Y-polarized light, respectively. Intensity maxima are denoted by white arrows and the electronic band structure of single-layer graphene is shown in the sketch. E_F is 0.4 eV above the Dirac point energy. (Reproduced with permission from Hwang *et al.* (2011). Copyright (2011) by the American Physical Society. By courtesy of Choonkyu Hwang.)

K point are also valid for 1D systems such as metallic nanotubes and wide armchair nanoribbons, as is demonstrated in the following sections. However, other symmetries result in semiconducting systems with varying gaps. Semiconducting nanotubes and ribbons with increasing diameter (or width) show a linear downscaling of their associated energy gaps. By using proper boundary conditions, the electronic band structure of both types of system can be analytically derived. One notes that this peculiar electronic band structure of graphene yields a specific behavior of the total density of states which can be written

$$\rho(E) = \int \frac{dk_x dk_y}{(2\pi)^2} \delta(E - \varepsilon_k) = \frac{2|E|}{(\pi \hbar^2 v_F^2)}, \tag{2.32}$$

while the carrier density is given by

$$n(E) = \frac{sgn(E)(E^2)}{(\pi \hbar^2 v_F^2)}. \tag{2.33}$$

2.2.3   Electronic properties of graphene beyond the linear approximation

The description of quasiparticles as massless Dirac fermions is accurate for low-energy excitations. However, it may be necessary to refine the model by including the deformation of the electronic band structure, which is particularly important for higher energies (trigonal warping). This can be captured by a development of the electronic dispersion up to second order in momentum shift with respect to the $K_\pm$ points. Finally, in very clean graphene, electron–electron interaction can induce some low-energy renormalization of the electronic bands and wavepacket velocity. All these effects are briefly reviewed in the following sections.

Trigonal warping corrections
It is important to know that deviations from the linear dispersion of the energy bands away from the Fermi level are designated as *trigonal warping*. Indeed, when expanding the full band structure close to one of the Dirac points ($\mathbf{k} = \mathbf{K}_\pm + \mathbf{p}/\hbar$ with $\mathbf{p}/\hbar \ll |\mathbf{K}_\pm|$), the energy dispersion is given by:

$$E_\pm(p) \simeq \pm v_F |\mathbf{p}| + \mathcal{O}\left[\left(\frac{p}{\hbar K_\pm}\right)^2\right], \tag{2.34}$$

where $\mathbf{p}$ is the momentum measured relatively to the Dirac point, and $v_F = \sqrt{3}\gamma_0 a/2\hbar$, the Fermi velocity. The expansion of the spectrum around the Dirac point up to the second order in p, and including second-nearest-neighbor interaction ($\gamma_0^{(2)}$) gives

$$E_\pm(\mathbf{p}) \simeq 3\gamma_0^{(2)} \pm v_F |\mathbf{p}| - \left(\frac{9\gamma_0^{(2)} a^2}{4} \pm \frac{3\gamma_0 a^2}{8} \sin(3\theta_p)\right) |\mathbf{p}|^2, \tag{2.35}$$

where $\theta_p = \arctan(p_x/p_y)$ is the angle in momentum space. Note that the presence of $\gamma_0^{(2)}$ shifts the position of the Dirac point in energy, thus breaking the electron–hole symmetry. Consequently, up to order $(p/\hbar K_\pm)^2$, the dispersion depends on the p direction in momentum space and has a threefold symmetry. This is the so-called "trigonal warping" of the electronic spectrum (Ando, Nakanishi & Saito, 1998, Dresselhaus & Dresselhaus, 2002) (see also Problem 2.10).

Effects beyond nearest neighbors and comparison with first-principles calculations

Most *tight-binding* (TB) studies use a first-nearest-neighbors π-π^* scheme to describe graphene electronic properties. However, when compared to *ab initio* calculations, this simplified TB approach predicts the electronic energies correctly in a limited energy range. A better, but more complicated, π-π^* parameterization has been proposed in the literature for both graphene (Reich *et al.*, 2002) and few-layer graphene (Grüneis *et al.*, 2008).

When interactions are included up to the third nearest-neighbors (3*nn*), the resulting two centers 3rd nearest-neighbors π-π^* orthogonal TB model turns out to be much more efficient (Lherbier *et al.*, 2012), accurately describing first-principles results over the entire Brillouin zone. In contrast to the 1st *nn* π-π^* model which produces a totally symmetric band structure, the 3rd *nn* TB model recovers the asymmetry between valence and conduction van Hove singularities, and the agreement with *ab initio* band structures is quite satisfactory, as illustrated in Fig. 2.6. The 3rd *nn* parameters of the TB model used to construct these band structures (Fig. 2.6) are only composed of a single on-site energy term ε_{p_z} and three hopping terms $\gamma_0^{(1)} = \gamma_0$, $\gamma_0^{(2)}$ and $\gamma_0^{(3)}$ corresponding respectively to 1st, 2nd and 3rd *nn* interactions. The pristine graphene

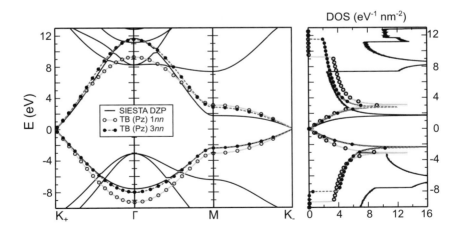

Figure 2.6 Electronic band structures and density of states (DOS) computed using the SIESTA code with a double-ζ polarized (DZP) basis set (full lines) along K-Γ-M-K path for a 1×1 supercell (unit cell). The TB band structures for a 1st nearest-neighbors model (1*nn*, lines with open circle symbols) and for a 3rd nearest-neighbors model (3*nn*, lines with filled circle symbols) are also plotted. Fermi energy is set to zero. (Adapted with permission from Lherbier *et al.* (2012). Copyright (2012) by the American Physical Society.)

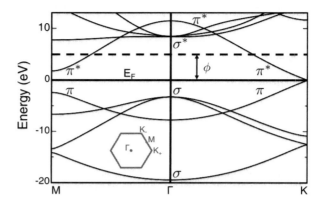

Figure 2.7 Electronic band structure of graphene. The bonding σ and the antibonding σ^* bands are separated by a large energy gap. The bonding π (highest valence band) and the antibonding π^* (lowest conduction band) cross at the K_+ (and K_-) points of the Brillouin zone. The Fermi energy (E_F) is set to zero and ϕ indicates the work function. Above the vacuum level (dashed horizontal line), the states of the continuum are difficult to describe and merge with the σ^* bands. The 2D hexagonal Brillouin zone is illustrated in the inset with the high-symmetry points Γ, M, K_+ and K_-. (Adapted with permission from Charlier, Blase & Roche (2007). Copyright (2007) by the American Physical Society.)

Hamiltonian then reads as

$$
\begin{aligned}
H = &\sum_i \varepsilon_{p_z} |\phi_i\rangle \langle \phi_i| \\
&+ \sum_{i,<j,k,l>} \left(\gamma_0^{(1)} |\phi_j\rangle \langle \phi_i| + \gamma_0^{(2)} |\phi_k\rangle \langle \phi_i| + \gamma_0^{(3)} |\phi_l\rangle \langle \phi_i| + h.c. \right),
\end{aligned}
\tag{2.36}
$$

with $\varepsilon_{p_z} = 0.6$ eV and $\gamma_0^{(1)} = -3.1$eV, $\gamma_0^{(2)} = 0.2$ eV, $\gamma_0^{(3)} = -0.16$ eV. The sum on index i runs over all carbon p_z orbitals. The sums over j, k, l indexes run over all p_z orbitals corresponding respectively to 1st, 2nd and 3rd nearest neighbors of the ith p_z orbital. In Fig. 2.6, the TB band structure (lines with symbols) is superimposed on the *ab initio* band structure (full lines). A good agreement is obtained especially for the valence bands. The conduction band side seems to be a little bit less accurate but this is uniquely due to the inability of the pristine graphene TB model to reproduce the conduction band along the K-M branch.

The *ab initio* electronic bands of graphene (Charlier, Blase & Roche, 2007) along the high-symmetry $M-\Gamma-K$ directions are presented in Fig. 2.7.[6] Its space group ($P3m$) contains a mirror symmetry plane, allowing symmetric σ and antisymmetric π states

---

[6] Note that in plots like the one in Fig. 2.7 the dispersion does not depend on the K point (K_+ or K_-) selected for the path in the Brillouin zone (horizontal axis in that figure). Therefore it is usual to call this point generically K. Notwithstanding one must remember that pseudospin is different on each valley.

to be distinguishable. In a 2D crystal, a parallel mirror symmetry operation separates the eigenstates for the whole Brillouin zone, and not only along some high-symmetry axis. The π and π^* bands cross at the vertices of the hexagonal Brillouin zone (vertices labeled by their momentum vector usually denoted by K_+ and K_- as mentioned above). *Ab initio* calculations confirm that the π and π^* bands are quasilinear (linear very close to K_+ or K_- and near the Fermi energy), in contrast with the quadratic energy–momentum relation obeyed by electrons at band edges in conventional semiconductors. When several interacting graphene planes are stacked as in few-layer graphite (nGLs) or in the perfect graphite crystal, the former antisymmetric π bands are split owing to bonding or antibonding patterns, whereas the σ bands are much less affected by the stacking, as explained in the next section.

Interaction-driven distortions at the Dirac point

Although *ab initio* density-functional theory (DFT) calculations confirm the *tight-binding* linear dispersion picture, the estimation of the Fermi velocity v_F was found to be smaller by 15–20% than the experimental value (Calandra & Mauri, 2007). Consequently, the role of electron–electron self-energy effects in the quasiparticle (QP) band structures and the Fermi velocity has been clarified (Trevisanutto *et al.*, 2008, Siegel *et al.*, 2011). With respect to the density-functional theory within the local-density approximation (Fig. 2.8(a)), the Fermi velocity is renormalized with an increase of 17%, such that it corrects the DFT underestimation and leads to a value of $1.12 \times 10^6 \ \mathrm{m\,s^{-1}}$ (Trevisanutto *et al.*, 2008), in good agreement with accurate magnetotransport

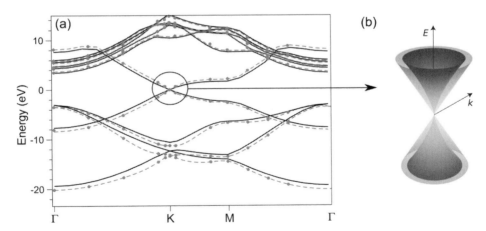

Figure 2.8 Many-body effects in graphene: (a) electronic band structures calculated within both DFT-LDA (solid lines) and GW (circles and dashed lines) approaches; (b) reshaping of the Dirac cone due to the interaction-driven renormalization (increase) of the Fermi velocity at low momenta. The outer cone represents the linear Dirac spectrum without many-body effects. ((a) is reproduced from Trevisanutto *et al.*, 2008. Copyright (2008) by the American Physical Society. (b) is reprinted with permission from Macmillan Publishers Ltd: *Nature Physics* (Elias *et al.*, 2011), copyright (2011)).

measurement of 1.1×10^6 m s^{-1} (Zhang *et al.*, 2005). Furthermore, the nearly linear DFT band dispersion is in GW considerably distorted. Close to the Dirac point the self-energy results in an unusual negative GW band-gap correction and the appearance of a kink in the band structure (Fig. 2.8(b)) which is due to a coupling with the π plasmon at $\sim$5 eV and the low-energy $\pi \rightarrow \pi^*$ single-particle excitations.

By measuring the cyclotron mass in suspended graphene with carrier concentrations which varied by three orders of magnitude, Elias *et al.* (2011) showed departures from linear behavior due to electron–electron interactions (with increasing v_F near the Dirac point). Interestingly, no gap was found even at energies as close as 0.1 meV to the Dirac point and no new interaction-driven phases were observed. This suggests that if there is a gap in graphene it is not larger than 0.1meV.

Consequently, the quasiparticle properties of graphene are modified by the presence of long-range Coulomb interactions. Their effects are especially pronounced when the Fermi energy is close to the Dirac point, and can result in strong renormalization of the Dirac band structure (the Fermi velocity v_F) and a reconstruction of the Dirac cone structure near the charge neutrality point. Consequently, many electronic characteristics and transport phenomena are strongly affected by these many-body effects that are sensitive to the value of the Coulomb interaction constant in graphene.

In the preceding pages the electronic properties of two-dimensional graphene have been described either using an effectively massless Dirac fermion model in the vicinity of the charge neutrality point or using a *tight-binding* approach within the nearest-neighbors approximation. Refinements of the electronic band structure have been obtained by going beyond the linear approximation, including trigonal warping effects, extending the *tight-binding* model to third nearest neighbors, or by performing full first-principles calculations, including GW corrections. Actually, both DFT and GW simulations confirm the low-energy linear energy dispersion close to the charge neutrality point, although renormalized Fermi velocity is obtained owing to many-body effects (long-range Coulomb interactions) at the Dirac point. In the next section the effect of the stacking on the electronic properties of few-layer graphene is highlighted.

2.3   Electronic properties of few-layer graphene

Bulk 3D graphites are semi-metallic materials that exhibit very peculiar electronic properties. Indeed, the first semi-empirical models for Bernal (or *ABAB*-stacking) (McClure, 1957, Slonczewski & Weiss, 1958) and rhombohedral (or *ABC*-stacking) graphites (Haering, 1958, McClure, 1969) have demonstrated that the shape of the Fermi surface, and consequently the nature of the charge carriers, are strongly dependent upon the geometry of the stacking between layers. The Fermi surfaces of these two graphite structures are represented in Fig. 2.9, and compared to the "ideal" simple hexagonal case (*AAA*-stacking) where all graphene planes are piled up exactly on top of each other. Fermi surfaces are located along the $H - K - H$ edge of the 3D Brillouin zones and exhibit a complex shape, due to the coexistence of holes and electrons at the charge

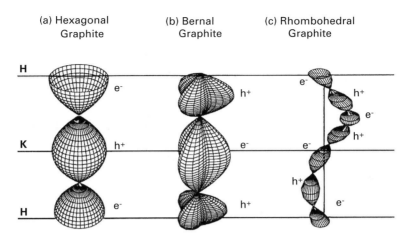

Figure 2.9 Electronic properties of graphite(s) with various stacking: Fermi surfaces of (a) simple hexagonal graphite (AAA-stacking), (b) Bernal graphite ($ABAB$-stacking), and (c) rhombohedral graphite (ABC-stacking). The three Fermi surfaces are centered on the vertical $H-K-H$ edge of the 3D Brillouin zone. Electron and hole pockets are labeled using e^- and h^+, respectively. (Reprinted from Charlier, Gonze & Michenaud, 1994a, Copyright (1994), with permission from Elsevier.)

neutrality point (Charlier *et al.*, 1991, Charlier, Michenaud & Gonze, 1992, Charlier, Gonze & Michenaud, 1994a).

Analogously, since few-layer graphenes are intermediate quasi-2D crystals between bulk graphite(s) and graphene, their electronic structures will be reminiscent of both of them. The weak interlayer interaction that creates the band dispersion out of the basal plane in graphite(s) is now responsible for the band mixing between isolated graphene bands occurring in few-layer graphene. Since coexistence of carriers is only possible when different bands are present in the same energy range, the number of layers and the geometry dependence of the interlayer interaction are key parameters influencing the transport properties in these quasi-2D graphene-based systems.

In bilayer graphene (AB-stacking), due to symmetry reasons, the $P\bar{3}m1$ group does not contain the horizontal mirror plane. Consequently, the valence band and conduction band only exhibit two contact points since they are not degenerated, except along high-symmetry axes, thus avoiding any deep domain of coexistence of electrons and holes (Fig. 2.10(a)). The close-up of the overlapping region clearly demonstrates the loss of the linear dispersion of the kinetic energy of the charge carriers ($E \propto k$) previously obtained for graphene (Latil & Henrard, 2006, Varchon *et al.*, 2007). Indeed, the band structures of the bilayer system present a parabolic shape ($E \propto k^2$) along the high-symmetry axes. In addition, both band extrema are actually saddle points (delimiting a pseudogap $\varepsilon_{psg} \sim 3$ meV) and the real overlap between the two touching points (K itself and one point along the $K - \Gamma$ axis) is $\delta\varepsilon \sim 1$ meV (Latil & Henrard, 2006). Unfortunately, this domain of coexistence is far too narrow to be visible experimentally. Nevertheless, the important feature that is preserved is the absence of a band gap, since the upper valence band touches the lower conduction band at the K point of the Brillouin

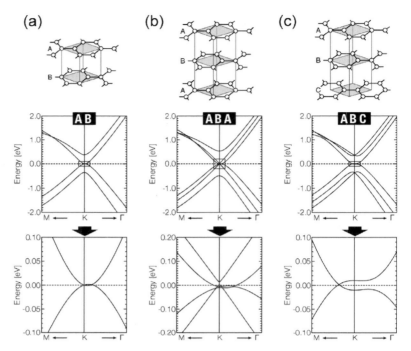

Figure 2.10 Electronic properties of few-layer graphene: band structures in the vicinity of K, and near the Fermi level (zero energy), for: (a) bilayer graphene (ABC-stacking); (b) trilayer graphene (ABA-stacking); and (c) trilayer graphene (ABC-stacking). (Adapted with permission from Latil & Henrard (2006). Copyright (2006) by the American Physical Society.)

zone. However, a significant band gap could be theoretically induced by lowering the symmetry of the system through the application of a perpendicular electric field (Castro *et al.*, 2007). Indeed, a bilayer graphene-based material has been produced experimentally, exhibiting an electrically tunable band gap, a phenomenon of great significance for both basic physics and applications (Mak *et al.*, 2009).

Regarding the trigonal warping corrections discussed before for monolayer graphene, we note that this effect turns out to be strong only close to the charge neutrality point in this case. *Tight-binding* models developed for graphite can be easily extended to the bilayer structure which presents the AB-stacking as in 3D bulk graphite. The set of hopping parameters for graphene has to be completed with $\gamma_1 \simeq 0.4$ eV (hopping energy between A_1 and A_2 atoms from the two layers), $\gamma_4 \simeq 0.04$ eV (hopping energy between A_1 (A_2) and B_2 (B_1) atoms from the two layers), and $\gamma_3 \simeq 0.3$ eV (hopping interaction between B_1 and B_2; see Fig. 2.11(a). The hopping γ_4 leads to a k-dependent coupling between the sublattices. The same role is played by the inequivalence between sublattices within a layer. However, in this approximation, the γ_3 hopping term qualitatively changes the spectrum at low energies since it introduces a trigonal distortion, or warping, of the bands. Unlike the one introduced by large momentum in Eq. (2.35), such trigonal warping drastically modifies the parabolic dispersions at the Dirac point at low energies ($\varepsilon < 5$ meV). The electron–hole symmetry is preserved but, instead of

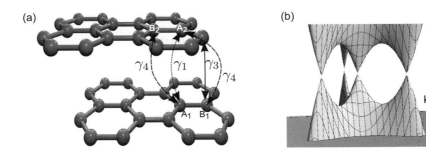

Figure 2.11 (a) Model of bilayer graphene where the hopping parameters connecting different layers mentioned in the text are indicated. (b) The dispersion relation of bilayer graphene very close to the charge neutrality point splits into four pockets when γ_3 is considered (this contrasts dramatically with the parabolic dispersion that it is found when this term is ignored).

two bands touching at $k = 0$, three sets of Dirac-like linear bands are obtained. The Dirac point is thus split in four pockets (see Fig. 2.11(b)), one Dirac point remaining at $\varepsilon = 0$ and $k = 0$, while the three other Dirac points, also at $\varepsilon = 0$, lie at three equivalent points with a finite momentum (McCann & Fal'ko, 2006, McCann, Abergel & Fal'ko, 2007) (see also Problem 2.12).

From the AB bilayer, trilayer graphene is constructed with an additional layer, keeping either the Bernal ABA-stacking pattern (Fig. 2.10(b)) or the rhombohedral ABC-stacking pattern (Fig. 2.10(c)). Indeed, when adding this third layer, electronic properties change, and two different situations arise depending on how this extra layer is stacked on the others.

The first possibility involves stacking the third layer so that it mirrors the first layer (Fig. 2.10(b)). Also referred to as Bernal or ABA-stacking, this arrangement has an electrical structure of overlapping linear and quadratic bands. The band structure of the ABA trilayer is characterized by band crossings in the vicinity of the Fermi level (Fig. 2.10(b)). The $P\bar{6}m2$ space group of the ABA trilayer contains the horizontal mirror symmetry, allowing the separation of the antisymmetric states and the symmetric ones. Moreover, the two symmetric bands exhibit a quasi-linear dispersion (*massless fermions*); however, unlike monolayer graphene, a gap opens due to the nonequivalence of carbon atoms in the same layer ($\varepsilon_{\mathrm{gap}} \sim 12$ meV). The band overlap between the top of the quasi-massless holes band and the electrons band is predicted to be of the order of a few meV (Latil & Henrard, 2006).

The second possibility, known as rhombohedral or ABC-stacking, involves displacing the third layer with respect to the second layer in the same direction again as the second with respect to the first. As the first precursor of the "rhombohedral family," the electronic structure of ABC trilayer graphene has cubic dispersion ($E \propto k^3$). This means that for low carrier concentrations (which correspond to low momentum states) the relative kinetic energy of the particles in rhombohedral graphene's cubic bands will be less than that in bilayer graphene's quadratic bands, which is less again than monolayer graphene's linear bands. The band structure of the ABC trilayer ($P\bar{3}m1$ space group) exhibits a single crossing point between valence and conduction bands, located along the K–M axis (Fig. 2.10(c)). Consequently, any coexistence of charge carriers is strictly forbidden in this specific stacking. However, a graphene *quasi-massless* dispersion is preserved and bounded by a pseudogap ε_{psg} ~18 meV. The group velocities vary from 1.9–2.6 10^5 m s^{-1} (Latil & Henrard, 2006).

But what happens when the number of layers is increased? N layers of graphene have approximately $2^{(N-2)}$ possible arrangements (Yacoby, 2011). One of these arrangements is the natural extension of the *ABC*-stacked trilayers that consists of a multi-layer with cyclic arrangement given by *ABCABCA*, and so on. The dispersion of such multilayers is predicted to have even lower kinetic energy ($E \propto k^N$). The reason that such electronic behavior has not yet been seen in graphite, the macroscopic form of multilayer graphene, is because natural graphite usually exhibits Bernal stacking where such effects would be absent. However, if effective ways of growing artificial few-layer graphene with rhombohedral stacking are found, *ABC* trilayer graphene might just be a new playground to tailor the electronic properties of few-layer graphene-based nano-structures.

Moreover, *ab initio* calculations on bilayer and trilayer graphene (Latil, Meunier & Henrard, 2007) suggest that the massless fermion behavior, a typical signature of single-layer graphene, is preserved in incommensurate multilayered graphitic systems. Indeed, the linear dispersion is conserved in turbostratic multilayer systems despite the presence of adjacent layers (Fig. 2.12), thus predicting the presence of Dirac carriers in disori-ented few-layer graphene. More generally, the electronic properties (and consequently the optical, vibrational, and transport properties) of a given FLG film are found to be controlled mainly by the misorientation of the successive layers rather than their num-ber (Latil, Meunier & Henrard, 2007). Recent experiments have shown that a rotation between the different graphene layers can generate van Hove singularities which, inter-estingly, can be brought arbitrarily close to the Fermi energy when the angle of rotation is changed (Li *et al.*, 2010), thereby opening promising opportunities for tuning the role of interactions in the material.[7]

To add even more thrill to this area, recent experiments and simulations in bilayer graphene (Kim *et al.*, 2013) show that even tiny imperfections (stacking and twist angle) can dramatically change the electronic structure. The ARPES data presented in Kim *et al.* (2013) show the co-existence of massive and massless Dirac fermions due to a distribution of twists as small as 0.1 degrees.

In conclusion, the electronic properties of few-layer graphene (FLG) are quite com-plex and present exotic electronic states. Indeed, depending on the stacking geometry and on the number of layers, an FLG can be either metallic (with single or mixed carri-ers) or an extremely narrow-gap semiconducting 2D system.

In the next two sections, confinement effects will be described when the graphene sheet is cut in strips as for graphene nanoribbons or rolled up in cylinders such as car-bon nanotubes. The remarkable properties of graphene derived close to the K_+ and K_-

---

[7] The problem of rotated or twisted graphene layers is fascinating and many questions are still open. For example, from which rotation angle does a twisted bilayer behave as a monolayer? In Suárez Morell *et al.* (2010) it was suggested that this transition occurs at a finite angle of 1.5 degrees when decoupling is achieved.

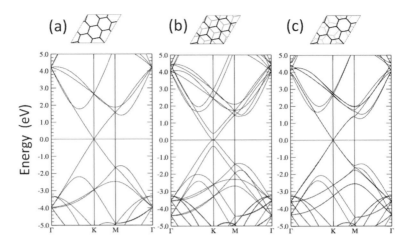

Figure 2.12 Electronic properties of single and bilayer graphenes: band structures of (a) single-layer; (b) Bernal bilayer; and (c) turbostratic bilayer in the vicinity of K point of the Brillouin zone and the Fermi energy. The corresponding supercells are also represented. (Adapted with permission from Latil, Meunier & Henrard, 2007. Copyright (2007) by the American Physical Society.)

points are found to remain valid for 1D systems such as metallic nanotubes and wide armchair nanoribbons. However, other symmetries result in semiconducting systems with varying gaps. Semiconducting nanotubes and ribbons with increasing diameter (or width) show a linear downscaling of their associated energy gaps. By using proper boundary conditions, the electronic band structure of both types of system can be analytically derived, as illustrated in the following.

2.4 Electronic properties of graphene nanoribbons

As mentioned above, the combination of high charge carrier mobilities and long coherent lengths makes graphene an outstanding material for nanoscale electronics. However, this wonder material has an Achilles heel. The electric conduction cannot be turned off by, for example, changing a gate voltage as is usual in field effect transistors. The ability to "turn off" graphene is crucial for achieving the control of the current flow needed in active electronic devices. Therefore, opening a band gap in graphene is an important problem and many different creative ways of doing it have been proposed.[8] One possible solution is to use narrow strips of graphene, also called graphene nanoribbons (GNRs).

Graphene nanoribbons can be obtained by cutting a graphene sheet as shown in Fig. 2.13. If one follows a certain direction when cutting, two typical shapes are

---

[8] Alternatives in the bulk material include growing epitaxial graphene on a SiC substrate (Zhou *et al.*, 2007). In that case the interaction with the substrate breaks the symmetry between the A and B sublattices, thereby opening a bandgap of about 0.26 eV. In bulk bilayer graphene, a gap can be opened by applying an electric field perpendicular to its surface (Zhang *et al.*, 2009).

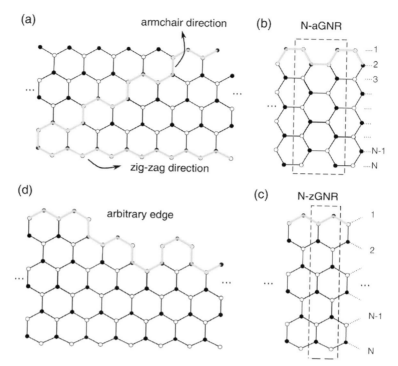

Figure 2.13 (a) Honeycomb lattice of graphene showing both armchair and zigzag directions. Ribbons with *armchair* and *zigzag* edges are shown in the schemes in (b) and (c) respectively, where the shape of the edges at the top of each scheme is highlighted with a grey shadow. Their corresponding 1D unit cells are marked with dashed boxes. A ribbon with a more general edge shape is shown in (d).

basically possible: the armchair edge (Fig. 2.13(b)) and the zigzag edge (Fig. 2.13(c)), having both edges with a difference of 30° between them. More complex shapes other than these two "ideal" cases are a combination of armchair and zigzag shaped pieces (Kobayashi *et al.*, 2005, Enoki, Kobayashi & Fukui, 2007). This last case is commonly found in experiments where obtaining atomically precise ribbons is difficult. This may jeopardize the achievement of clean bandgaps even for ribbons a few nanometers wide. Fortunately, strategies involving top-down[9] and bottom-up[10] approaches have allowed for important progress. Furthermore, unzipping carbon nanotubes, which could be termed a theorist's dream, has also been experimentally demonstrated (Jiao *et al.*, 2010, Roche, 2011, Shimizu *et al.*, 2011, Wang *et al.*, 2011).

The atomic structure of nanoribbons with armchair and zigzag edges is represented in Fig. 2.13(b) and (c), along with their corresponding unit cells. Here we follow previous

---

[9] See for example Han *et al.*, 2007, Chen *et al.*, 2007, Li *et al.*, 2008, Tapaszto *et al.*, 2008, Datta *et al.*, 2008, Ci *et al.*, 2008, Jiao *et al.*, 2009, Kosynkin *et al.*, 2009, Jiao *et al.*, 2010, Roche, 2011, Shimizu *et al.*, 2011, Wang *et al.*, 2011.

[10] See Campos-Delgado *et al.* 2008, Sprinkle *et al.* 2010, Cai *et al.* 2010, Kato & Hatakeyama 2012. In this last the authors demonstrated GNR devices with a transport gap of ~60 meV and high on/off ratios (>10^4).

conventions (Nakada *et al.*, 1996, Wakabayashi *et al.*, 1999, Miyamoto, Nakada & Fujita, 1999, Kawai *et al.*, 2000, Okada & Oshiyama, 2001, Lee *et al.*, 2005, Ezawa, 2006, Brey & Fertig, 2006, Sasaki, Murakami & Saito, 2006, Abanin, Lee & Levitov, 2006, Son, Cohen & Louie, 2006*a*, Son, Cohen & Louie, 2006*b*) and GNRs with armchair (zigzag) edges on both sides are classified by the number of dimer lines (zigzag lines) across the ribbon width. We will denote with N-aGNR and N-zGNR such armchair and zigzag GNRs, N being respectively the number of dimer and zigzag lines. In addition, if not otherwise stated, in the following the dangling bonds on the edge sites of GNRs will be assumed to be terminated by hydrogen atoms, although dangling bonds would not make any contribution to the electronic states near the Fermi level. Of course, these ideally shaped edges do not correspond to the experimental observations (Ritter & Lyding, 2009, Liu *et al.*, 2009, Girit *et al.*, 2009), where GNRs currently have a high degree of edge roughness. A more complex edge is illustrated in Fig. 2.13(d). Topological aspects of this edge disorder will be presented in the sections dedicated to the study of quantum transport in disordered GNRs.

The edges in graphene nanoribbons confine the electronic wavefunctions along the direction perpendicular to the ribbon axis. Their electronic properties can be obtained by imposing the appropriate boundary conditions on the Schrödinger's equation within the simple *single-band tight-binding approximation* based on π-states of graphene (Nakada *et al.*, 1996, Wakabayashi *et al.*, 1999, Ezawa, 2006) or on the two-dimensional Dirac's equation with an effective speed of light ($\sim 10^6$ m/s) (Brey & Fertig, 2006, Sasaki, Murakami & Saito, 2006, Abanin, Lee & Levitov, 2006).

As we will see below, the presence of edges introduces new states not present in bulk 2D graphene (Nakada *et al.*, 1996). These states appear because of the hard boundary conditions at the edges. Indeed, if instead of a vanishing wavefunction at the edges one considers periodic boundary conditions (this is essentially the zone-folding approximation for carbon nanotubes introduced later in Section 2.5), one finds that the wavevector along the direction perpendicular to the ribbon axis k_{perp} is quantized, thereby defining a set of "cutting" lines in the Brillouin zone. Within this approximation the band structure results from cutting the dispersion for bulk graphene along those lines. Figure 2.14 illustrates the basic picture of the zone-folding scheme: starting from a four-atom unit cell (Fig. 2.14(a)) one gets the corresponding Brillouin zone (gray rectangle in Fig. 2.14(b)). It is easy to see that for this Brillouin zone the K points for zigzag ribbons fold at $\pm 2\pi/(3a)$ along the vertical axis. For armchair ribbons the K points fold directly onto the Γ point. Projecting the bulk dispersion of graphene on each of the zigzag and armchair directions gives the shaded areas in Fig. 2.14 (c) and (d) respectively. While this approximation may give the correct overall shape of the band structure and even the states at the "bulk" of the ribbon (which can be formed by superposition of states with k_{perp} and $-k_{\text{perp}}$), it fails dramatically for low energies where it misses edge states. We note that edge states are generically present in all nanoribbons, even those with irregular edges, except the armchair ones (Akhmerov, 2011).

The specific edge symmetry of zigzag and armchair nanoribbons is shown in Fig. 2.13. When using the Dirac equation, appropriate boundary conditions need to be applied at the edges (vanishing wavefunction). For zigzag nanoribbons, given that

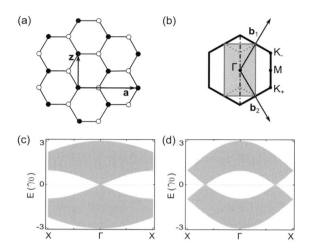

Figure 2.14 (a) A four-atom unit cell used in the zone-folding procedure. (b) The corresponding Brillouin zone in shaded grey together with the unit vectors and the Brillouin zone of bulk graphene. Depending on the ribbon termination K_+ and K_- can be folded at $\pm 2\pi/(3a)$ along the vertical axis (zigzag ribbons) or onto the Γ point (armchair ribbons). (c) and (d) The projections of graphene's bulk dispersion onto the zigzag and armchair directions respectively.

one edge is formed entirely of A-type atoms while the other edge contains only B-type atoms, the boundary conditions can be imposed separately on each sublattice (Brey & Fertig, 2006). For armchair graphene nanoribbons this is no longer the case (Brey & Fertig, 2006). Here, we take a different path, and consider armchair and zigzag graphene nanoribbons within a *tight-binding* model following Cresti *et al.* (2008) and Dubois *et al.* (2009).

2.4.1   Electronic properties of armchair nanoribbons (aGNRs)

By analogy with what has been done previously for graphene, the *tight-binding* Hamiltonian of the ribbons can be written as:

$$\mathcal{H} = \sum_i \varepsilon_i \hat{c}_i^\dagger \hat{c}_i - \sum_{i,j} \gamma_{ij} \hat{c}_i^\dagger \hat{c}_j, \tag{2.37}$$

where ε_i represent the onsite energies which can be chosen as the reference energy ($\varepsilon_i = \varepsilon = 0$), γ_{ij} are the transfer integrals between the jth and ith π orbitals, and $\hat{c}_i^\dagger$ ($\hat{c}_i$) is an operator which creates (annihilates) an electron at the orbital localized around site i. Within the first nearest-neighbor approximation, $\gamma_{ij} = \gamma_0$ for i, j neighboring sites and zero otherwise. (This is equivalent to assuming that the ribbon edges are passivated in such a way that bulk graphene is reproduced.)

Thanks to the periodicity along the axis of the aGNR, the $\hat{c}_i$ and $\hat{c}_i^\dagger$ operators may be expressed as Bloch sums,

$$\hat{c}_i = \frac{1}{\sqrt{N}} \sum_k e^{ikR_i} \hat{c}_k(i), \tag{2.38}$$

where R_i is the position of the ith site, and $\hat{c}_k(i)$ is one of the $\{\hat{c}_k^{1A}, \hat{c}_k^{1B}, \hat{c}_k^{2A}, \hat{c}_k^{2B}, \ldots, \hat{c}_k^{NB}\}$ operators, named by reference to the unit cell sites $\{1_A, 1_B, \ldots, N_A, N_B\}$ as represented in Fig. 2.13(b).

Replacing Eq. (2.38) into Eq. (2.37), the *tight-binding* Hamiltonian of *armchair* GNRs can be expressed in terms of the basis set $\{\hat{c}_k^{1A}, \hat{c}_k^{1B}, \hat{c}_k^{2A}, \hat{c}_k^{2B}, \ldots, \hat{c}_k^{NB}\}$,

$$\mathcal{H} = \sum_{k \in BZ} \mathcal{H}_k , \qquad (2.39)$$

with

$$\mathcal{H}_k = \gamma_0 \hat{\phi}_k^\dagger \begin{pmatrix} 0 & e^{ik\frac{a}{2}} & 0 & 1 & 0 & 0 & \cdots \\ e^{-ik\frac{a}{2}} & 0 & 1 & 0 & 0 & 0 & \cdots \\ 0 & 1 & 0 & e^{ik\frac{a}{2}} & 0 & 1 & \cdots \\ 1 & 0 & e^{-ik\frac{a}{2}} & 0 & 1 & 0 & \cdots \\ 0 & 0 & 0 & 1 & 0 & e^{ik\frac{a}{2}} & \cdots \\ 0 & 0 & 1 & 0 & e^{-ik\frac{a}{2}} & 0 & \cdots \\ \cdots & \cdots & \cdots & \cdots & \cdots & \cdots & \cdots \end{pmatrix} \hat{\phi}_k , \qquad (2.40)$$

where

$$\hat{\phi}_k = \left[\hat{c}_k^{1A}, \hat{c}_k^{1B}, \hat{c}_k^{2A}, \hat{c}_k^{2B}, \ldots, \hat{c}_k^{NA}, \hat{c}_k^{NB} \right]^T . \qquad (2.41)$$

By diagonalizing the Hamiltonian of Eq. (2.40), the band structure is obtained. Figure 2.15 shows the results for three different nanoribbons (15-aGNR, 16-aGNR and 17-aGNR). We note that the typical Dirac-like linear dispersion or the direct gap always appears at $k = 0$ for the *armchair* configuration, a fact that can also be predicted by using the zone-folding approximation. (An alternative approach based on a mode decomposition in *real* space is presented in Section 4.2.2.)

From these results we can see that some ribbons exhibit semiconducting behavior while others are metallic. An analytical calculation (Cresti *et al.*, 2008) of the eigenvalues of the *tight-binding* Hamiltonian at $k = 0$, shows that the energy gap (Δ_N) is width dependent:

$$\Delta_N = \begin{cases} \Delta_{3\ell} = |\gamma_0| \left(4\cos\frac{\pi\ell}{3\ell+1} - 2 \right), \\[2mm] \Delta_{3\ell+1} = |\gamma_0| \left(2 - 4\cos\frac{\pi(\ell+1)}{3\ell+2} \right), \\[2mm] \Delta_{3\ell+2} = 0, \end{cases} \qquad (2.42)$$

with $\Delta_{3\ell} > \Delta_{3\ell+1} > \Delta_{3\ell+2} = 0$, where N (related to the ribbon width) and ℓ are integers.

Therefore, the *tight-binding* model predicts that N-aGNRs are metallic for every $N = 3\ell + 2$ (where ℓ is a positive integer), and semiconducting otherwise.[11]

More precise DFT calculations, however, reveal that even for the $3\ell + 2$-GNRs there is a small gap at $k = 0$. According to DFT, all armchair ribbons remain semiconducting

---

[11] An alternative to the path described above is described in Section 4.2.2, where a mode decomposition in real space, obtained by straightforward parallelization of the *tight-binding* Hamiltonian for GNRs and carbon nanotubes, is introduced. This scheme is useful especially when carrying out transport calculations and therefore is introduced later on when dealing with ballistic transport in carbon nanostructures.

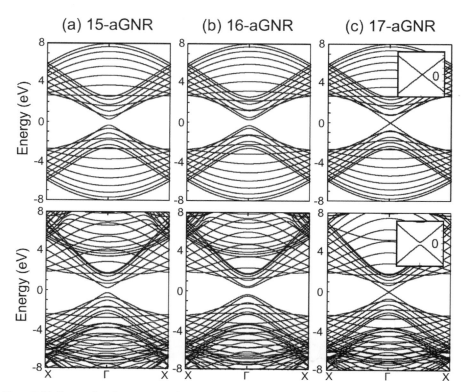

Figure 2.15 Energy band structures of three armchair nanoribbons (*N*-aGNRs) of various widths (*N* = 15, 16, 17). The *tight-binding* band structures (top) computed using a constant hopping energy (γ_0 = 2.7eV) between nearest neighbors are compared to *ab initio* band structures (bottom). Reproduced from Dubois, 2009.

(Son, Cohen & Louie, 2006*a*, Son, Cohen & Louie, 2006*b*) with gaps which decrease as the width of the aGNR increases, reaching the zero-gap value of graphene for infinite ribbon width. For a ~5 nm wide ribbon with metallic behavior as predicted by the *tight-binding* model, the *ab initio* band gap has a magnitude of only ~0.05 eV.

The origin of this gap opening can be attributed to edge effects (Son, Cohen & Louie, 2006*a*) not taken into account in the simple *tight-binding* model. Indeed, the edge carbon atoms of the aGNR are passivated by hydrogen atoms, by some foreign atoms, or by molecules in general. Therefore, one may generally expect the σ bonds between hydrogen and carbon as well as the onsite energies of the carbons at the edges and their bonding distances to be different from those in the middle of the ribbon. The bonding distances between carbons at the edges decrease from 1.44 Å (Fig. 2.16) leading to an increase of ~ 15% in the hopping integral between π orbitals. This physically explains the emergence of a band gap for all the aGNRs (Son, Cohen & Louie, 2006*a*, Dubois, 2009). By introducing these modified onsite and hopping energies within an improved nearest-neighbors *tight-binding* model one obtains an improved agreement with the DFT band structure.

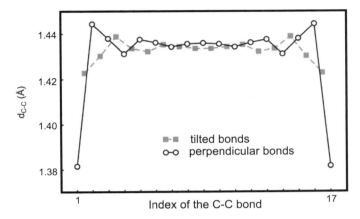

Figure 2.16 Evolution of the carbon–carbon bond length (d_{C-C}) across the 17-aGNR ribbon width, computed using *ab initio* structural optimization techniques. Both the length of the carbon–carbon parallel (empty circles) and tilted (filled squares) bonds with respect to the ribbon axis are illustrated. Reproduced from Dubois, 2009.

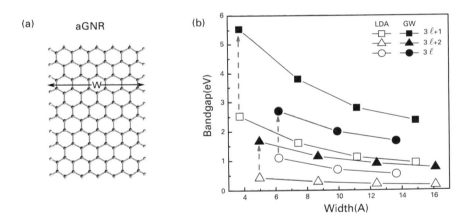

Figure 2.17 (a) Ball-stick model for an 11-aGNR which exhibits 11 C–C dimer lines making up its width *w*. Hydrogen atoms (light colored balls on the left and right edges) are used to passivate the edge σ-dangling bonds. (b) Width-dependence of aGNRs band gaps. The three families of aGNRs are represented by different symbols. The values of the same family of aGNRs are connected by solid lines as a guide to the eyes. The open symbols are LDA band gaps while the solid symbols are the corresponding quasiparticle band gaps. Dashed arrows are used to indicate the self-energy correction for the smallest width ribbon of each of the three aGNRs families. (Adapted with permission from Yang *et al.* (2009). Copyright (2007) by the American Physical Society. By courtesy of Li Yang.)

Finally, it has to be noted that the *ab initio* energy gaps are further increased when the electron–electron correlations are taken into account by means of GW corrections (Yang *et al.*, 2007). DFT-LDA and quasiparticle band gaps for armchair GNRs of various width are compared in Fig. 2.17. In agreement with DFT-LDA calculations, the quasiparticle band structure has a direct band gap at the zone center for all investigated aGNRs. In addition, the band gaps of the three families of N-aGNRs, which are classified according to whether $N = 3\ell+1$, $3\ell+2$, or 3ℓ (N being the number of dimer chains as explained earlier in Fig. 2.13(b), and ℓ being an integer), present qualitatively the same hierarchy as those obtained in DFT-LDA ($E_g^{3\ell+1} > E_g^{3\ell} > E_g^{3\ell+2} \neq 0$). Although including electron–electron interaction, the energy gaps are also found to decrease as the widths of the aGNR increase, reaching the zero-gap value of graphene for the infinite width.

2.4.2   Electronic properties of zigzag nanoribbons (zGNRs)

Taking into account a periodicity along the ribbon length, one rewrites the Hamiltonian as a Bloch sum ($\mathcal{H} = \sum_{k \in BZ} \mathcal{H}_k$). For zGNRs, $\mathcal{H}_k$ finally reads

$$
\mathcal{H}_k = \gamma_0 \hat{\phi}_k^\dagger
\begin{pmatrix}
0 & 2\cos(\frac{ka}{2}) & 0 & 0 & 0 & \cdots \\
2\cos(\frac{ka}{2}) & 0 & 1 & 0 & 0 & \cdots \\
0 & 1 & 0 & 2\cos(\frac{ka}{2}) & 0 & \cdots \\
0 & 0 & 2\cos(\frac{ka}{2}) & 0 & 1 & \cdots \\
0 & 0 & 0 & 1 & 0 & \cdots \\
\cdots & \cdots & \cdots & \cdots & \cdots & \cdots
\end{pmatrix}
\hat{\phi}_k, \quad (2.43)
$$

where $\hat{\phi}_k = [\hat{c}_k^{1A}, \hat{c}_k^{1B}, \hat{c}_k^{2A}, \hat{c}_k^{2B}, \ldots, \hat{c}_k^{NA}, \hat{c}_k^{NB}]^t$.

By diagonalizing the Hamiltonian of Eq. (2.43), the band structure shown in Fig. 2.18 is obtained. One notes that a dispersion relation reminiscent of the Dirac cones develops around $k = \pm 2\pi/(3a)$. Another salient feature is the formation of a sharp peak in the density of states at E_F, resulting from the formation of partially flat and degenerate bands with zero energy (between the Dirac points and the border of the Brillouin

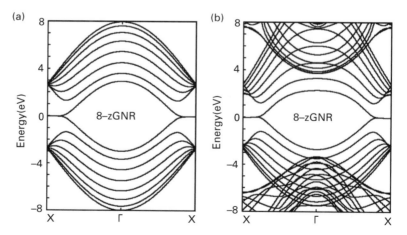

Figure 2.18 Band structure of zigzag nanoribbons (8-zGNR): (a) *tight-binding* band structures using a constant nearest-neighbors hopping energy ($\gamma_0 = -2.7\text{eV}$); (b) *ab initio* band structures. Reproduced from Dubois, 2009.

zone ($2\pi/(3a) \leq |k| \leq \pi/a)$), and which represents the highest valence and lowest conduction bands.

First-principles calculations of the electronic structure confirm that the zero-energy states are mainly confined along the ribbon edges, but progressively spread along the ribbon lateral dimension as the wavevector is shifted from π/a to $2\pi/(3a)$ (see Fig. 2.18(b)). A slight dispersion of those states (which is width dependent) develops due to the overlap between opposite edge states and the formation of bonding and anti-bonding states. Actually, simple calculations show that an edge shape with three or four zigzag sites per sequence is enough to form the edge state (Nakada *et al.*, 1996). The presence of such remarkably confined electronic edge states has been confirmed by STM and STS measurements (Kobayashi *et al.*, 2005).

In contrast with the aGNR case, the modification of the carbon–carbon bonding distance at the edges is found not to affect the low-energy electronic structures (Fig. 2.19). The partially flat bands at the Fermi energy are actually "topologically protected," and thus insensitive to the precise hopping energy across the ribbon width.

The peculiar edge states of zGNRs are furthermore evidence of some local magnetic ordering, although the ribbon as a whole has a nonmagnetic ground state, with ferro-magnetic ordering at each zigzag edge and antiparallel spin orientation between the two edges (Fig. 2.20(a)) (Wakabayashi *et al.*, 1999, Okada & Oshiyama, 2001, Lee *et al.*, 2005, Son, Cohen & Louie, 2006*a*, Son, Cohen & Louie, 2006*b*). One notes, however, that the difference in total energy per edge atom between non-spin-polarized and spin-polarized edge states is only of the order of a few tens of meV (Son, Cohen & Louie, 2006*a*). Additionally, zGNRs exhibit a Curie-like temperature dependence of the Pauli paramagnetic susceptibility, and a crossover is predicted from high-temperature diamagnetic to low-temperature paramagnetic behavior in the magnetic susceptibility (Wakabayashi *et al.*, 1999).

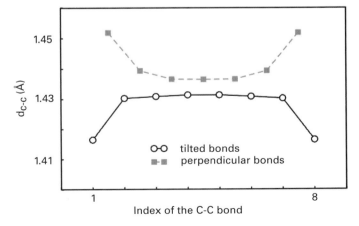

Figure 2.19 Evolution of the carbon–carbon bond length (d_{C-C}) across the 8-zGNR ribbon width, obtained after *ab initio* structural optimization. Both the length of the carbon–carbon parallel (empty circles) and tilted (filled squares) bonds with respect to the ribbon axis are shown. Reproduced from Dubois, 2009.

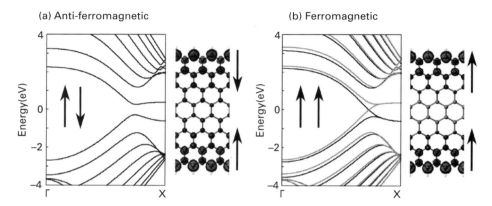

Figure 2.20 Spin-polarized electronic structure of the 8-zGNR: (a) ↑↓ configuration (ground state): a ferromagnetic order along its axis and antiparallel coupling between the edges; (b) ↑↑ configuration: a ferromagnetic order along its axis and parallel coupling between the edges. In each panel, the energy band structure (left) and the spin-polarized electronic densities ($\rho_\uparrow - \rho_\downarrow$, right) iso-surfaces are computed within the spin-dependent DFT framework. Adapted from Dubois, 2009.

The electronic ground state of the pristine 8-zGNR shows some antiparallel (↑↓) spin orientations between the edges (Fig. 2.20(a)), leading to a semiconducting band structure (0.5 eV band gap) with full spin degeneracy. Besides, the magnetic configuration with parallel (↑↑) spin orientations between the edges (Fig. 2.20(b)) is metastable (11 meV/edge-atom higher in energy). This configuration displays a metallic behavior, as the $\pi_\uparrow^*$ and $\pi_\downarrow$ bands cross at the Fermi energy with the formation of a total magnetic moment of 0.51 μ_B (per edge atom).

By performing spin-polarized calculations, the energy gap formation in zGNRs can be rationalized by the magnetic ordering-induced staggered sublattice potentials (Kane & Mele, 2005). This is unique to the edge symmetry since opposite spin states are forced to lie on different sublattices. Since the strength of the staggered potentials in the middle of the ribbon decreases with the ribbon width, the band gaps of zigzag GNRs are consequently inversely proportional to their width. The band structure of zGNRs is slightly modified when accounting for the GW correction (Yang *et al.*, 2007). The self-energy corrections are found to enlarge the energy gaps for all zGNRs and slightly increase the band dispersion of edge states (Fig. 2.21).

Differently to the band gap (Δ^0) located around three-fourths of the way to the Brillouin zone edge (Fig. 2.21), the zGNR energy gap at the zone boundary (Δ^1) is width-insensitive because of its dominant edge-state character (Yang *et al.*, 2007). The dependence of the GW correction on the wave vector slightly affects the band dispersion of zGNRs (Fig. 2.21(b–c)). The GW correction is calculated to be about 1 eV for Δ^0 for a ribbon width between 1–2.5 nm (Yang *et al.*, 2007). Since the Δ^1 gap is width independent, the corresponding GW correction remains of the order of 1.5 eV.

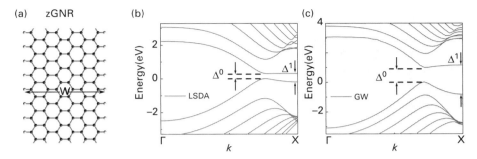

Figure 2.21 (a) Ball-stick model for a 12-zGNR which exhibits 12 zigzag chains along the ribbon axis. Hydrogen atoms (light colored balls on the left and right edges) are used to passivate the edge σ-dangling bonds. (b and c) Calculated band structure (and energy gap) of a 12-zGNR within the LSDA (b) and GW (c) approximations. The up and down spin states are degenerated for all the bands in the $\uparrow\downarrow$ ground state configuration, and the top of the valence band is set at zero. The symbols Δ^0 and Δ^1 denote the direct band gap and the energy gap at the zone boundary. (Adapted with permission from Yang *et al.*, 2007. Copyright (2007) by the American Physical Society.)

To conclude, we have seen the fundamental role played by edge symmetries, for determining the precise values and the width dependence of energy band gaps in both *armchair* and *zigzag* GNRs. The enhanced electron–electron interaction in these quasi-one-dimensional systems yield to significant self-energy correction in both *armchair* and *zigzag* GNRs. The states near the band gap of zGNRs are sensitive to the wavevector, giving rise to a larger band width and smaller effective mass (Yang *et al.*, 2007). The computed band gaps lie in the range 1 to 3 eV for GNRs with widths from 3 to 1 nm.

2.5 Electronic properties of carbon nanotubes

2.5.1 Structural parameters of CNTs

In 1991, helped by state-of-the-art transmission microscopy, Sumio Iijima from NEC laboratories in Japan discovered and first characterized *"Helical microtubules of graphitic carbon"* (Iijima, 1991). The microtubules were made of concentric cylindrical shells with a spacing between them of about 3.4 Å, the same as usually found in conventional graphite materials. Their diameter ranged from a few nanometers for the inner shells to several hundred nanometers for the outer shells and they constituted what today we call carbon nanotubes. A few years later, arc discharge methods with transition metal catalysts were used to successfully synthesize carbon nanotubes made of a *single* graphene layer rolled into a hollow cylinder (Iijima & Ichihashi, 1993, Bethune *et al.*, 1993). In contrast to the multiwall carbon nanotubes (MWNTs) obtained earlier, these structures, called single wall carbon nanotubes (SWNTs), had diameters of about one nanometer and an impressively perfect crystalline structure. They were considered the "ultimate" carbon-based 1D systems.

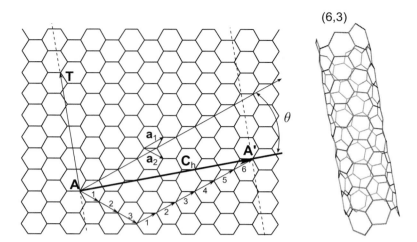

Figure 2.22 The graphene network. The lattice vectors are indicated by $\mathbf{a}_1$ and $\mathbf{a}_2$. In this example the chiral vector is $\mathbf{C}_h = 6\mathbf{a}_1 + 3\mathbf{a}_2$. The direction perpendicular to $\mathbf{C}_h$ is the tube axis (dashed lines) where the translational vector $\mathbf{T}$ is indicated. The angle between $\mathbf{C}_h$ and the $\mathbf{a}_1$ "zigzag" direction of the graphene lattice defines the chiral angle θ. The resulting $(6, 3)$ nanotube is shown on the right. The unit cell for this nanotube is a rectangle bounded by $\mathbf{C}_h$ and $\mathbf{T}$.

As shown in Fig. 2.22, the structure of single-wall carbon nanotubes is that of a rolled graphene strip (Saito, Dresselhaus & Dresselhaus, 1998). Their structure can be specified by the chiral vector ($\mathbf{C}_h$) that connects two equivalent sites (A and A' in Fig. 2.22) on a graphene sheet. Therefore, the chiral vector can be specified by two integer numbers (n and m), $\mathbf{C}_h = n\mathbf{a}_1 + m\mathbf{a}_2$, and represents the relative position of the pair of atoms on the graphene network which form a tube when rolled. The (n, m) pair uniquely labels SWNTs.

Since the chiral vector $\mathbf{C}_h$ defines the circumference of the tube, its diameter can be estimated as $d_t = |\mathbf{C}_h|/\pi = \frac{a}{\pi}\sqrt{n^2 + nm + m^2}$, where a is the lattice constant of the honeycomb network ($a = \sqrt{3} \times a_{cc}$ and $a_{cc} \simeq 1.42$ Å , the C–C bond length). The chiral vector $\mathbf{C}_h$ uniquely defines a particular (n, m) tube, as well as its chiral angle θ which is the angle between $\mathbf{C}_h$ and $\mathbf{a}_1$ ("zigzag" direction of the graphene sheet, see Fig. 2.22).

The chiral angle θ can be calculated from

$$\cos\theta = \frac{\mathbf{C}_h \cdot \mathbf{a}_1}{|\mathbf{C}_h||\mathbf{a}_1|} = (2n + m)/(2\sqrt{n^2 + nm + m^2}), \qquad (2.44)$$

and lies in the range $0 \leq |\theta| \leq 30°$, because of the hexagonal symmetry of the graphene lattice. Nanotubes of the type $(n, 0)$ ($\theta = 0°$) are called *zigzag* tubes, because they exhibit a zigzag pattern along the circumference. Such tubes display carbon–carbon bonds parallel to the nanotube axis. On the other hand, nanotubes of the type (n, n) ($\theta = 30°$) are called *armchair* tubes, because they exhibit an armchair pattern along the circumference. Such tubes display carbon–carbon bonds perpendicular to the nanotube axis. Both zigzag and armchair nanotubes are achiral tubes, in contrast with general $(n, m \neq n \neq 0)$ chiral tubes (compare for example the structure of the tubes shown in Fig. 2.23(a)).

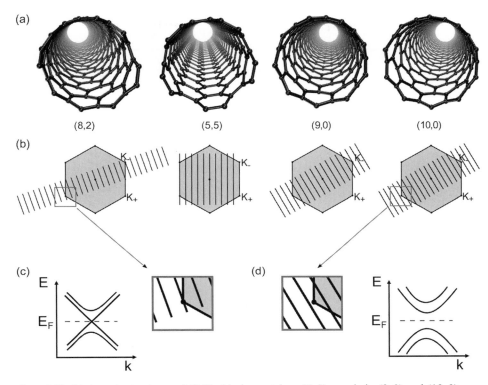

Figure 2.23 (a) Atomic structures of $(8, 2)$ chiral nanotubes, $(5, 5)$ armchair, $(9, 0)$ and $(10, 0)$ zigzag. (b) Allowed k-vectors for the same nanotubes mapped onto the graphene Brillouin zone. The areas within the grey rectangles in (b) are zoomed for better visualization in (c) and (d). For metallic nanotubes, the allowed k-vectors include the K point of the graphene Brillouin zone as shown in (c). The corresponding dispersion relations are linear and exhibit a metallic behavior. In contrast, for semiconducting nanotubes the K point is not an allowed vector and there is an energy gap, as shown in (d). A sketch with the dispersion relations for these two cases is also shown in (c) and (d).

Besides the tube diameter, the chiral vector also determines the unit cell. The translational period t along the tube axis is given by the smallest graphene lattice vector $\mathbf{T}$ perpendicular to $\mathbf{C}_h$. The translational vector $\mathbf{T}$ can be written as a linear combination of the basis vectors $\mathbf{a}_1$ and $\mathbf{a}_2$ as $\mathbf{T} = t_1 \mathbf{a}_1 + t_2 \mathbf{a}_2$. Using the orthogonality relation $\mathbf{C}_h \cdot \mathbf{T} = 0$, one obtains $t_1 = (2m + n)/N_R$ and $t_2 = -(2n + m)/N_R$, where N_R is the greatest common divisor of $(2m + n)$ and $(2n + m)$. The length of the translational vector is given by $t = |\mathbf{T}| = \sqrt{3}a\sqrt{n^2 + nm + m^2}/N_R$. The nanotube unit cell is thus a cylindrical surface with height t and diameter d_t. The number of carbon atoms per unit cell is $N_C = 4(n^2 + nm + m^2)/N_R$. All this information is condensed in Table 2.1.

2.5.2 Electronic structure of CNTs within the zone-folding approximation

Now that we have completely defined the structure of the nanotubes (diameter, chirality, unit cell, etc.) from the pair (n, m) let us turn to their electronic properties. We start

Table 2.1 Structural parameters for (n, m) carbon nanotubes. (Adapted with permission from Charlier et al., 2007. Copyright (2007) by the American Physical Society.)

Symbol	Name	Formula/value
a	graphene lattice constant	$a = \sqrt{3} \times a_{cc} \simeq 2.46$ Å
		$a_{cc} \simeq 1.42$ Å
$\mathbf{a}_1, \mathbf{a}_2$	graphene basis vectors	$\left(\frac{\sqrt{3}}{2}; \frac{1}{2}\right) a, \left(\frac{\sqrt{3}}{2}; -\frac{1}{2}\right) a$
$\mathbf{b}_1, \mathbf{b}_2$	graphene reciprocal lattice vectors	$\left(\frac{1}{\sqrt{3}}; 1\right) \frac{2\pi}{a}, \left(\frac{1}{\sqrt{3}}; -1\right) \frac{2\pi}{a}$
$\mathbf{C}_h$	chiral vector	$\mathbf{C}_h = n\mathbf{a}_1 + m\mathbf{a}_2 \equiv (n, m)$
		$(0 \leq \mid m \mid \leq n)$
d_t	tube diameter	$d_t = \frac{\mid \mathbf{C}_h \mid}{\pi} = \frac{a}{\pi}\sqrt{n^2 + nm + m^2}$
θ	chiral angle	$0 \leq \mid \theta \mid \leq \frac{\pi}{6}$
		$\sin\theta = \frac{\sqrt{3}m}{2\sqrt{n^2+nm+m^2}}$
		$\cos\theta = \frac{2n+m}{2\sqrt{n^2+nm+m^2}}$
		$\tan\theta = \frac{\sqrt{3}m}{2n+m}$
$\mathbf{T}$	translational vector	$\mathbf{T} = t_1\mathbf{a}_1 + t_2\mathbf{a}_2 \equiv (t_1, t_2)$
		$\gcd(t_1, t_2) = 1^{(*)}$
		$t_1 = \frac{2m+n}{N_R}, t_2 = -\frac{2n+m}{N_R}$
		$N_R = \gcd(2m + n, 2n + m)^{(*)}$
N_C	number of C atoms per unit cell	$N_C = \frac{4(n^2+nm+m^2)}{N_R}$

In this table, n, m, t_1, t_2 are integers.

$^{(*)}$ $\gcd(n, m)$ denotes the greatest common divisor of the two integers n and m.

from the single-band *tight-binding* model of graphene in a nearest-neighbor approximation as introduced before. Simplicity is a big advantage of this approximation which, together with a zone folding approach, allows for the prediction of the electronic properties (Hamada, Sawada & Oshiyama, 1992, Saito *et al.*, 1992a). As made clear in the next paragraphs, the zone folding neglects curvature effects, thereby giving a good approximation for tubes of large enough radii ($d_t > 1$nm). The zone-folding approach considers a nanotube as a piece of graphene sheet with periodic boundary conditions along the circumferential direction. This can be expressed by the condition

$$\Psi_{\mathbf{k}}(\mathbf{r} + \mathbf{C}_h) = e^{i\mathbf{k}\cdot\mathbf{C}_h}\Psi_{\mathbf{k}}(\mathbf{r}) = \Psi_{\mathbf{k}}(\mathbf{r}), \tag{2.45}$$

where vectors $\mathbf{r}$ and $\mathbf{C}_h$ lie on the nanotube surface. The first equality in the last equation is a result of applying Bloch's theorem. These boundary conditions impose a quantization of the allowed wavevectors "around" the nanotube circumference, $\mathbf{k} \cdot \mathbf{C}_h = 2\pi q$

(q integer). In contrast, the wavevectors along the nanotube axis remain continuous.[12] Therefore, when plotting the allowed wavevectors in reciprocal space we are left with a set of parallel lines whose direction and spacing depend on the indices (n, m) (see the scheme in Fig. 2.23(b)). The dispersion for each allowed wavevector in the circumferential direction is then obtained by cutting the dispersion relation of 2D graphene along these cutting lines. Superposition of these curves gives the electronic structure of the (n, m) nanotube.

Given graphene's peculiar dispersion relation, a nanotube will be metallic whenever one of the cutting lines crosses the $\mathbf{K}$ (either $\mathbf{K}_+$ or $\mathbf{K}_-$) point. Based on this fact, a rule for metallicity follows from imposing that $\mathbf{K}$ is an allowed wavevector for the given (n, m) nanotube, i.e. $\exp(i\mathbf{K} \cdot \mathbf{C}_h) = 1$. Using $\mathbf{K} = |\mathbf{K}|\mathbf{a}_2/a$ with $|\mathbf{K}| = 4\pi/(3a)$ we find that $n + 2m$ must be a multiple of 3 or, in other terms, $n + 2m \equiv 0 \pmod{3}$. Since $3m \equiv 0 \pmod{3}$ for any m, it follows that $n - m \equiv 0 \pmod{3}$. *Therefore, a nanotube defined by the (n, m) indices will be metallic if $n - m = 3\ell$, with ℓ an integer, or semiconducting if $n - m = 3\ell \pm 1$.*

As a result of the previous rule, most nanotubes are semiconductors and only a fraction (1/3) are metallic. Furthermore, (n, n) armchair nanotubes are always metallic whereas $(n, 0)$ zigzag nanotubes are metallic whenever n is a multiple of 3. For metallic nanotubes, in the vicinity of E_F ($\mathbf{k} = \mathbf{K} + \delta\mathbf{k}$) the dispersion relation is

$$E_\pm(\delta\mathbf{k}) \simeq \pm\frac{\sqrt{3}a}{2}\gamma_0|\delta\mathbf{k}|, \tag{2.46}$$

presenting a linear energy–momentum relation (Fig. 2.23(c)).

For semiconducting nanotubes, the K point is not included and the conduction and valence bands emerge from states with k vectors located on the allowed line(s) closest to the K point (see Fig. 2.23(d)). Choosing (n, m) such that $n - m = 3\ell \pm 1$ gives a gap opening at the Fermi level with a magnitude that can be estimated by (see Problem 2.8)

$$\Delta E_g^1 = \frac{2\pi a\gamma_0}{\sqrt{3}|\mathbf{C}_h|} = \frac{2a_{cc}\gamma_0}{d_t}. \tag{2.47}$$

The value of ΔE_g^1 decreases with the inverse of the tube diameter d_t ($d_t = |\mathbf{C}_h|/\pi$) (White & Mintmire, 1998).[13] In the large-diameter limit one gets a zero-gap semiconductor, as is expected since graphene is recovered. For a realistic $(17, 0)$ tube with a diameter of 1.4 nm, one gets $\Delta E_g^1 \simeq 0.59$ eV.

An instructive exercise is to determine the Brillouin zone of a carbon nanotube, a task that we leave for Problem 2.7. Given that nanotubes are essentially one-dimensional, their Brillouin zone is one-dimensional as well. The zone edges are denoted with X and X', with $X' = -X$ due to time-reversal symmetry. The case of band folding for a $(5, 5)$ armchair nanotube is shown in Fig. 2.24 within a band-folding scheme. Note that the

---

[12] Strictly speaking this is true for infinite nanotubes. However, for typical nanotube sizes the discrete nature of the states may become evident at low enough temperatures.

[13] This $1/d_t$ dependence of the gap on the diameter relies on the assumption of a linear dispersion of the bands around E_F (White & Mintmire, 1998). Away from E_F the dispersion deviates from linear, an effect called trigonal warping (Saito, Dresselhaus & Dresselhaus, 2000) which induces a dependence of the band gap not only on the diameter, but also on the (n, m) indices.

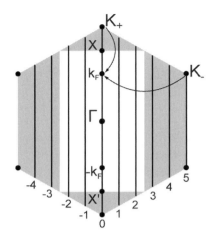

Figure 2.24 Brillouin zone for graphene (grey shaded hexagon) together with the Brillouin zone (white rectangle) for a four-atom unit cell with lattice vectors $\mathbf{T} = \mathbf{a}_1 - \mathbf{a}_2$ (translational vector for armchair tubes) and $\mathbf{a}_1 + \mathbf{a}_2$ (equal to $\mathbf{C}_h/n$ for armchair tubes). The allowed k vectors for a $(5, 5)$ nanotube lie on the black lines depicted in the figure. To compute the $(5, 5)$ band structure, fold the the corners of the hexagonal Brillouin zone onto the rectangular cell (white) and superimpose the bands calculated for bulk graphene along the black lines of length $2\pi/|\mathbf{T}|$.

K points are folded at a distance of $\pm 2\pi/(3a)$ from the Γ point whereas for zigzag nanotubes they are folded onto the Γ point itself.

In Fig. 2.25(a), the dispersion relations $E(k)$ for an $(8, 2)$ chiral nanotube are illustrated. Since $n - m$ is a multiple of 3, this nanotube exhibits metallic behavior with a band crossing at $k = \pm 2\pi/3\mathbf{T}$. Other chiral nanotubes, like the $(9, 6)$ (not shown), display a zero energy gap at $k = 0$. The DOS of chiral nanotubes (see Fig. 2.25(a)) displays van Hove singularities as for the achiral tubes (Charlier & Lambin, 1998) shown in the other panels.

The electronic band structure of an armchair $(5, 5)$ carbon nanotube is presented in Fig. 2.25(b). Six bands for the conduction states, and an equal number for the valence, are observable. However, four of them are degenerate, leading to ten electronic levels in each case, consistent with the ten hexagons around the circumference of the $(5, 5)$ nanotube. For all armchair nanotubes, the energy bands exhibit a large degeneracy at the zone boundary, where $k = \pm \pi/a$ (X point), so that Eq. (2.16) becomes $E(k = \pm \pi/a) = \pm \gamma_0$. This comes from the absence of dispersion along the segments connecting the neighboring centers of the BZ sides (the M points), an effect that yields the so-called trigonal warping of the bands as already discussed. The valence and conduction bands for armchair nanotubes cross at $k = k_F = \pm 2\pi/(3a)$, a point that is located at two thirds of ΓX (Fig. 2.25(b)). This means that the original K vertices of the original graphene hexagonal BZ are folded at two thirds of the ΓX line (or its inversion symmetry image). As discussed above, the $(5, 5)$ armchair nanotube is thus a zero-gap semiconductor which will exhibit metallic conduction at finite temperatures, since only infinitesimal excitations are needed to promote carriers into the conduction bands.

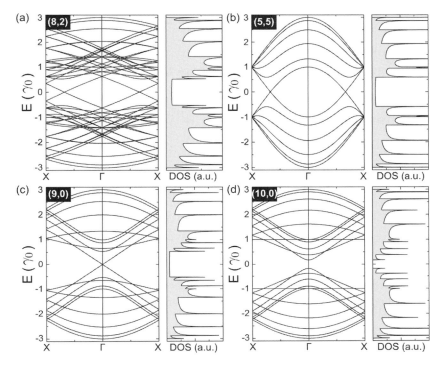

Figure 2.25 Band structure and density of states for: (a) an $(8,2)$ chiral nanotube; (b) a $(5,5)$ armchair nanotube; (c) a $(9,0)$ zigzag nanotube; and (d) a $(10,0)$ zigzag nanotube within the zone-folding model. The 1D energy dispersion relations are presented in the $[-3\gamma_0; 3\gamma_0]$ energy interval in units of γ_0, the nearest-neighbor C–C *tight-binding* hopping parameter (~ 2.9 eV). The energy bands are plotted along the $X - \Gamma - X$ direction. The Fermi level is located at zero energy.

The density of states (DOS) $\Delta N/\Delta E$ represents the number of available states ΔN for a given energy interval ΔE ($\Delta E \to 0$). This DOS is a quantity that can be measured experimentally under some approximations. The shape of the density of states is known to depend dramatically on dimensionality. In 1D, as shown below, the density of states diverges as the inverse of the square root of the energy ($1/\sqrt{E}$) close to band extrema. These "spikes" in the DOS are called van Hove singularities (vHs) and manifest confinement properties in the directions perpendicular to the tube axis. As carbon nanotubes are one-dimensional, their corresponding DOS exhibit such a spiky behavior at energies close to band edges (see Fig. 2.25). For all metallic nanotubes, the density of states per unit length along the nanotube axis is a constant at the Fermi energy (E_F), and can be expressed analytically (Mintmire & White, 1998):

$$\rho(\varepsilon_F) = 2\sqrt{3}a_{cc}/(\pi\gamma_0|\mathbf{C}_h|). \tag{2.48}$$

The calculated 1D dispersion relations $E(k)$ for the $(9,0)$ and the $(10,0)$ zigzag nanotubes are illustrated in Fig. 2.25(c,d), respectively. As expected, the $(9,0)$ tube is metallic, with the Fermi surface located at Γ, whereas the $(10,0)$ nanotube exhibits a finite energy gap at Γ. In particular, in the case of the $(10, 0)$ nanotube, there is a dispersionless energy band at $E/\gamma_0 = \pm 1$, which gives a singularity in the DOS at these particular energies. For a general $(n, 0)$ zigzag nanotube, when n is a multiple of 3, the energy gap at $k = 0$ (Γ point) becomes zero. However, when n is *not* a multiple of 3, an energy gap opens at Γ. The corresponding densities of states have a zero value at the Fermi energy for the semiconducting nanotube, and a small nonzero value for the metallic one.

Note that the k values for the band crossing at E_F in metallic nanotubes are $k = \pm 2\pi/3|\mathbf{T}|$ or $k = 0$ for armchair or zigzag tubes, respectively. These k values are also the locations of the band gaps for the semiconducting zigzag nanotubes. The same k values also denote the positions of the energy gaps (including zero energy gaps) for the general case of chiral nanotubes.

In semiconducting zigzag or chiral nanotubes, the band gap (as expressed in Eq. (2.47)) is independent of the chiral angle and varies inversely with the nanotube diameter: $\Delta E_g^1 = 2\gamma_0 a_{cc}/d_t$ (in the linear bands approximation). Density of states measurements by scanning tunneling spectroscopy (STS) provide a powerful tool for probing the electronic structure of carbon nanotubes. It can be shown, indeed, that under some assumptions the voltage–current derivative dI/dV is proportional to the DOS. These experiments (Wilder *et al.*, 1998, Odom *et al.*, 1998) confirmed that the energy band gap of semiconducting tubes is roughly proportional to $1/d_t$, and that about 1/3 of nanotubes are conducting, while the other 2/3 are semiconducting. Resonances in the DOS have also been observed experimentally (Wilder *et al.*, 1998, Odom *et al.*, 1998) on both metallic and semiconducting nanotubes whose diameters and chiral angles were determined using a scanning tunneling microscope (STM) (Venema *et al.*, 1999). Several other experimental techniques such as resonant Raman scattering (Jorio *et al.*, 2001), optical absorption and emission measurements (O'Connell *et al.*, 2002, Bachilo *et al.*, 2002, Lefebvre, Homma & Finnie, 2003), have also confirmed this structure in van Hove singularities of the electronic densities of states in single-wall carbon nanotubes.

2.5.3   Curvature effects: beyond the zone-folding model

In the previous section, the electronic properties of CNT are directly deduced from confinement of the electrons around the tube circumference through the restriction of the allowed k Bloch vectors, which neglects any curvature effects. However, such curvature effects become increasingly important as the nanotube diameter is further reduced. To account for the cylindrical geometry, one considers that carbon atoms are placed onto a cylindrical wall which implies that: (I) the C-C bonds perpendicular and parallel to the axis become different, so that the $\mathbf{a}_1$ and $\mathbf{a}_2$ have different lengths; (II) as a result, the formation of an angle for the two p_z orbitals located on bonds renormalize the hopping terms γ_0 between a given carbon atom with its three neighbors; (III) the broken planar symmetry induces a mixing between π and σ, forming hybrid orbitals that exhibit partial sp^2 and sp^3 character, all effects which are neglected in the zone-folding model of graphene.

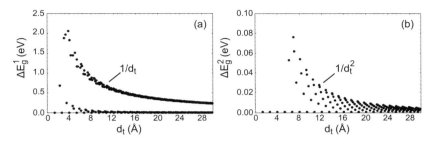

Figure 2.26 Magnitude of both primary (ΔE_g^1) and secondary (ΔE_g^2) gaps in carbon nanotubes with radii less than 15 Å. The primary gap (ΔE_g^1) scales as $1/d_t$: (a) the secondary gap (ΔE_g^2 or curvature induced gap) scales as $1/d_t^2$; (b) the dots at $\Delta E_g^1 = \Delta E_g^2 = 0$ correspond to the *armchair* nanotubes which always preserve their metallic character. (Adapted with permission from Kane & Mele, 1997. Copyright (1997) by the American Physical Society.)

Here, we briefly summarize the effect of finite curvature on the electronic properties of nanotubes. The above mentioned modifications indexed (I) and (II) change the conditions at which occupied and unoccupied bands are crossing (at k_F), which shifts this Fermi vector k_F away from the Brillouin zone corners (K point) of the graphene sheet (Kane & Mele, 1997). Taking curvature into account for *armchair* nanotubes shifts the Fermi wavevector along an allowed line of the graphene Brillouin zone. However, for symmetry reasons, the metallic nature of *armchair* tubes remains insensitive to finite curvature. In contrast, for *non-armchair* metallic nanotubes, k_F is found to shift away from the K point perpendicularly to the allowed k-lines, which produces the formation of a small band gap at E_F (see Fig. 2.26).

Thus, in the presence of curvature effects, the sole zero-band gap tubes are the (n, n) *armchair* nanotubes, whereas (n, m) tubes with $n - m = 3\ell$ (ℓ is a nonzero integer) all fall into the category of tiny-gap semiconductors. *Armchair* tubes are usually labeled "type I" metallic tubes, while the others are of "type II." Remaining nanotubes belong to the intermediate-gap semiconductors (with gaps a few tenths of an eV). Tiny-gap semiconducting nanotubes also present a secondary gap induced by curvature, which depends on the tube diameter (as $1/d_t^2$) and chiral angle (Kane & Mele, 1997). The secondary gap in quasi-metallic *zigzag* nanotubes (chiral angle=0) is found to be

$$\Delta E_g^2 = \frac{3\gamma_0 a_{cc}^2}{4d_t^2}, \tag{2.49}$$

which is vanishingly small so that one generally considers that all the $n - m = 3\ell$ tubes are metallic at room temperature (see Fig. 2.26). Measurements of the density of states using scanning tunneling spectroscopy have nicely confirmed the predicted $1/d_t^2$ dependence for three zigzag nanotubes (Ouyang *et al.*, 2001b), together with the true metallic nature of armchair nanotubes. The band-folding picture based on the single-band *tight-binding* approach is therefore highly reasonable for large enough tube diameter (above 1 nm) (Hamada, Sawada & Oshiyama, 1992, Saito *et al.*, 1992a, Mintmire, Dunlap & White, 1992).

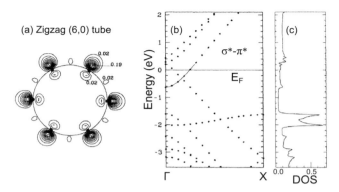

Figure 2.27 Electronic structure of a *zigzag* $(6, 0)$ nanotube. Contour plot of the charge density for state (a) at Γ. The contours are in a plane perpendicular to the axis of the tube containing six carbon atoms. The numbers quoted are in units of $e/(a.u.)^3$. The circle represents a cross section of the cylinder on which the atoms lie. Electronic band structure (b) and density of states (c). The energies are in eV, the reference zero energy is at the Fermi level, and the DOS is in states/eV atom. The new band $(\sigma^*-\pi^*)$ around the center of the Brillouin zone is traced as a guide to the eye. (Reproduced with permission from Blase *et al.*, 1994. Copyright (1994) by the American Physical Society.)

2.5.4   Small-diameter nanotubes: beyond the *tight-binding* approach

The effect of curvature is significant for very small tube diameter, when σ and π states are strongly rehybridized (effect (III)). The zone-folding picture ceases to be correct, demanding that *ab initio* calculations be achieved (Blase *et al.*, 1994) (Fig. 2.27). Strongly modified low-lying conduction band states are introduced into the band gap of insulating tubes because of hybridization of the σ^* and π^* states, which reduces the energy gaps of some nanotubes by more than 50%. For example, the $(6, 0)$ tube, predicted to be a semi-metal in the band-folding scheme, becomes a true metal within LDA, with a density of states at the Fermi level equal to 0.07 state/eV atom (Fig. 2.27(b) and (c)). The $\sigma^*-\pi^*$ hybridization is confirmed by drawing the charge density associated with the states around the Fermi level, as shown in Fig. 2.27(a). Such states are no longer antisymmetric with respect to the tube wall, with a charge spilling out of the tube. Nanotubes with diameters above 1 nm evidence no $\sigma-\pi$ rehybridization.

Electronic properties of small-diameter nanotube have also been explored within the GW approximation, which makes it possible to account for many-body corrections (Miyake & Saito, 2003). The GW calculations found that the energy of the $\sigma^*-\pi^*$ state is reduced with decaying diameter (as drawn by a * in Fig. 2.28), which affects the $1/d$ law in a particularly strong fashion as soon as $d_t \sim 0.8$ nm. For instance, the gap of the $(7, 0)$ tube becomes 0.6 eV (Fig. 2.28(c)). Such a considerable many-body correction is actually compensated by a lattice relaxation (Miyake & Saito, 2003). The lowered state crosses the Fermi level in the $(5, 0)$ and $(6, 0)$ tubes which both become metallic (Fig. 2.28(a–b)). Therefore GW corrections are not strongly renormalizing the DFT results, which are reasonably good for varying tube diameters.

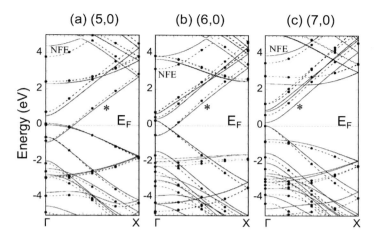

Figure 2.28 Electronic band structure of *zigzag*: (a) $(5, 0)$; (b) $(6, 0)$; and (c) $(7, 0)$ nanotubes. *Ab initio* (DFT-LDA) results (solid line) are compared to GW calculations (circles) for each relaxed geometry. Energy is measured from the Fermi level. Dashed lines are guides for the eyes. (Adapted with permission from Miyake & Saito, 2003. Copyright (2003) by the American Physical Society.)

The analysis of optical spectra of both semiconducting and metallic tubes crucially needs to account for electron–hole interaction effects, which become prominent in small-diameter single-walled carbon nanotubes, but require *ab initio* calculations (Spataru *et al.*, 2004). Finally, note that ultrasmall tube diameters (diameter of about 4 Å) have been obtained by performing tube growth inside AlPO$_4$-5 zeolite channels (with inner diameter of about 7.3 Å) (Wang *et al.*, 2000). Some specific (but controversial) superconductivity has been reported in such ultrasmall tubes (Tang *et al.*, 2001), which given their reported diameter distribution around 4 Å, limits the possible geometries to the $(3, 3)$, $(4, 2)$, and $(5, 0)$ nanotubes.

Such ultrasmall tubes have been studied using *ab initio* simulations by Connetable and coworkers (Connétable *et al.*, 2005), who reported that the $(5, 0)$ tube (predicted as a semiconductor in band-folding representation) becomes metallic with two bands (one doubly degenerate) crossing the Fermi level (yielding two different k_F), a curvature effect related to the Peierls distortion (Fig. 2.29(a)). In these calculations the armchair $(3, 3)$ remains semi-metallic, but with a $\pi - \pi^*$ band crossing at E_F that is displaced off its ideal $\frac{2}{3}$ ΓX position (Fig. 2.29(b)). To conclude, the zone-folding model is certainly valid as long as the tube diameter remains larger than 1 nm. When this condition is not satisfied, the single-band *tight-binding* approach can be misleading, requiring recourse to more accurate calculations, either through a sophisticated *tight-binding* approach, *ab initio* DFT-LDA, or even GW approximation depending on the tube under study.

2.5.5   Nanotubes in bundles

In the previous sections, only a special achiral subset of carbon tubes known as *armchair* nanotubes was predicted to exhibit true metallic behavior. These single-wall

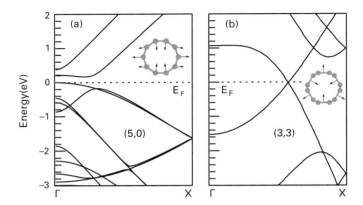

Figure 2.29 Band structures of: (a) the distorted D_{2h} *zigzag* $(5, 0)$ nanotube; and (b) the *armchair* $(3, 3)$ tube. The zero of energy has been set to the top of the valence bands and at the Fermi level, respectively. Inset are symbolic representations of the out-of-plane acoustical and optical modes at Γ for the phonon bands driving the Peierls instability in the (a) $(5, 0)$ and (b) $(3, 3)$ tubes, respectively. (Adapted with permission from Connétable *et al.*, 2005. Copyright (2005) by the American Physical Society.)

(n, n) nanotubes are the only real 1D cylindrical conductors with only two open conduction channels (energy subbands in a laterally confined system that cross the Fermi level). Hence, with increasing length, their conduction electrons ultimately become localized owing to residual disorder in the tube, which is inevitably produced by interactions between the tube and its environment. However, theoretical calculations (White & Mintmire, 1998) have demonstrated that, unlike normal metallic wires, conduction electrons in *armchair* nanotubes experience an effective disorder averaged over the tube's circumference, leading to electron mean free paths that increase with nanotube diameter. This increase should result in exceptional ballistic transport properties and localization lengths of 10 μm or more for tubes with the diameters that are typically produced experimentally. These transport properties of *armchair* nanotubes are described in detail in Chapter 4.

Although the close-packing of individual nanotubes into ropes does not significantly change their electronic properties, *ab initio* calculations predicted that broken symmetry of the $(10, 10)$ tube caused by the interactions between tubes in a rope induces a pseudogap of about 0.1 eV at the Fermi level (Delaney *et al.*, 1998) (Fig. 2.30). Consequently, this pseudogap strongly modifies many of the fundamental electronic properties of the *armchair* tubes, explaining particularly a semi-metallic-like temperature dependence of the electrical conductivity, as well as the presence of a finite gap in the infrared absorption spectrum for bundles of nanotubes.

As mentioned earlier, the electronic properties of isolated (n, n) *armchair* nanotubes are dictated by their geometrical structure, which impose the two linear π–π^* bands to cross at the Fermi energy (Fig. 2.30(a)). These two linear bands give rise to constant density of states near the Fermi level and to true metallic behavior. The atomic structure of an isolated (n, n) nanotube exhibits n mirror planes containing the tube

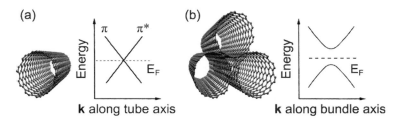

Figure 2.30 Energy band structures for: (a) an isolated (10, 10) nanotube; and (b) a bundle of (10, 10) nanotubes. The two diagrams exhibit: (a) the crossing of the two linear $\pi-\pi^*$ bands for the isolated tube (π^*) character; and (b) band gap opening due to the breaking of the mirror symmetry. E_F is the Fermi energy and k is the wavevector. (Adapted by permission from Macmillan Publishers Ltd: *Nature*, Delaney *et al.*, 1998, copyright (1998).)

axis. The π-bonding state is *even* (the wavefunction has no sign change) while the π-antibonding state is *odd* (sign change) under these symmetry operations. The band crossing is thus allowed and the *armchair* nanotube is metallic, as illustrated schematically in Fig. 2.30(a). Note that it is precisely this symmetry of the isolated (n, n) tube that induces the intrinsic metallic behavior of the tube and its extraordinary ballistic conduction (White *et al.*, 1998). Breaking of this symmetry, however, completely alters this picture. If the tubes in the rope are separated enough to eliminate any nanotube interactions, the band structure will remain identical. However, the inter-tube distances in the bundle are small enough that each nanotube can feel the potential due to all neighboring tubes (Delaney *et al.*, 1998). As a consequence of this perturbation, the Hamiltonian at any point k where the two $\pi-\pi^*$ bands used to cross becomes

$$H_k = \begin{pmatrix} \varepsilon_0 + \delta_{11} & \delta_{12} \\ \delta_{21} & \varepsilon_0 + \delta_{22} \end{pmatrix},$$

where ε_0 is the unperturbed energy. The diagonal matrix elements δ_{11} and δ_{22} merely act to shift the energy and location in k-space of the band crossing. The off-diagonal elements (δ_{12} and δ_{21}) represent the quantum-mechanical level repulsion, thus opening a gap as illustrated schematically in Fig. 2.30(b). If the vertical line through k has high symmetry, the off-diagonal matrix elements may still be zero and a crossing may persist. However, at a general k point, the inter-tube interactions will dramatically change the physics of the ropes. If the symmetry of the nanotube is not broken in the bundle (i.e. for (6, 6) armchair nanotubes), the crossing is preserved (Charlier, Gonze & Michenaud, 1995). These inter-tube interactions, which break the rotational symmetry of *armchair* (n, n) tubes due to the local environment, have been measured experimentally using low-temperature scanning tunneling spectroscopy (Ouyang *et al.*, 2001*b*), thus confirming that the magnitude of the pseudogap depends inversely on nanotube radius.

2.5.6 Multiwall nanotubes

Another mechanism based on the multi-shell concept may tailor the electronic properties of nanotubes. Indeed, the weak interaction between the concentric shells in a

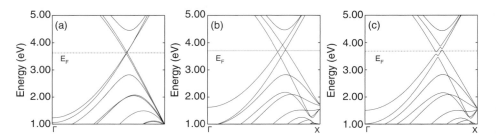

Figure 2.31 Energy band structures of a double-wall nanotube consisting of two aligned coaxial *armchair* nanotubes: $(5,5)@(10,10)$. Near degenerate bands with no gap characterize the $(5,5)@(10,10)$ double-wall nanotube without inter-tube interaction (a). In the presence of inter-tube interaction, depending on the mutual tube orientation, the $(5,5)@(10,10)$ system may exhibit zero gap (b) in the most symmetric (point group symmetry: D_{5h}), or four pseudogaps (c) in a less symmetric and stable configuration (point group symmetry: C_5). (Adapted with permission from Kwon & Tománek, 1998. Copyright (1998) by the American Physical Society. By courtesy of Young-Kyun Kwon.)

multiwall nanotube may induce significant modifications of the electronic properties of the pristine individual nanotubes (Lambin *et al.*, 1994, Kwon & Tománek, 1998). The inter-wall coupling, which has already been mentioned as opening a pseudogap in a bundle of single-wall nanotubes due to symmetry lowering, may periodically open and close four such pseudogaps near the Fermi energy (E_F) in a metallic double-wall nanotube during its rotation normal to the nanotube axis. Indeed, Fig. 2.31 illustrates the intriguing interplay between geometry and electronic structure during the rotation of the inside $(5,5)$ armchair nanotube into the outside $(10,10)$ nanotube, sharing the common axis.

The individual $(5,5)$ and $(10,10)$ tubes are both metallic and present the conventional "graphitic" inter-wall separation of 3.4 Å when nested. To determine the electronic properties of the double-wall nanotube, a *tight-binding* technique with parameters extracted from *ab initio* calculations for simpler structures has been used (Lambin *et al.*, 1994, Kwon & Tománek, 1998). Due to the relatively high symmetry of the coaxial system consisting of a D_{5d} $(5,5)$ nanotube nested inside the D_{10h} $(10,10)$ nanotube, the dependence of the inter-tube interaction on the tube orientation presents an 18° period-icity. In the absence of inter-tube interaction, the band structure of each isolated tube is preserved, and characterized by two crossing linear $\pi-\pi^*$ bands near E_F, one for the "left" and one for the "right" moving electrons. The band structures of a pair of decoupled $(5,5)$ and $(10,10)$ coaxial nanotubes are illustrated in Fig. 2.31(a) as a mere superposition of the individual band structures. Switching on the inter-tube interaction in the $(5,5)@(10,10)$ double-wall tube removes the near degeneracy of the bands near E_F as well (see Fig. 2.31(b,c)). In the most stable orientation, the double-wall system is still characterized by the D_{5d} symmetry of the inner tube. The four bands cross, with a very small change in the slope (Fig. 2.31(b)). While the same argument also applies to a least stable configuration, markedly different behavior is found at any other tube orientation that lowers the symmetry, giving rise to four band crossings (Fig. 2.31(c)).

This translates into four pseudogaps in the density of states near E_F (Kwon & Tománek, 1998). At the Fermi level, the density of states of double-wall nanotubes is thus affected by the mutual orientation of the two constituent nanotubes, since the positions of the four pseudogaps depend significantly on it. The opening and closing of pseudogaps during the libration motion of the double-wall tube is a unique property that cannot be observed in single-wall nanotube ropes (Kwon & Tománek, 1998). Finally, self-consistent charge density and potential profiles for double-wall nanotubes, considering various chiralities, have been obtained (Miyamoto, Saito & Tománek, 2001), and demonstrate that the atomic structure of the inner tube modifies the charge density associated with states near E_F, even outside the outer tube, so that it could even be probed experimentally using an STM. A significant amount of charge, originating mainly from the π electron system of the tubes is transferred mainly into a new interwall state, related to the interlayer state in graphite (Miyamoto, Saito & Tománek, 2001).

2.6   Spin–orbit coupling in graphene

The role of spin–orbit coupling in graphene is an important question especially in regards to the reported spin diffusion lengths and in connection with graphene spintronics (see Section 2.8 for discussion). We cover here the basic elements of spin–orbit coupling in graphene. Spin–orbit coupling is a relativistic interaction between electrons and a local electric field (which can be derived from the Dirac equation). In a generic form it is written as

$$\mathcal{H}_{so} = \frac{\mu_B}{2mc^2}\mathbf{E}.(\hat{\sigma} \times \mathbf{p}). \tag{2.50}$$

It can be seen as a relativistic effect occurring since the moving electrons sense an effective magnetic field in their rest frame, triggering an effective Zeeman effect, which is given by $\mathcal{H}_{so} = \frac{g\mu_B}{2}\mathbf{B}_R.\hat{\sigma}$. The effective magnetic field $\mathbf{B}_R \sim \mathbf{E}.(\mathbf{p} \times \hat{z})$ is named a Rashba field, and is thus parallel to the graphene plane. Such a field induces a precession of the electron spin around the field, defined by a spin precession frequency (Larmor frequency), given by $\Omega = \frac{\mu_B|\mathbf{B}_R|}{\hbar}$. For a tight-binding Hamiltonian, setting $\mathbf{E} = E_0\hat{z}$ and $\mathbf{p}_{ij} = \langle \mathbf{p}_z^i|\mathbf{p}|\mathbf{p}_z^j \rangle$, one can rewrite the spin–orbit Hamiltonian as

$$\mathcal{H}_{so} = \frac{E_0}{4m^2c^2}\sum_{ij}c_{i\sigma}^{\dagger}\left[(\hat{\sigma} \times \mathbf{p}_{ij}).\hat{z}\right]c_{j\sigma'}. \tag{2.51}$$

The effective magnetic field can also be rewritten introducing the electronic potential at the atomic scale which generates local electric fields, $\mathbf{E}(\mathbf{r}) = -\nabla V(\mathbf{r})$. In the two-center approximation, the potential $V(\mathbf{r})$ is approximated by the spherically symmetric atomic potential with $V(\mathbf{r}) = V(|\mathbf{r}|)$ and $\nabla V(\mathbf{r}) = \frac{\mathbf{r}}{r}\frac{dV}{dr}$. In this approximation, the SOC operator is rewritten as a term which couples the spin and the angular momentum operators, $\mathcal{H}_{so} = \xi(r)\mathbf{L}.\mathbf{S}$, where the function $\xi(r)$ contains the entire radial dependence of the SOC Hamiltonian operator. The scalar product of the momentum and spin operators can be rewritten using the identity $\mathbf{L}.\mathbf{S} = \frac{1}{2}(\hat{L}_+\hat{S}_- + \hat{L}_-\hat{S}_+) + \hat{L}_z\hat{S}_z$, with standard definitions for $\hat{L}_\pm = \hat{L}_x \pm i\hat{L}_y$, and commutation relations (for instance $[\hat{L}_+, \hat{L}_-] = 2\hbar\hat{L}_z$). The atomic orbital wave functions, given by spherical harmonics $Y_{l,m}(\hat{r}) = \langle \hat{r}|l,m \rangle$, are eigenfunctions of the angular momentum. On this basis, the kinetic part of the TB Hamiltonian is rewritten applying the properties of the angular momentum algebra to the atomic orbital, that is $\hat{L}_z|l,m\rangle = m\hbar|l,m\rangle$, and $\hat{L}_\pm|l,m\rangle = \hbar\sqrt{l(l+1) - m(m\pm1)}|l,m\pm1\rangle$. Using the orthogonality of the atomic orbitals, a set of nonzero onsite expectation values of the SOC Hamiltonian is obtained:

$$\langle l,m,\mathbf{R}|\mathcal{H}_{so}|l,m,\mathbf{R}\rangle = \xi_l\delta_{l,l}\langle l,m,\mathbf{R}|\mathbf{L}.\mathbf{S}|l,m,\mathbf{R}\rangle, \qquad (2.52)$$

where the strength of the atomic SOC is defined by the TB parameter $\xi_l = \int_0^\infty dr R_l^2(r)\xi(r)$ with the angular momentum quantum number l. Neither the radial part of the orbital wavefunctions $R_l(r)$ nor $\xi(r)$ is known explicitly for carbon atoms in graphene. Therefore the SOC parameters ξ_l are arbitrary and must be fitted to reproduce

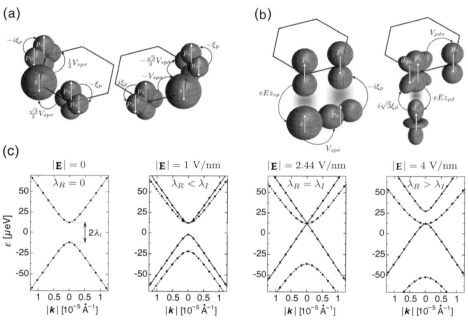

Figure 2.32 (a) Two of the possible nnn hopping paths through the s, p orbitals, arrows, with a corresponding spin, shown by arrows on the orbitals. The opposite sign for the clockwise (left) and the anticlockwise (right) effective hopping is determined by the signs of the two SOCs of the p orbitals. (b) A representative leading hopping path, arrows, which is responsible for the Rashba SOC effect, by coupling states of different spins, illustrated by arrows on the orbitals. The effective hopping is between nearest neighbors. Left: dominant p orbital contribution. Right: negligible d orbital contribution. For clarity the orbitals of the same atoms are separated vertically, according to their contribution either to the σ bands (bottom) or to the π bands (top). (c) Band structure of graphene with SOC as a function of an external transversal electric field with varying intensity. The first-principles results are represented by circles, and the Fermi level is at zero. The curves are fits to the analytical model. The spin branch $\mu = 1$ is shown in solid and $\mu = -1$ in dashed. The calculated Fermi velocity is $v_F = 0.833 \times 10^6$ m/s. (Adapted with permission from Gmitra et al., 2009. Copyright (2009) by the American Physical Society. Images courtesy of Jaroslav Fabian and Martin Gmitra.)

the SOC effects in the band structure obtained by the first-principles calculations. The onsite matrix elements of the dimensionless angular part of the SOC Hamiltonian have been computed in Gmitra *et al.* (2009), where the hopping matrix elements of the SOC Hamiltonian between different atoms have been assumed to be zero, because the spin–orbit interaction has its largest effect on electrons at the nucleus. Explicitly, in the Slater Koster two-center approximation, one gets (Gmitra *et al.*, 2009, Konschuh, Gmitra & Fabian, 2010)

$$\lambda_I \sim \frac{2(\varepsilon_p - \varepsilon_s)}{9V_{sp\sigma}^2}\xi_p^2 + \frac{9V_{pd\pi}^2}{2(\varepsilon_d - \varepsilon_p)^2}\xi_d \qquad (2.53)$$

for the intrinsic spin–orbit coupling, whereas the Rashba term (zero in the absence of an electric field), tunable with an external electric field, is found to be

$$\lambda_R \sim \frac{2eEz_{sp}}{3V_{sp\sigma}}\xi_p + \sqrt{3}\frac{eEz_{sp}}{(\varepsilon_d - \varepsilon_p)}\frac{3V_{pd\pi}}{(\varepsilon_d - \varepsilon_p)}\xi_d. \qquad (2.54)$$

Figure 2.32 shows the evolution of the band structure computed using an effective two-band Hamiltonian defined by $\mathcal{H}_0 = -\hbar v_F(\tau\sigma_x k_x + \sigma_y k_y)$ (clean massless Dirac fermions), together with $\mathcal{H}_I = \lambda_I \tau \sigma_z s_z$ for the intrinsic SOC (Eq. 2.53) and $\mathcal{H}_R = \lambda_R(\tau\sigma_x s_y - \sigma_y s_x)$ for the Rashba term (Eq. 2.54), with σ_z and s_z the pseudospin and real spin Pauli matrices.

2.7 Magnetic field effects in low-dimensional graphene-related materials

2.7.1 Short historical perspective

In 1993, Ajiki and Ando (Ajiki & Ando, 1993, 1996) theoretically predicted that an axial magnetic field should tune the nanotube band structure between a metal and a semiconductor, owing to the modulation of the electronic wavefunctions through the Aharonov–Bohm phase (Aharonov & Bohm, 1959). For too strongly disordered nanotubes (Bachtold *et al.*, 1999, Roche & Saito, 2001), such a phenomenon is actually masked by the $\Phi_0/2$ Altshuler–Aronov–Spivak magnetoresistance oscillations (Altshuler, Aronov & Spivak, 1981), so it requires clean nanotubes and a ballistic regime to be observed. After a series of initial magnetotransport experiments (Stojetz *et al.*, 2005, Strunk, Stojetz & Roche, 2006), the fascinating possibility of turning a metallic nanotube to a semiconducting one and vice versa has finally been nicely confirmed in a careful experiment (Fedorov *et al.*, 2007).

Besides, the application of a perpendicular magnetic field was also predicted to generate Landau levels (LL) with peculiar features (Saito, Dresselhaus & Dresselhaus, 1994, Roche *et al.*, 2000). The confirmation of such an interesting structure of LL in nanotubes has been revealed in clean nanotubes (ballistic regime), through the B-dependent modulation of a Fabry-Perot cavity (Raquet *et al.*, 2008). The peculiarities of LL in graphene nanoribbons have also recently been revealed in experiments (Poumirol *et al.*, 2010, Ribeiro *et al.*, 2011). For two-dimensional graphene, strong perpendicular magnetic fields produce the quantum Hall effect, which is addressed in Section 5.3. Below we provide the essential theory for understanding magnetic-field dependences of electronic structure and transport in carbon nanotubes.

2.7.2   Peierls substitution

The application of an external magnetic field to carbon nanotubes or graphene nanoribbons can be captured through the Aharonov–Bohm effect. To examine such phenomena, let us first rewrite the Hamiltonian in the presence of the B-field as $\mathcal{H} = (\mathbf{p}/2m - e\mathbf{A})^2 + V(\mathbf{r})$, including the vector potential $\mathbf{A}$ (with $\mathbf{B} = \mathbf{rot}\,\mathbf{A}$). The Bloch function in the static $\mathbf{B}$ is simply

$$\Psi(\mathbf{k}, \mathbf{r}) = \frac{1}{N} \sum_{\mathbf{R}} e^{i\mathbf{k}.\mathbf{R}} e^{i\varphi_R} \Phi(\mathbf{r_R}),$$

with $\mathbf{R}$ a lattice vector and φ_R the field-dependent phase factor, which is straightforwardly derived as (Luttinger, 1951)

$$\varphi_R(\mathbf{r}) = \int_{\mathbf{R}}^{\mathbf{r}} \mathbf{A}(\mathbf{r}').d\mathbf{r}' = \int_0^1 (\mathbf{r} - \mathbf{R}).\mathbf{A}(\mathbf{R} + \lambda[\mathbf{r} - \mathbf{R}])d\lambda. \qquad (2.55)$$

In a *tight-binding* representation of the Hamiltonian, the effect of the vector potential is thus introduced through extra phase factors (Eq. (2.55)), in the hopping terms between π orbitals as

$$\gamma_{ij}(\mathbf{B}) = \gamma_0 \exp\left(\frac{2i\pi}{\phi_0} \int_{\mathbf{r}_i}^{\mathbf{r}_j} \mathbf{A}(\mathbf{r}).d\mathbf{r}\right) = \gamma_0 e^{-i\varphi_{ij}}, \qquad (2.56)$$

where $\phi_0 = h/e$ is the flux quantum. This is known as the Peierls substitution (Peierls, 1933).

To explore the **B**-dependent bandstructure effects, it is technically convenient to use the Cartesian basis $(\mathbf{e}_x, \mathbf{e}_y, \mathbf{e}_z)$ where $\mathbf{e}_z$ and $\mathbf{e}_x$ are respectively taken along and perpendicular to the tube axis. In the Landau gauge, $\mathbf{A} = B(0, u_z x, u_x y)$, the Aharonov–Bohm phase acquired during an electronic motion between an orbital located at (x_i, y_i, z_i) and another located at (x_j, y_j, z_j) can be easily derived:

$$\varphi_{ij} = \frac{2\pi}{\phi_0} B\left[(y_j - y_i)u_z \frac{x_i + x_j}{2} + (z_j - z_i)u_x \frac{y_i + y_j}{2}\right], \qquad (2.57)$$

which can be used regardless of the direction of **B** with respect to the tube axis. In the axial configuration $u_x = 0$, $u_z = 1$, whereas in the perpendicular configuration $u_x = 1$, $u_z = 0$.

2.7.3   Parallel field, Aharonov–Bohm gap opening and orbital degeneracy splitting

When the B-field is applied parallel to the tube axis, a spectacular field-dependent bandgap is generated, as well as symmetry breaking of the orbital degeneracy. To deepen this effect, let us consider the two-dimensional Cartesian coordinates $\tilde{\mathbf{r}} = (\tilde{x}, \tilde{y})$ in the basis defined by $(\mathbf{C}_h, \mathbf{T})$. The vector potential is thus rewritten as $\mathbf{A} = (\phi/|\mathbf{C}_h|, 0)$,

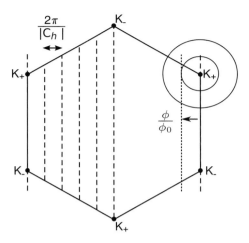

Figure 2.33 Brillouin zone of a graphene sheet together with allowed states for an armchair tube (dashed lines) at zero magnetic flux. Allowed electronic states near the **k**-points under an axial magnetic field are given as dotted lines, while circles give the equipotentials close to the Dirac point.

while the phase factor between two π orbitals located at $\tilde{\mathbf{r}}_i = (\tilde{x}_i, \tilde{y}_i)$ and $\tilde{\mathbf{r}}_j = (\tilde{x}_j, \tilde{y}_j)$ becomes $\varphi_i - \varphi_j = i\phi(\tilde{x}_i - \tilde{x}_j)/|\mathbf{C}_h|$. As a consequence, the periodic boundary conditions on the quantum phase are changed following

$$\Psi_k(\mathbf{r} + |\mathbf{C}_h|) = e^{i\mathbf{k}.\mathbf{C}_h} e^{i\frac{2\pi}{\phi_0} \int_{\mathbf{r}}^{\mathbf{r}+\mathbf{C}_h} \mathbf{A}(\mathbf{r}').d\mathbf{r}'} \Psi_k(\mathbf{r}), \qquad (2.58)$$

and the additional magnetic phase factor thus reduces to $2\pi\phi/\phi_0$, so that the change in the quantum momentum becomes

$$\kappa_\perp \rightarrow \kappa_\perp(\phi) = \frac{2\pi}{|\mathbf{C}_h|}\left(q \pm \frac{\alpha}{3} + \frac{\phi}{\phi_0}\right), \qquad (2.59)$$

with $\alpha = 0$ for metallic tubes, whereas $\alpha = \pm 1$ for semiconducting tubes. In Fig. 2.33 the modification of available electronic states is illustrated in reciprocal space. Using Eq. (2.59), the field-dependent gap oscillation for an initially metallic tube is

$$\Delta E_B = E^+_{q=0}(k_\parallel, \phi/\phi_0) - E^-_{q=0}(k_\parallel, \phi/\phi_0) = 3\Delta E_0 \phi/\phi_0, \qquad (2.60)$$

if $\phi \leq \phi_0/2$ while $\Delta E_0 = 2\pi a_{cc}\gamma_0/|\mathbf{C}_h|$ denotes the gap at zero flux. If $\phi_0/2 \leq \phi \leq \phi_0$ then $\Delta E_B = 3\Delta E_0 |1 - \phi/\phi_0|$, so that the bandgap exhibits an oscillation between 0 and $2\pi a\gamma_0/|\mathbf{C}_h|$ with period ϕ_0 (Ajiki & Ando, 1993, 1996). For example, $\Delta E_B \approx 75$ meV at 50 T for a $(22, 22)$ tube (diameter ≈ 3 nm), while $\Delta E_B \approx 40$ meV at 60 T for a $(10, 10)$ tube (diameter = 1.4 nm). To obtain a magnetic field equivalent to $\phi = \phi_0$ in nanotubes with diameters of 1 nm, 10 nm, 20 nm, and 40 nm, magnetic fields of 5325 T, 53 T, 13 T, and 3 T, are respectively needed.

Besides the ϕ_0-periodic bandgap oscillation, the Aharonov–Bohm effect influences the whole subband structure through the field-dependent energy splitting of the van Hove singularities (Roche *et al.*, 2000). Indeed, in the absence of a magnetic field, each

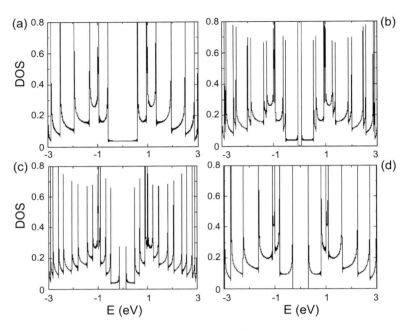

Figure 2.34 Density of states of a $(5, 5)$ nanotube under increasing magnetic flux values: (a) $\phi/\phi_0 = 0$; (b) $\phi/\phi_0 = 0.1$; (c) $\phi/\phi_0 = 0.2$; and (d) $\phi/\phi_0 = 0.5$.

energy level is fourfold degenerate (including spin and orbital degeneracy). The orbital degeneracy is attributed to the symmetry between clockwise $(+)$ and counterclockwise $(-)$ electronic motions around the tube. In the presence of the axial magnetic field, electrons in degenerate $(+)$ and $(-)$ eigenstates acquire opposite orbital magnetic moment $\pm\mu_{orb}$, which thus yields an upshift of the energy of $(+)$ and a downshift of the energy of $(-)$, lifting the orbital degeneracy (van Hove singularity splitting) (Roche *et al.*, 2000).

This mechanism is illustrated in the DOS plots for a $(5, 5)$ tube in Fig. 2.34, in which the tube is metallic at zero magnetic field. The calculation has been performed using a simple *tight-binding* model in the π orbital approximation, while the nearest neighbors hopping integrals are renormalized in the presence of the magnetic field using Eq. (2.57). As predicted, by applying a finite magnetic flux ϕ threading the tube, the bandgap opens and increases linearly with ϕ, to reach a maximum value at half flux quantum $(\phi_0/2)$. Further, the bandgap is linearly reduced until it finally closes again when the field reaches a flux quantum (not shown here). For all armchair (n, n) metallic tubes, the magnitude of the field-dependent splitting of the qth vHs can actually be derived analytically as

$$\Delta E_B(q, \phi/\phi_0) = 2\gamma_0\left[\sin\frac{\pi}{q}(\cos\frac{\pi\phi}{q\phi_0} - 1) - \cos\frac{\pi}{q}\sin\frac{\pi\phi}{q\phi_0}\right]. \qquad (2.61)$$

Semiconducting tubes [i.e., (n, m) tubes with $n - m = 3\ell \pm 1$ (ℓ being an integer)] are affected in a similar way, but the gap expression is slightly different. One

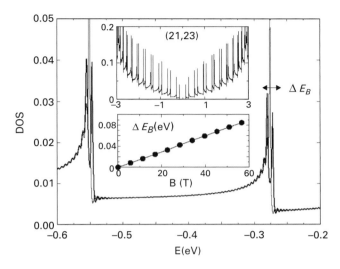

Figure 2.35 Density of states of a $(21, 23)$ tube at zero and finite flux. Top inset: expanded plot of the DOS. Bottom inset: evolution of the vHs splitting ΔE_B as a function of magnetic field. (Reproduced with permission from Charlier, Blase & Roche, 2007. Copyright (2007) by the American Physical Society.)

finds $\Delta E_B = \Delta E_0 |1 - 3\phi/\phi_0|$ if $0 \leq \phi \leq \phi_0/2$, and $\Delta E_B = \Delta E_0 |2 - 3\phi/\phi_0|$ when $\phi_0/2 \leq \phi \leq \phi_0$. Hence, the initial zero-field energy gap (ΔE_0) continuously decays with ϕ, reaching zero at $\phi = \phi_0/3$. The gap further opens as ϕ increases from $\phi_0/3$, reaching a local maximum $(\Delta E_0/2)$ at $\phi = \phi_0/2$, before closing again at $\phi = 2\phi_0/3$, and finally recovering its original value ΔE_0 at $\phi = \phi_0$. Figure 2.35 shows the DOS of a 3 nm diameter semiconducting single-walled tube with and without magnetic flux (main panel) near a van Hove singularity as well as the evolution of the vHs splitting with the field (bottom inset).

The van Hove singularity splitting has been observed by spectroscopic experiments (Zaric *et al.*, 2004), while magnetoresistance oscillations were first studied in disordered and large-diameter multiwall carbon nanotubes (Bachtold *et al.*, 1999), first revealing Altshuler–Aronov–Spivak $\phi_0/2$-periodic magnetoresistance oscillations driven by the quantum interferences (Altshuler, Aronov & Spivak, 1981). Several years afterwards, the joint contribution of field-modulated bandstructure features was confirmed (Stojetz *et al.*, 2005, Strunk, Stojetz & Roche, 2006). Three-terminal devices with conduction channels formed by quasi-metallic carbon nanotubes were shown to operate as nanotube-based field-effect transistors under strong magnetic fields (Fedorov *et al.*, 2007).

Figure 2.36 shows the transfer characteristics of two samples measured at different magnetic fields (Fedorov *et al.*, 2007). At zero field, the $G(V_g)$ curve exhibits ambipolar behavior, indicating the presence of a small 10 meV gap in the electronic spectrum, typical of non-armchair metallic nanotubes. A magnetic field in the axial direction strongly affects the transfer characteristics (Fig. 2.36(a) and (c)). The effect of an axial magnetic field is most pronounced at the gate voltage V_g^* corresponding

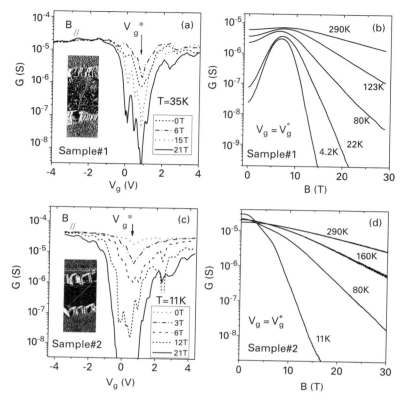

Figure 2.36 (a, b) Transfer characteristics $G(V_g)$ of CNTFETs at axial magnetic fields from 0 to 21 T. At $V_g = V_g^*$, a strong suppression of conductivity is observed at high magnetic fields. The insets show the AFM images of CNTFETs. (b, d) Magnetoconductance curves $G(B)$ measured at $V_g = V_g^*$ at different temperatures. (Courtesy of G. Fedorov.)

to the minimum value of conductance. The exact value of V_g^* changes with temperature, but not with magnetic field for $B > 10$ T. When the magnetic field is changed between 6 and 15 T, the conductance $G(V_g^*)$ (sample 1, Fig. 2.36(a)) drops by about three orders of magnitude at 35 K. A similarly strong effect is observed in another sample (Fig. 2.36(c)). The off-state conductance of the devices is actually found to exponentially decrease with the magnetic flux intensity, confirming the gap opening driven by the Aharonov–Bohm effect (Fedorov *et al.*, 2007). Remarkably, intrinsic properties of a quasi-metallic CNT, such as the helical symmetry, as well as the characteristics of the Schottky barriers formed at the metal–nanotube contacts, can also be obtained by using temperature-dependent magnetoresistance measurements (see Fedorov *et al.*, 2007, for details).

2.7.4   Perpendicular field and Landau levels

Landau levels develop when the magnetic field is applied perpendicular to the tube axis. In this configuration, the vector potential within the Landau gauge for the nanotube

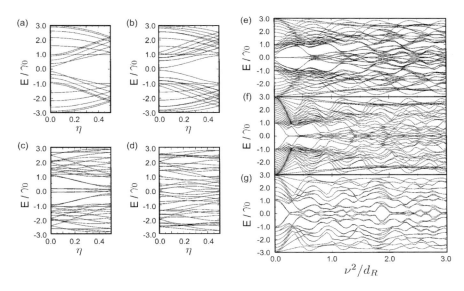

Figure 2.37 Energy dispersion for $(10, 0)$ tube as a function of $\eta = k_\parallel |\mathbf{T}|/2\pi$ for several values of the (dimensionless) inverse magnetic length ν: (a) 0.0; (b) 1.0; (c) 2.0; (d) 3.0. Curves (e), (f), and (g) give the energy at $k_\parallel = 0$ as a function of ν^2/d_R (d_R is the highest common divisor of $(2n + m, n + 2m)$) for tubes $(20, 0)$, $(20, 20)$ and $(9, 9)$ respectively. (Reproduced with permission from Saito *et al.*, 1994, 1996*a*. Copyright (2006) by the American Physical Society. Courtesy of Riichiro Saito.)

surface is rewritten as $\mathbf{A} = (0, (|\mathbf{C}_h|B/2\pi) \sin(2\pi \tilde{x}/|\mathbf{C}_h|))$, keeping $(\tilde{x}, \tilde{y})$ as the coordinates along the circumferential and nanotube axis directions. For symmetry reasons the net magnetic flux threading the tube is now zero. The phase factors can be computed either using this basis, or the Cartesian basis defined earlier. The energy dispersion can then be evaluated as a function of the dimensionless wave vector $k_\parallel |\mathbf{T}|/2\pi$, for several values of the dimensionless inverse magnetic length $\nu = |\mathbf{C}_h|/2\pi \ell_B$, where $\ell_B = \sqrt{\hbar/eB}$ is the magnetic length.

As seen in Fig. 2.37, an increase of the magnetic field results in a reduction of the subband dispersion, with a particularly strong effect in the vicinity of the charge neutrality point, where a zero-energy Landau level forms. Using the **k.p** method, the expression of the eigenstates under magnetic field close to the **K** points can actually be derived analytically for metallic tubes. The **k.p** equation at the Dirac point ($\delta \mathbf{k} = 0$) in the presence of a perpendicular magnetic field can be decoupled into two equations (Ando & Seri, 1997):

$$\left(-\frac{\partial}{\partial \tilde{x}} + \frac{|\mathbf{C}_h|}{2\pi \ell_B^2} \sin\left(\frac{2\pi \tilde{x}}{|\mathbf{C}_h|}\right) \right) \mathcal{F}_A^k(\tilde{x}) = 0, \qquad (2.62)$$

$$\left(+\frac{\partial}{\partial \tilde{x}} + \frac{|\mathbf{C}_h|}{2\pi \ell_B^2} \sin\left(\frac{2\pi \tilde{x}}{|\mathbf{C}_h|}\right) \right) \mathcal{F}_B^k(\tilde{x}) = 0, \qquad (2.63)$$

from which two independent solutions can be obtained:

$$\Psi_A(\tilde{x}) = \begin{pmatrix} 1 \\ 0 \end{pmatrix} \mathcal{F}_A(\tilde{x}), \quad \Psi_B(\tilde{x}) = \begin{pmatrix} 0 \\ 1 \end{pmatrix} \mathcal{F}_B(\tilde{x}),$$

$$\mathcal{F}_A(\tilde{x}) = \frac{1}{\sqrt{|\mathbf{C}_h|I_0(2\nu^2)}} \exp\left(-\nu^2 \cos\left(\frac{2\pi x}{|\mathbf{C}_h|}\right)\right),$$

$$\mathcal{F}_B(\tilde{x}) = \frac{1}{\sqrt{|\mathbf{C}_h|I_0(2\nu^2)}} \exp\left(+\nu^2 \cos\left(\frac{2\pi x}{|\mathbf{C}_h|}\right)\right),$$

where $I_0(2\nu^2)$ is a modified Bessel function of the first kind. Note that for sufficiently large magnetic field ($\nu \gg 1$), these wavefunctions become strongly localized in the circumference direction; that is, $\Psi_A(\tilde{x})$ is a wavefunction localized around $\tilde{x} = \pm|\mathbf{C}_h|/2$ at the bottom side of the cylinder, whereas $\Psi_B(\tilde{x})$ is localized around the top side $\tilde{x} = 0$. As a result, the boundary condition on the wavefunction becomes irrelevant and the resulting band structures, starting from an initially metallic or semiconducting nanotube, become identical (Ajiki & Ando, 1993). For a small $\delta\mathbf{k}$ around the $\mathbf{K}$ points, the low-energy properties are described by an effective Hamiltonian that can be determined by the two degenerate states Ψ_A and Ψ_B as

$$\mathcal{H}_{\text{eff}} = \begin{pmatrix} 0 & -i\gamma_0\delta k I_0^{-1}(2\nu^2) \\ +i\gamma_0\delta k I_0^{-1}(2\nu^2) & 0 \end{pmatrix}, \tag{2.64}$$

whose eigenvalues are $E_{q=0}^{\pm} = \pm\gamma_0|\delta\mathbf{k}|/I_0(2\nu^2)$, with a group velocity given by $v = \gamma_0/\hbar I_0(2\nu^2)$, while the density of states becomes $\rho(E_F) \sim I_0(2\nu^2)/\pi\gamma_0 \sim e^{\nu^2}/\sqrt{4\pi\nu^2}$ ($\nu \gg 1$) (Ando & Seri, 1997). The DOS at the charge neutrality point thus diverges

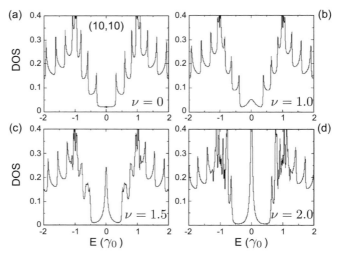

Figure 2.38 (a-d) Density of states (in arbitrary units) of a (10, 10) tube in a perpendicular magnetic field for several field strengths. The field strengths are expressed in terms of the dimensionless parameter $\nu = |\mathbf{C}_h|/2\pi\ell_B$, where $\ell_B = \sqrt{\hbar/eB}$ is the magnetic length.

exponentially with increasing magnetic field. This is shown in Fig. 2.38 for a $(10, 10)$ tube for several magnetic field strengths.

Given the diameter-dependent scaling property of the band structure, the larger the tube diameter, the smaller the required value of the magnetic field to fully develop such a Landau subband at CNP. As already seen in the field-dependent band structures (Fig. 2.37), the whole DOS also progressively degrades as the Landau subbands form, although a strong mixing between high-energy subbands remains (Fig. 2.38).

As soon as $v = |\mathbf{C}_h|/2\pi \ell_B \geq 1$ is satisfied, the electronic spectrum actually becomes fully dominated by Landau levels. One finds that for tubes with diameters of 1 nm, 10 nm, 20 nm, and 40 nm, the condition $v = 1$ corresponds to magnetic field strengths of 2635 T, 26 T, 6.6 T, and 1.6 T, respectively. For all cases, $\ell_B \ll \ell_e$ (or $\omega_c \tau_e \gg 1$) is satisfied for clear observation of Landau quantization (ℓ_e is the mean free path, $\omega_c = eB/m$ is the cyclotron frequency, and τ_e is the scattering time). Compelling evidence of the formation of Landau levels in carbon nanotubes was reported by Raquet and coworkers (Raquet *et al.*, 2008).

2.7.5 Landau levels in graphene

The massless Dirac fermion nature of electronic excitations in monolayer graphene is actually spectacularly manifested in the high magnetic field regime, through the energy spectrum which splits up into nonequidistant Landau levels (LL) (McClure, 1956, Goerbig, 2011). The modification of the electronic spectrum in the presence of an external perpendicular magnetic field B can be technically described by the Peierls substitution, which is applicable as long as the characteristic magnetic length $l_B = \sqrt{\hbar/eB}$ remains much larger than the lattice spacing. This is actually the case for experimentally accessible magnetic fields since $a/l_B \simeq 0.005 \times \sqrt{B[\mathrm{T}]}$.

The Peierls substitution consists of replacing the wavevector $\mathbf{q}$ in the linearized Dirac Hamiltonian by the gauge-invariant kinetic momentum, $\mathbf{q} \to \Pi/\hbar \equiv -i\nabla + e\mathbf{A}(\mathbf{r})/\hbar$, where $\mathbf{A}(\mathbf{r})$ is the vector potential that generates the magnetic field $B\mathbf{e}_z = \nabla \times \mathbf{A}(\mathbf{r})$. Using the commutation relation $[x_\mu, p_\nu] = i\hbar\delta_{\mu,\nu}$ between the components x_μ of the position operator and those $p_\mu = -i\hbar\partial/\partial x_\mu$ of the canonical momentum operator ($\mu, \nu = x, y$ for the 2D plane), one obtains the noncommutativity between the components of the kinetic momentum

$$[\Pi_x, \Pi_y] = -i\frac{\hbar^2}{l_B^2}, \tag{2.65}$$

such that these components may be viewed as conjugate. Let us introduce the convenient ladder operators

$$\hat{a} = \frac{l_B}{\sqrt{2}\hbar}\left(\Pi_x - i\Pi_y\right) \text{ and } \hat{a}^\dagger = \frac{l_B}{\sqrt{2}\hbar}\left(\Pi_x + i\Pi_y\right), \tag{2.66}$$

which satisfy the usual commutation relation $[\hat{a}, \hat{a}^\dagger] = 1$, as in the case of the harmonic oscillator. In terms of these ladder operators, the linearized Hamiltonian, which in the absence of magnetic field is given by $H_{\mathbf{q}}^{\text{eff},\xi} = \xi\hbar v_F(q_x\sigma_x + \xi q_y\sigma_y)$ (see Section 2.2), becomes for a finite magnetic field (Goerbig, 2011)

$$H_B^\xi = \xi\sqrt{2}\frac{\hbar v_F}{l_B}\begin{pmatrix} 0 & \hat{a} \\ \hat{a}^\dagger & 0 \end{pmatrix}, \qquad (2.67)$$

where (when contrasted to the zero-field Hamiltonian) the A and B components have interchanged in the spinors describing electrons around the K_- point ($\xi = -$).

The solution of the equation $H_B^\xi\psi_n = E_{\lambda n}\psi_n$, in terms of the two-spinors,

$$\psi_n = \begin{pmatrix} u_n \\ v_n \end{pmatrix}, \qquad (2.68)$$

provides the Landau level spectrum of electronic excitations in graphene

$$\hat{a}^\dagger\hat{a}\,v_n = \left(\frac{E_{\lambda n}}{\sqrt{2}\hbar v_F/l_B}\right)^2 v_n, \qquad (2.69)$$

which indicates that $v_n \propto |n\rangle$ is an eigenstate of the number operator $\hat{a}^\dagger\hat{a}$, $\hat{a}^\dagger\hat{a}|n\rangle = n|n\rangle$, with energies

$$E_{\lambda n} = \lambda\frac{\hbar v_F}{l_B}\sqrt{2n}, \qquad (2.70)$$

where $\lambda = \pm$ denotes the levels with positive and negative energy, respectively (Goerbig, 2011). Furthermore, substitution of this result in the eigenvalue equation yields $u_n \propto \lambda\hat{a}|n\rangle$. For example, the Landau level spectra for magnetic fields of 10 T and 50 T are shown in Fig. 2.39. These calculations have been performed using the Lanczos recursion method (see Appendix D).

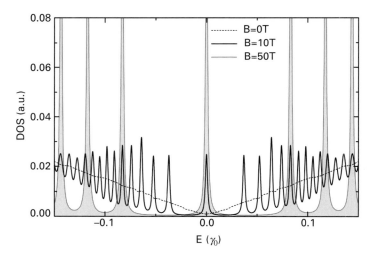

Figure 2.39 Numerical calculation of the density of states showing the Landau levels in pristine graphene in the absence of magnetic field (dashed lines), for $B = 10\,\text{T}$ (black solid line) and $B = 50\,\text{T}$ (grey solid line). The DOS is expressed in arbitrary units. Courtesy of Dinh van Tuan.

Note that the term *relativistic* is used to distinguish the $\lambda\sqrt{Bn}$ dispersion of the levels from the conventional (non-relativistic) Landau levels, which disperse linearly in Bn. A remarkable difference with respect to nonrelativistic Landau levels in metals (with parabolic bands) is the presence of a zero-energy Landau level with $n = 0$. This level needs to be treated separately, and indeed the solution of the eigenvalue equation yields an eigenvector

$$\psi_{\xi,n=0} = \begin{pmatrix} 0 \\ |n = 0\rangle \end{pmatrix}, \qquad (2.71)$$

with a single nonvanishing component. As a consequence, zero-energy states at the K_+ point are restricted to the B sublattice, whereas those at the K_- have a nonvanishing weight only on the A sublattice. For Landau levels with $n \neq 0$, the eigenstates

$$\psi^{\xi}_{\lambda,n\neq 0} = \frac{1}{\sqrt{2}} \begin{pmatrix} |n - 1\rangle \\ \xi\lambda|n\rangle \end{pmatrix} \qquad (2.72)$$

are spinors in which both sublattices are equally populated, but the components correspond to different nonrelativistic Landau states. More information about the extension to the situation of the strongly interacting case can be found in Goerbig (2011).

Recently, by measuring high field magnetotransport properties of graphene on top of a boron-nitride substrate, the long sought-after *Hofstadter's butterfly*[14] (Hofstadter, 1976) was finally unveiled (Ponomarenko *et al.*, 2013, Dean *et al.*, 2013). This was made possible thanks to the Moiré pattern created by the interaction between graphene and boron-nitride and related long superlattices periodicities (on the order of tens of nanometers). Besides providing the first direct evidence of the fractal spectrum predicted by Hofstadter, this demonstrated the enormous potential of layered structures.

2.8 Defects and disorder in graphene-based nanostructures

In the first section of the present chapter, the fascinating electronic properties of graphene are described, suggesting the latter to be a material of choice for future nanoelectronics. Indeed, graphene exhibits an extremely high stability, huge carrier mobilities and a thermal conductivity predicted to be nearly twice that of diamond. Graphene also presents a high response to perpendicular electric fields, making it possible to tune the type and concentration of charge carriers. Its high carrier mobilities and long phase coherence length (Miao *et al.*, 2007) suggest potential applications in integrated circuits, high-mobility transistors, or single-molecule gas sensors. Eventually, graphene can also be patterned using existing lithographic techniques and meets many requirements for the process and design of nanoelectronic devices.

However, the lack of an electronic (or transport) gap in pristine graphene is an issue that has to be overcome in order to achieve a high I_{on}/I_{off} current ratio in

---

[14] Hofstadter's butterfly is a fractal pattern in the energy spectrum of a 2D system in a magnetic field. It arises due to the interplay between the lattice periodicity and the one imposed by the magnetic field and leads to a butterfly-like shape when looking to the spectrum (energy versus magnetic flux). The interested reader can find more on this fascinating subject and the related experiments in graphene in our website.

graphene-based field-effect devices (Xia *et al.*, 2010). In this context, controlled defects engineering in sp^2 carbon-based materials has become a topic of great excitement (Krasheninnikov & Banhart, 2007). Indeed, the electronic (and transport) properties of carbon nanotubes (Charlier, Blase & Roche, 2007) and graphene-based nanomaterials (Suenaga *et al.*, 2007, Cresti *et al.*, 2008) can be considerably enriched by chemical modifications, including substitution and molecular doping (Latil *et al.*, 2004, Lherbier *et al.*, 2008) as well as functionalization. Another approach to tuning the electronic properties of graphene consists in using ion or electron beam irradiation in order to introduce structural point defects (e.g. vacancies, Stone–Wales defects, adatoms, etc.) in sp^2 carbon-based nanostructures. Indeed, convincing room-temperature signatures of an Anderson regime in irradiated carbon nanotubes (Gómez-Navarro *et al.*, 2005) and graphene (Nakaharaim *et al.*, 2013) or low-T saturating conductivities in graphene samples (Chen *et al.*, 2009) have been reported. Consequently, it is crucial not only to understand the influence of defects on the electronic properties of graphene-based nanostructures in order to conquer their detrimental effects, but also because controlled defect introduction may be used to tune the carbon nanosystem properties in a desired direction (e.g. gas sensor, etc.)

In the following, a few important defects in sp^2 carbon nanostructures are illustrated, and their specific effect on the electronic properties of the host material are discussed.

2.8.1   Structural point defects in graphene

Structural point defects exist in various geometrical forms in graphene and more generally in all sp^2 carbon-based nanomaterials (Banhart, Kotakoski & Krasheninnikov, 2011, Kotakoski *et al.*, 2011). For example, the Stone–Wales (SW) defect is a well-known and common planar defect in sp^2 carbon nanostructures, which consists in a 90° rotation of a carbon–carbon bond (Stone & Wales, 1986). This topological transformation yields to the formation of two heptagons connected with two pentagons (Fig. 2.40(a,b) and Fig. 2.41(a)).

Vacancies are missing carbon atoms in the honeycomb lattice and can be created by irradiating graphene with ions such as Ar^+ for instance. By removing one C atom from the graphene plane, three C atoms are left with an unsaturated bond. When the system is relaxed, this monovacancy undergoes a Jahn–Teller distortion: two of the unsaturated carbon atoms form a weak covalent bond resulting in a pentagonal rearrangement (Fig. 2.40(c)). The third unsaturated carbon atom moves radially out of the plane, modifying the initial D_{3h} symmetry of the hexagonal network into the favored C_s symmetry (Amara *et al.*, 2007). The monovacancy is a magnetic defect since the localized orbitals of the unsaturated carbon atom exhibit a net local magnetic moment (Lehtinen *et al.*, 2004*a*, 2004*b*), which could explain the issue of induced magnetization in some carbon nanomaterials (Yazyev & Helm, 2007, Yazyev, 2008, 2010). However, these single vacancies migrate easily in the graphene plane and are stabilized when they recombine with another, thus reconstructing into various divacancy defects (Krasheninnikov & Banhart, 2007, Lee *et al.*, 2005, Kim *et al.*, 2011).

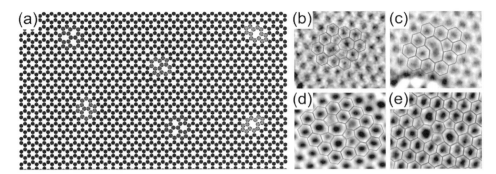

Figure 2.40 Structural point defects in graphene: (a) model for various realistic defects randomly oriented and distributed in graphene. HRTEM images of defects observed in graphene: (b) Stone–Wales; (c) vacancy; (d,e) more complex topologies containing pentagon–heptagon pairs. ((a) is adapted from Lherbier *et al.*, 2012. Copyright (2008) American Physical Society. (b–e) are adapted with permission from Meyer *et al.*, 2008. Copyright (2008) American Chemical Society.)

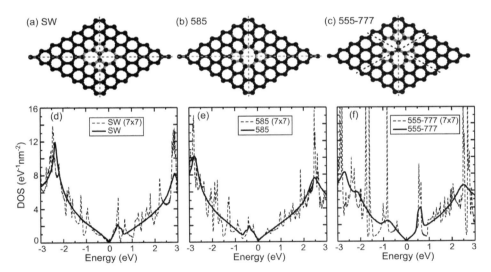

Figure 2.41 Topological defects in graphene and their corresponding effect on its electronic properties. Model of the three structural defects: (a) Stone–Wales; (b) 585; and (c) 555-777 divacancies. Symmetry axes are drawn in dashed lines, *tight-binding* densities of states for a single defect in a 7×7 graphene supercell (dashed lines – concentration of $\sim 1\%$) and for a large plane of graphene containing randomly distributed and oriented divacancies (thick lines – concentration of 1%) for the defects illustrated above: (d) Stone–Wales; (e) 585; and (f) 555-777 divacancies. (Adapted with permission from Lherbier *et al.*, 2012. Copyright (2008) American Physical Society.)

For example, the reconstruction can lead to formation of two pentagons and one octagon (so-called 585, Fig. 2.41(b)), or also to three pentagons and three heptagons (so-called 555-777, Fig. 2.41(c)). According to *ab initio* calculations, the formation energy of the 555-777 divacancy is smaller than that of the 585 divacancy by about

0.9 eV (Lherbier *et al.*, 2012). Such a stabilization of the 555-777 divacancy with regards to the 585 divacancy in graphene contrasts to the case of carbon nanotubes (Lee *et al.*, 2005). A third kind of divacancy has also been reported (Banhart, Kotakoski & Krasheninnikov, 2011, Meyer *et al.*, 2008), exhibiting a larger extension and involving four pentagons, one hexagon and four heptagons (so-called 5555-6-7777 – not illustrated here). In contrast with monovacancies that cannot be considered as a reversible geometrical modification of the ideal graphene plane, divacancies and Stone–Wales defects belong to the class of topological defects. One also notes that the 585 divacancy as well as the SW defect possesses a D_{2h} symmetry since two orthogonal symmetry axes can be defined (Fig. 2.41(a, b)), whereas the 555-777 divacancy possesses a D_{3h} symmetry (Fig. 2.41(c)). Observation of these three structural point defects has already been reported in graphene by means of STM experiments (Suenaga *et al.*, 2007, Ugeda *et al.*, 2012) or transmission electron microscopy (TEM) images (Meyer *et al.*, 2008) and their influence on transport properties has also deserved in-depth inspection (Lherbier *et al.*, 2011).

In order to investigate the effect of these structural defects on graphene, two centers 3rd nearest-neighbors π–π^* orthogonal *tight-binding* models for both pristine and defective graphene can be computed using *ab initio* calculations. To extract optimized TB parameters, a set of points $E(k)$ is chosen in the *ab initio* band structure and used as constraints in a fit procedure (Lherbier *et al.*, 2012). These TB parameters are usually fitted to reproduce as accurately as possible the full band structure (as described previously). An alternative strategy can also be applied when the *ab initio* Hamiltonian is expressed in a localized orbitals basis set. Indeed, the TB parameters can thus be directly *extracted* by performing successive operations on this *ab initio* Hamiltonian. In particular, the basis set has to be reduced to a single p_z orbital.

Local TB parameters corresponding to a given disorder potential are obtained using the same fitting technique as for the pristine graphene. For the impurity potential, only onsite modifications are considered, but the new arrangement of neighbors for carbon atoms in the core of the defects is carefully taken into account. For the SW defect, the rotation of the carbon–carbon bond leads to a modification of first, second, and third nearest neighbors for carbon atoms in the vicinity of the rotated bond. The validity of the TB parameterization for the defects is checked by comparing the *ab initio* and the TB band structures for a 7×7 supercell containing a single defect. In Fig. 2.41, the TB band structure (lines with symbols) is superimposed on the *ab initio* band structure (full lines). A good agreement is obtained, especially for the valence bands. The conduction band side seems to be less accurate but this is uniquely due to the inability of the pristine graphene TB model to reproduce a conduction band along a K-M branch.

The density of states of random distribution of structural point defects in the honeycomb lattice (large graphene planes) can then be estimated, revealing the salient features that persist after taking into account the randomness character of the disorder. Figure 2.41(d–f) presents the total DOS of large graphene samples containing 1% of SW, 585 and 555-777 divacancies, computed using the recursion method (see Appendix D), and compared with the total DOS calculated for a 7×7 supercell containing a single defect (more or less the same concentration, $\sim$ 1%). A first observation is that the DOS of random disordered systems is much smoother than the one corresponding to a single defect in a supercell. In the random disorder case, most of the peaks have disappeared except the ones close to the Dirac point (set to zero). The broadening due to the distribution disorder is more efficient in energy regions containing several bands. Close to the Dirac point, the number of bands is smaller, which preserves the defect-induced resonances. Secondly, the position of resonance energy peaks is consistent with the *ab initio* supercell band structures obtained for defective graphene (Lherbier *et al.*, 2012). The DOS of randomly disordered graphene suggest that the electron transport in an energy region around $E = 0.35$ eV should be mainly damaged by SW defects, whereas hole transport should be altered around $E = -0.35$ eV for 585 divacancies, and finally that 555-777 divacancies exhibit several resonance energies around $E = 0.6, -0.8, -2.1$ eV, which should also lead to reduced transport performances. Such an in-depth analysis of the transport properties of graphene containing a random distribution of structural point defects is presented in Chapter 7.

2.8.2 Grain boundaries and extended defects in graphene

Single-atom-thick graphene sheets are presently produced by chemical vapor deposition (Li, Luican & Andrei, 2009) on macroscopic scales (up to meters (Bae *et al.*, 2010)), making their polycrystallinity almost unavoidable. This polycrystalline nature of graphene samples at micrometer length scales induces the presence of intrinsic topological defects of polycrystalline materials, such as grain boundaries and dislocations. Theoretically, graphene grain boundaries are predicted to inevitably affect all kinds of physical properties of graphene, but these drastic modifications strongly depend on their atomic arrangement. Using atomic-resolution imaging, experimentalists have been able to determine the location and identity of every atom at a grain boundary, discovering that different grains stitch together predominantly through pentagon–heptagon pairs (Huang *et al.*, 2011) (see Fig. 2.42(a,b)). By correlating grain imaging with scanning probe and transport measurements, these grain boundaries were found to severely

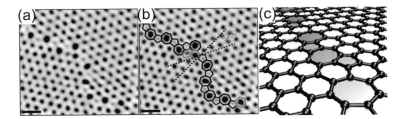

Figure 2.42 Atomic-resolution STEM images of polycrystalline graphene: (a) two grains (bottom left, top right) intersect with a 27° relative rotation. An aperiodic line of defects stitches the two grains together (scale bars 5Å). This grain boundary is composed of pentagons, heptagons, and distorted hexagons as outlined in (b). (c) Model of the atomic structure of a tilt grain boundary in graphene separating two crystalline domains. ((a) and (b) are adapted with permission from Macmillan Publishers Ltd: *Nature* from Huang *et al.*, 2011, copyright (2011). (c) By courtesy of Oleg Yazyev.)

weaken the mechanical strength of graphene, but they do not so drastically alter their electrical properties.

From a theoretical point of view, grain boundaries (GB) with large-angle symmetric configurations were found to be energetically favorable using *ab initio* calculations (Yazyev & Louie, 2010*a*, Yazyev & Louie, 2010*b*) (see Fig. 2.42(c)). Drastic stabilization of small-angle configuration GBs via out-of-plane deformation has also been predicted (Yazyev & Louie, 2010*b*), which is a remarkable feature of graphene as a truly two-dimensional material. Grain boundaries are expected to markedly alter electronic transport in graphene. Indeed, charge-carrier transport across periodic grain boundaries is primarily governed by a simple momentum conservation law (Yazyev & Louie, 2010*a*). Two distinct transport behaviors have been predicted – either perfect reflection or high transparency for low-energy charge carriers depending on the grain boundary atomic structure (see Fig. 2.42(c)). Furthermore, engineering of periodic grain boundaries with tunable transport gaps has also been suggested (Yazyev & Louie, 2010*a*), allowing for controlling charge currents without the need to introduce bulk bandgaps in graphene. Tailoring electronic properties and quantum transport by means of grain boundary engineering may pave a new road towards practical digital electronic devices based on graphene at a truly nanometer scale.

The controlled engineering of extended defects represents a viable approach to creation and nanoscale control of one-dimensional charge distributions with widths of several atoms. By growing graphene on two non-equivalent threefold hollow sites of Ni(111) substrate, termed fcc (face-centered cubic) and hcp (hexagonal close-packed) sites (Fig. 2.43(a)), a one-dimensional extended line of defects can be formed without any unsaturated dangling bonds by restructuring the two graphene half-lattices that are translated by a fractional unit cell vector $1/3(a_1 + a_2)$ (where a_1 and a_2 are the two graphene unit cell vectors) (Lahiri *et al.*, 2010) (Fig. 2.43(a)). The two graphene domains can be joined at their boundary so that every carbon has a threefold coordination, forming a one-dimensional topological defect consisting of a pair of pentagons and one octagon periodically repeated along the dislocation line (Fig. 2.43(b)).

The atomic locations identified from scanning tunneling microscopy (Fig. 2.43(c)) confirm that the defect is composed of one octagon surrounded by a pair of pentagons, and a period of twice the unit cell vector of graphene along the defect line has been measured (Lahiri *et al.*, 2010). Although STM constitutes an excellent experimental tool for identifying defects in graphene (Amara *et al.*, 2007), the interpretation of the images can sometimes be extremely complicated. Therefore, in order to overcome the problem of defect identification, simulated images have been calculated from the *ab initio* local density of states of an extended line of 585 defects embedded in GNRs using the Tersoff–Hamann approximation (Botello-Mendez *et al.*, 2011*b*) (Fig. 2.43(d)). LDOS are computed between 0.2–0.3 eV with respect to the Fermi energy in order to account for an n-type doping substrate (i.e. Ni in the experiment). It is noteworthy that the *ab initio* STM image for the extended line of 585 defects (Fig. 2.43(d)) exhibits a quite good agreement with the experimental one (Fig. 2.43(c)).

Various architectures of an extended line of defects embedded in graphene, and exhibiting pentagonal, heptagonal, and octagonal rings of carbon, were explored using

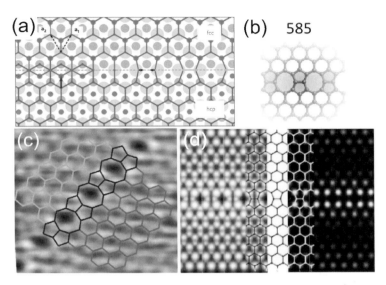

Figure 2.43 Extended one-dimensional defect in graphene. (a) Structural model and schematic formation: the two half-lattices can be joined along the $a_1 - a_2$ direction (indicated by the dashed line) by restructuring the graphene lattice. The domain boundary can be constructed as illustrated, by joining two carbon atoms, indicated by the two arrows, along the domain boundary line. This reconstructed domain boundary forms a periodic structure consisting of octagonal and pentagonal carbon rings. The underlying Ni(111) structure illustrates how the extended defect is formed by anchoring two graphene sheets to a Ni(111) substrate at slightly different adsorption sites. If one graphene domain has every second carbon atom located over an fcc-hollow site (upper part) and the other domain over a hcp-hollow site (lower part), then the two domains are translated by $1/3(a_1 + a_2)$ relative to one another. The light, medium, and dark grey spheres correspond to Ni atoms in the 1st, 2nd, and 3rd layers, respectively. (b) Schematic model based on the periodic repetition of a 585 (pentagon–octagon–pentagon) defect. (c) Experimental STM images. (d) *Ab initio* STM images simulated at constant current (left) and constant height (right) for the extended 585 one-dimensional extended line with the superimposed defect model. ((a) and (c) are reprinted by permission from Macmillan Publishers Ltd: *Nat. Nanotech.*, Lahiri *et al.*, 2010, copyright 2010. (b) and (d) are from Botello-Mendez *et al.*, 2011*b*. Adapted by permission of The Royal Society of Chemistry.)

first-principles simulations (Botello-Mendez *et al.*, 2011*b*). Three different stable atomic configurations were predicted to arise from the reconstruction of periodic divacancies.

Indeed, different divacancy defects with various orientations relative to the zigzag direction of graphene could be formed (Fig. 2.44). A first option consists in removing carbon dimers oriented perpendicularly to the zigzag chains (Fig. 2.44(a)). Intuitively, it is expected that after geometrical relaxation, the structure would be composed of two pentagons separated by an octagon, also perpendicular to the zigzag orientation. However, the energy needed to achieve such a large strain prevents the octagons from being formed. Instead, a C–C bond rotation (as in a Stone–Wales defect) is needed to relieve the strain, as illustrated by the d5d7 structure in Fig. 2.44(a). The second option consists in removing the carbon dimers tilted 30° from the zigzag orientation

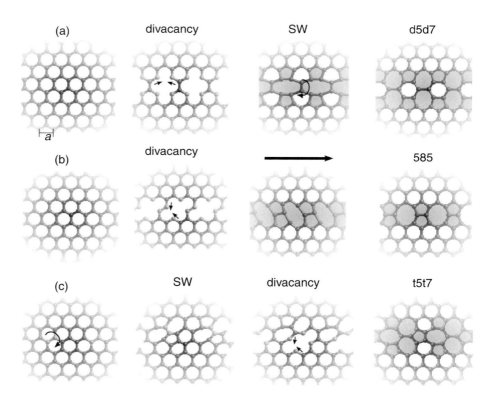

Figure 2.44 Formation of extended lines of defects in graphene through the reconstruction of divacancies. Lines of defects are formed after the removal of carbon dimers, either (a) perpendicular or (b) with a 30° deviation from the zigzag direction. These extended arrays of defects are called (a) d5d7 (double-5 double-7 structure) and (b) 585 (pentagon–octagon–pentagon structure), respectively. Further displacement of one graphene side by $\frac{1}{2}a$ (black arrow in b) is required to relax the 585 grain boundary as observed in Lahiri *et al.*, 2010. An extended array of defects composed of a series of three pentagons and three heptagons (t5t7 – triple-5 triple-7 structure) is also topologically possible when the divacancy reconstruction occurs along with a SW transformation (c). (Images from Botello-Mendez *et al.*, 2011*b*. Adapted by permission of The Royal Society of Chemistry.)

of graphene, resulting in an alternated series of octagons and two pentagons sharing a same side, as represented by the 585 structure in Fig. 2.44(b).

A less strained 585 structure could be obtained by displacing one of the two graphene domains connected to the grain boundary by $\frac{1}{2}a$, where a is the lattice parameter of graphene. Such a grain boundary structure is the one observed during the epitaxial growth of graphene (Lahiri *et al.*, 2010) (Fig. 2.43(c)). An alternative array of defects could be reconstructed from divacancies by means of a SW transformation, thus leading to a triple-pentagon triple-heptagon (t5t7) structure as depicted in Fig. 2.44(c). Such a defect shape has already been suggested as a stable topology for the reconstruction of an isolated divacancy in graphene (Amorim *et al.*, 2007).

The energetic stability and the ground-state properties of these lines of defects arising from the reconstruction of divacancies have also been investigated using the density functional theory formalism (Botello-Mendez *et al.*, 2011*b*). LDA calculations predict that the most stable reconstruction of an array of divacancies is a line of t5t7 defects (Fig. 2.44(c)), whereas the GGA calculations predict that the most stable reconstruction is the 585 array of defects (Fig. 2.44(b)). Such a discrepancy can be easily explained by the fact that LDA calculations tend to underestimate the lattice parameters of graphene, while GGA calculations tend to overestimate these values. It is also noteworthy that the experimental observation of the 585 reconstruction is constrained by the specific synthesis conditions (Lahiri *et al.*, 2010). However, in a top-down approach, e.g. vacancy creation through irradiation, the reconstruction would be either with the t5t7 or d5d7 line of defects, and would be most probably driven by the kinetics and interaction with the substrate.

In order to verify the potential advantages of these 1D arrays of defects in nanoelectronics, electronic band structure calculations have been performed (Lahiri *et al.*, 2010, Botello-Mendez *et al.*, 2011*b*). The presence of 5- and 7-member rings embedded into the sp^2-hybridized carbon network is found to induce an unexpected always-metallic behavior (Terrones *et al.*, 2000). Indeed, an almost flat band, similar to that of zigzag-edged GNR, is present close to the Fermi energy (Fig. 2.45(a)), resulting in a spike in the DOS at the Fermi level. The corresponding electronic states from the band close to the Fermi level produce a local doping in a narrow stripe along the line defect, thus creating a perfect one-dimensional metallic wire embedded in the perfect graphene sheet. Such a well-defined atomic structure of a nanowire embedded in an atomically perfect graphene sheet can help to address practically the formation of well-controlled contacts at the atomic level, as required for the future development of molecular electronics.

In the case of the extended line of t5t7 defects, extra conduction channels are induced (Botello-Mendez *et al.*, 2011*b*) and localized states could enhance the chemical reactivity of graphene. This extended defect opens the possibility of arranging molecules or

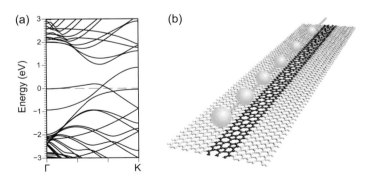

Figure 2.45 *Ab initio* electronic band structure of the 585 extended one-dimensional defect (a), exhibiting a flat band close to the Fermi level in the first half of the Brillouin zone. (b) Model illustrating the enhancement of chemical reactivity along the extended line of t5t7 defects allowing new molecular-sensing self-assembling possibilities. ((a) is reprinted by permission from Macmillan Publishers Ltd: *Nat. Nanotech.*, Lahiri *et al.*, 2010, copyright 2010. Image (b) by courtesy of Andrés Botello-Méndez (Botello-Mendez *et al.*, 2011*b*).)

atoms in a linear fashion, thus behaving as a 1D template (Fig. 2.45(b)). Such quasi-one-dimensional carbon-based metallic wires could have a big impact on the future development of smaller functional devices and may form building blocks for atomic-scale, all-carbon electronics.

2.8.3   Structural defects at graphene edges

As illustrated in the previous sections, the atomic structure of the edges is responsible to a large extent for the electronic properties of graphene nanoribbons. In many nanoscale materials, the surfaces and the edges fix the symmetry inside the "bulk" and determine the corresponding low-energy electronic structure. In GNRs, the edges also turn out to rule the appearance of flat π bands at the Fermi level and ferromagnetic ordering of π electrons along the ribbon axis. Aiming at future applications of graphene in electronic and/or spintronic devices, the precise control of the edges is thus crucial.

Although the physics of the graphene edges has been intensively investigated in the literature, only very few works consider the possibility for the edge to relax towards other geometries than the standard mono-hydrogenated zigzag and armchair patterns (Koskinen, Malola & Häkkinen, 2008, Wassmann et al., 2008). In addition, experimental studies of the graphene edge are even more scarce due to the difficulty in resolving atomically the terminations without perturbing their intrinsic structures (Girit et al., 2009, Meyer et al., 2008a). Such a lack of edge characterization continues to prevent a deep understanding of many experimental results. Indeed, despite theoretical calculations that predict nearly flat bands around the Fermi level for the zigzag edges, the experimental measurements reveal semiconducting behavior for all tested GNRs. Besides, the measured transport properties of graphene ribbons seem to be independent of their crystallographic orientation (Han et al., 2007, Li et al., 2008, Wang et al., 2008). Many reasons related to sample preparation (e.g. presence of adsorbates, substrate effects, etc.) or to the appearance of mobility gaps induced by edge disorder have been put forward to explain this disagreement between theory and experiments. Presently, the graphene edge structure is still a widely debated issue.

Many edge patterns can be considered, just by modifying the number of hydrogen connecting carbon atoms at the graphene edge. Pristine zigzag and armchair patterns with various hydrogen coverage on the edges are labeled as $z_{n_1 n_2 ... n_x}$ and $a_{m_1 m_2 ... m_x}$ respectively, where n_i and m_i refer to the number of hydrogen atoms covalently bonded to the ith edge side, and x is the number of adjacent edge-sites within the unit cell of length a (see Figs. 2.46 and 2.47). For example, z_{211} refers to a GNR with zigzag-shaped edges whose unit cell is built from a sequence of three elementary zigzag units. The carbon atoms on the edges of the first zigzag unit are doubly hydrogenated, the others are saturated by one hydrogen per carbon site.

In addition to these conventional zigzag and armchair patterns, other geometrical reconstructions of the edges can be considered. The $z(57)_{ij}$ structure depicted in Fig. 2.48 is a reconstruction of the zigzag edge where two adjacent hexagons of the edge transform in a pentagon and a heptagon. This edge pattern corresponds to a cut through a line of Stone–Wales defects (Stone & Wales, 1986) and should make it possible to

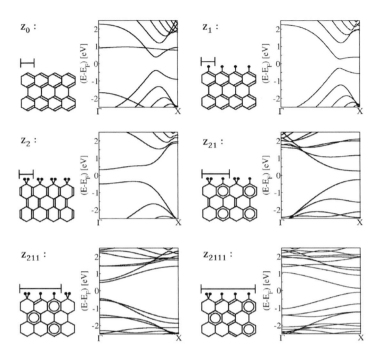

Figure 2.46 Schematic representation and electronic band structure of various possible edge passivations using hydrogen for zigzag graphene ribbons. The atomic structures depicted are periodic along the ribbon edge. The length of the unit cell is represented by a line segment. The carbon network is represented within the Clar's sextet notation (black lines), the hydrogen atoms are represented using filled circles. The electronic band structures are represented along the one-dimensional unit cell between the Γ and X special points. Reproduced from Dubois, 2009.

connect the graphene plane with a Haeckelite-type structure (Terrones *et al.*, 2000). Such a stable spontaneous Stone–Wales-type reconstruction of the z_{00} edge has been suggested by Koskinen, Malola & Häkkinen (2008). The $z(600)_{ijkl}$ structure (see Fig. 2.48) has also been reported in the literature (Wassmann *et al.*, 2008). Its geometry corresponds to a zigzag edge from which two thirds of the carbon atoms on the edges have been removed, giving rise to a saw-blade pattern.

The $a(677)_{ij}$ edge pattern (Fig. 2.48) has been proposed recently as a probable structure competing with the standard armchair edge (Koskinen, Malola & Häkkinen, 2008). This pattern is similar in nature to the $a(57)_{ij}$ (Fig. 2.48) since they both correspond to a reconstruction of the edge by a Stone–Wales-like mechanism (i.e. the 90° rotation of a carbon–carbon σ bond). Note that the $a(57)$ structure is especially interesting since its specific topology would allow connecting armchair and zigzag ribbons side by side. Finally, the $a(56)_i$ structure (Fig. 2.48) is slightly different, since its formation requires diffusion of carbon atoms in order to obtain alternation of hexagons and pentagons along the edge.

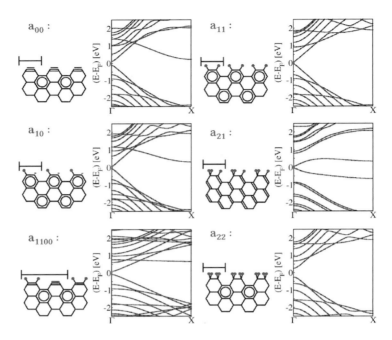

Figure 2.47 Schematic representation and electronic band structure of various possible edge passivations using hydrogen for armchair graphene ribbons. The atomic structures depicted are periodic along the ribbon edge. The length of the unit cell is represented by a line segment. The carbon network is represented within the Clar's sextet notation (black lines), the hydrogen atoms are represented using filled circles. The electronic band structures are represented along the one-dimensional unit cell between the Γ and X special points. Reproduced from Dubois, 2009.

From an electronic point of view, a general conclusion can be drawn from the data presented here (Figs. 2.46–2.48). Depending on their edge profile, zigzag and armchair GNRs may exhibit either semiconducting, semi-metallic or metallic behavior. Consequently, a knowledge of the atomic structure of the graphene edge is crucial to understand its corresponding electronic and transport properties.

In order to compare the energetic stability of the different passivated and reconstructed edges, edge-formation energies have to be computed using conventional *ab initio* techniques (Dubois, 2009). These calculations reveal that the most stable edge is the $z(600)_{2222}$ pattern, suggesting that neither the standard zigzag nor the standard armchair edges presents the lowest edge-formation energy. The stability of the $z(600)$ configuration is followed directly by the a_{22} pattern ($\sim$ 10 meV greater in energy), and then by the a_{11}, z_{21}, z_{211}, and z_{2111} configurations (more than 100 meV greater in energy), which are standard armchair and zigzag shapes with various hydrogen coverages.

The concept of aromaticity can bring additional clarification to the close connections between the ribbon topology and its relative stability. Indeed, such an aromaticity concept has been intensively used in the chemical literature, as first introduced by Clar in

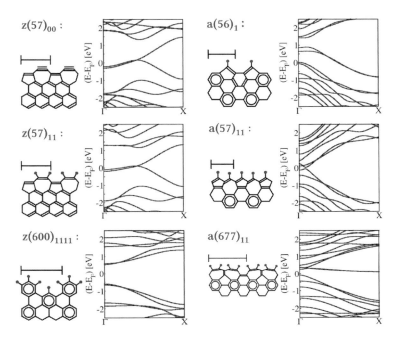

Figure 2.48 Schematic representation and electronic band structure of various possible edge reconstructions for zigzag (left) and armchair (right) graphene ribbons. The atomic structures depicted are periodic along the ribbon edge. The length of the unit cell is represented by a line segment. The carbon network is represented within the Clar's sextet notation (black lines), the hydrogen atoms are represented using filled circles. The electronic band structures are represented along the one-dimensional unit cell between the Γ and X special points. Reproduced from Dubois, 2009.

the 1970s (Clar, 1964, 1972). Clar proposed an interpretation of the electronic structure of polycyclic aromatic hydrocarbons (PAHs) based on the formal arrangement of localized π electrons in aromatic sextets. Here, an aromatic sextet (sometimes called a benzenoid aromatic ring) is defined as the benzene-like arrangement of six localized π electrons on the ring. In terms of resonance theory, it can be viewed as the resonance between two hexagonal rings with alternating single and double bonds (see Fig. 2.49). Clar's theory describes the π-electronic system of hydrocarbons on the basis of the resonance structure that maximizes the number of aromatic sextets drawn for the system. That is, according to Clar, the most stable isomers of a given hydrocarbon are those with the highest numbers of aromatic sextets. While Clar's concepts originate from an intuitive interpretation of experimental facts, there have been many papers grounding their theoretical justification within the Hückel framework (Li & Jiang, 1995).

Similarly to large PAHs, various properties of carbon nanosystems can be understood to a large extent by analysis of the π-electronic structure. Recently, Clar's rules have been shown to provide a satisfactory description of the electronic and chemical properties of carbon nanotubes (Matsuo, Tahara & Nakamura, 2003, Ormsby & King, 2004). Such simple concepts of organic chemistry have also been used to rationalize

(a)

(b)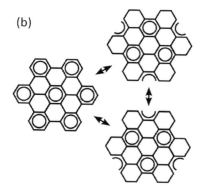

Figure 2.49 (a) Representation of the benzenoid aromatic ring within the standard Clar's sextet scheme. (b) Three equivalent representations of bulk graphene as the ideal case of aromaticity. All the π electrons belong to a benzenoid ring and the sextet density $\rho_{sx} = \frac{1}{3}$.

both the stability and the electronic properties of GNRs (Baldoni, Sgamellotti & Mercuri, 2008, Wassmann *et al.*, 2008). Within Clar's theory, graphene is the ideal 2D case of localized π electrons. Indeed, graphene can be viewed as the resonance between three structures where all π electrons belong to an aromatic sextet (represented by circles in Fig. 2.49). In each resonance structure, one third of the hexagonal rings is benzenoid and no double bonds have to be drawn. In the following, the sextet density $\rho_{sx} = $ (# sextets/#hexagons) is used to characterize the benzenoid (aromatic) character of GNR edges. An example of full benzenoid structure is given by the graphene plane where $\rho_{sx} = \frac{1}{3}$. In Figs. 2.46–2.48, the edge profiles are drawn in a representation that maximizes the number of aromatic sextets. The Clar's representations of the standard armchair (a_{11}) and zigzag (z_1) edge profiles appear to be completely different, illustrating the two extreme cases of fully benzenoid ($\rho_{sx}(a_{11}) = \frac{1}{3}$) and non-benzenoid ($\rho_{sx}(z_1) = 0$) systems.

According to Clar's rule, the z_1 edge profile is thus expected to be less stable than other edge shapes compatible with a fully benzenoid GNR. Consequently, the z_1 edge shape is not a favorable boundary profile for the hexagonal network of π electrons and may be seen as a good candidate for reconstruction. As shown in Fig. 2.46, periodic hydrogenation of the zigzag edge modifies the aromatic character of the zGNRs. The z_{21}, z_{211}, and z_{2111} edge profiles restore a partial aromaticity within the ribbon ($\rho_{sx}(z_{21}) = \frac{1}{4}$, $\rho_{sx}(z_{211}) = \frac{1}{3}$, $\rho_{sx}(z_{21}) = \frac{1}{4}$). According to Clar's rule, the z_{211} profile is then expected to be the most stable zigzag edge profile.

In contrast with the z_1 edge profile, the a_{11} and a_{22} profiles are compatible with the bonding pattern of the infinite graphene plane. Consequently, one may expect the a_{11} and a_{22} edge to be stable and relatively inert. This is globally true for all the edge structures that preserve the *armchair* boundary of the $\pi - \pi^*$ electronic network (i.e. a_{22},

a_{11}, a_{1100}, a_{00}). Note that while the $a(56)_1$ edge is compatible with the fully benzenoid character of the GNR, the large edge strains introduced by the pentagonal terminations make this edge profile less stable. These qualitative predictions are in good agreement with the *ab initio* computed edge-formation energies (Dubois, 2009). In addition, the *"aromatic"* edge structures (those with a sextet density $\rho_{sx} = \frac{1}{3}$) do not produce edge states and consequently do not give rise to magnetic order along the edges, as pointed out in Wassmann *et al.* (2008). Indeed, when the sextet density in the vicinity of the edge is smaller than that of bulk graphene, the low-energy electronic structure of the ribbon results from a competition between the bulk, where the aromaticity prevails, and the edge, where all carbon atoms tend to form four saturated bonds. It is precisely such a competition which, by forcing some edge atoms to be over- or under-saturated, gives rise to these electronic defects, i.e. the edge states.

Finally, Clar's structures also allow us to rationalize the dependence of the electronic properties of GNRs with respect to their width. As mentioned in the previous sections, the electronic properties of standard a_{11}-GNRs (i.e. the energy gap) fluctuate periodically with the ribbon width. On one hand, the monotonic decrease of the energy gap for increasing widths is the direct manifestation of the electronic confinement in the direction perpendicular to the ribbon axis. On the other hand, the periodic pattern superimposed on this monotonic decrease is due to a periodic variation of the electronic structure at the ribbon termination as highlighted by the Clar's representation.

The Clar's representations of GNRs with z_1, z_{21}, and a_{11} edge profiles are depicted for different ribbon widths in Fig. 2.50. Note that Clar's representation of the (partially) benzenoid GNRs (i.e. z_{21} and a_{11}) depends on the ribbon width. Hence, varying the width of these GNRs implies a modification of the aromaticity along the ribbon edges. On the contrary, Clar's representation of the nonbenzenoid GNRs (e.g. z_1) is independent of their width. The ribbons keep a nonbenzenoid character for all widths. As illustrated in Fig. 2.51, the variations of the edge aromaticity have an important impact on the electronic properties of the ribbons. The energy gap of the $[a_{11}]$-GNRs and $[z_{21}]$-GNRs exhibits an oscillating behavior with respect to changes in the ribbon width.

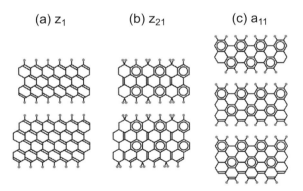

(a) z_1 (b) z_{21} (c) a_{11}

Figure 2.50 Clar's representations for $[z_1]$-, $[z_{21}]$-, and $[a_{11}]$-GNRs of various widths. (a) $[z_1]$-GNRs with $N = 4$ and $N = 5$ widths. (b) $[z_{21}]$-GNRs with $N = 4$ and $N = 5$ widths. (c) $[a_{11}]$-GNRs with $N = 6$, $N = 7$ and $N = 8$ widths.

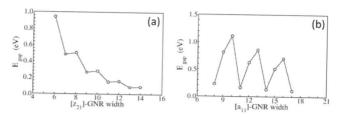

Figure 2.51 Energy gap with dependence in GNRs. The abscissa refers to the number of zigzag/dimer lines of carbon atoms across: (a) a $[z_{21}]$-GNR; and (b) a $[a_{11}]$-GNR with increasing width. Reproduced from Dubois (2009).

These results demonstrate that simple binding theories, such as the Clar's sextet theory, can be easily extended in order to bring some clarification of the stability and electronic properties of graphene edges. Though they cannot be used to make quantitative predictions, these concepts of aromaticity would probably help in the search for the most stable edge patterns in various chemical environments.

In summary, the electronic and energetic properties of various *reconstructed* edge patterns have been unveiled by means of *ab initio* calculations. Though the interest of the scientific community has been focused on the properties of the standard (a_{11} and z_1) edge patterns, the computed edge-formation energies reveal that these latter are stable at very low hydrogen concentrations (Dubois, 2009), suggesting that, in most experimental conditions, the standard edges do reconstruct. Though the most stable edge structures are likely to be dependent on the actual chemical environment, z_{211}, a_{22} and $z(600)$ edge patterns are suggested to be probable candidates for the edge reconstruction, presenting no magnetic order along the GNR edges.

2.8.4   Defects in carbon nanotubes

In the previous sections, only the geometrical aspect and the local environment of carbon nanotubes have been investigated. As in graphene, the intrinsic honeycomb network of carbon nanotubes is probably not as perfect and ideal as previously considered. Indeed, defects like pentagons, heptagons, vacancies, or dopants could certainly also be found, thus modifying dramatically the electronic properties of these 1D nanosystems. Introducing defects in the carbon network is thus an interesting way to tailor the intrinsic properties of the tube, in order to propose novel potential applications in nanoelectronics.

Thanks to the sensitivity of the metallic/semiconducting character of carbon nanotubes to their chirality, they can be used to form all-carbon metal–semiconductor, semiconductor–semiconductor, or metal–metal junctions. These junctions have great potential for applications since they are of nanoscale dimensions and made entirely of carbon. In constructing this kind of on-tube junction, the key is to join two half-tubes of different helicity seamlessly with each other, without too much cost in energy or disruption in structure. The introduction of pentagon–heptagon pair defects into the hexagonal network of a single wall carbon nanotube has been shown to change the helicity of the carbon nanotube and fundamentally alter its electronic structure (Dunlap, 1994, Lambin *et al.*, 1995, Saito, Dresselhaus & Dresselhaus, 1996*b*, Charlier, Ebbesen & Lambin, 1996, Chico *et al.*, 1996). Both the existence of such atomic-level structures and the measurement of their respective electronic properties have already been resolved experimentally (Yao *et al.*, 1999, Ouyang *et al.*, 2001*a*).

The defects, however, must induce zero net curvature to prevent the tube from flaring or closing. The smallest topological defect with minimal local curvature (hence less energy cost) and zero net curvature is a pentagon–heptagon pair. When the pentagon is attached to the heptagon, as in the aniline structure, it only creates topological changes (but no net disclination), which can be treated as a single local defect. Such a pair will create only a small local deformation in the width of the nanotube, and may also generate a small change in the helicity, depending on its orientation in the hexagonal network. Figure 2.52 depicts the connection, using a single 5–7 pair, between two nanotubes exhibiting different electronic properties. As mentioned above, the $(8, 0)$ nanotube has a 1.2 eV gap in the *tight-binding* approximation, and the $(7, 1)$ tube is a metal (although a small curvature-induced gap is present close to the Fermi energy).

Joining a semiconducting nanotube to a metallic one, using a pentagon–heptagon 5–7 pair incorporated in the hexagonal network, can thus be proposed as the basis of a nanodiode (or molecular diode) for nanoelectronics. The system illustrated in Fig. 2.52 forms a quasi-1D semiconductor–metal junction, since within the band-folding picture the $(7, 1)$ half-tube is metallic and the $(8, 0)$ half-tube is semiconducting. This led to the prediction that these defective nanotubes would behave as the desired nanoscale metal–semiconductor Schottky barriers, semiconductor heterojunctions, or metal–metal junctions with novel properties, and that they could act as building blocks in future nanoelectronic devices.

The beam of a transmission electron microscope can be used to irradiate nanostructures locally. Covalently connected crossed single-wall carbon nanotubes can thus be created using electron beam welding at elevated temperatures (Terrones *et al.*, 2000, Terrones *et al.*, 2002). These molecular junctions of various geometries ("X","Y", and

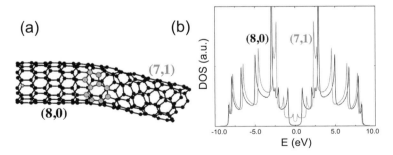

Figure 2.52 (a) Atomic structure of an $(8, 0)/(7, 1)$ intramolecular carbon nanotube junction. The large light-gray balls denote the atoms forming the heptagon–pentagon pair. (b) The electron density of states related to the two perfect $(8, 0)$ and $(7, 1)$ nanotubes is illustrated with thick black and thin grey lines, respectively. (Adapted with permission from Chico *et al.*, 1996. Copyright (1996) American Physical Society. By courtesy of Leonor Chico.)

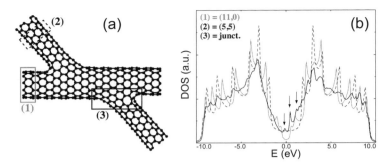

Figure 2.53 Atomic structure (a) and electronic properties (b) of an ideal X-junction, created by intersecting an $(5, 5)$ tube with an $(11, 0)$ tube. (b) shows the one-dimensional electronic densities of states of a semiconducting $(11, 0)$ nanotube (light curve), a metallic $(5, 5)$ nanotube (dashed curve) and the average over the intersecting region of the molecular junction (black curve). The Fermi level is positioned at the zero energy. Localized states due to the presence of defects are indicated by arrows. (Adapted with permission from Terrones *et al.*, 2002. Copyright (2002) by the American Physical Society.)

"T") are found to be stable after the irradiation process. To study the relevance of some of these nanostructures, various models of ideal molecular junctions can be generated. The presence of heptagons is found to play a key role in the topology of nanotube-based molecular junctions. Figure 2.53 depicts an ideal "X" nanotube connection, where a $(5, 5)$ armchair nanotube intersects an $(11, 0)$ zigzag tube. In order to create a smooth topology at the molecular junctions, six heptagons are introduced at each crossing point (Terrones *et al.*, 2002).

The local densities of states of the metallic $(5, 5)$ nanotube and the (semiconducting) $(11, 0)$ nanotube are illustrated in Fig. 2.53. The LDOS of the regions where the two nanotubes cross reveals an enhancement of the electronic states at the Fermi level. It is also notable that the presence of localized donor states in the conduction band (as indicated by arrows) is caused by the presence of heptagons. The novel small peak on the valence band (also shown by an arrow), close to the Fermi energy, can probably be attributed to the high curvature of the graphitic system (Terrones *et al.*, 2002). The van Hove singularities present in the LDOS of the two achiral nanotubes are dramatically less pronounced in the junction region (Fig. 2.53), thus illustrating a clear loss of the one-dimensional character. Local density of states of CNT-based junction models suggests their importance in electronic device applications and paves the way towards controlled fabrication of nanotube-based molecular junctions and network architectures exhibiting exciting electronic and mechanical behavior.

To close this subsection we note that following the previous idea of introducing pentagons and heptagons into hexagonal networks, a novel class of perfect crystals, consisting of layered sp^2-like carbon and containing periodic arrangements of pentagons, heptagons and hexagons, has been suggested theoretically (Terrones *et al.*, 2000). These sheets are rolled up so as to generate single-wall nanotubes (Fig. 2.54), which resemble locally the radiolaria drawings of Ernst Haeckel (1862).

(a)　　　　　　　(b)　　　　　　　(c)

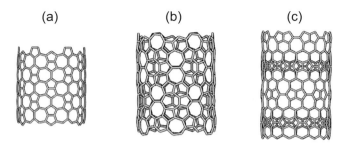

Figure 2.54 Nonchiral Haeckelite nanotubes of similar diameter (1.4 nm): (a) nanotube segment containing only heptagons and pentagons paired symmetrically; (b) nanotube segment exhibiting repetitive units of three agglomerated heptagons, surrounded by alternating pentagons and hexagons; (c) nanotube segment containing pentalene and heptalene units bound together and surrounded by six-membered rings. (Adapted with permission from Terrones *et al.*, 2000. Copyright (2000) by the American Physical Society.)

These ideally defective tubes exhibit intriguing electronic properties: local density of states calculations of Haeckelite tubes reveal an important enhancement of electronic states close to the charge neutrality point, independent of orientation, tube diameter, and chirality. Considering the possible metallic properties of Hackelites, these new nanostructures should offer different advantages compared to carbon nanotubes in applications (i.e. no helicity selection for electronic interconnect applications). Our calculations also reveal that these Haeckelite structures are more stable than C_{60}, and present cohesive energies of the order of 0.3–0.4 eV/atom with respect to graphene, allowing the potential synthesis of this new class of nanotubes. Although, these ideal topologies have never been synthesized, the carbon coiled nanostructures have been explained by rolling up strips mainly made of heptagons, pentagons, and hexagons, with a predominance of nonhexagonal rings (Biró *et al.*, 2002).

2.9　Further reading and problems

- General suggested references on carbon nanotubes include Saito, Dresselhaus & Dresselhaus (1998) and Charlier, Blase & Roche (2007).
- For a very detailed review on the low-energy (**k.p**) approximation applied to graphene and carbon nanotubes see Marconcini & Macucci (2011).
- A very detailed account of the boundary conditions in a terminated graphene network is presented in Akhmerov (2011). Furthermore, a very nice presentation on graphene zigzag edge states can be found in Wimmer (2009).

Problems

2.1 *Non-orthogonal tight-binding scheme and graphene dispersion.*
Follow Section 2.2.1 and re-derive the dispersion relation for graphene without neglecting the overlap $s = \langle p_z^A | p_z^B \rangle$ between neighboring p_z orbitals. Compare the results obtained with and without the approximation.

2.2 *Electronic structure of graphene and boron-nitride: the $\pi-\pi^*$ model.*
Using a simple *tight-binding* approach (one π orbital per atom in a honeycomb lattice), the following TB Hamiltonian can be constructed:

$$H = \begin{pmatrix} \varepsilon_A & \gamma_0 f(k) \\ \gamma_0 f^*(k) & \varepsilon_B \end{pmatrix},$$

where ε_A and ε_B are the two onsite energies of the two corresponding atoms present in the unit cell, γ_0 is the hopping integral, and $f(k)$ is the sum of the nearest-neighbor phase factors and is equal to $e^{ik_x a/\sqrt{3}} + 2e^{-ik_x a/2\sqrt{3}} \cos{(k_y a/2)}$ (Wallace (1947)).

Solve the eigenvalue problem when $\varepsilon_A = \varepsilon_B$ (graphene) and when $\varepsilon_A \neq \varepsilon_B$ (boron-nitride sheet, Bernal graphite, etc.).

2.3 *Electronic structure of graphene: a tight-binding study.*
Most *tight-binding* studies use a first-nearest-neighbors $\pi-\pi^*$ model to describe the electronic properties of graphene. However, such an approximation produces a perfectly symmetric band structure (cf. previous exercise). In order to recover the existing asymmetry between valence (π) and conduction (π^*) bands, a third-nearest-neighbors $\pi-\pi^*$ model has to be used. Such a TB model is composed of a single onsite term ε_{p_z} and three hopping terms $\gamma_0^{(1)}$, $\gamma_0^{(2)}$, and $\gamma_0^{(3)}$ corresponding to the interaction between first-, second-, and third-nearest neighbors, respectively. The corresponding Hamiltonian can thus be expressed as follows:

$$H = \sum_i \varepsilon_{p_z} |\phi_i\rangle \langle\phi_i| + \sum_{i,<j,k,l>} \left(\gamma_0^{(1)} |\phi_j\rangle \langle\phi_i| + \gamma_0^{(2)} |\phi_k\rangle \langle\phi_i| + \gamma_0^{(3)} |\phi_l\rangle \langle\phi_i| + h.c. \right),$$

with $\varepsilon_{p_z} = 0.6\,\text{eV}$, $\gamma_0^{(1)} = -3.1\,\text{eV}$, $\gamma_0^{(2)} = 0.2\,\text{eV}$, and $\gamma_0^{(3)} = -0.16\,\text{eV}$. The sums on index i run over all carbon p_z orbitals. The sums over j, k, l indices run over all p_z orbitals corresponding, respectively, to first-, second-, and third-nearest neighbors of the ith p_z orbital (for more detailed information, see Lherbier *et al.* (2012)).

Compare the electronic band structures of graphene along high symmetry lines using both the first nearest-neighbors and the third nearest-neighbors $\pi-\pi^*$ models.

2.4 *Electron–hole symmetry in bipartite lattices.*
Consider a generic bipartite lattice, i.e. one that can be divided into two lattices where all the sites of one lattice have nearest neighbors that belong to the other.

(a) Arrange the basis vectors spanning the Hilbert space in such a way that all the orbitals corresponding to sublattice A come first. By assuming that only hoppings connecting sites of different sublattices are allowed, write a generic form for the Hamiltonian in block-matrix form.

(b) Use the previously obtained form of the Hamiltonian to show that if E is an eigenvalue of the Schrödinger equation, then $-E$ is also an eigenvalue and that therefore there is particle–hole symmetry.

(c) Consider a graphene network with a line defect as shown in Fig. 2.43(b). Say if the bipartite nature of the graphene lattice is preserved by this defect. What can you conclude on the electron–hole symmetry of the system with the defect?

2.5 *Electron–hole symmetry in finite systems.*
As noted in Section 2.2.1, the Hamiltonian is said to have electron–hole symmetry if there is a transformation, $\mathcal{P}$, such that $\mathcal{P}^\dagger \mathcal{H} \mathcal{P} = -\mathcal{H}$. This implies that if Ψ is an eigenstate of $\mathcal{H}$ with a positive energy E (electron function), then $\mathcal{P}\Psi$ is also an eigenstate with energy $-E$. For the case of a finite size system such as a ribbon, the transformation must also be compatible with the boundary conditions. Can you mention one transformation that satisfies both requirements, independently of the edge termination? (For further reading see Akhmerov & Beenakker (2008).)

2.6 *Edge states of zigzag graphene nanoribbons.*
(a) Consider a semi-infinite graphene sheet with a zigzag edge. Within a simple π orbitals Hamiltonian show that there is a normalizable solution for $E = 0$ which corresponds to a state localized at the edges. How does the decay length of this edge state (in the direction perpendicular to the edge) depend on the k vector?

(b) Repeat the previous point for a zigzag graphene nanoribbon of finite width.

(c) Find an alternative solution to the previous two points by starting from the effective Dirac Hamiltonian and by imposing appropriate boundary conditions.

2.7 *Structural parameters and Brillouin zone of carbon nanotubes.*
(a) Demonstrate the expressions for the structural parameters of CNTs given in Table 2.1.

(b) Consider $(5, 5)$ and $(6, 3)$ nanotubes. Determine their structural parameters, reciprocal basis vectors and their Brillouin zones.

2.8 *Zone-folding approximation for carbon nanotubes.*
(a) Re-derive by your own method the rule for metallicity of an SWNT within the zone-folding approximation.

(b) Analyze the case of semiconducting nanotubes and derive an expression for their bandgap (Eq. 2.47).

(c) Choose a set of four different nanotubes (including zigzag, armchair, and chiral), determine their structural parameters and draw the associated cutting lines in reciprocal space. Are they metallic or semiconducting?

2.9 *Ratio of metallic to semiconducting carbon nanotubes.*
Within the zone-folding approximation one third of carbon nanotubes are metallic. To demonstrate this statement:
(a) Determine the number of all the possible nanotubes up to (n, n), N.

(b) Determine the ratio between the total number of metallic nanotubes up to (n, n), N_m, and N. Assuming a random distribution of chiralities and large enough n, this gives the ratio of metallic to semiconducting nanotubes.

2.10 *Flat bands in carbon nanotubes and trigonal warping corrections.*
If you examined carefully the dispersion relation for a $(10, 0)$ nanotube in Fig. 2.25 you might have noticed the appearance of flatbands located at $\pm\gamma_0$. Here, we examine this in more detail.

(a) Plot the dispersion relation of bulk graphene (3D plot of $E(k_x, k_y)$) and visualize the constant energy lines.

(b) Plot graphene's isoenergy lines in k-space for different values of E from zero up to a few eV. Observe how these lines deviate from circles as the energy is shifted away from the Dirac point and try to conclude why these flat bands appear only for some CNTs.

2.11 *Electronic structure of graphene: an* ab initio *study.*
In order to obtain the *ab initio* electronic properties of graphene, the simulation has to be performed in three successive steps using the ABINIT code (for more detailed information, tutorials are available on the website *www.abinit.org* and also in Appendix A).

(a) Choose the pseudopotential for carbon (suggestion: LDA pseudopotential of Troullier–Martins or Goedecker–Teter–Hutter, also available on *www.abinit.org*.)

(b) A self-consistent calculation has to be performed in order to optimize the structural lattice parameters of graphene and estimate the electronic density using the following input file for the ABINIT code:

acell	2.5	2.5	10	(Å)
rprim				
	$\sqrt{0.75}$	0.5	0.0	
	0.0	1.0	0.0	
	0.0	0.0	1.0	
ntypat	1			
znucl	6			
natom	2			
typat	1	1		
xred				
	1/3	1/3	0.0	
	2/3	2/3	0.0	
kptopt	1			
nshiftk	1			
shiftk	0.0	0.0	0.0	
ngkpt	12	12	1	
ionmov	3			
ntime	100			
tolmxf	10^{-6}			
nband	10			
toldfe	10^{-10}			
nstep	20			
ecut	40			

A description of each variable is available on the ABINIT website.

(c) Finally, a non-self-consistent calculation has to be performed in order to estimate the electronic band structure along a specific path in the reciprocal space using the following input file for the ABINIT code:

acell	4.6165	4.6165	13.228	*(Bohr)*
rprim				
	$\sqrt{0.75}$	0.5	0.0	
	0.0	1.0	0.0	
	0.0	0.0	1.0	
ntypat	1			
znucl	6			
natom	2			
typat	1	1		
xred				
	1/3	1/3	0.0	
	2/3	2/3	0.0	
kptopt1	1			
nshiftk1	1			
shiftk1	0.0	0.0	0.0	
ngkpt1	12	12	1	
prtden1	1			
iscf2	-2			
getden2	-1			
kptopt2	-3			
ndivk2	20	20	20	
kptbounds2				
	2/3	1/3	0.0	*(K)*
	2/6	1/6	0.0	*(Γ)*
	1/2	1/2	0.0	*(M)*
	1/3	2/3	0.0	*(K′)*
tolwfr2	10^{-12}			
enunit2	1			
nband2	10			
toldfe1	10^{-10}			
nstep	20			
ecut	40			

A description of each variable is available on the ABINIT website.

2.12 *Bilayer graphene and trigonal warping effects.*
In this exercise we consider an effective Hamiltonian for bilayer graphene with Bernal stacking. The unit cell contains two inequivalent sites labeled as $A1$, $B1$ on the top layer and $A2$, $B2$ on the bottom layer. Their arrangement is such that atom $B1$ is on top of atom $A2$. By indexing the wavefunctions $\Psi = (\psi_{A1}, \psi_{B2}, \psi_{A2}, \psi_{B1})^T$ for the K_+ valley

and $\Psi = (\psi_{B2}, \psi_{A1}, \psi_{B1}, \psi_{A2})^T$ for the K_- valley, one can write a low-energy effective Hamiltonian as (McCann *et al.*, 2006)

$$\mathcal{H}(\mathbf{k}) = \xi \begin{pmatrix} 0 & v_3\pi & 0 & v\pi^\dagger \\ v_3\pi^\dagger & 0 & v\pi & 0 \\ 0 & v\pi^\dagger & 0 & \xi\gamma_1 \\ v\pi & 0 & \xi\gamma_1 & 0 \end{pmatrix},$$

where $\xi = 1(-1)$ for valley $K_+(K_-)$, $\pi = p_x + ip_y$, $v = (\sqrt{3}/2)a\gamma_0/\hbar$, $v_3 = (\sqrt{3}/2)a\gamma_3/\hbar$, and a is the graphene lattice constant. The hopping parameters take the values $\gamma_0 = 3.16$ eV, $\gamma_1 = 0.39$ eV and $\gamma_3 = 0.315$ eV.

(a) Calculate and plot the energy dispersion for this model of bilayer graphene with and without the parameter γ_3. The hopping parameter γ_3 is responsible for the trigonal warping effects commented on in Section 2.2.3.

(b) Repeat the previous exercise for the density of states. In which energy range can trigonal warping be neglected?

(c) What would happen if you break the symmetry between the two graphene sheets?

** Additional exercises and solutions are available at our website.

3   Quantum transport: general concepts

The previous sections have been devoted to the electronic structure of carbon-based materials. The rest of the book is now focused on their transport properties. This part is meant as a nexus, providing a brief reminder on quantum transport with a focus on the tools that are needed later in the book. After a discussion of the most relevant length scales and the different transport regimes, three different formalisms are reviewed, namely Landauer theory, the Kubo formalism and the semiclassical Boltzmann transport equation. More technical details concerning the use of Green's functions methods and the Lanczos method for computing the density of states and wave-propagation are discussed in Appendixes C and D respectively.

3.1   Introduction

3.1.1   Relevant time and length scales

Electron transport through a device is a phenomenon that takes place in time and space and as such there are relevant time *and* length scales. Given a device with characteristic dimensions L_x, L_y, and L_z, if the system is metallic then one has the Fermi wavelength $\lambda_F = 2\pi/k_F$ associated with its Fermi wave-number k_F. The elastic mean free path ℓ_{el} can be defined as the distance that an electron travels before getting elastically backscattered (off impurities for example); $\ell_{el} = v_F \tau_{el}$, where τ_{el} is the mean time between those elastic scattering events which are usually produced by defects or imperfections in the crystal structure. In disordered systems, when the disorder strength is such that $\ell_{el} \sim \lambda_F$, the wavefunctions become localized on a length scale ξ, the localization length.

Analogously to ℓ_{el}, one can define the inelastic mean free path $\ell_{in} = v_F \tau_{in}$ as the mean distance between inelastic scattering events such as those due to electron–phonon or weak electron–electron interactions. Generically, it is usual to speak of the electronic mean free path ℓ, without discerning the specific source, elastic or inelastic.[1] The phase coherence length ℓ_ϕ (and corresponding coherence time τ_ϕ) is defined as the length over which the phase of the single-electron wavefunction is preserved (within an independent electrons approximation), which limits the scale of quantum phase interferences. Typical values for graphene, carbon nanotubes and other materials are given in Table 3.1.

---

[1]   In graphene the main sources of scattering include charged impurities, defects in the crystal structure and microscopic corrugations of the graphene sheet (also called ripples). Their relative importance is still debated.

Table 3.1 Typical magnitudes of the charge density (n), the mean free path ℓ, Fermi wavelength (λ_F) and the coherence length (L_ϕ) at 4 K in various materials.

	GaAs-AlGaAs	Metals	Graphene	SWNT	MWNT
n	4×10^{11} cm^{-2}	$10^{21} - 10^{23}$ cm^{-3}	$10^{11} - 10^{12}$ cm^{-2}	10^{11} cm^{-2}	,,
ℓ	$100 - 10^4$ nm	$1 - 10$ nm	50 nm to 3 μm^a	$1\ \mu$m	$10 - 40$ nm
λ_F	40 nm	0.5 nm	$2\sqrt{\pi/n}$	0.74 nm	,,
L_ϕ	100 nm	0.5 μm	0.5 μm^b	3 μm^c	100 nm

^{a} In suspended graphene, mean free paths of about 100 nm were found at 4 K for $n \sim 10^{11}$ cm^{-2} (and about 75 nm at 300 K) in Du *et al.* (2008), while Bolotin *et al.* (2008) estimate ℓ of up to 1.2 μm for $n \sim 2 \times 10^{11}$ cm^{-2}. On the other hand, reported values for devices made from graphene sandwiched in between hBN crystals go up to 3 μm (Mayorov *et al.*, 2011).
^{b} See Tikhonenko *et al.* (2009).
^{c} See Stojetz *et al.* (2005).

3.1.2   Coherent versus sequential transport

Coherent or sequential? is probably one of the most crucial questions, since it dictates the general framework that better suits a particular system under investigation in a particular experimental condition (Weil & Vinter, 1987, Jonson & Grincwajg, 1987, Luryi, 1989, Foa Torres, Lewenkopf & Pastawski, 2003). Note, however, that the answer most probably lies in between these two extreme situations (see also Section 3.5).

Let us imagine that we start with the sample (nanotube, graphene ribbon, etc.) decoupled from the electrodes. As the coupling between them is turned on, there is an increasing escape rate which determines the intrinsic width (Γ_α) of the levels (ε_α) corresponding to the isolated sample. The more isolated is the sample from the electrodes, the longer the lifetime τ_D of an electron in any of those levels and the smaller the intrinsic level width $\Gamma_\alpha = \hbar/\tau_D$. If the lifetime associated with the intrinsic level width is longer than the coherence time (τ_ϕ), then the electrons will spend enough time inside the sample to suffer phase breaking events leading to a decoherent regime.

In the decoherent limit, one may use a *sequential picture* for transport, in which the electronic motion is divided, as in a theater play, into different parts:

1. *Tunneling in.* The electron is transmitted from the left electrode into the sample;
2. *Dwelling.* The electron dwells in the sample, eventually interacting with other electrons or with phonons/vibrational degrees of freedom;
3. *Tunneling out.* The electron tunnels into the right electrode or is reflected back to the left one.

A sometimes implicit assumption of such a picture is that transport is decoherent. Therefore, the description can be at a semiclassical level where only the occupation *probabilities* (and not the amplitudes) are taken into account into a set of rate equations. Typically, these rate equations take into account the different possible processes (tunneling in and out of the sample, electron–electron and inelastic interactions) through a Fermi golden rule for the associated transition rates. By solving these equations one gets the occupation probabilities, from which the current and other quantities of interest can be computed. The widely used Boltzmann equation belongs to this class of schemes, and is introduced later in Section 3.3.

Transport in the Coulomb blockade regime (see also Section 5.8.1) is usually described by such a sequential picture (Beenakker, 1991). In this regime, the contacts to the electrodes are weak enough such that the charge inside the sample is well defined and quantified. One says that the transport is suppressed (or blocked) and is only possible at precise energies, which can be tuned by varying the gate voltage (conductance peaks). The energy scale governing such peaks is the charging energy (E_c): the energy necessary to compensate for the electron repulsion and add one more electron to the system.

When the coherence time is longer than the residence time in the sample, the tunneling processes through the contacts and dwell inside the sample cannot be treated in a separate fashion anymore. The picture is that of a coherent transport mechanism and the theater play becomes a weird quantum game. This is the realm where quantum interference effects and even more exotic phenomena involving correlated motion between electrons like the Kondo effect may take place.[2] The Landauer–Büttiker theory and the Kubo formalism, which are briefly introduced in Sections 3.2 and 3.4.4, provide an appropriate framework for coherent, noninteracting electrons.

A crucial magnitude controlling the transition between these regimes is the intrinsic energy level width Γ_α of the sample connected to outside world. As one moves from the coherent to the sequential regime, Γ_α is reduced until it becomes the smallest energy scale in the problem (the sample being more and more disconnected from the electrodes). Simultaneously, the value of the charging energy increases from zero to a value where it dominates over the mean level spacing Δ and dictates a sequential and discretized transfer of charges from a source to a drain electrode.

A beautiful experiment showing this transition is reproduced in Fig. 3.1 (Babic & Schönenberger, 2004). The coupling with the leads changes as the gate voltage V_g is varied, thereby producing a crossover from low transparency to high transparency contacts and allowing observation of the transition from coherent (lower V_g region in Fig. 3.1 (a) and (b)) to sequential tunneling (high V_g region in the figures, where isolated resonances are observed). The conductance accordingly exhibits a wealth of phenomena which includes, from higher to lower gate voltage: Coulomb blockade peaks, strong cotunneling, and Kondo effect, and destructive interference which is manifested as Fano resonances.[3] This experiment illustrates in a magnificent way that the occurrence and the

---

[2] The Kondo effect is one of the most studied many-body phenomena in condensed matter physics (for a review see Kouwenhoven & Glazman (2001)), and is also an active topic in graphene physics, both theoretically (Cornaglia, Usaj & Balseiro, 2009, Cazalilla et al., 2012) and experimentally (Chen et al., 2011).

[3] Fano resonances, also known as anti-resonances in the context of electronic transport (Guinea & Vergés, 1987, D'Amato, Pastawski & Weisz, 1989), are a coherent effect of destructive interference pioneered by Fano (1935) in spectroscopy and observed since then in many contexts in different nanostructures (Miroshnichenko, Flach & Kivshar, 2010).

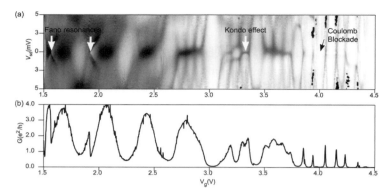

Figure 3.1 (a) Density plot of the differential conductance versus bias voltage V_{sd} and gate voltage V_g (high conductance in black and low conductance in white). (b) Linear response conductance versus gate voltage. The coupling to the leads strongly depends on the gate voltage, allowing for the observation of very different phenomena in the same experiment, namely, Coulomb blockade, Kondo effect and Fano resonances. (Adapted from Babic & Schönenberger (2004). Copyright (2004) by the American Physical Society. Courtesy of Christian Schönenberger.)

nature of quantum transport phenomena through a mesoscopic sample strongly depend on the conditions, the measurement setup, and the dominant energy scales of the system under study.

3.2   Landauer–Büttiker theory

One of the most influential frameworks for the study of quantum transport is Landauer theory, pioneered originally by Rolf Landauer in the early fifties (Landauer, 1957, 1970) and generalized later on by Büttiker and others (Büttiker *et al.*, 1985) for multi-lead systems. The simplicity of Landauer's picture for transport boosted it as a driving force in the field of nanoscale transport. As will be shown later, several reasons make it particularly useful in the context of graphene-based devices and therefore we dedicate the next pages to a brief presentation of its main points while trying to clarify the underlying assumptions and limitations.

Let us consider a sample or device that is connected through leads to reservoirs. A particular case with two leads is represented in Fig. 3.2. Within Landauer's approach conductance through a device is seen as a scattering process where electrons injected from the reservoirs are incident onto the device and then scattered back into the reservoirs. Landauer's theory relates the conductance, measuring the ease with which the electrons flow, with the transmission probability through the device. The current through electrode j (I_j) is given by

$$I_j = \frac{2e}{h} \int \sum_{i=1}^{N} \left[T_{j,i}(\varepsilon) f_i(\varepsilon) - T_{i,j}(\varepsilon) f_j(\varepsilon) \right] d\varepsilon, \tag{3.1}$$

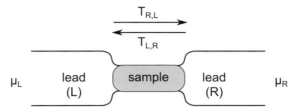

Figure 3.2 A setup where a small conductor is sandwiched between two leads (L and R). The leads are connected to reservoirs kept at equilibrium chemical potentials μ_L and μ_R which in turn determine their equilibrium occupation probabilities (Fermi functions) denoted by f_L and f_R.

where $T_{j,i}(\varepsilon)$ is the transmission probability of an electron of energy ε from lead i to lead j, and the integral is over all available energies. The factor 2 on the right hand side of the previous equation comes from considering the electron spin, which here just duplicates the number of available channels. Therefore, the current is given by a balance between the contributions from the different leads weighted by their corresponding transmission probabilities.[4] These transmission probabilities can be conveniently expressed in terms of the Green's functions within the device region through the *trace formula*. The reader not familiar with this formalism is encouraged to follow Appendix C and complete the brief overview given here.

For a two-terminal setup in the low bias, zero temperature limit ($V \to 0$, $k_B T \to 0$), Eq. (3.1) can be simplified to

$$I = \frac{2e^2}{h} T_{RL}(\varepsilon_F) V, \tag{3.2}$$

where T_{RL} is the total transmission probability from left to right, ε_F is the Fermi energy and $I_L = I_R = I$ is the current through the device. The conductance is essentially given by the transmission probability in this limit.

It is important to note that there are no blocking factors multiplying the occupation probabilities in the leads (such as $f_i(1 - f_j)$) in the previous equation. Although for systems with time-reversal symmetry the additional terms would just cancel out leaving the expression unchanged; the difference is a conceptual one. The blocking factors would be an attempt to ensure that the final states are empty and therefore not blocked by the Pauli principle (which would somehow mean coming back to a sequential view of transport!). However, within Landauer's scattering viewpoint the scattering states are occupied according to their asymptotic occupation probability given by the Fermi functions

---

[4] Note that one can also interpret Eq. (3.1) as the balance between two terms, each one being the product of the probability that a state of energy ε is occupied in lead j (i) times the density of states of that channel times velocity of the corresponding state times the probability that it is transmitted T_{ij} to the other lead (T_{ji}). For one-dimensional channels the density of states is proportional to the inverse group velocity, canceling it out and leading to the referred formula.

in the leads. These states are orthogonal and extend from the leads and throughout the whole device. Therefore, there is no need of an actual transition to pass from one side of the device to the other.[5]

To get more insight, both a heuristic (Section 3.2.1) and a more formal derivation (Section C.1) will be presented. But before that let us briefly discuss the current fluctuations. Besides the average current, its fluctuations from the average value are also of interest. They can be characterized by a current–current correlation function $\langle \Delta I(t) \Delta I(t') \rangle$, where $\Delta I(t) = I(t) - \langle I(t) \rangle$. Its statistical moments give the full counting statistics (Blanter & Büttiker, 2000). Most often one is interested in the low frequency current noise, which is given by the zeroth order moment (Blanter & Büttiker, 2000). At zero temperature, when there are no fluctuations in the distribution of the incident electrons, the zero-frequency noise is given by

$$S_I = \frac{2e^3 \, |V|}{\pi \hbar} Tr\left(r^\dagger r t^\dagger t\right),$$ (3.3)

where $|V|$ is the bias voltage applied between the electrodes, and t and r are matrices containing the transmission and reflection amplitudes (evaluated at the Fermi energy) between the different channels. If we denote the scattering matrix by S, its element $S_{i,j}$ being the probability amplitude of going from channel j to channel i, then the transmission matrix t contains the off-diagonal elements of S while the reflection matrix r contains the diagonal elements of S. The matrix $t^\dagger t$ can be diagonalized and its real eigenvalues give the transmission probabilities that we denote with T_q. In this basis of eigenchannels, the last equation can be written as (Blanter & Büttiker, 2000)

$$S_I = \frac{2e^2}{\pi \hbar} \sum_{q=1}^{N} T_q (1 - T_q).$$ (3.4)

More generally, within Landauer's theory, the zero-frequency noise for a two-terminal conductor in thermal equilibrium at a temperature T is

$$S_I = \frac{2e^2}{\pi \hbar} \sum_{q=1}^{N} \int \left(T_q[f_L(1 - f_L) + f_R(1 - f_R)] + T_q(1 - T_q)(f_L - f_R)^2 \right) d\varepsilon,$$ (3.5)

which contains contributions from: (i) the fluctuations of the incident electron beams (encoded in the Fermi functions), and (ii) the noise due to charge quantization. The zero-temperature Eq. (3.3) represents the pure shot noise contribution. An important remark emphasized by Blanter & Büttiker (2000) is that while the zero-temperature conductance can be expressed fully in terms of the transmission *probabilities* independently of the choice of the basis, S_I cannot. *The scattering amplitudes (and not the probabilities) are then the crucial quantities ruling both the conductance and noise.*

---

[5] See for example the discussion in Chapter 2 of Datta (1995).

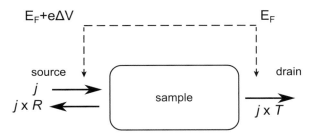

Figure 3.3 Scattering through a system with a single incoming and outgoing channel and an applied voltage difference ΔV. A current density j is injected from the source. The carriers are then scattered and a fraction R is reflected and a fraction T transmitted.

3.2.1 Heuristic derivation of Landauer's formula

Let us consider a one-dimensional metallic system coupled to two 1D electronic leads which drive incoming and outgoing currents as sketched in Fig. 3.3. The temperature of the system is set to zero, so only electrons at the Fermi level participate in the electronic current. This current originates from a potential difference between right and left leads with $e\Delta V \ll E_F$. Such potential difference is related to a density gradient $\delta n = n(E_F + e\Delta V) - n(E_F)$ ($n(E)$ being the electron density) which can be approximated, including spin degeneracy, by

$$\frac{\partial n}{\partial E}|_{E_F}.e\Delta V = 2e\Delta V/(\pi \hbar v_F). \tag{3.6}$$

On the other hand, this electron density difference δn can also be written in terms of the current densities in steady state as

$$\delta n = \frac{j + jR}{ev_F} - \frac{jT}{ev_F}, \tag{3.7}$$

introducing R and T, the reflection and transmission probabilities respectively. From Eq. (3.7) one infers that $\Delta V = [j(1 + R - T)/(ev_F)] \times \pi \hbar v_F/(2e)$. The total current $I = Tj$, so that the resistance of the system reads finally

$$\mathcal{R} = \Delta V/I = \frac{h}{2e^2}\frac{R}{T}. \tag{3.8}$$

Based on this derivation, the quantum conductance becomes $G = 1/\mathcal{R} = 2e^2/h \times T/R$, which has an ill-defined value in the limit of perfect transmission ($T = 1$). Indeed, given current conservation, $R + T = 1$, a perfect transmission through the system means $T = 1$ and $R = 0$, or $G \to \infty$ which is clearly unphysical. The reason for such a singularity comes from the neglect of contact effects. This problem was pioneered by Rolf Landauer (Landauer, 1970, Imry & Landauer, 1999) who demonstrated that in the situation of ballistic transport across a given (low-dimensional) system, the resistance and dissipation will take place at the interface between the measured system and the metallic electrodes acting as charge reservoirs. This problem can be solved by rewriting the Landauer formula as

$$\mathcal{R} = \frac{h}{2e^2} \frac{1}{T} \tag{3.9}$$

$$= \frac{h}{2e^2} + \frac{h}{2e^2} \frac{1-T}{T}, \tag{3.10}$$

so that the quantum resistance is then seen to split into two parts for the single conducting channel case. The first term of Eq. (3.10) gives the contact resistance between the disorder-free one-dimensional metallic conductor and an electron reservoir with many more electron channels, whereas the second term of Eq. (3.10) actually provides information about the intrinsic resistance of the system, which can dominate the total resistance when the intrinsic transmission is very low. This second term is physically connected to the so-called four-points transport measurements, which allow access to such intrinsic resistance, by excluding contact effects.

3.3   Boltzmann semiclassical transport

The Boltzmann equation describes the transport properties of quantum particles (electrons, phonons) driven by a semiclassical dynamics. It determines how the particles of the system are accelerated in external fields, losing part of their accumulated energy through scattering-induced momentum relaxation. Scattering processes are determined by static (impurities, defects) as well as dynamical (phonons) disorders. The Boltzmann transport equation describes the dynamics of the distribution function $f_k(r,t)$, which gives the *probability* (and not the probability amplitude) of finding a particle in momentum-state $|\mathbf{k}\rangle$ in the neighborhood of $|\mathbf{r}\rangle$ and at time t. Its most general form states

$$\frac{\partial f_k(\mathbf{r},t)}{\partial t} + \mathbf{v}_k \cdot \nabla_{\mathbf{r}} f_k(\mathbf{r},t) + \mathbf{F} \cdot \nabla_{\mathbf{k}} f_k(\mathbf{r},t) = \left. \frac{\partial f_k(\mathbf{r},t)}{\partial t} \right|_{\text{coll}}, \tag{3.11}$$

with $\mathbf{F}$ describing external (Lorentz) forces acting on the particles, $\mathbf{v}$ denoting their velocity, and where $\frac{\partial f_k(\mathbf{r},t)}{\partial t}|_{\text{coll}}$ is the collision term which drives the system towards equilibrium, and depends on the sources of scattering and dissipation. The wave nature of electrons is accounted for in the collision term, as well as in the particles energetics (with $E(k)$ and $\mathbf{v}_k = \frac{1}{\hbar} \nabla_k E(k)$ given by the crystalline band structure of the clean system), but the particle dynamics is treated classically in the sense that quantum interferences between multiple scattering events are disregarded. The Boltzmann transport theory is therefore invalidated when localization phenomena enter into play and should instead be replaced by the Kubo approach (described in Section 3.4). In the regime of high charge density and high temperatures, the Boltzmann transport theory applies reasonably well, however, even in low-dimensional materials such as graphene.

In Eq. (3.11), the collision term describes the abrupt changes of momentum due to scattering of the particles. To keep the calculation simple, we hereafter consider only elastic scattering (particle momentum is changed in the scattering process but energy is conserved), but electron–phonon coupling can be treated similarly (Hwang & Sarma, 2008, Muñoz, 2012). The rate of change of the distribution function f_k due to scattering can be written as

$$\frac{\partial f_k(\mathbf{r}, t)}{\partial t}\Big|_{\text{coll}} = \sum_{k'} f_{k'}(1 - f_k)p_{k',k} - \sum_{k'} f_k(1 - f_{k'})p_{k,k'} \tag{3.12}$$

$$= \sum_{k'} (f_{k'} - f_k)p_{k,k'}. \tag{3.13}$$

The first term on the right hand side of Eq. (3.12) accounts for scattering events from all other states $|k'\rangle$ to $|k\rangle$. The probability of one scattering event is proportional to the probability that the state $|k'\rangle$ (resp. $|k\rangle$) is occupied (resp. vacant), and to the transition probability $p_{k,k'}$. The second term on the right hand side of Eq. (3.12) denotes the contribution of the scattering from $|k\rangle$ to $|k'\rangle$. As these processes decrease f_k, the second term gets a minus sign. To obtain Eq. (3.13), the detailed balance property $p_{k,k'} = p_{k',k}$ has been used. The transition probability from state $|k\rangle$ to state $|k'\rangle$ is given by Fermi's golden rule:

$$p_{k,k'} = \frac{2\pi}{\hbar}|\langle\mathbf{k'}|V|\mathbf{k}\rangle|^2\delta(\varepsilon_{k'} - \varepsilon_k), \tag{3.14}$$

introducing the $\delta(\varepsilon_{k'} - \varepsilon_k)$ function ensuring energy conservation.

3.3.1 The relaxation time approximation and the Boltzmann conductivity

To solve Eq. (3.11) one needs further approximation. The most straightforward is the relaxation time approximation (RTA) which introduces a single timescale for the non-equilibrium distribution function $f_k(r, t)$ to relax to the Fermi–Dirac function $f_k^0 = 1/(e^{\beta(\varepsilon_k - E_F)} + 1)$ ($\beta = 1/(k_B T)$ and E_F is the Fermi energy). One defines the deviation to its equilibrium state as $g_k = f_k - f_k^0$. The relaxation time approximation assumes that the system is driven back to its equilibrium position as

$$\frac{\partial f}{\partial t}\Big|_{\text{coll}} = -\frac{g_k}{\tau_k}, \tag{3.15}$$

with τ_k the relaxation time, which measures how fast the system relaxes to the equilibrium distribution after turning off the external field. The calculation of τ_k is related to

$$-\frac{g_k}{\tau_k} = \sum_{k'} (f_{k'} - f_k)p_{k,k'} \tag{3.16}$$

$$= \sum_{k'} (g_{k'} - g_k)p_{k,k'}, \tag{3.17}$$

assuming that the energy and the relaxation time do not depend on the direction of the k vector and considering only scattering with $k = k'$. Within the RTA, the Bloch–Boltzmann equation becomes

$$\frac{\partial f_k}{\partial t} = -\mathbf{v}_k \cdot \nabla_{\mathbf{r}} f_k - \frac{e\mathbf{E}}{\hbar} \cdot \nabla_{\mathbf{k}} f_k - \frac{g_k}{\tau_k}. \tag{3.18}$$

For a homogeneous electrical field and absence of temperature gradient, the steady-state solution can be written

$$\frac{e\mathbf{E}}{\hbar} \cdot \nabla_{\mathbf{k}} f_k + \frac{g_k}{\tau_k} = 0, \tag{3.19}$$

and using $\nabla_{\mathbf{k}} f_k = \hbar \mathbf{v}_k \partial f_k^0 / \partial \varepsilon_k + \partial g_k / \partial \mathbf{k}$, one finally obtains

$$f_k = f_k^0 - \left(\frac{\partial f_k^0}{\partial \varepsilon_k} \right) e \tau_k \mathbf{v}_k \cdot \mathbf{E}. \tag{3.20}$$

The relaxation time can be then calculated using Eq. (3.17) as

$$\frac{1}{\tau_k} = \sum_{k'} \left(1 - \frac{g_{k'}}{g_k} \right) p_{k,k'} \tag{3.21}$$

$$= \sum_{k'} \left(1 - \frac{\left(-\frac{\partial f_{k'}^0}{\partial \varepsilon_{k'}} \right) e \tau_{k'} \mathbf{v}_{k'} \cdot \mathbf{E}}{\left(-\frac{\partial f_k^0}{\partial \varepsilon_k} \right) e \tau_k \mathbf{v}_k \cdot \mathbf{E}} \right) p_{k,k'} \tag{3.22}$$

$$= \sum_{k'} \left(1 - \frac{\hat{\mathbf{k}}' \cdot \mathbf{E}}{\hat{\mathbf{k}} \cdot \mathbf{E}} \right) p_{k,k'}. \tag{3.23}$$

In the last step the relation $\mathbf{v}_k = v_k \mathbf{k}/k = v_k \hat{\mathbf{k}}$ is used. Without loss of generality the vector $\mathbf{k}$ may be chosen to point in the x direction, so $\hat{\mathbf{k}} = \mathbf{e}_x$, and in two dimensions wavevectors are decomposed as $\hat{\mathbf{k}}' = \cos\theta_{k'} \mathbf{e}_x + \sin\theta_{k'} \mathbf{e}_y$, while $\mathbf{E} = E_x \mathbf{e}_x + E_y \mathbf{e}_y$. Therefore

$$\frac{1}{\tau_k} = \sum_{k'} \left(1 - \frac{E_x \cos\theta_{k'} + E_y \sin\theta_{k'}}{E_x} \right) p_{k,k'} \tag{3.24}$$

$$= \sum_{k'} (1 - \cos\theta_{k'}) p_{k,k'}. \tag{3.25}$$

Note that the term proportional to $p_{k,k'} \sin\theta_{k'}$ cancels by summation, since $\sin\theta_{k'}$ is an odd function of $\theta_{k'}$ while $p_{k,k'}$ is symmetric with respect to $\theta_{k'}$. This result holds for isotropic scattering in k-space and if all transition probabilities $p_{k,k'} = 0$ for $k \neq k'$. The inverse relaxation time is thus obtained by summing up the probabilities of all scattering events, weighted with the transport factor $(1 - \cos\theta_{k'})$ which favors large-angle scattering. This result makes sense physically as the large-angle scattering events more strongly alter the distribution function, hence controlling the behavior of the relaxation time. Calculation of the Boltzmann conductivity is derived from the current density $\mathcal{J}$, defined as (using Eq. (3.20))

$$\mathcal{J} = \frac{e}{\Omega} \sum_k v_k f_k \tag{3.26}$$

$$= -\frac{e^2}{\Omega} \sum_k \left(\frac{\partial f_k^0}{\partial \varepsilon_k} \right) \tau_k \mathbf{v}_k (\mathbf{v}_k \cdot \mathbf{E}). \tag{3.27}$$

This gives the conductivity in the RTA as

$$\sigma_{xx} = -\frac{e^2}{2\pi} \int k dk \left(\frac{\partial f_k^0}{\partial \varepsilon_k}\right) \tau_k v_k^2. \tag{3.28}$$

3.4   Kubo formula for the electronic conductivity

The conductivity of a bulk material is defined at finite frequency ω as the tensorial ratio between the applied electric field and the resulting electronic current: $\mathbf{J}(\omega) = \sigma(\omega)\mathbf{E}(\omega)$. We assume that the transport measurement direction is along the (Ox) axis, so that only diagonal elements are taken into account: $\mathcal{J}_x(\omega) = \sigma(\omega)E_x(\omega)$. The Kubo approach is a technique to calculate linear response in materials (optical, electric, etc.). It is based on the *fluctuation–dissipation theorem* that establishes a correspondence between the *dissipative* out-of-equilibrium response (namely, the conductivity) and the *fluctuations* at the equilibrium (the correlation function of the charge carrier velocities).

We provide here a comprehensive derivation of the Kubo formula for electronic conductivity (Roche, 1996, Triozon, 2002, Lherbier, 2008), which is suitable for studying quantum transport phenomena in disordered graphene-based materials, based on numerical simulations. It is inspired by a derivation by Nevill Mott which calculates the absorbed power driven by electronic transitions induced by the exchanges between the system and the electromagnetic field (P).

Let us assume an electronic system described by the Hamiltonian $\hat{\mathcal{H}}_0 = \frac{\hat{\mathbf{P}}^2}{2m} + \hat{\mathcal{V}}$, where $\hat{\mathcal{V}}$ gives the crystal potential which can also include the effect of crystal imperfections. Then assume that its electronic spectrum is given by $\varepsilon_k, |\Psi_k\rangle$. By applying an external (weak) electric field, the system will undergo internal fluctuations, which are usually well captured by electronic transition between states of the system at equilibrium. To compute σ, we start with the equation $P = \mathcal{J} \cdot \mathbf{E}$ with $\mathcal{J} = \sigma \mathbf{E}$. The electric field $\mathbf{E}(t)$ is given by $E_0 \cos(\omega t)\mathbf{u}_x$, but for computational convenience we use an oscillatory field throughout the derivation, while the limit to the static case is taken at the end $(\mathbf{E}(t) = E_0\mathbf{u}_x)$ with $\omega \to 0$. The associated vector potential $\mathbf{A}(t)$ in the Coulomb gauge is

$$\mathbf{A}(t) = -\frac{E_0}{2i\omega} \left(e^{i\omega t} - e^{-i\omega t}\right) \mathbf{u}_x, \tag{3.29}$$

while the total power absorbed per unit time is

$$P_{\text{tot abs}} = \sum_{n,m} P_{\text{abs}}^{n \to m} - P_{\text{diss}}^{m \to n}. \tag{3.30}$$

The average power absorbed (P_{abs}) and dissipated (P_{diss}) per unit time can be estimated from the transition probabilities $\tilde{p}_{n \to m}$ from electronic states n to m (and inversely $(m \to n)$) and Fermi–Dirac distribution $f(E)$:

$$P_{\text{abs}}^{n \to m} = \left[\hbar\omega f(E_n)(1 - f(E_m))\right]\tilde{p}_{n \to m}, \tag{3.31}$$

$$P_{\text{diss}}^{m \to n} = \left[\hbar\omega f(E_m)(1 - f(E_n))\right]\tilde{p}_{m \to n}. \tag{3.32}$$

Such transition probabilities per unit time are derived from a first-order perturbation theory in the electric field as

$$\tilde{p}_{n \to m} = \frac{p_{n \to m}(t)}{t} = \frac{1}{\hbar^2 t} \left| \int_0^t dt' e^{i(E_m - E_n)t'/\hbar} \langle m | \delta \hat{\mathcal{H}}(t') | n \rangle \right|^2, \tag{3.33}$$

with $\delta \hat{\mathcal{H}}$ being the time-dependent perturbation of the total Hamiltonian. At first order it directly relates to the velocity operator $\hat{V}$ and vector potential $\mathbf{A}$ through

$$\delta \hat{\mathcal{H}}(t') = e \, \hat{\mathbf{V}} \cdot \mathbf{A}(t'), \tag{3.34}$$

$$\delta \hat{\mathcal{H}}(t') = e \hat{V}_x A_x(t') \quad \text{(for the 1D case).} \tag{3.35}$$

Using Eqs. (3.29–3.35) we obtain

$$P_{\text{tot abs}} = \frac{\pi \hbar e^2 E_0^2}{2\hbar \omega} \sum_{n,m} |\langle m | \hat{V}_x | n \rangle|^2 \delta(E_m - E_n - \hbar \omega) \big[f(E_n) - f(E_m) \big], \tag{3.36}$$

and finally the total power absorbed per unit time and volume $P = \frac{P_{\text{abs}}}{\Omega}$ (Ω being the sample volume) is related to the conductivity by

$$P = \frac{P_{\text{tot abs}}}{\Omega} = \sigma \langle \mathbf{E} \cdot \mathbf{E} \rangle = \frac{\sigma E_0^2}{2}. \tag{3.37}$$

Using Eq. (3.37), where $\langle \cos^2(\omega t) \rangle$ has been replaced by its average value $1/2$, one gets the Kubo conductivity

$$\sigma(\omega) = \frac{\pi \hbar e^2}{\Omega} \sum_{n,m} |\langle m | \hat{V}_x | n \rangle|^2 \delta(E_m - E_n - \hbar \omega) \frac{f(E_n) - f(E_m)}{\hbar \omega}. \tag{3.38}$$

Using the properties of $\delta(x)$ functions and rewriting the expression as a trace of operators the general expression becomes

$$\sigma(\omega) = \frac{\pi \hbar e^2}{\Omega} \int_{-\infty}^{+\infty} dE \frac{f(E) - f(E + \hbar \omega)}{\hbar \omega} \text{Tr} \left[\hat{V}_x^\dagger \delta(E - \hat{\mathcal{H}}) \hat{V}_x \delta(E + \hbar \omega - \hat{\mathcal{H}}) \right]. \tag{3.39}$$

It is also instructive to rewrite this formula introducing the autocorrelation function of velocity ($C(E, t)$), together with the mean square spreading of wavepackets defined as ($\Delta X^2(E, t)$). Using

$$\delta \left(E + \hbar \omega - \hat{\mathcal{H}} \right) = \frac{1}{2\pi \hbar} \int_{-\infty}^{+\infty} dt \, e^{i \left(E + \hbar \omega - \hat{\mathcal{H}} \right) t/\hbar} \tag{3.40}$$

inside the trace, which is further denoted by $\mathbb{A}_1$:

$$\mathbb{A}_1 = \text{Tr} \left[\hat{V}_x^\dagger \delta \left(E - \hat{\mathcal{H}} \right) \hat{V}_x \delta \left(E + \hbar \omega - \hat{\mathcal{H}} \right) \right], \tag{3.41}$$

$$\mathbb{A}_1 = \frac{1}{2\pi \hbar} \int_{-\infty}^{+\infty} dt \, e^{i\omega t} \, \text{Tr} \left[\hat{V}_x^\dagger \delta \left(E - \hat{\mathcal{H}} \right) \hat{V}_x \, e^{i \left(E - \hat{\mathcal{H}} \right) t/\hbar} \right], \tag{3.42}$$

$$\mathbb{A}_1 = \frac{1}{2\pi \hbar} \int_{-\infty}^{+\infty} dt \, e^{i\omega t} \, \text{Tr} \left[\hat{V}_x^\dagger \delta \left(E - \hat{\mathcal{H}} \right) e^{i\hat{\mathcal{H}}t/\hbar} \hat{V}_x \, e^{-i\hat{\mathcal{H}}t/\hbar} \right]. \tag{3.43}$$

The velocity operator in its Heisenberg representation being

$$\hat{V}_x(t) = \left(e^{i\hat{\mathcal{H}}t/\hbar} \, \hat{V}_x \, e^{-i\hat{\mathcal{H}}t/\hbar} \right), \tag{3.44}$$

we get

$$\mathbb{A}_1 = \frac{1}{2\pi\hbar} \int_{-\infty}^{+\infty} dt \, e^{i\omega t} \, \mathrm{Tr}\left[\hat{V}_x^\dagger(0) \, \delta(E - \hat{\mathcal{H}}) \, \hat{V}_x(t) \right]. \tag{3.45}$$

Then, one uses the general definition of quantum average for a given energy E, from which any operator $\hat{Q}$ has

$$\langle \hat{Q} \rangle_E = \frac{\mathrm{Tr}\left[\delta(E - \hat{\mathcal{H}})\hat{Q} \right]}{\mathrm{Tr}\left[\delta(E - \hat{\mathcal{H}}) \right]}. \tag{3.46}$$

Replacing $\hat{Q}$ by the product $\hat{V}_x(t)\hat{V}_x^\dagger(0)$,

$$\langle \hat{V}_x(t)\hat{V}_x^\dagger(0) \rangle_E = \frac{\mathrm{Tr}\left[\hat{V}_x^\dagger(0)\delta(E - \hat{\mathcal{H}})\hat{V}_x(t) \right]}{\mathrm{Tr}\left[\delta(E - \hat{\mathcal{H}}) \right]}, \tag{3.47}$$

and using this result to rewrite $\mathbb{A}_1$,

$$\mathbb{A}_1 = \frac{1}{2\pi\hbar} \int_{-\infty}^{+\infty} dt \, e^{i\omega t} \, \mathrm{Tr}\left[\delta(E - \hat{\mathcal{H}}) \right] \langle \hat{V}_x(t)\hat{V}_x^\dagger(0) \rangle_E, \tag{3.48}$$

$$\mathbb{A}_1 = \frac{1}{2\pi\hbar} \mathrm{Tr}\left[\delta(E - \hat{\mathcal{H}}) \right] \int_{-\infty}^{+\infty} dt \, e^{i\omega t} \, \langle \hat{V}_x(t)\hat{V}_x^\dagger(0) \rangle_E, \tag{3.49}$$

$$\mathbb{A}_1 = \frac{1}{2\pi\hbar} \mathbb{A}_2 \, \mathbb{A}_3, \tag{3.50}$$

with $\quad \mathbb{A}_2 = \mathrm{Tr}\left[\delta(E - \hat{\mathcal{H}}) \right], \quad$ and $\quad \mathbb{A}_3 = \int_{-\infty}^{+\infty} dt \, e^{i\omega t} \, \langle \hat{V}_x(t)\hat{V}_x^\dagger(0) \rangle_E. \tag{3.51}$

Two interesting quantities emerge, with $\mathbb{A}_2$ the total density of states. The second quantity can be reformulated as ($\mathbb{A}_3$) using the definition of velocity autocorrelation function $C(E, t) = \langle \hat{V}_x(t)\hat{V}_x^\dagger(0) \rangle_E$, so that

$$\mathbb{A}_3 = \int_{-\infty}^{+\infty} dt \, e^{i\omega t} \, C(E, t), \tag{3.52}$$

$$\mathbb{A}_3 = \int_{-\infty}^{0} dt \, e^{i\omega t} \, C(E, t) + \int_{0}^{+\infty} dt \, e^{i\omega t} \, C(E, t), \tag{3.53}$$

$$\mathbb{A}_3 = \int_{0}^{+\infty} dt \, e^{-i\omega t} \, C(E, -t) + \int_{0}^{+\infty} dt \, e^{i\omega t} \, C(E, t), \tag{3.54}$$

and using $C(E, -t) = \langle \hat{V}_x(-t)\hat{V}_x^\dagger(0) \rangle_E = \langle \hat{V}_x(0)\hat{V}_x^\dagger(t) \rangle_E = C(E, t)^\dagger$, one gets

$$\mathbb{A}_3 = \int_{0}^{+\infty} dt \, e^{-i\omega t} \, C(E, t)^\dagger + e^{i\omega t} \, C(E, t), \tag{3.55}$$

$$\mathbb{A}_3 = \int_{0}^{+\infty} dt \, 2\mathfrak{Re}\left(e^{i\omega t} \, C(E, t) \right). \tag{3.56}$$

One can easily show that the real part of the velocity autocorrelation function is proportional to the second derivative of the mean squared spread

$$\frac{\partial^2}{\partial t^2} \Delta X^2(E, t) = 2\Re e \ C(E, t), \tag{3.57}$$

with $\Delta X^2(E, t)$ defined as

$$\Delta X^2(E, t) = \langle |\hat{X}(t) - \hat{X}(0)|^2 \rangle_E. \tag{3.58}$$

One can consequently rewrite $\mathbb{A}_1$ as follows:

$$\mathbb{A}_1 = \frac{1}{2\pi\hbar} \mathbb{A}_2 \int_0^{+\infty} dt \ 2\Re e \left(e^{i\omega t} \ C(E, t) \right), \tag{3.59}$$

and $\mathbb{A}_1$ can be replaced in Eq. (3.39) to get another formulation of the Kubo conductivity (Roche, 1996, Triozon, 2002, Lherbier, 2008):

$$\sigma(\omega) = \frac{e^2}{2} \int_{-\infty}^{+\infty} dE \frac{f(E) - f(E + \hbar\omega)}{\hbar\omega} \frac{\text{Tr}\left[\delta(E - \hat{\mathcal{H}})\right]}{\Omega} \int_0^{+\infty} dt \ 2\Re e \left(e^{i\omega t} \ C(E, t) \right). \tag{3.60}$$

This last Eq. (3.60) is the total density of states per volume unit $\rho(E) = \text{Tr}\left[\delta(E - \hat{\mathcal{H}})\right]/\Omega$. This is a general form for σ, which can now be simplified taking two limits. First, let us go to the static electric field limit $\omega \mapsto 0$,

$$\sigma_{DC} = -\frac{e^2}{2} \int_{-\infty}^{+\infty} dE \frac{\partial f(E)}{\partial E} \rho(E) \int_0^{+\infty} dt \ 2\Re e \left(C(E, t) \right), \tag{3.61}$$

$$\sigma_{DC} = -\frac{e^2}{2} \int_{-\infty}^{+\infty} dE \frac{\partial f(E)}{\partial E} \rho(E) \int_0^{+\infty} dt \frac{\partial^2}{\partial t^2} \Delta X^2(E, t), \tag{3.62}$$

$$\sigma_{DC} = -\frac{e^2}{2} \int_{-\infty}^{+\infty} dE \frac{\partial f(E)}{\partial E} \rho(E) \lim_{t \to \infty} \frac{\partial}{\partial t} \Delta X^2(E, t), \tag{3.63}$$

while the zero-temperature limit $(T \mapsto 0)$ implies that $-\frac{\partial f(E)}{\partial E} \mapsto \delta(E - E_F)$, so that

$$\sigma_{DC}(E_F) = \frac{e^2}{2} \int_{-\infty}^{+\infty} dE \ \delta(E - E_F) \ \rho(E) \lim_{t \to \infty} \frac{\partial}{\partial t} \Delta X^2(E, t), \tag{3.64}$$

$$\sigma_{DC}(E_F) = \frac{e^2}{2} \rho(E_F) \lim_{t \to \infty} \frac{\partial}{\partial t} \Delta X^2(E_F, t). \tag{3.65}$$

This last expression means that $\frac{\partial}{\partial t} \Delta X^2(E_F, t)$ should converge in the limit $t \mapsto \infty$, to define a meaningful conductivity. The propagation of the wavepacket thus needs to establish a saturation regime before conductivity can be safely calculated. However, as shown in other chapters, the time-dependent scaling on the conductivity can be followed and allowed to follow localization phenomena as long as phase coherence is maintained. This formula, known as the Kubo conductivity (Kubo, 1966), is the most general starting point to study quantum (or classical) transport in any type of disordered materials, provided that electron–electron interaction can be described as a perturbation with respect to the initial electronic structure, introducing additional transitions (inelastic scattering), but preserving the independent electron description of transport quantities.

3.4.1   Illustrations for ballistic and diffusive regimes

The behavior of $\Delta X^2(t)$ and related diffusion coefficient $D_x(t)$ defined by

$$D_x(t) = \frac{\Delta X^2(t)}{t} \tag{3.66}$$

is easily determined in two important transport regimes. Below we outline some consequences of the transport regime on the scaling property of the quantum conductivity, as computed from the Kubo formula.

Ballistic regime

First, in the absence of any structural imperfection, the electronic propagation remains ballistic with the mean square spread just defined by the initial velocity of the wavepacket $\Delta X^2(t) = v_x^2(0)t^2$, with $v_x(0)$ the velocity at $t = 0$. The diffusion coefficient is then linear in time, $D_x(t) = v_x^2(0)t$, while the Kubo conductivity is given by

$$\sigma_{DC}(E)_{\text{bal}} = \frac{e^2}{2}\rho(E) \lim_{t\to\infty} \frac{\partial}{\partial t}\Delta X^2(E,t) = e^2\rho(E) \lim_{t\to\infty} v_x^2(0,E)t, \tag{3.67}$$

so that $\sigma_{DC}(E)_{\text{bal}}$ diverges in the long time limit. This singularity is inherent to the fact that when deriving the linear response theory, a finite dissipation source, intrinsic to the sample, is introduced both physically and mathematically. The ballistic limit is therefore not well defined in this formalism, although as shown below a complete equivalence exists with the Landauer–Büttiker formulation, and the quantization of the conductance can be obtained from the Kubo formula with some extra assumptions. The conductance of the materials can indeed be derived from the conductivity through $G = \sigma L^{d-2}$, with d the space dimension. For one-dimensional systems $G = \sigma/L$. Dividing Eq. (3.67) by the relevant length scale L, we can recover a quantized conductance expected in a ballistic regime (when reflectionless contacts are assumed). By replacing L by $2v_x t$ (since the length propagated during t is $2\sqrt{\Delta X^2(t)} = 2v_x t$), the conductance then becomes

$$G(E) = e^2 \rho_{1D}(E) \lim_{t\to\infty} \frac{v_x^2(E)t}{L} = e^2 \rho_{1D}(E) \lim_{t\to\infty} \frac{v_x^2(E)t}{2v_x(E)t}, \tag{3.68}$$

$$G(E) = \frac{e^2}{2}\rho_{1D}(E)v_x(E) = \frac{2e^2}{h} = G_0, \tag{3.69}$$

using $\rho_{1D}(E) = 2/\pi\hbar v_x(E)$ and with G_0 the conductance quantum (spin degeneracy included). So even in the most unfavorable transport regime, the quantization of the conductance can be recovered and identified to the situation of perfect transmission through reflectionless contacts (Landauer–Büttiker approach, Section 3.2)

Diffusive regime

The velocity autocorrelation function in the time relaxation approximation is given by $\langle v_x(0)v_x(t)\rangle = v_x^2(0)e^{-t/\tau}$ (introducing the transport time τ and restricting the discussion to elastic scattering events), which yields

$$\lim_{t\to\infty} \Delta X^2(t) = \lim_{t\to\infty} 2\tau v_x^2(0)\,[t-\tau] \longmapsto 2\tau v_x^2(0)t. \tag{3.70}$$

Similarly (using Eq. (3.66)) one gets $\lim_{t\to\infty} D_x(t) \longmapsto 2\tau v_x^2(0)$. The Kubo formula for a diffusive regime then gives access to the semiclassical conductivity (σ_{sc}):

$$\sigma_{sc}(E) = \sigma_{DC}(E)_{\text{diff}} = \frac{e^2}{2}\rho(E) \lim_{t\to\infty} \frac{\partial}{\partial t}\Delta X^2(E,t), \tag{3.71}$$

$$\sigma_{sc}(E) = e^2\rho(E)\tau(E)v_x^2(0,E), \tag{3.72}$$

$$\sigma_{sc}(E) = e^2\rho(E)v_x(0,E)\ell_e(E), \tag{3.73}$$

where the mean free path $\ell_e(E)$ is introduced. For the diffusive regime,

$$\sigma_{sc}(E) = \frac{e^2}{2}\rho(E) \lim_{t\to\infty} D_x(E,t) = \frac{e^2}{2}\rho(E)D_x^{\text{max}}(E), \tag{3.74}$$

where D_x^{max} corresponds to the maximum value ($D_x^{\text{max}} = 2\tau v_x^2(0)$). In this regime, by defining the charge density as $n(E) = \int dE\rho(E)$, the mobility μ is given by

$$\mu(E) = \frac{\sigma_{sc}(E)}{n(E)e}. \tag{3.75}$$

For free electrons $E(k) = (\hbar k)^2/2m$ and $v(k) = \hbar k/m$, with $\rho_{1D}(E) = \frac{2}{\pi\hbar}\left(\frac{m}{2E}\right)^{1/2}$ and $n_{1D}(E) = \frac{2}{\pi\hbar}(2mE)^{1/2}$, so that using Eq. (3.72) and Eq. (3.75), the mobility finally is given by

$$\mu(E) = \frac{e^2\rho_{1D}(E)\tau(E)v^2(E)}{en_{1D}(E)} = \frac{e\tau(E)v^2(E)}{2E}, \tag{3.76}$$

$$\mu(E) = \frac{e\tau(E)\hbar^2 k^2}{2\left(\frac{\hbar^2 k^2}{2m}\right)m^2} = \frac{e\tau(E)}{m}, \tag{3.77}$$

which are familiar expressions for semiclassical transport (absence of quantum interferences). One notes that estimation of the mobility becomes problematic for graphene-based materials for plenty of reasons. First, for clean graphene-based materials (nanotubes, graphene ribbons or two-dimensional graphene), the mean free path might become longer than the electrode spacing, so that the use (or even the definition) of Eq. (3.75) becomes inappropriate since it neglects contact effects. Additionally, in the presence of intrinsic disorder (vacancies, adsorbed adatoms, etc.), strong scattering and a significant contribution of quantum interferences occur, which again invalidate the use of Eq. (3.75). Quantum interferences up to 100 K have been measured experimentally in disordered graphene materials (see for instance Moser *et al.* (2010)), so even if inelastic scattering restores in principle the validity of Eq. (3.75), the experimental estimations have to be scrutinized with care. One general assumption is that the quality of the sample can be appreciated by estimating the mobility at a charge density of (typically) 10^{11} cm^{-2} with varying temperature, and that the absolute value allows comparison of sample quality. In the numerical calculations (using the Kubo formula) that are discussed later, the estimations of mobility using Eq. (3.75) are made using the semiclassical conductivity computed at zero temperature.

3.4.2 Kubo versus Landauer

The Kubo approach is a quantum generalization of the semiclassical Bloch–Boltzmann approach for studying electron transport in materials, which includes all multiple scattering effects driven by disorder. The Kubo–Greenwood formalism (Kubo, 1966) is well suited for exploring the intrinsic transport properties of a given disordered material of high dimensionality. It mainly applies to the study of weakly or strongly disordered systems, characterized by a diffusive regime and localization phenomena in the low temperature limit. It gives all information on the intrinsic quantum conductivity which can be accessed experimentally by four-points transport measurements (meaning two electrodes for generating voltage drop and two others for measuring induced current). With this formalism, when the system is translational invariant, no scattering takes place, and the "intrinsic" mean free path is infinite. Differently, the Landauer–Büttiker transport formalism is directly linked with two-points transport measurements (meaning two identical electrodes for generating voltage drop and measuring induced current) and is proportional to the transmission probability for charges to be transfered through a given system connected to external electrodes. A connection between Kubo and Landauer can be made by rewriting the two-points resistance (computed with the Landauer–Büttiker method) as e.g. $R = R_0/T = R_0 + R_{int}$, making explicit the "intrinsic resistance" $R_{int} = R_0(1 - T)/T$, which could be derived applying the Kubo–Greenwood approach.

Within this formalism, when the system is free of scattering, or when the density of impurities is sufficiently low such that $\ell_e \gg L$ (L is the distance between source/drain electrodes) the transport regime is ballistic, with a transmission probability at energy E entirely proportional to the number of propagating modes, that is $G(E) = G_0 N_\perp(E)$.

In the situation of a large amount of scatterers (such as chemical impurities), i.e. when $\ell_e \ll L$, the transport regime becomes diffusive and the conductance scales as $G(E) = G_0 N_\perp(E)\ell_e(E)/L$. An interpolation formula allows covering of the so-called quasiballistic regime with $T = N_\perp(E)/(1 + L/\ell_e)$. If the quantum transmission at the system/electrode interface is perfect (induces no scattering), then both Kubo and Landauer formalisms are totally equivalent, although some geometrical factors differentiate them if computed with the different formalisms (Akkermans & Montambaux, 2007). The extracted Landauer mean free path ℓ_e^L and Kubo mean free path ℓ_e^K are expected to be proportional, $\ell_e^L = \kappa \ell_e^K$ ($\kappa = 2$ for $d = 1$, $\kappa = \frac{\pi}{2}$ ($d = 2$), $\kappa = \frac{4}{3}(d = 3)$) (Akkermans & Montambaux, 2007). In the case of a rectangular waveguide, the κ coefficient depends on the dimensionality of the system (Datta, 1995).

For instance, to determine exactly the κ coefficient for a finite nanotube, one needs to solve the diffusion equation for the specified geometry and given boundary conditions (Datta, 1995). $\kappa = 2$ at the charge neutrality point. This can also be shown using the

Einstein relationship for conductivity, $\sigma_F = e^2 \rho_F D_F$, where $\rho_F = 4/\pi \hbar v_F$ is the total density of states at CNP, and $D_F = \ell_e v_F$ is the diffusivity coefficient at CNP. The total conductivity for the quasi-1D system is obtained by using Ohm's law, $G = G_0 N_\perp 2\ell_e/L$.

3.4.3   Validity limit of Ohm's law in the quantum regime

Ohm's law in the classical regime can be easily derived using the 1D formula for the conductance of a diffusive system, i.e. $G = \sigma_{sc} L^{d-2} = e^2 \rho(E) D/L$, with $\rho(E) = 2/h v_F$ and $D = \ell_e v_F$. Then $G = 2e^2/h\frac{\ell_e}{L}$ which uses the additivity rule of resistance, i.e. $\mathcal{R}(L_1 + L_2) = \mathcal{R}(L_1) + \mathcal{R}(L_2)$. In the quantum regime, if one uses the Landauer expression for the conductance/resistance, one demonstrates that the resistance $\mathcal{R}(L_1 + L_2) > \mathcal{R}(L_1) + \mathcal{R}(L_2)$ because of multiple scattering phenomena.

3.4.4   The Kubo formalism in real space

An efficient real space implementation of the Kubo formula was first developed by Roche and Mayou in 1997 for the study of quasiperiodic systems (quasicrystals) (Roche & Mayou, 1997). It was then adapted by Roche and coworkers to allow exploration of mesoscopic (magneto)-transport in complex and disordered mesoscopic systems including carbon nanotubes, semiconducting nanowires, and graphene-based materials (Roche, 1999, Roche & Saito, 2001, Roche et al., 2005, Latil, Roche & Charlier, 2005, Lherbier et al., 2008, Ishii et al., 2009). The typical disordered samples studied with such methodology already contain several tens of millions of orbitals, and with the use of high performance computing resources, the simulation of samples with 1 billion atoms can be envisioned in the next decade. This numerical transport method therefore offers unprecedented exploration possibilities of complex quantum transport phenomena, not only in realistic models of disordered graphene-based materials, but also in any other types of materials of exciting scientific and technological interest (silicon nanowires (Persson et al., 2008), organic crystals (Ortmann & Roche, 2011), topological insulators, etc).

We present here the basic ingredients of the numerical implementation and provide in further sections extensive illustrations of its use in the study of disordered graphene-based materials (Roche, 1996, Triozon, 2002, Lherbier, 2008). Appendix D provides an extensive technical derivation of such a real space (and order N) implementation using the Lanczos method, which is also reviewed in detail. We present here a summary of such a derivation, since it will help us to explore most quantum transport regimes in complex forms of graphene-based materials. We start again with the general form of the Kubo conductivity:

$$\sigma(\omega) = \frac{2\pi e^2 \hbar}{\Omega} \int_{-\infty}^{+\infty} \frac{f(E) - f(E + \hbar\omega)}{\hbar\omega} \mathrm{Tr}\left[\hat{V}_x \, \delta(E - \hat{\mathcal{H}}) \, \hat{V}_x \, \delta(E - \hat{\mathcal{H}} + \hbar\omega)\right] dE, \quad (3.78)$$

where $\hat{\mathcal{H}}$ is the Hamiltonian operator, $\hat{V}_x$ is the operator for the electronic velocity along the x axis and $f(E)$ is the Fermi distribution function. The DC conductivity corresponds to the limit $\omega = 0$. Using the property

$$\lim_{\omega \to 0} \frac{f(E) - f(E + \hbar\omega)}{\hbar\omega} = -\frac{\partial f}{\partial E} = \delta(E - E_F) \qquad (3.79)$$

and after a Fourier transform, the diagonal conductivity can be simplified to

$$\sigma_{DC} = e^2 \rho(E_F) \lim_{t \to \infty} \left[\frac{1}{t} \left\langle \Delta X^2(t) \right\rangle_E \right], \qquad (3.80)$$

where $\rho(E_F)$ is the density of states per unit of volume and $\left\langle \Delta X^2(t) \right\rangle_E$ measures the electronic quadratic spread at energy E, defined as

$$\left\langle \Delta X^2(t) \right\rangle_E = \frac{\mathrm{Tr}\left[\delta(E - \hat{\mathcal{H}}) \left(\hat{X}(t) - \hat{X}(0) \right)^2 \right]}{\mathrm{Tr}\left[\delta(E - \hat{\mathcal{H}}) \right]}, \qquad (3.81)$$

where $\hat{X}(t)$ is the position operator along the x axis, written in the *Heisenberg representation* for the time t. We note that Eq. (3.80) is slightly different from Eq. (3.65), which is the most general starting point, but which is more computationally demanding. However, in most cases the simplification of using Eq. (3.80) is sufficient to extract the main physics. We refer the reader to Lherbier *et al.* (2012), for a numerical comparison between both formulas. We then modify Eq. (3.81), using the time-reversal symmetry and the properties of the trace. One can demonstrate that

$$\mathrm{Tr}\left[\delta(E - \hat{\mathcal{H}}) \left(\hat{X}(t) - \hat{X}(0) \right)^2 \right] = \mathrm{Tr}\left[A^\dagger(t)\, \delta(E - \hat{\mathcal{H}})\, A(t) \right], \qquad (3.82)$$

$$A(t) = \left[\hat{X}, \hat{u}(t) \right] = \hat{X}\hat{u}(t) - \hat{u}(t)\hat{X}, \qquad (3.83)$$

where $\hat{X}$ is the position operator in the Schrödinger representation and $\hat{u}(t) = \exp(-i\hat{\mathcal{H}}t/\hbar)$ is the usual evolution operator. Secondly, the trace in Eq. (3.81) is approximated by expectation values on random phase states. Random phase states are expanded on all the orbitals $|n\rangle$ of the basis set and defined thus:

$$|wp\rangle = \frac{1}{\sqrt{N}} \sum_{n=1}^{N} \exp(2i\pi\,\alpha(n))\, |n\rangle, \qquad (3.84)$$

where $\alpha(n)$ is a random number in the $[0, 1[$ range. An average over few tens of random phases states is usually sufficient to calculate the expectation values,

$$\mathrm{Tr}[\ldots] \longrightarrow \langle wp| \ldots |wp\rangle, \qquad (3.85)$$

and the spread (3.81) can finally be rewritten:

$$\left\langle \Delta X^2(t) \right\rangle_E = \frac{\langle wp|\, A^\dagger(t)\, \delta(E - \hat{\mathcal{H}})\, A(t)\, |wp\rangle}{\langle wp|\, \delta(E - \hat{\mathcal{H}})\, |wp\rangle}. \qquad (3.86)$$

Equation (3.86) is now suitable for order $O(N)$ numerical techniques and calculation of the transport properties is possible. We leave additional technical details to Appendix D and turn now to a physical discussion of this approach. Actually, the quadratic spread

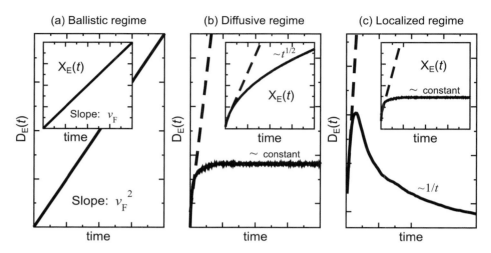

Figure 3.4 Typical behaviors of the diffusion and spread coefficients for the three characteristic regimes: (a) ballistic; (b) diffusive; and (c) localized. Courtesy of Dinh Van Tuan.

(Eq. 3.86) is a key quantity as it is directly related to the diffusion coefficient (or diffusivity) through

$$D_E(t) = \left\langle X^2(t) \right\rangle_E \cdot \frac{1}{t}, \tag{3.87}$$

whose time dependence fully determines the transport mechanism. It is worthwhile also to define the electronic spread

$$X_E(t) = \sqrt{\left\langle X^2(t) \right\rangle_E} = \sqrt{t D_E(t)}. \tag{3.88}$$

The three main transport mechanisms which can be generically followed through the time evolution of the wavepacket dynamics (illustrated in Fig. 3.4) are:

- *Ballistic regime.* Electrons travel through the systems without suffering any scattering, so that $D_E(t)$ and $X_E(t)$ remain linear functions in time, with slopes respectively equal to v_F^2 and v_F. Figure 3.4(a) shows a typical ballistic motion in clean 2D graphene at the Dirac point for metallic armchair nanotubes, whereas Fig. 3.5 shows the extracted energy-dependent velocity in clean two-dimensional graphene from the linear regime, which agrees perfectly with the analytical (exact) result. In particular, at the Dirac point $v_F \sim 2.1\gamma_0$ Å/$\hbar$.
- *Diffusive regime.* Behavior in weakly disordered graphene is characterized by a saturation of $D_E(t \to \infty)$. The saturation value identifies the elastic relaxation (or transport) time τ (see Fig. 3.4(b)).
- *Localized regime.* Behavior in strongly disordered graphene is manifested by an increasing contribution of quantum interference which reduces the diffusion coefficient, roughly following a $\sim 1/t$ decay. Spreading $X_E(t)$ reaches an asymptotic value that is related to the localization length $\xi(E)$ (see Fig. 3.4(c)).

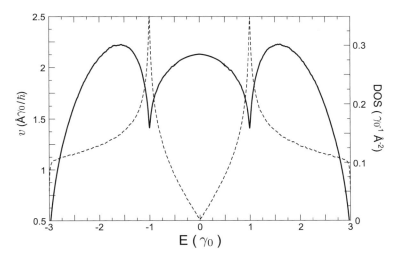

Figure 3.5 Velocity versus energy (solid line) and density of states (dashed line) for pristine graphene.

All the dynamics of the electronic system is actually conveyed by the $\hat{\mathcal{H}}$ operator. Since the Hamiltonian accounts for the presence of static disorder (e.g. randomly located defects), the time-dependent quantum dynamics of electronic wavepackets capture all multiple scattering phenomena including those accessible within the semiclassical transport regime (Bloch–Boltzmann) such as the elastic mean free path, or the localization length defining the Anderson insulator state. Inelastic scattering phenomena cannot be captured rigorously with such an approach, except in some indirect manner by coupling molecular dynamics with the time-dependent wavepacket approach.

3.4.5 Scaling theory of localization

The scaling theory of localization in disordered systems was developed in the early 1980s, initiated earlier by P. W. Anderson and further consolidated by Abrahams *et al.* (1979), who established comprehensive foundations of transport theory in disordered systems. Theoretical predictions have been confirmed by decades of experimental work. In one and two dimensions, any metallic system is predicted to be continuously driven to an (Anderson) insulating state as temperature decays to zero, and all states are localized at zero temperature. From a general perspective, the conductance of a system can be viewed as the sum ($\mathcal{P}_{P \to Q}$) over all probability amplitudes of propagating trajectories starting from one location P and going to another one Q in real space, or more explicitly

$$G = \frac{2e^2}{h} \mathcal{P}_{P \to Q}, \tag{3.89}$$

$$\mathcal{P}_{P \to Q} = \sum_i |\mathcal{A}_i|^2 + \sum_{i \neq j} \mathcal{A}_i \mathcal{A}_j e^{i(\alpha_i - \alpha_j)}, \tag{3.90}$$

defining $|\mathcal{A}_i| e^{i\alpha_i}$ as the probability amplitude of trajectory i. The conductance for a disordered system is obtained by averaging over an ensemble of random configurations,

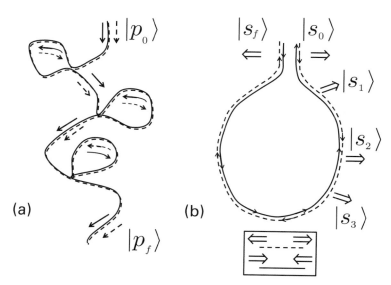

Figure 3.6 (a) Two clockwise and counterclockwise scattering trajectories which interfere constructively at various locations in real space where paths cross again. (b) Zoom-in of one particular loop where in addition to momentum shift upon scattering off impurities, adiabatic rotation of spin degree of freedom (conveyed by a weak spin–orbit scattering) is also pinpointed by arrows (two for up spin and down spin) on both trajectories.

which suppresses almost all interfering terms, simplifying Eq. (3.90) to the sum of probabilities of individual trajectories, as expected from a classical calculation, $\langle G \rangle_{dis.} = \frac{2e^2}{h} \sum_i |\mathcal{A}_i|^2$.

This is, however, not completely correct since there exists a class of scattering trajectories which contain geometrical loops returning to an initial point in real space (illustrated in Fig. 3.6(a)). In the presence of time-reversal symmetry, the probability amplitude associated with the closed trajectory ($\mathcal{C}_+$) moving in a clockwise manner along the loop turns out to interfere constructively with the one ($\mathcal{C}_-$) topologically identical, but moving counterclockwise (illustrated in Fig. 3.6(b)). Indeed, if we denote $\alpha_+ = \oint_{\mathcal{C}_+} \mathbf{p}.d\mathbf{r}$, it is clear that $\alpha_- = \alpha_+$.

Accordingly, the quantum interferences driven by the contribution of all such families of trajectories are not canceled by disorder averaging and eventually enhance the probability of return to some origin, $\mathcal{P}_{O \to O} = 4|\mathcal{A}_O|^2$ (being twice the classical result in the absence of interferences ($|\mathcal{A}_O|^2 + |\mathcal{A}_O|^2$)). Such a contribution of quantum interferences which yields increase of the quantum resistance is also known as the Cooperon contribution (see Appendix C).

This weak localization phenomenon was first theoretically described in a seminal paper by Altshuler & Aronov (1985). An interesting point is that by applying an external magnetic field, this interference effect is actually tunable by extra phase factors related to the vector potential. Indeed, since $\mathbf{p} \to \mathbf{p} + \frac{e}{c}\mathbf{A}$, the phase factors along clockwise and counterclockwise trajectories have a sign difference, $\alpha_{\pm} = \pm \frac{e}{\hbar} \oint \mathbf{A}.d\mathbf{r}$. The total probability of return to some origin is then modified as (defining $\phi_0 = h/e$, the magnetic flux quantum)

$$|\mathcal{A}_0|^2 \left|1 + e^{i(\alpha_+ - \alpha_-)}\right|^2 = 2|\mathcal{A}_0|^2 \left(1 + \cos\frac{2\pi\varphi}{\phi_0/2}\right) \le 4|\mathcal{A}_0|^2, \qquad (3.91)$$

which means that the Cooperon contribution will always be reduced compared to its zero-field value, resulting in a resistance decrease upon switching on the field (negative magnetoresistance) (Bergman, 1984). The quantum (Cooperon) correction $\delta\sigma(L)$ can be actually derived as a perturbative correction of the semiclassical result $\sigma(L) = \sigma_{sc} + \delta\sigma(L)$ (see Appendix C), and rewritten in real space as

$$\delta\sigma(L) = -\frac{2e^2 D}{\pi\hbar\Omega}\int_0^\infty dt \mathcal{Z}(t)\left(e^{-t/\tau_\varphi} - e^{-t/\tau_{el}}\right), \qquad (3.92)$$

with $\mathcal{Z}(t) = \int d^d r P(\mathbf{r}, \mathbf{r}, t)$ the space integral of the total probability of returning to some origin, which obeys a diffusion law, of which a solution is $P(\mathbf{r}, \mathbf{r}, t) = 1/(4\pi Dt)^{d/2}$ (in D dimensions). In the integral, the factor $(e^{-t/\tau_\varphi} - e^{-t/\tau_{el}})$ restricts the size of the loops which contribute to interferences. Indeed, since trajectories develop in the diffusive regime ($t \ge \tau_{el}$, τ_{el} is the elastic scattering time), and within the coherent regime ($t \le \tau_\varphi$, with τ_φ the coherent time), all trajectories accomplished in a timescale shorter than τ_{el} and longer than τ_φ do not contribute to the total interference. In two dimensions, it is found that $\delta\sigma(L) = -(2e^2/\pi h)\ln(L/\ell_{el})$ (Lee & Ramakrishnan, 1985). The transition to the insulating state is continuous and reached when the quantum correction is of the same order as the semiclassical conductivity, that is when $\Delta\sigma(L = \xi) \simeq \sigma_{sc}$. This simple criterion allows us to establish a universal relationship (Thouless, 1973) between the two transport length scales of both (metallic and insulating) regimes, as $\xi = \ell_{el}\exp(\pi\sigma_{sc}/G_0)$ (Lee & Ramakrishnan, 1985).

Spin/pseudospin effect on the Cooperon

It turns out that the presence of additional degrees of freedom such as the spin (or pseudospin in graphene) brings new features in the interference pattern. This was first theoretically established by Hikami and coworkers (Hikami, Larkin & Nagaoka, 1980), and then confirmed experimentally (Bergman, 1984) on thin metallic film functionalized with deposited magnetic atoms. It is usually assumed that if $|s\rangle$ is the initial spin state, it can generally be rewritten as a superposition of the spin up $|\Uparrow\rangle$ and spin down $|\Downarrow\rangle$ states. In principle, there exist two main possibilities of how the spin orientation can be changed on the scattering path, in the presence of spin–orbit coupling (defined by a Hamiltonian $\mathcal{H}_{so}$). The first mechanism initially derived for metals, and known as the Elliot–Yafet mechanism, assumes that the presence of spin–orbit coupling results in a spin rotation each time the electron is scattered at the impurities. In contrast, the Dyakonov–Perel mechanism describes spin precession while the electron propagates between the scattering centers. The origin of the spin precession has been generally described in terms of lack of inversion symmetry (i.e. in zinc blende crystals) or of an asymmetric potential shape of the quantum well forming a 2D electron gas (the Rashba effect).

Regardless of the underlying mechanism, if an electron propagates along a closed loop, its spin orientation is changed. The modification of the spin orientation can be expressed by a rotation matrix U. For propagation along the loop in a forward (f) direction, the final state $|s_f\rangle$ can be expressed by $|s_f\rangle = U|s\rangle$, where U is the corresponding rotation matrix. For propagation along the loop in a backwards directions (b), the final spin state is given by $|s_b\rangle = U^{-1}|s\rangle$. Here, use is made of the fact that the rotation matrix of the counterclockwise propagation is simply the inverse of U. For interference between the clockwise and counterclockwise electron waves, not only the spatial component is relevant but also the interference of the spin component: $\langle s_b|s_f\rangle = \langle s|U^2|s\rangle$, making use of the fact that U is a unitary matrix. Weak localization, and thus constructive interference, is recovered if the spin orientation is conserved in the case that U is the unit matrix. However, if the spin is rotated during electron propagation along a loop, in general no constructive interference can be expected. Moreover, for each loop a different interference will be expected. Interestingly, averaging over all possible trajectories even leads to a reversal of the weak localization effect. The generalization of the Cooperon factor in the presence of spin–orbit coupling has been demonstrated to be as illustrated in Fig. 3.6(b) (Chakravarty & Schmid, 1986):

$$\delta\sigma = -\frac{2e^2 D}{\pi\hbar\Omega}\int_0^\infty dt \mathcal{Z}(t)\langle\mathcal{Q}_{s.o}(t)\rangle(e^{-t/\tau_\varphi} - e^{-t/\tau}). \tag{3.93}$$

The amplitude terms get extra phase factors related to the spin rotation which accumulates along the scattering trajectory as $|s_{n+1}\rangle = e^{-i\Delta\theta S_z/\hbar}|s_n\rangle$, assuming that $\Delta\theta$ is some finite rotation angle while $|s_n\rangle$ denotes the spin state along the trajectory and $S_z = \frac{\hbar}{2}\sigma_z$. Then along the full trajectory the total spin-dependent accumulated phase factor can be described by introducing the time ordering operator $\mathcal{T}$ $\mathcal{R}_t = \mathcal{T}e^{-\frac{i}{\hbar}\int_0^t \mathcal{H}_{so}dt}$, with $\mathcal{H}_{so} = \frac{\hbar}{4m^2c^2}\sigma.(\nabla V(\mathbf{r}) \times \mathbf{p})$, the spin–orbit component of the total Hamiltonian, $\sigma = (\sigma_x, \sigma_y, \sigma_z)$ with Pauli matrices. The additional term entering into the Cooperon contribution is actually

$$\langle\mathcal{Q}_{so}(t)\rangle = \sum_\pm \langle s_0|\mathcal{R}_{-t}^\dagger|s_f\rangle\langle s_f|\mathcal{R}_t|s_0\rangle. \tag{3.94}$$

In Figure 3.6, clockwise and counterclockwise trajectories returning to some origin are shown together with some schematics of the adiabatic rotation of spin degree of freedom (assuming a weak spin–orbit coupling). Up and down spin are adiabatically rotated along the path yielding a $\pm\pi$ extra phase. The total Berry phase adds up to 2π, which yields sign reversal of the Cooperon contribution evidenced by the extra factor $\langle\mathcal{Q}_{so}(t)\rangle = -1/2$ (see Chakravarty & Schmid, 1986). In two dimensions and in the absence of magnetic field, the weak antilocalization correction to the conductivity can be generally recast as

$$\delta\sigma(L,\ell_e) = +\frac{2e^2}{\pi h}\ln\left(\frac{L}{\ell_{el}}\right), \tag{3.95}$$

and the Kubo (quantum) conductivity can be rewritten as $\sigma = \sigma_{BB} + |\frac{2e^2}{\pi h} \ln(\frac{L}{\ell_{el}})|$, which means that the conductivity is increased, with respect to its semiclassical value, by the quantum interference contribution.

3.5 Quantum transport beyond the fully coherent or decoherent limits

Beyond fully coherent or decoherent transport, the intermediate regime where decoherence partially suppresses quantum effects is also of high relevance. Such an intermediate regime is out of reach of the technical tools presented so far, and innovative theoretical approaches are crucially needed. A detailed presentation of new ideas and emerging methods to tackle such problem is beyond the scope of this book, but a brief discussion and useful references are provided for interested readers.

The key question is how to model decoherent effects by keeping a simple one-body description. An early insight into this issue was given by Büttiker (1988a), who proposed that decoherent effects could be simulated by the action of an imaginary voltmeter attached to the sample; see the scheme in Fig. 3.7. The electrons that get absorbed by the voltmeter are reinjected so as to keep the zero-current condition on the imaginary electrode. Since no phase memory is retained between the incoming and reinjected electrons, phase coherence is steadily lost. This appealing picture has been used in many situations and has the advantage that it can be readily incorporated within Landauer's formalism.

For the simplest case of a single resonant level E_0 coupled to leads through escape rates Γ_L, Γ_R and a voltmeter through Γ_ϕ, a simple calculation shows that the effective transmission probability entering the Landauer conductance is

$$\widetilde{T}_{R,L}(\varepsilon) = \frac{\Gamma_R \, \Gamma_L}{(\varepsilon - E_0)^2 + (\Gamma_L + \Gamma_R + \Gamma_\phi)^2} \left\{ 1 + \frac{\Gamma_\phi}{\Gamma_L + \Gamma_R} \right\}. \tag{3.96}$$

Therefore, we can see that the conductance is the sum of a coherent term, where electrons do not suffer from any phase-breaking event, plus a decoherent term (second term

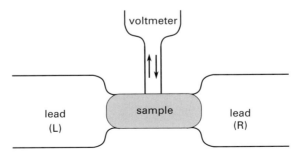

Figure 3.7 Scheme of Büttiker's imaginary voltage probe model for decoherence: decoherence can be modeled by the action of a voltmeter whose effect is to randomize the phase of the carriers. Each carrier entering the voltmeter is reinjected to keep the current balance but without a phase correlation, thereby randomizing the electronic phase.

in the curly brackets of the previous equation). Note that even the first term, which one would associate with the coherent component, is modified by the decoherent processes that change the escape rates in the denominator.

An important question at this point is whether this scheme can be implemented within a Hamiltonian model. D'Amato & Pastawski (1990) proposed a tight-binding model where the imaginary probes were included through complex self-energies with a self-consistent chemical potential adjusted in order to fulfill the voltmeter condition. Further generalizations were also presented in Pastawski (1991). This technique has been further used in a variety of systems (Nozaki, Girard & Yoshizawa, 2008, Maassen, Zahid & Guo, 2009) where the input values can be modeled so as to represent an effective decoherence rate from acoustic phonons or other sources.

3.6   Further reading and problems

- A general textbook on transport concepts in mesoscopic physics is provided by Datta (1995).
- For a very nice discussion on the role of symmetries in transport see Büttiker (1988b).
- For a review on shot noise beyond the brief discussion given in Section 3.2 see Blanter & Büttiker (2000).
- On decoherence, the loss of interference and the quantum-classical transition, we recommend Stern, Aharonov & Imry (1990) and Zurek (2003).

Problems

3.1 *Landauer conductance of pristine carbon-based materials.*[6] In this computational exercise you are encouraged to implement a calculation of the conductance for carbon nanotubes and graphene nanoribbons using the Landauer formula introduced in Section 3.3.

(a) Consider a simple system where an infinite CNT or GNR (modeled through a tight-binding Hamiltonian) is divided into three regions: the left and right will take the role of leads and the central one will be the sample where you may change the site energies to emulate a bias voltage.

 Hint: In the process you may take advantage of the details provided in Appendix C for calculation of the self-energies due to the leads and the Green's functions of the central region. You may use the sample codes provided in our website as well as the datasets available for different systems.

(b) As a byproduct you should also compute the local and total density of states for different systems and rationalize it.

3.2 *Shot noise and exchange interference effects.* For a two terminal conductor, prove that the zero-frequency noise in the independent particle approximation is given by Eq. (3.3). Convince yourself that it cannot be written independently of the chosen basis in terms of transmission and reflection probabilities, as is the case with the conductance.

---

[6] Problems 3.1 and 3.3 may also be solved after reading Chapter 4.

This shows that the noise depends critically on the interference among the carriers from different channels and could therefore serve to probe it. It is also predictable for the shot noise to be particularly sensitive to interaction effects.

3.3 *Conductance through carbon-based materials using the Kubo formula.*[6]

(a) Let us reconsider the previous exercise, but now using the Kubo formula introduced in Section 3.4.4. To do so, take advantage of the recursion methods introduced in Appendix D. You may use the sample codes provided in our website.

(b) For almost the same cost you can compute the local and total density of states.

** Additional exercises and solutions are available at our website.

4   Klein tunneling and ballistic transport in graphene and related materials

In this chapter we start with a presentation of the so-called Klein tunneling mechanism, which is one of the most striking properties of graphene. Later we give an overview of ballistic transport both in graphene and related materials (carbon nanotubes and graphene nanoribbons). After presenting a simple real-space mode-decomposition scheme, which can be exploited to obtain analytical results or to boost numerical calculations, we discuss Fabry-Pérot interference, contact effects, and the minimum conductivity in the 2D limit.

4.1   The Klein tunneling mechanism

The Klein tunneling mechanism was first reported in the context of quantum electrodynamics. In 1929, physicist Oskar Klein (Klein, 1929) found a surprising result when solving the propagation of Dirac electrons through a single potential barrier. In non-relativistic quantum mechanics, incident electrons tunnel a short distance through the barrier as evanescent waves, with exponential damping with the barrier depth. In sharp contrast, if the potential barrier is of the order of the electron mass, $eV \sim mc^2$, electrons propagate as antiparticles whose inverted energy–momentum dispersion relation allows them to move freely through the barrier. This unimpeded penetration of relativistic particles through high and wide potential barriers has been one of the most counterintuitive consequences of quantum electrodynamics, but despite its interest for particle, nuclear, and astro-physics, a direct test of the *Klein tunnel effect* using relativistic particles still remains out of reach for high-energy physics experiments.[1]

In 1998, Ando, Nakanishi, and Saito deduced a full suppression of backscattering in metallic carbon nanotubes for long range disorder, as a consequence of Berry's phase and electron–hole symmetry (Ando, Nakanishi & Saito, 1998). In 2006, Katsnelson, Novoselov and Geim demonstrated that massless Dirac fermions in graphene offer a unique test of Klein's *gedanken* experiment (Klein, 1929). This remarkable prediction was then confirmed by a series of finely tuned experiments (Stander, Huard & Goldhaber-Gordon, 2009, Young & Kim, 2009, 2011). A collimating effect on ballistically transmitted carriers was revealed through analysis of the conductance oscillations (and phase shift in the conductance fringes at low magnetic

---

[1]   We emphasize that despite its name, the Klein tunnel effect does not involve tunneling as usually meant in quantum mechanics because it does not rely on evanescent waves. It is the roles of pseudospin conservation and electron–hole symmetry that are crucial in this case.

fields) in a graphene-based p-n junction, providing compelling evidence of the perfect transmission of carriers normally incident on the junctions (Cheianov & Fal'ko, 2006, Shytov, Rudner & Levitov, 2008). In the following we provide a complete derivation of the analytical calculation for both a single layer and bilayers.

4.1.1 Klein tunneling through monolayer graphene with a single (impurity) potential barrier

Let us consider a system such as the one represented in Fig. 4.1: two normal regions with a potential barrier in between. To obtain the transmission probability, we follow a simple scheme and solve for the wavefunctions in each of the regions and then match them.

The 2D wavefunction of an electron in graphene
As discussed in Section 2.2.2, the graphene electronic structure at low energy can be obtained from the two-component effective equation $\mathcal{H}\Psi = E\Psi$, where

$$\mathcal{H} = \hbar v_F \begin{pmatrix} 0 & \hat{k}_x - i\hat{k}_y \\ \hat{k}_x + i\hat{k}_y & 0 \end{pmatrix} = \hbar v_F(\sigma_x \hat{k}_x + \sigma_y \hat{k}_y), \tag{4.1}$$

where σ_x, σ_y are Pauli spin matrices and $\hat{\mathbf{k}} = (\hat{k}_x, \hat{k}_y) = -i\nabla$ is a wavevector operator. In the previous equation we have dropped the subindex K_+ in the Hamiltonian; the

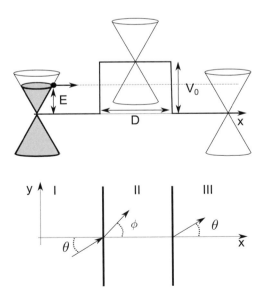

Figure 4.1 Scattering of Dirac electron by a square potential.

calculations for the other valley proceed in an analogous way. The 2D spinor wavefunction is written as

$$\Psi_{s\mathbf{k}}(\mathbf{r}) = \begin{pmatrix} \Psi_1 \\ \Psi_2 \end{pmatrix} \exp(i\mathbf{kr}), \tag{4.2}$$

Substituting Eq. (4.2) into (4.1) yields

$$\hbar v_F \begin{pmatrix} 0 & k_x - ik_y \\ k_x + ik_y & 0 \end{pmatrix} \begin{pmatrix} \Psi_1 \\ \Psi_2 \end{pmatrix} = E \begin{pmatrix} \Psi_1 \\ \Psi_2 \end{pmatrix}. \tag{4.3}$$

To solve this equation, we use the band dispersion

$$E_{s\mathbf{k}} = s\hbar v_F |k|, \tag{4.4}$$

where $s = +1$ and -1 denote the conduction band and valence bands, respectively. By substituting into Eq. (4.3), we obtain

$$\Psi_2 = \frac{k_x + ik_y}{s|k|} \Psi_1 = se^{i\phi} \Psi_1, \tag{4.5}$$

with

$$k_x = |k| \cos \phi, \qquad k_y = |k| \sin \phi, \tag{4.6}$$

so the corresponding momentum space pseudospinor eigenfunction becomes

$$\Psi_{s,\mathbf{k}} = \begin{pmatrix} \Psi_1 \\ \Psi_2 \end{pmatrix} = \begin{pmatrix} 1 \\ se^{i\phi} \end{pmatrix} \Psi_1, \tag{4.7}$$

and the wavefunction is

$$\Psi(x,y) = \begin{pmatrix} \Psi_1(x,y) \\ \Psi_2(x,y) \end{pmatrix} = \begin{pmatrix} 1 \\ se^{i\phi} \end{pmatrix} \Psi_1 e^{ik_x x + ik_y y}. \tag{4.8}$$

Using the normalization condition of the wavefunction, we finally get $\Psi_1 = \frac{1}{\sqrt{2}L}$, where L^2 is the area of system; this is

$$\Psi(x,y) = \frac{1}{\sqrt{2}L} \begin{pmatrix} 1 \\ se^{i\phi} \end{pmatrix} e^{ik_x x + ik_y y}. \tag{4.9}$$

Now that we have the wavefunctions for the different regions, let us proceed by appropriately matching them.

Wavefunction matching

Let us calculate the wavefunction in different regions of the Klein tunneling model and then evaluate the transmission probability. We consider a potential barrier that has a rectangular shape with width D and is infinite along the y axis,

$$V(x) = \begin{cases} V_0 & \text{in region II,} \\ 0 & \text{in regions I, III.} \end{cases} \tag{4.10}$$

Wavefunctions in different regions

We assume that the incident electron wave propagates at an angle ϕ in region I and then diffracts into region II with angle θ with respect to the x axis. As we showed above, the wavefunction in region I is given by Eq. (4.9). Because the potential does not change along the y axis, the y component of momentum is conservative,

$$k_y^I = k_y^{II} = k_y^{III} = k_y. \tag{4.11}$$

Furthermore, the potentials in regions I and III are the same so $k^I = k^{III}$, so that the electron wave in region III also propagates at an angle θ with respect to the x axis and

$$k_x^I = k_x^{III} = k_x. \tag{4.12}$$

Similarly to Eq. (4.3), the Schrödinger equation in region II becomes

$$\begin{pmatrix} V_0 & \hbar v_F(k_x^{II} - ik_y) \\ \hbar v_F(k_x^{II} + ik_y) & V_0 \end{pmatrix} \begin{pmatrix} \Psi_1^{II} \\ \Psi_2^{II} \end{pmatrix} = E \begin{pmatrix} \Psi_1^{II} \\ \Psi_2^{II} \end{pmatrix}. \tag{4.13}$$

From this, we deduce

$$k_x^{II} = \sqrt{(E - V_0)^2 / (\hbar v_F)^2 - k_y^2}, \tag{4.14}$$

$$\tan \theta = \frac{k_y}{\sqrt{(E - V_0)^2 / (\hbar v_F)^2 - k_y^2}}. \tag{4.15}$$

The wavefunction in the different regions can be written in terms of incident and reflected waves. This is:

$$\Psi^I(x, y) = \frac{1}{\sqrt{2L}} \left\{ \begin{pmatrix} 1 \\ se^{i\phi} \end{pmatrix} e^{i(k_x x + k_y y)} + r \begin{pmatrix} 1 \\ se^{i(\pi - \phi)} \end{pmatrix} e^{i(-k_x x + k_y y)} \right\}, \tag{4.16}$$

$$\Psi^{II}(x, y) = \frac{1}{\sqrt{2L}} \left\{ a \begin{pmatrix} 1 \\ s'e^{i\theta} \end{pmatrix} e^{i(k_x^{II} x + k_y y)} + b \begin{pmatrix} 1 \\ s'e^{i(\pi - \theta)} \end{pmatrix} e^{i(-k_x^{II} x + k_y y)} \right\}, \tag{4.17}$$

$$\Psi^{III}(x, y) = \frac{t}{\sqrt{2L}} \begin{pmatrix} 1 \\ se^{i\phi} \end{pmatrix} e^{i(k_x x + k_y y)}, \tag{4.18}$$

where $s = \text{sgn}(E)$ and $s' = \text{sgn}(E - V_0)$. This provides the whole set of wavefunctions for the different regions of the scattering problem.

The transmission probability and Klein tunnelling

The coefficients r, a, b, t are obtained by requiring continuity of the wavefunction,

$$\Psi^I(0, y) = \Psi^{II}(0, y), \tag{4.19}$$

$$\Psi^{II}(D, y) = \Psi^{III}(D, y). \tag{4.20}$$

Substituting the wavefunctions by their generic forms, we get for the first equation

$$\begin{cases} 1 + r = a + b, \\ s(e^{i\phi} + re^{i(\pi-\phi)}) = s'(ae^{i\theta} + be^{i(\pi-\theta)}). \end{cases} \tag{4.21}$$

Similarly, for the second we obtain

$$\begin{cases} ae^{iq_xD} + be^{-iq_xD} = te^{ik_xD}, \\ s'(ae^{i(\theta+q_xD)} + be^{i(\pi-\theta-q_xD)}) = tse^{i(\phi+k_xD)}, \end{cases} \tag{4.22}$$

where $q_x = k_x^{II}$. From Eq. (4.21), we have

$$b = -\frac{se^{i\phi} - s'e^{i\theta}}{se^{i\phi} + s'e^{-i\theta}} ae^{2iq_xD}. \tag{4.23}$$

Substituting Eq. (4.23) into the first expression in Eq. (4.21), we obtain

$$a = \frac{se^{i\phi} + s'e^{-i\theta}}{-2ise^{i\phi}\sin(q_xD) + 2s'\cos(\theta + q_xD)}(1 + r)e^{-iq_xD}. \tag{4.24}$$

Using Eq. (4.23) and Eq. (4.24), we obtain

$$b = -\frac{se^{i\phi} - s'e^{i\theta}}{-2ise^{i\phi}\sin(q_xD) + 2s'\cos(\theta + q_xD)}(1 + r)e^{iq_xD}. \tag{4.25}$$

Substituting Eq. (4.24) and Eq. (4.25) into the second expression of Eq. (4.21), we finally find that

$$s(e^{i\phi} - re^{-i\phi}) = \frac{(ss'e^{i(\phi+\theta)} + 1)e^{-iq_xD} + (ss'e^{i(\phi-\theta)} - 1)e^{iq_xD}}{-2ise^{i\phi}\sin(q_xD) + 2s'\cos(\theta + q_xD)}(1 + r), \tag{4.26a}$$

$$\Leftrightarrow \quad s(e^{i\phi} - re^{-i\phi}) = \frac{2ss'e^{i\phi}\cos(\theta - q_xD) - 2i\sin q_xD}{-2ise^{i\phi}\sin(q_xD) + 2s'\cos(\theta + q_xD)}(1 + r), \tag{4.26b}$$

$$\Leftrightarrow \quad r = \frac{ie^{i\phi}\sin(q_xD)(\sin\phi - ss'\sin\theta)}{\sin(q_xD) - ss'\left[\sin\phi\sin\theta\sin(q_xD) - i\cos\phi\cos\theta\cos(q_xD)\right]}. \tag{4.26c}$$

The transmission can then be obtained straightforwardly from $T(\phi) = tt^* = 1 - rr^* = 1 - R$:

$$T(\phi) = \frac{\cos^2\theta\cos^2\phi}{\cos^2(q_xD)\cos^2\theta\cos^2\phi + \sin^2(q_xD)(1 - ss'\sin\theta\sin\phi)^2}. \tag{4.27}$$

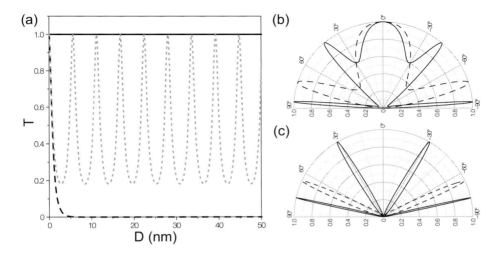

Figure 4.2 (a) Transmission probability $T(E)$ for normally incident electrons in single-layer and bilayer graphene (black solid and black dashed lines, respectively), and in a non-chiral zero-gap semiconductor (grey dotted line) as a function of width D of the tunnel barrier. In this plot the barrier for monolayer graphene is 450 meV and about 240 meV for the other two materials. $T(E)$ through a 100 nm-wide barrier as a function of the incident angle is shown for single-layer (b) and bilayer (c) graphene. In each of these plots two curves are shown; they correspond to a Fermi energy of ~ 80 (solid line) and 17 meV (dashed line), and $\lambda \sim 50$ nm. The barrier heights V_0 are 200 and 50 meV for the solid lines in (b) and (c) respectively, and 285 and 100 meV for the dashed lines in (b) and (c) respectively. (Adapted by permission from Macmillan Publishers Ltd: *Nature*, Katsnelson *et al.*, 2006, copyright (2006).)

In the limit $|V_0| \gg |E|$, the value of $\theta \to 0$ can be replaced in Eq. (4.27), which becomes

$$T(\phi) = \frac{\cos^2 \phi}{\cos^2(q_x D) \cos^2 \phi + \sin^2(q_x D)} = \frac{\cos^2 \phi}{1 - \cos^2(q_x D) \sin^2 \phi}. \qquad (4.28)$$

Equation (4.28) means that under resonance conditions $q_x D = \pi N, N = 0, \pm 1, \ldots$, the barrier becomes totally transparent ($T = 1$). Additionally, the barrier always remains perfectly transparent for angles close to normal incidence, $\varphi = 0$ (see Fig. 4.2), which stands as a feature unique to massless Dirac fermions and is directly related to the Klein paradox in QED. It is important to note that this perfect transmission for normal incidence is not a resonance effect. Indeed, this perfect tunneling can be rationalized in terms of the conservation of pseudospin. In the absence of pseudospin-flip processes, an electron moving to the right can be scattered only to a right-moving electron state or left-moving hole state (see also the discussion in Section 2.2.2). The matching between directions of pseudospin for quasiparticles inside and outside the barrier results in perfect tunneling. In the strictly one-dimensional case, such perfect transmission of Dirac fermions has also been discussed in the context of electron transport in carbon nanotubes (see Section 5.1).

4.1.2   **Klein tunneling through bilayer graphene with a single (impurity) potential barrier**

The 2D wavefunction of an electron in bilayer graphene

The Hamiltonian of an electron in bilayer graphene has the form

$$H = -\frac{\hbar^2}{2m} \begin{pmatrix} 0 & (\hat{k}_x - i\hat{k}_y)^2 \\ (\hat{k}_x + i\hat{k}_y)^2 & 0 \end{pmatrix}. \tag{4.29}$$

Similarly to the 2D spinor wavefunction of single-layer graphene, the wavefunction of bilayer graphene in the ith region satisfies

$$-\frac{\hbar^2}{2m} \begin{pmatrix} 0 & (\hat{k}_x - i\hat{k}_y)^2 \\ (\hat{k}_x + i\hat{k}_y)^2 & 0 \end{pmatrix} \Psi^i_{s\mathbf{k}}(\mathbf{r}) = (E - V_i)\Psi^i_{s\mathbf{k}}(\mathbf{r}), \tag{4.30}$$

where the wavefunction has the following form (the electrons are free in the y direction and the momentum in the y direction is unchanged):

$$\Psi^i_{s\mathbf{k}}(\mathbf{r}) = \left(a_i \begin{pmatrix} \Psi^1_1 \\ \Psi^1_2 \end{pmatrix}_i e^{ik_{ix}x} + b_i \begin{pmatrix} \Psi^2_1 \\ \Psi^2_2 \end{pmatrix}_i e^{-ik_{ix}x} \right.$$
$$\left. + c_i \begin{pmatrix} \Psi^3_1 \\ \Psi^3_2 \end{pmatrix}_i e^{\kappa_{ix}x} + d_i \begin{pmatrix} \Psi^4_1 \\ \Psi^4_2 \end{pmatrix}_i e^{-\kappa_{ix}x} \right) e^{ik_y y}. \tag{4.31}$$

Substituting such an expression for $\Psi^i_{s\mathbf{k}}(\mathbf{r})$ into Eq. (4.30), and following the same steps as for the monolayer case, we find that

$$\begin{pmatrix} \Psi^1_1 \\ \Psi^1_2 \end{pmatrix}_i = \begin{pmatrix} 1 \\ s_i e^{2i\phi_i} \end{pmatrix} \Psi^1_{1i}, \tag{4.32a}$$

$$\begin{pmatrix} \Psi^2_1 \\ \Psi^2_2 \end{pmatrix}_i = \begin{pmatrix} 1 \\ s_i e^{-2i\phi_i} \end{pmatrix} \Psi^2_{1i}, \tag{4.32b}$$

$$\begin{pmatrix} \Psi^3_1 \\ \Psi^3_2 \end{pmatrix}_i = \begin{pmatrix} 1 \\ -s_i h_i \end{pmatrix} \Psi^3_{1i}, \tag{4.32c}$$

$$\begin{pmatrix} \Psi^4_1 \\ \Psi^4_2 \end{pmatrix}_i = \begin{pmatrix} 1 \\ -s_i/h_i \end{pmatrix} \Psi^4_{1i}, \tag{4.32d}$$

where

$$s_i = \text{sgn}(V_i - E), \qquad \hbar k_{ix} = \sqrt{2m|E - V_i|} \cos\phi_i,$$
$$\hbar k_y = \sqrt{2m|E - V_i|} \sin\phi_i = const,$$
$$\kappa_{ix} = \sqrt{k_{ix}^2 + 2k_y^2}, \qquad h_i = \left(\sqrt{1 + \sin^2\phi_i} - \sin\phi_i \right)^2.$$

The way to obtain the first two pseudospinors is the same as for single-layer graphene, whereas the two others can be derived as follows:

$$-\frac{\hbar^2}{2m}\begin{pmatrix} 0 & (\hat{k}_x - i\hat{k}_y)^2 \\ (\hat{k}_x + i\hat{k}_y)^2 & 0 \end{pmatrix}\begin{pmatrix} \Psi_1^3 \\ \Psi_2^3 \end{pmatrix}_i e^{\kappa_{ix}x + ik_y y} = (E - V_i)\begin{pmatrix} \Psi_1^3 \\ \Psi_2^3 \end{pmatrix}_i e^{\kappa_{ix}x + ik_y y},$$

(4.33a)

$$-\frac{\hbar^2}{2m}\begin{pmatrix} 0 & (\hat{k}_x - i\hat{k}_y)^2 \\ (\hat{k}_x + i\hat{k}_y)^2 & 0 \end{pmatrix}\begin{pmatrix} \Psi_1^4 \\ \Psi_2^4 \end{pmatrix}_i e^{-\kappa_{ix}x + ik_y y} = (E - V_i)\begin{pmatrix} \Psi_1^4 \\ \Psi_2^4 \end{pmatrix}_i e^{-\kappa_{ix}x + ik_y y}.$$

(4.33b)

From this, we obtain

$$\begin{pmatrix} 0 & (\kappa_{ix} + k_y)^2 \\ (\kappa_{ix} - k_y)^2 & 0 \end{pmatrix}\begin{pmatrix} \Psi_1^3 \\ \Psi_2^3 \end{pmatrix}_i = \frac{2m(E - V_i)}{\hbar^2}\begin{pmatrix} \Psi_1^3 \\ \Psi_2^3 \end{pmatrix}_i,$$

(4.34a)

$$\begin{pmatrix} 0 & (\kappa_{ix} - k_y)^2 \\ (\kappa_{ix} + k_y)^2 & 0 \end{pmatrix}\begin{pmatrix} \Psi_1^4 \\ \Psi_2^4 \end{pmatrix}_i = \frac{2m(E - V_i)}{\hbar^2}\begin{pmatrix} \Psi_1^4 \\ \Psi_2^4 \end{pmatrix}_i,$$

(4.34b)

so we have the relations

$$\kappa_{ix}^2 = k_y^2 + \frac{2m|E - V_i|}{\hbar^2} = k_y^2 + k_i^2 = k_{ix}^2 + 2k_y^2,$$

(4.35a)

$$\Psi_{2i}^3 = -s_i \frac{(\kappa_{ix} - k_y)^2}{k_i^2}\Psi_{1i}^3 = -s_i h_i \Psi_{1i}^3,$$

(4.35b)

$$\Psi_{2i}^4 = -s_i \frac{k_i^2}{(\kappa_{ix} - k_y)^2}\Psi_{1i}^4 = -\frac{s_i}{h_i}\Psi_{1i}^4.$$

(4.35c)

Therefore, the wavefunction in the ith region is

$$\Psi_{s\mathbf{k}}^i(\mathbf{r}) = \begin{pmatrix} \psi_1^i(x, y) \\ \psi_2^i(x, y) \end{pmatrix},$$

(4.36)

where

$$\psi_1^i = \left(a_i e^{ik_{ix}x} + b_i e^{-ik_{ix}x} + c_i e^{\kappa_{ix}x} + d_i e^{-\kappa_{ix}x}\right)e^{ik_y y},$$

$$\psi_2^i = s_i(a_i e^{i(k_{ix}x + 2\phi_i)} + b_i e^{-i(k_{ix}x + 2\phi_i)} - c_i h_i e^{\kappa_{ix}x} - \frac{d_i}{h_i}e^{-\kappa_{ix}x})e^{ik_y y}.$$

This gives the general form of the wavefunctions for the bilayer case.

Transmission probability and chiral tunneling in bilayer graphene

Equation (4.36) gives the form of wavefunctions in the different regions. To avoid divergence of the wavefunction as $x \to -\infty$ ($x \to +\infty$), $d_1 = 0$ ($c_3 = 0$) in region I (III). There is no reflected wave in region III so $b_3 = 0$. Using the continuity conditions for both components of the wavefunction and their derivatives, for the case of an electron beam that is incident normally ($\phi = 0$) and low barrier $V_0 < E$:

$$a_1 + b_1 + c_1 = a_2 + b_2 + c_2 + d_2, \tag{4.37}$$

$$a_1 + b_1 - c_1 = a_2 + b_2 - c_2 - d_2,$$

$$k_1(ia_1 - ib_1 + c_1) = k_2(ia_2 - ib_2 + c_2 - d_2),$$

$$k_1(ia_1 - ib_1 - c_1) = k_2(ia_2 - ib_2 - c_2 + d_2),$$

$$a_2e^{ik_2D} + b_2e^{-ik_2D} + c_2e^{k_2D} + d_2e^{-k_2D} = a_3e^{ik_1D} + d_3e^{-k_1D},$$

$$a_2e^{ik_2D} + b_2e^{-ik_2D} - c_2e^{k_2D} - d_2e^{-k_2D} = a_3e^{ik_1D} - d_3e^{-k_1D},$$

$$k_2(ia_2e^{ik_2D} - ib_2e^{-ik_2D} + c_2e^{k_2D} - d_2e^{-k_2D}) = k_1(ia_3e^{ik_1D} - d_3e^{-k_1D}),$$

$$k_2(ia_2e^{ik_2D} - ib_2e^{-ik_2D} - c_2e^{k_2D} + d_2e^{-k_2D}) = k_1(ia_3e^{ik_1D} + d_3e^{-k_1D}),$$

where $k_1 = k_{1x} = \kappa_{1x} = \frac{\sqrt{2mE}}{\hbar}$ and $k_2 = k_{2x} = \kappa_{2x} = \frac{\sqrt{2m|E-V_0|}}{\hbar}$. From these equations we have $\psi_1 = -\psi_2$ both inside and outside the barrier. For the case of an electron beam that is incident normally ($\phi = 0$) and high barrier $V_0 > E$ we have similar equations to Eq. (4.38), but with $s_2 = -s_1 = -s_3 = -1$, from which we obtain the transmission coefficient t:

$$t = \frac{a_3}{a_1} = \frac{4ik_1k_2}{(k_2 + ik_1)^2e^{-k_2D} - (k_2 - ik_1)^2e^{k_2D}}. \tag{4.38}$$

The transmission probability T is given by

$$T = |t|^2 = \frac{4k_1^2k_2^2}{(k_1^2 + k_2^2)^2 \sinh^2(k_2D) + 4k_1^2k_2^2}. \tag{4.39}$$

Therefore, T in this case decays exponentially with the height and width of the barrier. This is in striking contrast with monolayer graphene where transmission is unity. For bilayer graphene, pseudospin conservation does not forbid backscattering. The results for both cases are shown in Fig. 4.2, where they are also compared with those for a non-chiral zero-gap semiconductor.

4.2   Ballistic transport in carbon nanotubes and graphene

One of the main problems hindering molecular electronics has been the poor quality of the contacts between the molecular sample and the metallic electrodes. This leads to low conductance values and usually takes us out of the coherent regime analyzed in this section (see Section 3.1.2). The outstanding quality of the contacts achieved for carbon nanotubes and graphene devices changed this picture radically allowing, for example, for the observation of ballistic transport and Fabry-Pérot interference (Liang *et al.*, 2001, Kim *et al.*, 2007, Herrmann *et al.*, 2007, Wu *et al.*, 2007, Wu *et al.*, 2012), which is addressed in the following pages.

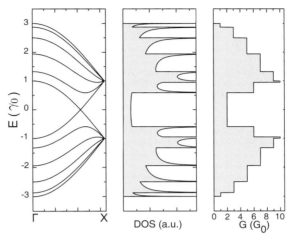

Figure 4.3 Band structure (left), density of states (middle) and conductance (right) for a $(5, 5)$ armchair nanotube.

4.2.1 Ballistic motion and conductance quantization

The intrinsic electronic transport along the nanotube axis is perfectly ballistic if the measured conductance is quantized (and L-independent), only varying with the number of available conducting channels at the considered energy ($N_\perp(E)$). In that regime $G(E) = 2e^2/h \times N_\perp(E)$ (including spin degeneracy). The quantized conductance profile for a given nanotube can actually be directly deduced from the band structure features, by counting the number of channels at a given energy. A metallic armchair nanotube presents only two quantum channels at the charge neutrality point (CNP), resulting in $G(E_F) = 2G_0$. At higher energies, the conductance increases as more channels become available for transport. Figure 4.3 shows the electronic bands and conductance of a clean $(5, 5)$ metallic tube.

This situation has been experimentally measured for very clean metallic nanotubes with ohmic contacts between the SWNT and metallic (palladium) voltage probes (Javey *et al.*, 2003, Javey *et al.*, 2004). Figure 4.4 (left panel) shows the $I_{DS}(V_{DS})$ for metallic single-walled carbon nanotubes (Pd Ohmic contacts) with lengths ranging from 700 nm down to 55 nm. The low-bias regime is clearly linear, and makes it possible to extract a corresponding conductance $G = dI/dV$ which turns out to be very close to the maximum quantized value $4e^2/h$ and (as expected) shows almost no temperature dependence (Fig. 4.4 (right panel)). Using semiclassical (Bloch–Boltzmann) transport simulations, mean free path acoustic phonon scattering is estimated to be in the order of 300 nm, whereas that for optical phonon scattering the inelastic length is estimated to be about 15 nm (see Section 5.1.1 for more details). Transport through very short (10 nm) nanotubes is free of significant acoustic and optical phonon scattering and thus ballistic and quasiballistic at the low- and high-bias voltage limits, respectively. High currents of up to 70 μA can flow ballistically through a short nanotube section in between Pd contacts (Javey *et al.*, 2003, Javey *et al.*, 2004).

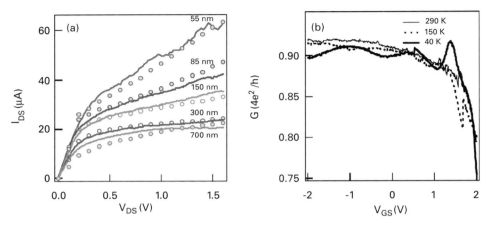

Figure 4.4 (a) Electrical properties of ohmically contacted metallic SWNTs of various lengths (diameters $d \in [2, 2.5]$ nm, oxide thickness is 10 nm). Solid lines are experimental $I_{DS} - V_{DS}$ curves, while symbols are Monte Carlo simulations. (b) Conductance versus gate voltage recorded (under a low bias of $V_{DS} \simeq 1mV$) at 290, 150, and 40 K. (Adapted from Javey *et al.*, 2004. Copyright (2004) by the American Physical Society. By courtesy of Hongjie Dai.)

These values are, however, the uppermost theoretical limits that can be experimentally accessible. In practical situations, lower values are found since quantum transmission is limited by interface symmetry mismatch, inducing Bragg-type backscattering. Additionally, topological and chemical disorders, as well as intershell coupling for multiwalled nanotubes, introduce intrinsic scattering along the tube, which also lowers the total transmission probability. Both effects will have an impact on the transmission through the different conducting channels or modes in the leads. If $T_n(E) \leq 1$ is the transmission probability through one of those channels at energy E, the conductance is given by $G(E) = G_0 \sum_{n=1,N_\perp} T_n(E)$ (Datta, 1995).

To make the last statements more concrete, in the following subsection we introduce a useful way of decomposing the system (be it a carbon nanotube or a graphene nanoribbon) into independent channels or modes, thereby giving a picture of what these conduction channels are.

4.2.2   Mode decomposition in real space

Solving the Hamiltonian by *brute force* to obtain the transport/electronic properties of pristine carbon nanotubes or graphene (armchair edge) nanoribbons, even by using a decimation procedure, is computationally demanding and much physical insight might be lost. In the following, a simple procedure to break the Hamiltonian into independent building blocks is described. The trick is simple but tremendously powerful: a suitable unitary transformation performs the desired decomposition which serves as a starting point for either more efficient computational codes or insightful analytical

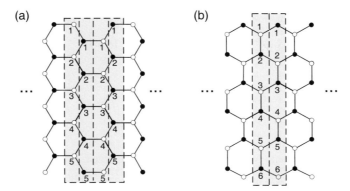

Figure 4.5 Representation of the atomic positions for different terminations of graphene nanoribbons: (a) armchair; and (b) zigzag. The dashed boxes separate different layers of carbon atoms. Note that zigzag and armchair carbon nanotubes can be obtained, respectively, from (a) and (b) by imposing periodic boundary conditions along the vertical direction. In the case of armchair nanotubes, one has to additionally arrange for the number of atoms in each layer to be an even number.

calculations. The following paragraphs follow previous works carried out for nanotubes (Mingo *et al.*, 2001) as well as graphene ribbons (Zhao & Guo, 2009; Rocha *et al.*, 2010).

Figure 4.5 shows arrangements of carbon atoms for nanoribbons of armchair and zig-zag edges. By adding a periodic boundary condition along the vertical direction, this represents as well the arrangement for the case of carbon nanotubes where curvature effects are neglected. In general, the Hamiltonian can be written in a block-matrix form where each block corresponds to the orbitals inside each layer (as depicted in Fig. 4.5) for a particular diameter/width of the nanotube/nanoribbon. The idea is to find a basis where all these block matrices have a diagonal form. As we see below, depending on the boundary conditions, this is sometimes possible thereby rendering a decomposition of the 2D lattice into several independent 1D lattices.

The nearest-neighbour π orbitals Hamiltonian is given by

$$
\mathcal{H} = \begin{pmatrix}
\ddots & & & & & \\
& \mathcal{H}_{l_1} & \mathcal{V}_1 & & & \\
& \mathcal{V}_1^\dagger & \mathcal{H}_{l_2} & \mathcal{V}_2 & & \\
& & \mathcal{V}_2^\dagger & \mathcal{H}_{l_3} & \mathcal{V}_1^\dagger & \\
& & & \mathcal{V}_1 & \mathcal{H}_{l_4} & \mathcal{V}_2^\dagger \\
& & & & \mathcal{V}_2 & \mathcal{H}_{l_5} & \ddots
\end{pmatrix},
\tag{4.40}
$$

where $\mathcal{H}_{l_i} = E_i \mathbb{I}_{m \times m}$ is the block matrix corresponding to the orbitals in the ith layer and $\mathcal{V}_1$ and $\mathcal{V}_2$ are the hopping matrices connecting layers of different type. The precise form of these matrices depends on the particular system at hand. In the following we consider zigzag and armchair nanotubes and armchair graphene nanoribbons.

Zigzag carbon nanotubes. Let us consider an $(n, 0)$ carbon nanotube. In this case $m = n$ and

$$\mathcal{V}_1 = \gamma_0 \begin{pmatrix} 1 & 0 & \cdots & & 1 \\ 1 & 1 & & & \\ 0 & 1 & 1 & \cdots & \\ \cdots & & \cdots & & \ddots \end{pmatrix}, \tag{4.41}$$

and $\mathcal{V}_2 = \gamma_0 \mathbb{I}_{n \times n}$. $\mathcal{V}_1$ can be easily diagonalized, i.e. there is an $(n \times n)$ matrix C such that $C^\dagger \mathcal{V}_1 C$ has a diagonal form. The eigenvectors of the matrix $\mathcal{V}_1$ are plane waves around the nanotube circumferential direction:

$$|\varphi_q\rangle = \frac{1}{\sqrt{n}} \sum_{j=1}^{n} \exp(iqj) |j\rangle, \tag{4.42}$$

where $q = 1, \ldots, n$ is the mode index, and $|j\rangle$ represents the π orbital localized at the jth atom in a given layer. Since $\mathcal{V}_2$ and $\mathcal{H}_{l_i}$ are proportional to the identity, they commute with $\mathcal{V}_1$ and the transformation defined by the previous equation (let us call it C) diagonalizes all the matrices simultaneously. Using a change of basis transformation of the form

$$\mathcal{U} = \begin{pmatrix} \cdots & & & \cdots \\ & C & 0 & 0 & \\ & 0 & C & 0 & \\ & 0 & 0 & C & \\ \cdots & & & \cdots \end{pmatrix}, \tag{4.43}$$

the transformed Hamiltonian is represented by n uncoupled chains with alternating hoppings γ_0 and $\gamma_q = 2\gamma_0 \exp(-i\pi q/n) \cos(q\pi/n)$ $(q = 1, \ldots, n)$ with q the mode index. Each of these modes is represented in Fig. 4.6.

Given that dimers with alternating hoppings like the ones above always have a gap unless the hoppings have equal absolute value, one can see that whenever n is a multiple

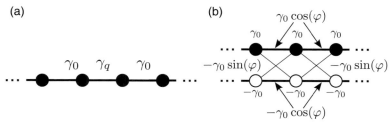

Figure 4.6 Representation of the modes/eigenchannels resulting after the mode decomposition described in the text for: (a) zigzag carbon nanotubes and armchair graphene nanoribbons; and (b) armchair carbon nanotubes.

of 3, $|\gamma| = |\gamma_q|$ is satisfied for $q = n/3, 2n/3$ and the system is metallic. The modes with $q = n/3, 2n/3$ are in this case the only ones which contribute to the density of states and the conductance in the vicinity of the charge neutrality point. The reader is encouraged to undertake Problem 4.3 at the end of this chapter.

Armchair graphene nanoribbons. In this case $m = n$ and $\mathcal{V}_2 = \gamma_0 \mathbb{I}_{n \times n}$, hence

$$
\mathcal{V}_1 = \gamma_0 \begin{pmatrix} 1 & 0 & \cdots & & 0 \\ 1 & 1 & & & \\ 0 & 1 & 1 & \cdots & \\ & & \cdots & \ddots & \end{pmatrix}.
\tag{4.44}
$$

Note that $\mathcal{V}_1$ differs from what is given in Eq. (4.41) only in the matrix element in the upper right corner (which gives the periodic boundary condition for carbon nanotubes). However, in contrast to the case of zigzag tubes, the matrix $\mathcal{V}_1$, Eq. (4.44), cannot be diagonalized. Therefore, a different strategy is required and a new basis set for armchair ribbons has to be obtained by imposing a "particle-in-a-box" assumption as described below.

Inspired by the geometrical arrangement of the A and B sublattices, an alternative block-diagonal change of basis transformation can be adopted:

$$
\mathcal{U} = \begin{pmatrix} \ddots & & & & & \\ & \mathcal{C}_1 & & & & \\ & & \mathcal{C}_2 & & & \\ & & & \mathcal{C}_2 & & \\ & & & & \mathcal{C}_1 & \\ & & & & & \ddots \end{pmatrix},
\tag{4.45}
$$

where the arrangement of the matrices $\mathcal{C}_1$ and $\mathcal{C}_2$ is periodically repeated with a four-layer periodicity (the same as the lattice). The matrix elements of $\mathcal{C}_1$ and $\mathcal{C}_2$ are chosen to satisfy hard boundary conditions:

$$
[\mathcal{C}_1]_{i,q} = \frac{2}{\sqrt{2n+1}} \sin\left(\frac{2iq\pi}{2n+1}\right),
\tag{4.46}
$$

$$
[\mathcal{C}_2]_{i,q} = \frac{2}{\sqrt{2n+1}} \sin\left(\frac{(2i-1)q\pi}{2n+1}\right).
\tag{4.47}
$$

Interestingly, the blocks of the transformed Hamiltonian $\mathcal{H}' = \mathcal{U}^\dagger \mathcal{H} \mathcal{U}$ are all diagonal. Indeed, the blocks proportional to the identity matrix remain invariant ($\mathcal{H}'_{l_i} = \mathcal{H}_{l_i}, \mathcal{V}'_2 = \mathcal{V}_2$), while $[\mathcal{V}'_1]_{i,q} = [\mathcal{C}_1^\dagger \mathcal{V}_1 \mathcal{C}_2]_{i,q} = 2\gamma_0 \delta_{i,q} \cos(q\pi/(2n+1))$. Therefore, the graphene armchair nanoribbon can also be represented as n independent one-dimensional chains with alternating hoppings γ_0 and $\gamma_q = 2\gamma_0 \cos(q\pi/(2n+1))$, $q = 1,\ldots,n$. A simple analysis of the electronic structure of each of these modes/eigenchannels shows that, for metallic armchair nanoribbons, there is a single mode with a nonvanishing density of states close to the charge neutrality point (see also Problem 4.3).

Armchair carbon nanotubes. Let us consider (m,m) armchair SWNTs. The carbon atoms are arranged into layers as shown in Fig. 4.5 (b), which in this case corresponds to a $(3,3)$ SWNT once periodic boundary conditions are taken along the vertical direction. In this case $n = 2m$, $\mathcal{V}_1 = \mathcal{V}_2^{\dagger} = \gamma_0 \mathbb{I}_{n \times n}$ and the only nontrivial matrices are the $\mathcal{H}_{l_i}$. By applying a change of basis transformation such as the one proposed in Problem 4.3, it is possible to obtain a set of decoupled circumferential modes (Mingo *et al.*, 2001). The Hamiltonian for each of these modes is represented in Fig. 4.6 and corresponds to ladders rather than 1D chains, $\varphi_q = \pi q/m$ with $q = 0, 1, \ldots m-1$. The circumferential mode contributing to the density of states at the charge neutrality point corresponds to $\varphi = 0$, leading to two decoupled 1D chains.

4.2.3   Fabry-Pérot conductance oscillations

Let us consider a device made up of a high quality sample which is connected to electrodes through *almost* perfect contacts. *Almost* is the crucial word here. It implies a departure from certainty – when the contacts are perfect transmission is perfect and therefore there is no uncertainty! – thereby giving room for interference. Indeed, the interference between the "paths" corresponding to different numbers of reflections at the interfaces may lead to a phenomenon similar to the Fabry-Pérot interference found in optics. But this time one has an *electrical* Fabry-Pérot interferometer working on the basis of *quantum* interference.

Several experiments have reported a successful realization of Fabry-Pérot interference both in carbon nanotube devices (Liang *et al.*, 2001, Kim *et al.*, 2007, Herrmann *et al.*, 2007, Wu *et al.*, 2007) and in graphene devices (Wu *et al.*, 2012, Oksanen *et al.*, undated). The evidence relies on the observation of oscillations in the conductance as the gate voltage is changed. In the following we comment on these results and outline a way to rationalize them through a minimal model. Though, throughout this book we have mostly been confronted with Hamiltonian models, this time scattering matrices are our starting point.

The experimental setup used, for example, in Liang *et al.* (2001) consists of a metallic carbon nanotube coupled to left and right electrodes and a gate (see scheme in Fig. 4.7(a)). The presence of Fabry-Pérot interference in this electrical measurement setup was verified by using the gate (V_g) and bias (V) voltages as control parameters in low-temperature experiments (Liang *et al.*, 2001) as shown in Fig. 4.7(c). At first sight this plot may seem reminiscent of the conductance pattern usually found in the Coulomb blockade regime. We emphasize, however, that this is not the case as can be appreciated by looking at the scale bar on the right. Indeed, the conductance minima (dark regions) do not show any blockade since the conductance remains close to $3e^2/h$. The maxima are not very far from the quantum limit of $4e^2/h$ for a metallic nanotube in the first conductance plateau. Furthermore, since the level spacing is

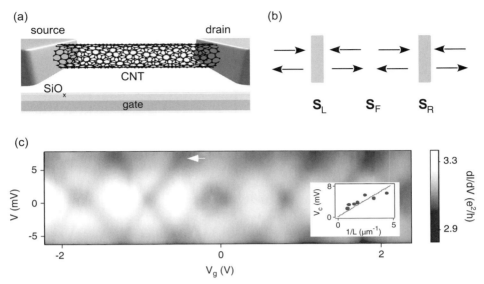

Figure 4.7 (a) A typical experimental setup used to observe Fabry-Pérot oscillations in the conductance of SWNTs. (b) The reflections and transmissions at the interfaces of the device which are modeled through the scattering matrices S_L (left lead), S_R (right lead) and S_F (for free propagation inside the sample). (c) Experimental results from Liang and coworkers (Liang *et al.*, 2001) showing the conductance as a function of the bias (V) and gate (V_g) voltages for a 220 nm long SWNT at 4K. (Reprinted by permission from Macmillan Publishers Ltd: *Nature*; Liang *et al.* (2001), copyright 2001.)

approximately constant close to the charge neutrality point, the regularity of the pattern gives access to a typical energy which in this case turns out to be the level spacing, whereas in the Coulomb blockade regime (see Section 5.8.1), the charging energy sets the dominant energy scale. The dependence of the spacing between the maxima on the inverse length of the device is plotted in the inset to Fig. 4.7(c) and shows the expected linear law.

Scattering matrix modeling. It is instructive to rationalize this experiment by resorting to the Landauer formalism in a minimal model as outlined below (details are left for Problem 4.5). To this end, we need to compute the transmission probability. Instead of using Green's functions for a model Hamiltonian, here we follow Liang and coworkers (Liang *et al.*, 2001) and propose a simple model for the scattering matrices corresponding to each of the processes involved in the interference: (i) partial reflections at the contacts, and (ii) free propagation through the sample. The composition of these scattering matrices makes possible calculation of the scattering matrix of the overall system.

The scattering matrix relates the incoming and outgoing probability amplitudes at a given scatterer. In this case, we have two active channels in the system, therefore S_α ($\alpha = L, R, F$) has dimension 4×4. $[S_L]_{ij}$ are the transmission and reflection amplitudes between the different channels $i, j = 1, 2, 3, 4$, where $i, j = 1, 2$ ($i, j = 3, 4$) correspond to the channels on the left (right).

Scattering matrices for free propagation through the sample. By assuming that the two propagating channels do not mix, the ballistic propagation inside the nanotube can be captured by

$$
S_F = \begin{pmatrix} 0 & 0 & e^{i\phi_1} & 0 \\ 0 & 0 & 0 & e^{i\phi_2} \\ e^{i\phi_1} & 0 & 0 & 0 \\ 0 & e^{i\phi_2} & 0 & 0 \end{pmatrix},
\tag{4.48}
$$

where ϕ_1 and ϕ_2 represent the phases accumulated during the propagation inside the sample in each of the two propagating channels.

Scattering matrices for the contacts. Let S_L and S_R be the scattering matrices for the left and right contacts respectively. The partial reflections at the contacts can be modeled in S_L and S_R by adding a suitable set of parameters. Here we show the alternative proposed by Liang *et al.* (2001), which consists in writing them as an exponential thereby ensuring unitarity:

$$
S_{L(R)} = \exp\left[i \begin{pmatrix} r_2 & r_1 \exp(\pm i\delta_1) & 0 & 0 \\ r_1 \exp(\pm i\delta_1) & r_2 \exp(\pm i\delta_2) & 0 & 0 \\ 0 & 0 & r_2 & r_1 \exp(\mp i\delta_1) \\ 0 & 0 & r_1 \exp(\mp i\delta_1) & r_2 \exp(\mp i\delta_2) \end{pmatrix} \right].
\tag{4.49}
$$

Furthermore, r_1 and r_2 can be assumed to be energy independent, leaving all the dependence on the bias and gate voltages in the phase shifts appearing in S_F. Using the above matrices, the total scattering matrix S_T can be obtained by composing them. In Problem 4.5 you are invited to continue with this calculation in detail. Once this is done, knowledge of S_T allows then for calculation of the Landauer conductance as a function of the bias and gate voltages. The Fabry-Pérot conductance maps simulated using this scattering matrix modeling compare very well with the experimental one as shown in Fig. 4.8 from Liang *et al.* (2001). Comparison with the experiment at hand may allow the extraction of useful information: Do the metal contacts introduce important intermode coupling? Are there other effects not considered here that may play a role? In the results shown in Fig. 4.8, the parameters used to adjust the experimental behavior did require an intermode coupling (i.e. a nonvanishing r_1).

Indeed, many issues which are beyond simple description above have been addressed over the past years, including the appearance of additional low frequency modulation of the interference pattern (Jiang, Dong & Xing, 2003) and interaction effects (Kim *et al.*, 2007).

We note that other very sensitive experiments have also probed the zero-frequency noise in the Fabry-Pérot regime (Kim *et al.*, 2007, Herrmann *et al.*, 2007, Wu *et al.*, 2007) for single-wall carbon nanotubes. Experiments suggest in some cases a good quantitative agreement with a coherent non-interacting picture (Herrmann *et al.*, 2007), while other results show moderate (Wu *et al.*, 2007) deviations which are attributed to

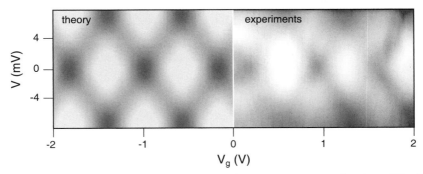

Figure 4.8 Calculated (left) and measured (right) conductance patterns as a function of V and V_g for a 220 nm SWNT device. Dark corresponds to 2.9 e^2/h and white to 3.2 e^2/h. (Reprinted by permission from Macmillan Publishers Ltd: *Nature* (Liang *et al.*, 2001), copyright 2001.)

either electron–electron interactions (Wu *et al.*, 2007, Herrmann *et al.*, 2007) or decoherence (Herrmann *et al.*, 2007).

Results compatible with a Tonomaga–Luttinger liquid were also found for weak backscattering at the contacts (Kim *et al.*, 2007). We mention that the presence of enhanced Coulomb interaction in low dimensionality was predicted to give rise to the formation of a Luttinger liquid (Luttinger, 1963), a phenomenon which was then applied to the case of metallic carbon nanotubes (Egger & Gogolin, 1997, Egger, 1999). The theoretical fingerprints for Luttinger liquid in nanotubes include peculiar power-law behavior of the temperature-dependent conductance, with an exponent depending on the contact geometry, or spin–charge separation. Transport evidence of such many-body states has been reported experimentally, but in very specific conditions, including a high quality metallic single-walled nanotube connected to external reservoirs with at least one poor (tunneling) contact (Bockrath *et al.*, 1999, Gao *et al.*, 2004). A more detailed presentation of the background and experiments is given in Charlier, Blase & Roche (2007).

We close this subsection by noting that, more recently, Fabry-Pérot oscillations have also been reported in sub-100 nm length graphene devices (Wu *et al.*, 2012) and also in outstanding experiments on suspended graphene devices (Grushina & Morpurgo, 2013; Rickhaus *et al.*, 2013).

4.2.4 Contact effects: SWNT-based heterojunctions and the role of contacts between metals and carbon-based devices

Up to now we have been mostly focused on intrinsic effects in carbon-based devices. However, it may be crucial to capture the contact effects, i.e. the contact resistance between the measured material (nanotubes, graphene) and the conducting electrodes. The electrodes are generally formed by other materials like palladium or gold, though there is also great interest in *all-carbon* devices (Anantram & Léonard, 2006).

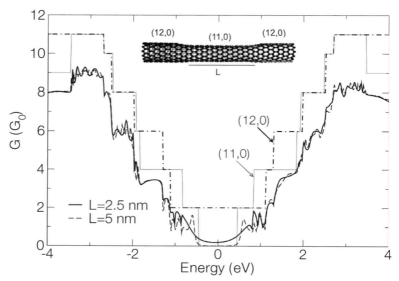

Figure 4.9 Conductance profile for a single semiconducting nanotube $(11,0)$ and a metallic nanotube $(12,0)$, together with the conductance of the double junction $(12,0)$–$(11,0)$–$(12,0)$ (a ball-and-stick structure is shown in the inset).

SWNT-based heterojunctions

The electronic structure of an intramolecular nanotube-based heterojunction was first investigated by Chico and coworkers (Chico *et al.*, 1996), who considered the case of the junction of two nanotubes of different diameter connected through pentagon–heptagon pairs. A different configuration (illustrated in the inset of Fig. 4.9) is that of a double heterojunction, where the leads are formed by $(12,0)$ nanotubes while the central region contains an $(11,0)$ semiconducting one. These junctions have been experimentally observed and measured by STM (Ouyang *et al.*, 2001*a*) (for a review see Odom, Huang & Lieber, 2002). The conductance between the STM tip and the device in different regions gives information on the local density of states. The different position of the peaks associated with the van Hove singularities allows characterization of the device. In this case, a good agreement has been found with the predictions of a simple π orbitals model.

What about the transport properties of such devices? Following a calculation within the Landauer–Büttiker formalism (Triozon, Lambin & Roche, 2005) one obtains the conductance shown in Fig. 4.9 as a function of the Fermi energy in the low temperature limit. Here, a simple *tight-binding* model for the π orbitals with a single hopping parameter for all carbon–carbon bonds is used. The solid and the dashed black lines are the conductance for an internal tube of 2.5 and 5.0 nm respectively. The results for the $(11,0)$ and $(12,0)$ SWNTs are also shown for reference. The overall decrease of the conductance is evident, with the conductance being limited by the smallest number of modes, $\min(N_{(12,0)}, N_{(11,0)})$, at the given energy. The small conductance for the shorter junction is due to tunneling through the gap and is already suppressed for 5.0 nm. The

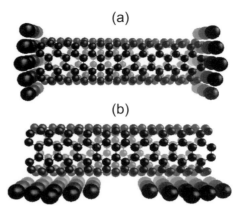

Figure 4.10 Contact types between a nanotube and electrode first layers: (a) end contacts; and (b) side or bulk contacts. Courtesy of Juan-José Palacios (Palacios *et al.*, 2003).

results for a longer system of 100 nm (not shown here) do not change much, with faster oscillations near the gap being the main difference.

Contacts between a metal and a metallic carbon-based material

The contact resistance of metallic interfaces mainly depends on the local atomic bonding and orbital rehybridization at the contact, and remains finite even for vanishing bias potential. As an illustration, let us consider the case of an M–MSWNT–M heterojunction. The scattering rate between the metal and the nanotube (which is also related to the self-energy of the contact Σ) can be estimated using a simple Fermi golden rule. Assuming that $|k_m\rangle = \sum_l e^{ik_m \cdot l}|\varphi_m^l\rangle$ (resp. $|k_F\rangle = \sum_l e^{ik_F \cdot l}|\varphi_{\text{NT}}^l\rangle$) are the propagating states with k_m (k_F) the wavevector in the metal (resp. nanotube), we take $|\varphi_{\text{NT}}^l\rangle$ the localized atomic basis orbitals (p_z-like) in cell (ℓ), that will have a finite overlap with $|\varphi_m^{l'}\rangle$ only for a few unit cells, defining the contact area ($(l - l')$ small).

The scattering rate between metal and nanotube is related to $\langle k_m|\mathcal{H}_{\text{contact}}|k_F\rangle$. This matrix element depends on the chemical nature of the interface bonding (covalent, ionic, etc.) and on the overlap $\langle \varphi_{\text{NT}}^l|\varphi_m^{l'}\rangle$ which changes depending on the interface geometry (end or side/bulk contacts, length of the contact as illustrated in Fig. 4.10), together with the momentum of the atomic orbitals contributing to $|k_m\rangle$. The optimization of the coupling is achieved when the wavevector conservation is maximally satisfied, i.e. $\sim \delta(k_m - k_F)$. For metallic armchair tubes, a larger coupling rate is obtained for $k_m \simeq 2\pi/3\sqrt{3}a_{cc}$, while a smaller metallic wavevector induces a smaller coupling rate. The tunneling rate from the metal to the nanotube can be effectively written as

$$1/\tau \sim \frac{2\pi}{\hbar}|\langle k_m|\mathcal{H}_{\text{contact}}|k_F\rangle|^2 \rho_{\text{NT}}(E_F)\rho_m(E_F), \qquad (4.50)$$

with $\rho_{\text{NT}}(E_F)$ ($\rho_m(E_F)$) the density of states of the nanotube (metal) at the Fermi level. Intriguingly, several experiments on metallic tubes have reported $G \simeq G_0$ at low bias, instead of the two theoretically predicted channels, assuming the π–π^* degeneracy at

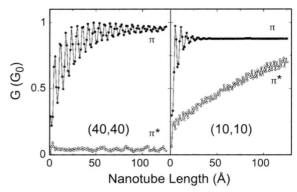

Figure 4.11 Contribution of π and π^* channels at CNP to the total nanotube conductance for armchair $(10, 10)$ and $(40, 40)$ tubes, and modeling the metal contact by a jellium. Adapted from Mingo & Han (2001). (Reprinted with permission from Mingo & Han (2001). Copyright (2001) by the American Physical Society.)

the charge neutrality point. This could be explained either by one channel becoming completely reflective or by a specific mismatch between the symetries of the incoming and outgoing states.

This issue was raised by Mingo and Han (Mingo & Han, 2001) who investigated such an imbalance in coupling strength between the π–metal (jellium) contact and the π^*–metal contact, by using quantum simulation with a Landauer–Büttiker approach. Figure 4.11 shows the contribution of the two channels to the conductance at the charge neutrality point as a function of nanotube length (for two different nanotube diameters). The transmission probability of the π^*–metal is clearly seen to almost vanish for a sufficiently large diameter, supporting the scenario of interface symmetry mismatch.

Metal/semiconducting nanotube/metal junctions
The different case of interfaces between metals and semiconducting nanotubes (M–SCSWNT–M junctions) deserves some particular consideration, given its central role in the operation of nanotube-based field effect transistors (see Section 4.2.4). Here, the formation of interface dipoles and Schottky barriers at the interface can produce very large contact resistance and a tunneling transport regime at low bias and low temperatures. The charge redistribution at the metal/semiconductor interface can be described by the band bending and existence of metal induced gap states. Those features strongly depend on the relative positions of the Fermi level and band edges of the metal and nanotube in contact, as discussed further below (Heinze *et al.*, 2002).

Several other papers have also emphasized the importance of the hybridization between carbon and metal orbitals at the contact (for instance Nemec, Tománek & Cuniberti, 2006), while other work has discussed the role of the Schottky barrier (Anantram & Léonard, 2006). The variations of the contact geometry (end, side, or melted), nanotube length, and metal type, can certainly explain why the experimental data are markedly scattered. Much effort is being devoted to controlling these contacts, as reported for example in Xia *et al.* (2011) where a transport efficiency of about 75% is achieved for palladium–graphene junctions. Moreover, while charge transfers, interface states, and the Schottky barrier physics are within the scope of *ab initio* simulations, computational limitations make it still impossible to compute any transport properties on μms long tubes (and in the presence of gate and bias voltages) at any degree of accuracy. Finally, for higher bias voltage between conducting probes, due to the potential drop profile along the tube, the modifications of bands along the tube axis produce additional backscattering (Anantram, 2000). This Bragg reflection is a fundamental point that could explain the experimental observation of limited turn-on current with increasing bias voltage (Poncharal *et al.*, 2002).

Schottky barriers and SWNT-based field effect transistors
Léonard and Tersoff pioneered theoretical studies on Schottky barriers in SWNT–metal contact interfaces, unraveling the fact that their peculiar nanoscale dimension and unconventional electrostatics (with poor screening effects) should result in a totally inefficient Fermi level pinning mechanism, a fact suggesting fine-tunability of Schottky-barrier height, eventually disappearing with the formation of a purely ohmic contact (Léonard & Tersoff, 1999, 2000*a*, 2000*b*, 2002). Nanoscale interfaces were thus envisioned as providing an unprecedented means to eliminate the inconvenient Schottky barrier hindering hole and electron injection. After tremendous efforts to precisely characterize the SB physics in carbon nanotube-based field effect transistors (SWNT-FETs) (Martel *et al.*, 2001), Javey and coworkers (Javey *et al.*, 2003) finally reported compelling experimental evidences of Schottky barrier suppression in SWNT-FET by using contacts between palladium and semiconducting nanotube with large enough tube diameter (Javey *et al.*, 2003, Kim *et al.*, 2005), a result which brought hope for the advent of all-carbon nanotube-device nanoelectronics (Tersoff, 2003).

Notwithstanding, there is still a fundamental need for in-depth understanding of nanoscale interfaces, since a large dispersion of experimental measurements is usually obtained (Léonard & Talin, 2011, Svensson & Campbell, 2011, Franklin & Chen, 2010, Anantram & Léonard, 2006). One problem lies in the experimental techniques commonly used to study contacts to bulk materials, which cannot be exploited at nanoscale, and poor statistics over devices shows a large discrepancy in the reported Schottky barrier heights as well as several contradictory conclusions (Martel *et al.*, 2001). In that perspective, accurate simulation of nanotube-transistor current–voltage characteristics are needed, for capturing the precise role of SWNT diameter and interface atomic structure of the metal/nanotube junction, as well as the contribution of chemical doping; since all those aspects eventually drive device control and performances. As argued by Tersoff (2003), the atomic-scale reasons for the disappearance of the SB-barrier for certain devices remain to be fully clarified. Many questions remain unsolved, such as: Why should palladium give a smaller Schottky barrier than platinum, gold, or titanium? The answer to this question should be sought in the chemical sticking properties of the metal to carbon structures, or in some combination of metal/carbon-dependent interfacial charging properties and long range electrostatics. To date, however, those questions are very challenging computationally, and would require self-consistent calculations fully based on first-principles methods to unveil the process of Schottky barrier formation for different types of nanoscale metal/nanotube interfaces. Besides, since ohmic contacts have been achieved for nanotubes with diameter above 2 nm (but with limited ON/OFF performances) (Kim et al., 2005), this theoretical knowledge would also be highly desirable for further monitoring and elimination of the SB in narrower tubes, whose electrical properties would be more favorable for practical devices.

4.3   Ballistic motion through a graphene constriction: the 2D limit and the minimum conductivity

The conductivity of a ballistic clean graphene has been discussed in the situation of a specific transport setup consisting of two heavily doped leads bridging a central (undoped) graphene region pinned at the Dirac point (charge neutrality point) and with geometry characterized by W and L, which are respectively the width and the length of the sandwiched graphene sample.

When considering a clean graphene sample in the ballistic regime where the electron mean free path $\ell_{el} \gg L$, the transport problem shares some similarities with Klein tunneling phenomena, where due to the chiral nature of the electrons, the transmission is always finite at specific angles through a p–n–p or n–n–n junction. One considers a graphene ribbon of length L and width W in the limit of short and wide armchair ribbons ($L < W$) with hardwall or smooth confining potential at the edges. The system is kept at energies around the CNP, and it is connected to two leads at high potential with a large number N of active conductive channels. By solving the noninteracting Dirac equation with such geometry, the conductivity is given by the number of evanescent modes, as $\sigma = (L/W)(4e^2/h) \sum_{n=1} T_n$, with $T_n = |1/\cosh(q_n L)|^2$, the transmission probability with q_n being the transverse wavevector, which can be rewritten as

$$\sigma = \frac{4e^2}{h} \frac{L}{W} \sum_{n=0}^{\infty} \frac{1}{\cosh 2[\pi (n + 1/2)L/W]} \longrightarrow \frac{4e^2}{\pi h} \quad (W \gg L). \quad (4.51)$$

Equation (4.51) can be derived using a twisted boundary condition $\Psi(y = 0) = \sigma_x \Psi(y = 0)$ and $\Psi(y = W) = -\sigma_x \Psi(y = W)$, where σ_x is a component of the Pauli matrix (Tworzydło et al., 2006). This particular boundary condition mimics the massless Dirac fermions inside the graphene sample but infinitely massive Dirac fermions in the leads. Now the above formula in the wide sample ($W \gg L$) limit converges to a universal value $\frac{4e^2}{\pi h}$, which is known as the *quantum limited conductivity* of graphene in the clean limit (as illustrated in Fig. 4.12). This theoretical result has been tested numerically by calculations based on a *tight-binding* model (Cresti, Grosso & Parravicini, 2007). This peculiar transmission property has been also confirmed experimentally (Miao et al., 2007).

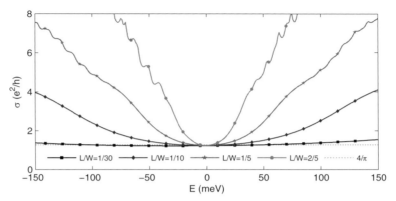

Figure 4.12 Conductivity versus energy for an armchair nanoribbon with lateral width $W = 150$ nm (and varying length in between highly doped source/drain contacts, $L = 5, 15, 30$, and 60 nm). The minimum conductivity $\frac{4e^2}{\pi h}$ is also drawn with a horizontal dashed line. Courtesy of A. Cresti.

4.4 Further reading and problems

- Regarding Klein tunneling we recommend the presentation in Katsnelson (2012). For a recent review on this subject, including the connection to phenomena in quantum optics, such as the Snell–Descartes law of refraction, total internal reflection, Fabry-Perot resonances and total internal reflection, we recommend Allain & Fuchs (2011).

Problems

4.1 *Potential step for Dirac fermions.* Consider Dirac fermions incident on a potential step of height V_0.
(a) Formulate three general conservations laws. (*Hint*: Remember that the Hamiltonian is time-independent and that it is translational invariant in one direction (y). What can you say about the current density along the direction perpendicular to the step?)
(b) Derive the transmission probability.

4.2 *Pseudospin conservation and the absence of backscattering.* Consider massless Dirac electrons entering a region where the potential is proportional to the identity matrix. Prove that for normal incidence they cannot be backscattered.

4.3 *Mode decomposition.*
(a) Following the lines of Section 4.2.2, deduce the mode decomposition for armchair SWNTs.
(b) Consider the mode-decomposition for armchair SWNTs and GNRs as well as zig-zag SWNTs given in Section 4.2.2. Determine the Hamiltonian for each of the independent modes that have (when possible) a nonvanishing DOS at the charge neutrality point. What can you say about the degeneracy close to this point?

(c) Compose the total density of states and the conductance by superposing the results for each of the independent modes for a $(5, 0)$ SWNT and for an armchair 8-aGNR.

4.4 *The zigzag challenge.* Starting from a simple *tight-binding* Hamiltonian and following the spirit of Section 4.2.2, you are challenged to obtain a suitable mode-decomposition for the case of zigzag GNRs. Can you do it? If not, explain why. You may also want to discuss your answer on our website.

4.5 *Fabry-Pérot conductance maps.* Following Section 4.2.3, we ask you to reconsider the setup of Fig. 4.7(a) and derive an expression for the Fabry-Pérot conductance map in Fig. 4.7(c). To this end we propose the following.
(a) Establish a simple model for the scattering matrices at the contacts and for the free propagation inside the sample. You may choose a model as the one proposed in the text or choose your own parameterization.
(b) Compose the partial scattering matrices in (a) to obtain the total scattering matrix for the device.
(c) Use these results to obtain the linear response conductance (Landauer) as a function of the gate and bias voltages.
(d) *Questions for further thought:* What would happen if the leads introduce inter-mode coupling? How would these results change if instead of a carbon nanotube one considers a graphene nanoribbon? What would be the effect of defects on the conductance patterns? (Suggested reading: Liang *et al.*, 2001, Jiang *et al.*, 2003.)

4.6 *Current noise for non-interacting electrons in the Fabry-Pérot regime.* Using the results of the previous exercise, determine the zero frequency noise for the setup of Fig. 4.7(a). (Suggested reading: Kim *et al.*, 2007, Herrmann *et al.*, 2007, Wu *et al.*, 2007.)

4.7 *Fabry-Pérot conductance maps revisited.* Let us reconsider the problem of the Fabry-Pérot conductance maps. This time you are encouraged to start from a simple *tight-binding* model for an infinite SWNT, which is divided into three regions; the central one of length L is going to be our resonant cavity and the rest the left and right leads. To simulate the weaker contact with the leads you may set a weakened carbon–carbon hopping between the leads and the central region, preserving the rotational symmetry of the system. Then by exploiting the mode decomposition of Section 4.2.2 try to obtain the conductance maps within the Landauer–Büttiker formalism. Discuss the limitations of the model.

** Additional exercises and solutions available at our website.

5   Quantum transport in disordered graphene-based materials

This chapter derives the main transport length scales and conduction regimes in disordered graphene-based materials, first focusing on model disorder (Anderson disorder, Gaussian impurities). Numerical implementations of the Kubo method presented earlier, and technically described in Appendix D, are validated by direct comparison with analytical results for both low-dimensional (nanotubes, nanoribbons) and two-dimensional graphene. Weak and strong localization phenomena are further studied, and some peculiar transport features in the quantum Hall regime are presented. Next, the effects of more specific defects, unique to graphene-based materials such as monovacancies, structural disorder, or grain boundaries are investigated, with a simplified *tight-binding* modeling. Finally, extensions to the study of phonon transport in disordered graphene, as well as some fundamental issues concerning Coulomb blockade are presented.

5.1   Elastic mean free path

The elastic mean free path (ℓ_{el}) is a key quantity in mesoscopic transport which dictates the crossover between ballistic and diffusive regimes. The behavior of ℓ_{el} in nanotubes and graphene nanoribbons exhibits unique scaling features, and can vary by orders of magnitude under a small Fermi level shift, owing to the close proximity of linearly dispersive bands and parabolic-like energy subbands. This allows spectacular tuning (using electrostatic gates or chemical doping) of transport regimes from ballistic to localization for the same sample.

It is first instructive to analyze the case of short range disorder, which allows illustration of common properties of transport length scales in all considered graphene-based low-dimensional materials. The Anderson potential for disorder is the most generic model for investigating localization phenomena in low dimension, being very convenient for both analytical derivations and numerical simulations. It was introduced by Anderson in the late fifties (Anderson, 1958). The Anderson disorder is a white noise (uncorrelated) disorder which is generally introduced through modulations of the onsite energies of a π orbital *tight-binding* Hamiltonian ($\varepsilon_\pi = \varepsilon_\pi + \delta\varepsilon_\pi$). The disorder strength is tuned by choosing randomly $\delta\varepsilon_\pi \in [-W/2, W/2]$ (with for instance a uniform probability distribution with $\mathcal{P} = 1/W$). For weak enough disorder, ℓ_{el} can then be derived analytically for both metallic carbon nanotubes and graphene nanoribbons

(Cresti *et al.*, 2008). We provide below the essential results. Let us first start with the case of 2D graphene, which offers the possibility for a straightforward analytical derivation, introducing the total density of states, approximated as

$$\rho(E) = \frac{2|E|}{\pi(\hbar v_F)^2}. \tag{5.1}$$

By writing $\ell_{\text{el}} = v_F \tau$, and using the Fermi golden rule (perturbation theory) to compute the elastic scattering time, τ ($\tau^{-1} = (2\pi/\hbar)\rho(E_F)W^2/12$), one finally readily obtains

$$\ell_{\text{el}} \propto 1/|E|, \tag{5.2}$$

which diverges when $|E| \rightarrow 0$. This crude estimation pinpoints a difficulty in calculating transport length scales when the Fermi level lies close to the Dirac point. A numerical calculation within the Kubo approach allows evaluation of ℓ_{el} at a quantitative level in 2D disordered graphene with Anderson scattering potential (numerical simulations are presented in Section 5.2.4).

In quasi-1D systems such as SWNTs and GNRs, scattering angles are restricted either to forward-scattering events at zero angle, which leads to momentum relaxation but does not affect the elastic transport length scale, or to backscattering events at an angle of π which thus monitor the behavior of ℓ_{el}. Using the Anderson disorder model, White and Todorov first derived an analytical formula for the low-energy elastic mean free path (ℓ_{el}) (White & Todorov, 1998, Roche *et al.*, 2000) using a two-bands model. A simple derivation is provided below, mainly following the path set out in White & Todorov (1998). For armchair metallic nanotubes, the scattering rate obtained in perturbation theory gives

$$\frac{1}{2\tau(E_F)} = \frac{2\pi}{\hbar} \left| \langle \Psi_{n1}(k_F)| \; \hat{U} \; |\Psi_{n2}(-k_F)\rangle \right|^2 \rho(E_F) \times N_c N_{\text{Ring}}, \tag{5.3}$$

with N_c and N_{Ring} the respective number of pair atoms along the circumference and the total number of rings taken in the unit cell (used for diagonalization), whereas the eigenstates at the Dirac point are given by

$$|\Psi_{n1,n2}(k_F)\rangle = \frac{1}{\sqrt{N_{\text{Ring}}}} \sum_{m=1,N_{\text{Ring}}} e^{imk_F} |\alpha_{n1,n2}(m)\rangle, \quad \text{with}$$

$$|\alpha_{n1}(m)\rangle = \frac{1}{\sqrt{2N_c}} \sum_{n=1}^{N_c} e^{\frac{2i\pi n}{N_c}} \left(|p_z^A(mn)\rangle + |p_z^B(mn)\rangle \right),$$

$$|\alpha_{n2}(m)\rangle = \frac{1}{\sqrt{2N_c}} \sum_{n=1}^{N_c} e^{\frac{2i\pi n}{N_c}} \left(|p_z^A(mn)\rangle - |p_z^B(mn)\rangle \right). \tag{5.4}$$

We consider here the simple case of an uncorrelated Anderson disorder defined by

$$\langle p_z^A(mn) | \; \hat{U} \; | p_z^A(m'n')\rangle = \varepsilon_A(m,n)\delta_{mm'}\delta_{nn'},$$

$$\langle p_z^B(mn) | \; \hat{U} \; | p_z^b(m'n')\rangle = \varepsilon_B(m,n)\delta_{mm'}\delta_{nn'},$$

$$\langle p_z^A(mn) | \; \hat{U} \; | p_z^A(m'n')\rangle = 0, \tag{5.5}$$

with $\varepsilon_B(m, n)$ and $\varepsilon_A(m, n)$ denoting the onsite energies of electron at atoms A and B in position (m, n), values which are taken at random within an interval $[-W/2, W/2]$ and with probability $\mathcal{P} = 1/W$. Replacing Eq. (5.4) in Eq. (5.3), and using Eq. (5.5), a simple calculation gives

$$\frac{1}{\tau(E_F)} = \frac{\pi \rho(E_F)}{\hbar} \left(\frac{1}{\sqrt{N_c N_{\text{Ring}}}} \sum_{N_c N_{\text{Ring}}} \varepsilon_A^2 + \frac{1}{\sqrt{N_c N_{\text{Ring}}}} \sum_{N_c N_{\text{Ring}}} \varepsilon_B^2 \right), \quad (5.6)$$

and finally

$$\ell_{\text{el}} = 18\sqrt{3}a_{cc}(\gamma_0/W)^2 N. \quad (5.7)$$

Such an expression shows that for fixed disorder strength, ℓ_{el} upscales linearly with the nanotube diameter, an unusual property suggesting a ballistic regime in the limit of very large diameter (or equivalently two-dimensional graphene). In Section 5.1.1, experimental data evidencing micrometers long mean free paths up to room temperature confirm such exceptional conduction capability of metallic carbon nanotubes.

As typical parameters, if considering an armchair $(5, 5)$ nanotube, with disorder $W = 0.2\gamma_0$, and after applying Eq. (5.7), $\ell_{\text{el}} \sim 550$ nm, a value much longer than the circumference, thus indicating a ballistic motion for long distances. Figure 5.1 shows ℓ_{el} in armchair metallic nanotubes with increasing diameters (from $(5, 5)$ to $(30, 30)$), using the Kubo approach implemented within the order N method (see Section 3.4.4).

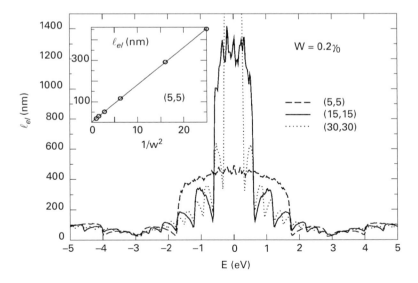

Figure 5.1 Energy-dependent mean free path as a function of diameter. Inset: ℓ_{el} versus W showing the $1/W^2$ scaling Fermi. (Reprinted with permission from Triozon *et al.* (2004). Copyright (2004) by the American Physical Society.)

The results fully validate the predicted scaling law with tube diameter in the vicinity of the Dirac point (with for instance $\ell_{el}(15, 15)/\ell_{el}(5, 5) = 3$).

An additional remarkable feature lies in the strong energy dependence of ℓ_{el}, particularly close to onsets of new subbands (or van Hove singularities). Besides, for higher energy subbands, the $1/W^2$ remains, but ℓ_{el} are found to be much smaller, without a linear scaling property with diameter. Such a large tunability of transport length scales upon small energy level shift was the origin of intense debate concerning the inherent transport mechanisms in single and multiwalled carbon nanotubes in the late nineties. The ballistic or diffusive nature of transport is here shown to be highly dependent on the energy-dependent transport mechanism, beyond the nature and strength of superimposed disorder. We note that multiwalled carbon nanotubes were shown to be particularly interesting, owing to intrinsic incommensurability between neighboring shells, allowing for the emergence of a diffusive regime, and quantum interference phenomena in the limit of ultraclean systems (Roche & Saito, 2001, Roche *et al.*, 2001). Concerning graphene nanoribbons, ℓ_{el} can also be derived in a similar fashion in metallic armchair graphene nanoribbons as (Areshkin, Gunlycke & White, 2007)

$$\ell_{el} = 12(\gamma_0/W)^2(N+1)a_{cc}, \qquad (5.8)$$

which also scales linearly with the ribbon width and follows $\ell_{el} \sim 1/W^2$.

In conclusion, low-dimensional (metallic) graphene-related systems exhibit mean free paths that may diverge with increasing diameter or ribbon width for a fixed disorder strength W. Notwithstanding, only armchair nanotubes truly behave as 1D massless Dirac fermions close to the charge neutrality point, since gaps form for all types of GNRs when edge boundary conditions are properly taken into account (chemical passivation of unsaturated dangling bonds, edge reconstruction, etc.). In that sense, armchair metallic nanotubes present the unique case of one-dimensional ballistic conductors up to room temperature.

5.1.1   Temperature dependence of the mean free path

The results presented in the section above have established the zero-temperature limit for ℓ_{el}. Although the coupling between electrons and acoustic vibrational degrees of freedom is weak at low temperatures, the temperature dependence of the mean free path can only be derived when properly taking into account inelastic scattering mediated by electron–phonon coupling. Besides, the temperature dependence of a nanotube resistance can also be driven by the electrical bias-induced excitation of phonon modes whose characteristics will depend on applied bias voltage, regardless of the temperature of the surrounding experimental setup.

In the low-temperature (and low-bias) regime, only acoustic modes play some role, which can be accounted for using perturbation theory (Fermi golden rule) and the Boltzmann transport equation (Lazzeri & Mauri, 2006). Alternatively, the effect of low-energy vibrational disorder (introducing time-dependent lattice distortions) can also be explored through the Kubo approach, using a time-dependent renormalization of the off-diagonal coupling matrix elements of the Hamiltonian (Ishii *et al.*, 2010). As a result, an inelastic (temperature-dependent) mean free path can be calculated: we named ℓ such a mean free path, which differs from ℓ_{el}, which gives the zero-temperature limit.

Technically, the time-dependent atomic displacements obtained by molecular dynamics can be transfered to renormalization of π–π off-diagonal coupling elements $\gamma_{ij}(t)$, which thus encode the electron–phonon interaction in both harmonic and anharmonic regimes, as discussed in Gheorghe, Gutiérrez & Ranjan (2005). A possible starting point is the empirical form $\gamma_{ij}(t) = \gamma_{ij}^0|\mathbf{R}_i^0 - \mathbf{R}_j^0|^2/|\mathbf{R}_i(t) - \mathbf{R}_j(t)|^2$ (Harrison, 1989), with $\gamma_{ij}^0 = 2.5\text{eV}$, where $\mathbf{R}_i(t)$ represents the atomic position at time t and $\mathbf{R}_i^0$ is at equilibrium. The phonon-vibration effects are accounted for from the molecular dynamics (MD) simulation using the Brenner–Tersoff potential for C–C bonds (Brenner, 1990). For fixed temperature T, the velocities of carbon atoms are normalized at each time step by the conditions of $\sum_{i=1}^{N_c} M_c\dot{\mathbf{R}}_i^2/2 = 3N_ck_BT/2$, where M_c and N_c are mass and number of carbon atoms, and k_B is the Boltzmann constant (Ishii et al., 2009).

Figure 5.2 shows the computed resistances of $(5, 5)$ SWNTs at $T = 0$ K and $T = 60$ K as a function of SWNT length L and for $W = 0.2$ (Anderson disorder). It is interesting to observe that the decay of the diffusion coefficient at zero temperature (inset) due to localization effects is fully suppressed at $T = 60$ K (as expected when introducing decoherence effects). Importantly, the length-dependent resistance pinpoints the crossover from a ballistic-like (length-independent) to a diffusive behavior, in which the increase of resistance scales linearly with tube length. The crossing point of the two asymptotic lines for ballistic and diffusive regimes enables estimation of the mean free path, $\ell \sim 0.4~\mu$m at $T = 60$ K (for $W = 0.2$) (Ishii et al., 2010).

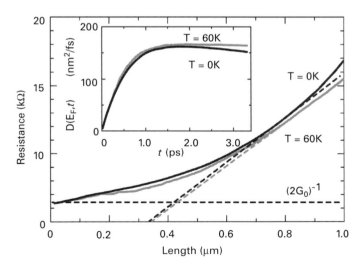

Figure 5.2 Length dependence of the total resistance for a $(5, 5)$ SWNT at 0 K (black line) and 60 K (gray line) with Anderson disorder potential $W = 0.2$. $(2G_0)^{-1}$ is also shown (horizontal dashed line), with $G_0 = 2e^2/h$. The other dashed line pinpoints the crossover from the ballistic to the diffusive regime. Inset: Time dependence of the diffusion coefficient for the same parameters. (Figure adapted from Ishii et al. (2010). Copyright (2010) by the American Physical Society.)

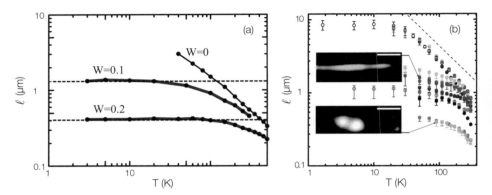

Figure 5.3 (a) Temperature dependence of the mean free paths for a $(5, 5)$ SWNT with both dynamical disorder and several strengths of the static disorder potential W. (Reproduced with permission from Ishii *et al.* (2010). Copyright (2010) by the American Physical Society.) (b) Mean free path for several nanotubes (metallic and semiconducting). Most metallic SWNTs (open circles) saturate at higher values than that of semiconductors (closed circles), T^{-1} (dashed line). Insets: scanning gate microscopy images taken on two different devices. Less current intensity is indicated by brighter color (grey). Defects are highlighted by the bright region (suppressed current). Scale bar is 500 nm. (Reproduced with permission from Purewal *et al.* (2007). Copyright (2007) by the American Physical Society. Courtesy of Philip Kim.)

In Fig. 5.3(a), the logarithm plot of temperature-dependent $\ell(T)$ is shown for $W = 0$, 0.1, and 0.2 with increasing temperature. The low-temperature behavior of $\ell(T)$ is clearly fixed by the static disorder strength ($T < 50$ K), whereas electron–phonon scattering events dominate in the high-temperature regime, whatever the static disorder strength. The scaling behavior $\ell(T) \sim T^{-1}$ obtained numerically is in perfect agreement with Fermi's golden rule (Suzuura & Ando, 2002) and with experimental data (Purewal *et al.*, 2007) shown in Fig. 5.3(b).

5.1.2   Inelastic mean free path in the high-bias regime

At very high temperature or in the high-bias voltage regime (that is for $V_{\text{bias}} \geq 0.2\text{V}$), the contribution of inelastic phenomena eventually yields strong current saturation. This was first reported by Yao and coworkers in metallic tubes (Yao, Kane & Dekker, 2000). In this experiment, a low-bias linear current–voltage characteristic is first observed, followed by a current saturation for $V_{\text{bias}} \sim 1$ volt, regardless of the temperature of the sample. The overall $I(V)$ response function is actually well described by a phenomenological law,

$$I = \frac{V}{R_0 + V/I_0},\tag{5.9}$$

with R_0 and I_0 two constants giving respectively a voltage-independent intrinsic resistance and a current saturation value in the order of $20 - 130\ \mu\text{A}$ (regardless of the tube diameter). The interpretation assumes that the saturation comes from the strong inelastic backscattering of electrons coupled to optic (or zone boundary) vibrational modes.

In this scenario, the inelastic mean free path is the propagating length needed for electrons to accumulate an additional energy of $\sim\hbar\Omega_{ph}$, the relevant optical phonon mode that will instantaneously produce electron backscattering. Assuming a linear potential drop between voltage probes separated by L, i.e. $V(l) = V_{\text{bias}}(1 - l/L)$, where l is the distance from the source, then the energy gain reads $e\int_0^{\ell_{ie}} \partial V/\partial l \, dl = \hbar\Omega_{ph}$, from which one finds $\ell_{ie} = (\hbar\Omega_{ph}L)/eV_{\text{bias}}$. Besides, assuming that the length-dependent resistance can be split into two contributions, by virtue of the Mathiessen rule, one can write

$$R(L) = \frac{h}{4e^2}\frac{L}{\ell_{\text{el}}} + \frac{h}{4e^2}\frac{L}{\ell_{ie}} = R_0 + \frac{h}{4e^2}\frac{eV_{\text{bias}}}{\hbar\Omega_{ph}} = R_0 + \frac{V_{\text{bias}}}{I_0}$$

with $R_0 = h/4e^2 L_{\text{tube}}/\ell_{\text{el}}$ the intrinsic resistance and ℓ_{el} the elastic mean free path, whereas $I_0 = (4e/h)\hbar\Omega_{ph}$, which is indeed in the range $20-30$ μA depending on the chosen phonon energy. Note that if ℓ_{el} is an intrinsic measure of the elastic disorder strength (defect density, etc.) which is a bias and temperature independent quantity, ℓ_{ie} is voltage dependent in this model, but at a fixed voltage it should remain inversely proportional to the nanotube length. Intriguingly however, by using a semiclassical Boltzmann approach, a fitting of the experimental data is achieved taking a fixed ℓ_{ie} (for a fixed tube length) independent of the voltage bias (Yao et al., 2000).

The observation of a length-dependent scaling of the resistance has been reported in several experimental works (Javey et al., 2004, Park et al., 2004). Lower resistance was measured for shorter tubes, and current saturation was shown to be reduced when decreasing L_{tube} from 700 nm to 50 nm, at which no saturation was observed for voltage bias up to 1.5 volts. Again, the analysis within the Fermi golden rule and Boltzmann approach allows extrapolation of some typical values for the inelastic scattering lengths that can be as short as 10 nm (Javey et al., 2004).

However, the discrepancy or strong fluctuations between theoretical estimates obtained by fitting procedures and the computed ℓ_{ie} (with perturbative theory), even when using ab initio calculations (Lazzeri et al., 2006, Lazzeri & Mauri, 2006), raises some fundamental questions about the applicability of Fermi's golden rule, the Mathiessen rule and semiclassical transport theory to tackle inelastic quantum transport in metallic carbon nanotubes, or within the context of carbon nanotubes-based field effect transistors (Appenzeller et al., 2004). The most probable scenario is that a scheme beyond the widely used Boltzmann equation is needed in situations like this, which lie between the fully decoherent and fully coherent regimes.

Furthermore, nonperturbative effects of electron–phonon interaction, which are beyond the validity of the Boltzmann approach, may also emerge. Indeed, although the Peierls distortion mechanism is ineffective in metallic SWNTs to produce a semiconducting state even at very low temperatures (Saito et al., 1992b, Mintmire, Dunlap & White, 1992) (except possibly for very short radius SWNTs (Connétable et al., 2005)), modifications in the phonon band structure due to a related mechanism, the Kohn anomaly (Kohn, 1959), are observable at moderate temperatures in both SWNTs (Farhat et al., 2007) and graphene (Pisana et al., 2007). A related Peierls-like mechanism was proposed at high bias, leading to non-equilibrium gaps at half the optical phonon energy above/below the charge neutrality point ($\pm\hbar\Omega/2$), which in turn would be observable as a small plateau in the current–voltage characteristics. The theoretical method adopted to tackle such phenomena employs the generalized Landauer–Büttiker formula expanded in a higher dimensional space (the electron–phonon Fock space) (Anda *et al.*, 1994, Bonča & Trugman, 1995). This approach is able to cope with the strong coupling of electrons to certain symmetry-selected phonon modes, a regime out of reach of perturbative methods and semiclassical transport concepts. We refer to Foa Torres & Roche (2006) and Foa Torres, Avriller & Roche (2008) for more details.

5.1.3 Quantum interference effects and localization phenomena in disordered graphene-based materials

A knowledge of the mean free path ℓ_{el} in disordered graphene-related systems is a first essential step since it allows identification of the frontier between the ballistic and the diffusive propagation of wavepackets. The localization length ξ is the other physical length scale that defines the transition towards the insulating regime, in which the conductance further decays exponentially with the system length as $G \sim G_0 \exp(-L/\xi)$ (Section 3.4.5).

Weak localization phenomena have been clearly observed in multiwalled carbon nanotubes with diameter ranging from ~ 3–20 nm (Stojetz *et al.*, 2005, Bachtold *et al.*, 1999) as well as in graphene nanoribbons with widths in the order of $\sim 200 - 500$ nm (Tikhonenko *et al.*, 2008). Weak antilocalization (WAL) has also been observed in graphene-based materials and relate to the pseudospin-related Berry's phase interferences, and induced sign reversal of the quantum correction (Section 3.4.5). Also, the transition from weak antilocalization to weak localization has also been reported for reduced ribbon width (Tikhonenko *et al.*, 2008). As the width of the graphene ribbons is reduced from ~ 20 nm down to ~ 5 nm, weak antilocalization is eventually suppressed, owing to an increasing contribution of edge defects and enhanced contribution of other disorder sources (topological, vacancies, adsorbed impurities, etc.).

It is instructive to analyze the variation of quantum transport features with varying dimensionality or defect-induced broken symmetries. Weak localization effects are generally revealed experimentally by tuning the strength of quantum corrections through the application of an external magnetic field (as explained in Section 3.4.5). However, for low-dimensional systems such as carbon nanotubes or graphene ribbons (with typical diameters or widths ≤ 10 nm), large magnetic fields also severely affect the electronic band structures, making the analysis of resulting magnetofingerprints much more difficult. Some illustrative examples have been presented in multiwalled carbon nanotubes (Stojetz *et al.*, 2005, Bachtold *et al.*, 1999), and in graphene nanoribbons (Ribeiro *et al.*, 2011, Poumirol *et al.*, 2010).

We provide here the most representative quantum localization effects in graphene nanoribbons. As seen in Chapter 2, zigzag-type GNRs display quite peculiar electronic properties, with low-energy wavefunctions sharply localized along the ribbon edges. Using the Landauer–Büttiker approach, the scaling properties of the quantum conductance of these systems can be numerically investigated. The typical energy dependence of conductance profiles for both zGNR and aGNR of width $\sim$20 nm are shown in Fig. 5.4(a) and Fig. 5.4(b), for the clean, weak disorder ($W = 0.5$) and strong disorder ($W = 2$) limits. For weak disorder ($W = 0.5$, Fig. 5.4(a)), transport in aGNR seems much less altered than the behavior displayed by zGNR. This contrasts with the case of stronger disorder ($W = 0.5$), where the conductance of both aGNR and zGNR are strongly reduced at low energies, with similar fluctuations indicating strong localization (Fig. 5.4(c)). Figures 5.4(b) and 5.4(d) show the exponential damping of averaged conductances (over $\sim$400 different disorder configurations), i.e. $\langle \ln G/G_0 \rangle \sim L/\xi$. By fitting these numerical results, ξ values in zigzag ribbons are found to be smaller than armchair ribbons by up to two orders of magnitude at high enough energies. In contrast, for larger disorder strength, such as $W = 2$ (Fig. 5.4(c)), the localization lengths for

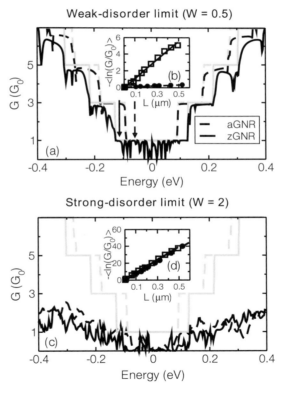

Figure 5.4 (a) Conductance for a single disorder configuration of a zigzag (solid black line) and an armchair (dashed black line) GNR with width $\sim$20 nm ($W = 0.5$). Grey lines correspond to ideal zigzag (solid line) and armchair (dashed lines) ribbons. (b) Configuration averaged (over $\sim$400 samples) normalized conductance as a function of GNR length for both zigzag and armchair GNRs. The solid (dashed) arrow shows the energy at which the calculations for the zGNR (aGNR) have been performed. (c) and (d) Same information as for (a) and (b) but for a larger disorder strength ($W = 2$). (Adapted from Lherbier *et al.* (2008) by courtesy of Blanca Biel.)

both types of ribbons become almost indistinguishable, showing that edge symmetry then loses its integrity.

An important result of mesoscopic physics (presented in Section 3.4.5) is the existence of a fundamental relationship between ℓ_{el} and ξ (referred to as the Thouless relation (Thouless, 1977)). In a strictly 1D system, it can be analytically demonstrated that $\xi = 2\ell_{el}$, whereas for quasi-1D systems (with $N_\perp(E)$ conducting channels), the relation is generalized as

$$\xi(E) = [\beta(N_\perp(E) - 1)/2 + 1]\ell_{el}(E), \qquad (5.10)$$

with β a factor dependent on the time-reversal symmetry (Beenakker, 1997). Avriller and co-workers (Avriller *et al.*, 2007) have extensively confirmed the applicability of such a foundational relation between transport length scales in chemically doped metallic carbon nanotubes, using the Landauer–Büttiker conductance method.

5.1.4   Edge disorder and transport gaps in graphene nanoribbons

Low-temperature conductance measurements in the Coulomb blockade regime display large fluctuations (Stampfer *et al.*, 2009) (see also Section 5.8.2) with an enhanced depletion of the conductance at low energy, referred to as a transport (or mobility) gap. The origin of such a transport gap has been debated theoretically, especially regarding the role of edge disorder induced localization effects (Areshkin *et al.*, 2007, Cresti & Roche, 2009, Evaldsson *et al.*, 2008, Wimmer *et al.*, 2008, Mucciolo, Castro Neto & Lewenkopf, 2009, Akhmerov & Beenakker, 2008). As shown below, the topological complexity of edge imperfections observed experimentally needs to be accounted for when analyzing the transport properties in edge-disorder GNRs.

In what follows, we discuss the transport properties in edge disordered GNRs using Green's function technique (Cresti, Grosso & Parravicini, 2007). As an illustration, we focus on the impact of various edge disorder configurations by scrutinizing the conductance properties of several 16-zGNRs (with length $L = 500$ nm), with randomly removed carbon edge atoms with equal probability 7.5%, and varying complexity of the edge defects topology (see Fig. 5.5). The disorder profile is defined by the probability P, controlling the number of defects on each edge to $P \times L/a$, where $a = 2.46$ Å is the lattice parameter (within a precision of 2%). The edge disorder profile is developed from a pristine zGNR and by removing edge carbon atoms randomly (as depicted in Fig. 5.5). The probability of removing atoms is chosen such that the total number of defects remains proportional to P. Transport properties for different defect types and comparable disorder strength can therefore be contrasted. Figure 5.5(a) denotes the richest edge defects profile which contains Klein defects (single dangling edge atoms, defect D1), together with missing hexagon defects, either one (defect D2), or two (defect D3).

Depending on the topology of edge defects, large transport fluctuations are obtained. Figure 5.5(a) shows the 16-zGNR conductance profile with the highest disorder complexity (largest variety of edge defects). The strong suppression of transmission in the

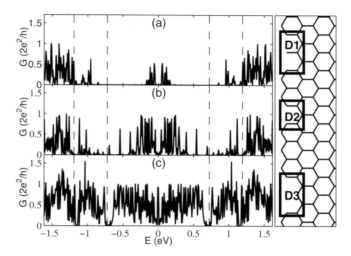

Figure 5.5 Left: conductance of a disordered 16-zGNR (with length $L = 500$ nm) with 7.5% of randomly removed edge carbon atoms. Case (a) includes dangling atom defects (D1), single (D2) and double (D3) missing hexagon defects, whereas D1 is prohibited in case (b) and D1 and D2 are disallowed for case (c). Right: a disordered ribbon edge with D1, D2, and D3 defects shown in boxes. (Reproduced with permission from Cresti & Roche (2009). Copyright (2009) by the American Physical Society.)

first plateau (region marked by two vertical lines close to $E = 0$), suggests an Anderson insulating regime with a localization length $\xi \lesssim L$. The conductance of a disordered ribbon in which Klein defects (D1) are discarded is shown in Fig. 5.5(b). In that case, the conductance remains large in the first plateau, but appears more suppressed at higher energies. Finally, by removing the possibility of both Klein defects and single missing hexagons (D1 and D2), a completely different transport regime forms (for the same length of the graphene ribbon). Figure 5.5(c) shows that the conductance is very close to $G_0 = 2e^2/h$, which gives the ballistic limit of the clean system (Wakabayashi *et al.*, 1999). The conductance can thus range from a localized to a quasi-ballistic regime depending on the local defect complexity and ribbon length. The structural property of local edge defects has therefore a genuine impact on resulting transport properties of GNRs.

The average conductance $G = \overline{T} \times (2e^2/h)$, where $\overline{T}$ is the averaged transmission coefficient over 1000 different configurations, is shown in Fig. 5.6, for 16-zGNR with single missing hexagons distributed with a probability $P = 7.5\%$. The quantized conductance of pristine GNRs sets the limit at $L = 0$. With increasing L, the transport regime evolves from a diffusive to an insulating state evidenced by the exponential decay of the conductance (Fig. 5.6). The crossover between diffusive and localization regimes is identified by comparing $\Delta(T)/\overline{T}$ and $\Delta(\ln T)/\overline{\ln T}$ (Δ stands for the standard deviation). The *transport (or mobility) gap* of the GNR is defined by an energy region for which $G < 0.01 \times 2e^2/h$. Here, such a transport gap is obtained when $L \geq 470$ nm

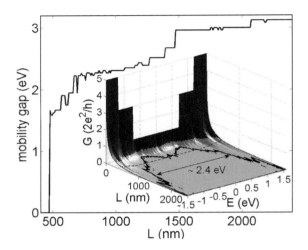

Figure 5.6 Transport gap evolution with L. Inset: Average conductance (1000 different configurations) of disordered 16-zGNR (defects are single missing hexagons with density $P = 7.5\%$). The bold line marks the (L, E) region for which $G < 0.01 \times 2e^2/h$. The L dependence of the transport gap width starts from $L \geq 470$ nm (dashed line). A particular value is shown (double arrow line) at $L = 1250$ nm. (Reproduced with permission from Cresti & Roche (2009). Copyright (2009) by the American Physical Society. Courtesy of Alessandro Cresti.)

(dotted line), but then further enlarges with the ribbon length, owing to the accumulation of quantum interferences. The existence of highly chemically reactive and disordered edges enhances the transport gap fluctuations, yielding a clear caveat against using nanoribbons in nanoelectronic devices, which would require perfectly controlled current–voltage characteristics, with very small sample-to-sample fluctuations.

5.2 Transport properties in disordered two-dimensional graphene

5.2.1 Two-dimensional disordered graphene: experimental and theoretical overview

The nature of disorder in graphene and its impact on transport properties remains a complex and debated issue. First, this is because there exist various sources of disorder introduced by the material and device fabrication techniques (graphite exfoliation, epitaxial or CVD growth, transfer to other substrate, electrical current cleaning, etc.). These methods produce either local or more long range lattice imperfections – defects, impurities, ripples and long range strain deformations – which impact on graphene electronic and transport properties in different ways, also depending on the charge densities and magnitude of screening phenomena. The possibility to observe unique transport features such as Klein tunneling, electron collimation or weak antilocalization phenomena is unique in condensed matter, but strongly sensitive to the nature of crystalline imperfections, and the atomic range of the corresponding disorder potential.

Defects in exfoliated graphene transfered to a silicon dioxide substrate generally range from local structural imperfections or glued adatoms to long range Coulomb scattering potentials produced by charged impurities (trapped in the oxide) or ripples quenched by substrate roughness after graphene deposition. In the high density limit, a nearly linear carrier-density-dependent conductivity $\sigma(n) \sim n$ has been observed experimentally (Yan & Fuhrer, 2011) (and convincingly interpreted with semiclassical physics, Section 3.3). Close to the Dirac point, when graphene is deposited onto SiO_2, the formation of so-called electron/hole puddles (which are spatially fluctuating charge densities) produces percolation transport which precludes the transition to the formation of the expected insulating state.

For clean enough graphene, the Dirac point conductivity usually remains finite even down to cryogenic temperatures, but its value is not universal and varies from sample to sample (depending on sample quality and precise preparation process). Significantly-low electron–phonon scattering has been experimentally reported (Chen *et al.*, 2008), implying that static disorder (especially close to the Dirac point) dominates low-temperature resistivity. Figure 5.7 (right inset) shows typical behavior for a pristine graphene sample in a large density range, which can be well fitted with $\sigma(n) = (1/(ne\mu_L) + \rho_S)^{-1}$ with $\mu_L \sim 26000$ cm^2/Vs, and $\rho_S = 53$ Ω, while the

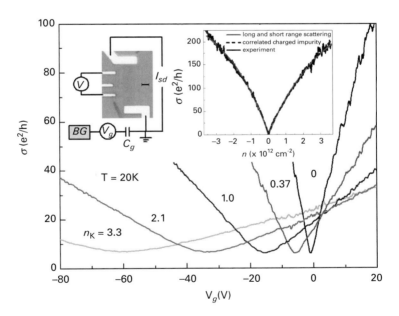

Figure 5.7 Conductivity versus gate voltage (at K) for pristine (undoped) graphene (right inset), and chemically doped graphene (main plot). Different curves are for potassium coverage density ranges within $[0.37, 3.3] \times 10^{12}$ cm^{-2}. Top-left inset is an optical microscope image of the monolayer graphene device, with a schematic of the measurement circuit. Courtesy of M. Fuhrer. (Reproduced with permission from Yan & Fuhrer (2011). Copyright (2011) by the American Physical Society.)

density is extracted from gate voltage V_g using $n = C_g(V_g - V_g^{min})/e$ (V_g^{min} being the value at which conductivity is minimum, while typical gate capacitance $C_g = 11\,nF/cm^2$) (taken from Chen *et al.*, 2008).

A popular quantity to characterize graphene structural quality is the charge mobility, which at zero temperature is given by $\mu(E) = \sigma_{sc}(E)/en(E)$, where $\sigma_{sc} = e^2\rho(E)v(E)\ell_{el}$ is the semi-classical conductivity deduced from the Einstein formula, with $\rho(E)$ the DOS, $n(E)$ the charge density at energy E, ℓ_{el} the elastic mean free path, and e the elementary charge (assuming zero temperature). Close to the charge neutrality (or Dirac) point, the measured experimental conductivities are mostly found to range within $\sim 2 - 5e^2/h$ although the charge mobility can vary by almost one order of magnitude (Zhang *et al.*, 2006, Oezyilmaz *et al.*, 2007, Jiang *et al.*, 2007). This effect has been attributed to the change of charge density due to the doping from the substrate and/or contacts.

On the theoretical side, the calculation of the Kubo conductivity for 2D graphene with short range disorder, and within the self-consistent Born approximation, yields $\sigma_{xx}^{min} = 4e^2/\pi h$ (h is the Planck constant) for the two Dirac nodes (Shon & Ando, 1998) (see Section 5.2.4), which is typically $1/\pi$ smaller than most of the experimental data. Numerical calculations using the Kubo formula confirm such a prediction (Nomura & MacDonald, 2006, Lherbier *et al.*, 2008, Lherbier *et al.*, 2011). Differently, by assuming that elastic scattering is dominated by a screened Coulomb potential (ionized impurities), Nomura & MacDonald (2006) have numerically reproduced the low-energy dependence of the electronic conductivity using a full quantum approach of the Kubo formula and performing a finite-size scaling analysis. They found that $\sigma_{xx}^{min} \sim e^2/h$ close to the Dirac point, in better agreement with most experiments. Other calculations have analyzed the effect of screened Coulomb potential on semiclassical Bloch–Boltzmann conductivity, taking into account the role of background zero potential fluctuations (electron–hole puddles) (Das Sarma *et al.*, 2011). These space-dependent charged inhomogeneities provide percolation paths for the propagating charges down to the zero density limit, which prohibit the exploration of transport at the Dirac point and yield a nonuniversal minimum conductivity which is sample dependent.

Unique transport features develop in graphene with low disorder. The Klein tunneling (KT) mechanism (see Section 4.1) is the first intrinsic and spectacular manifestation of massless Dirac fermion physics where backward reflection is partially or totally suppressed (depending on the incident angle of the incoming wavepacket and the height and width of the barrier) when charge crosses a local tunneling barrier. This result contrasts with usual behavior in Schrödinger physics, in which the propagation of an electronic wavepacket is exponentially damped when crossing a barrier of increasing width. Klein tunneling stands as an efficient mechanism to suppress localization effects, provided that the impurity potential is sufficiently long range to prohibit intervalley scattering between the two inequivalent Dirac cones. The KT mechanism is inherent to the symmetric electron–hole electronic band structure and the pseudospin degree of freedom associated with the AB sublattice degeneracy. In metallic carbon nanotubes (Ando, Nakanishi & Saito, 1998), a similar mechanism leads to a total suppression of backscattering, enforcing a ballistic motion of charges up to micrometers up to room temperature.

Pseudospin shares similar symmetries with the spin degree of freedom, and as such, the corresponding wavefunctions acquire extra phase factor which, when accumulated along a closed trajectory, produces an interference pattern dominated by a Berry's phase. The observation of the weak antilocalization (WAL) in graphene (Tikhonenko *et al.*, 2008, 2009) is such a manifestation of pseudospin effects on phase interferences. In 2006, McCann and coworkers provided a solid theoretical derivation of the Cooperons quantum correction in the presence of pseudospin-related Berry's phase factors (McCann *et al.*, 2006). The nature of disorder and contribution of intravalley versus intervalley scattering events is actually crucial for clarifying the predominance of WL versus WAL phenomena, as discussed later (see Section 5.2.6 for details). The possibility of strong (Anderson) localization is also inherent to the nature of disorder, and is expected to follow the weak localization regime, as dictated by the scaling theory of localization (Lee & Ramakrishnan, 1985).

Finally, intentional doping of graphene is a versatile way to tune electronic properties. The first experimental attempt was made using a controlled flux of atomic potassium, introducing physisorbed atoms to the graphene substrate, and resulting in a Fermi level shift and conductivity decrease. Figure 5.7 illustrates such a chemical doping effect driven by the physisorption of potassium atoms, which produce scattering centers and weak electron transfer to the graphene substrate. By varying the potassium coverage density ranges within $[0.37, 3.3] \times 10^{12}$ cm^{-2}, a continuous degradation of the conductivity is observed together with an energy downshift of the minimum conductivity (measured at the Dirac point) driven by charge transfer effects (Chen *et al.*, 2008, Yan & Fuhrer, 2011). In parallel, theoretical calculations suggested that boron and nitrogen impurities in substitution of carbon atoms would bring more spectacular impact in conduction, such as electron–hole transport asymmetry (Lherbier *et al.*, 2008), or mobility gaps in doped graphene nanoribbons (Biel *et al.*, 2009*a*).

In the next sections, we present the various possible sources of disorder in two-dimensional graphene and their specific impact on transport properties. We start from the more academic (but generic) short range potential such as the Anderson disorder or long range potential profile (Coulomb potential), and continue with more structurally and chemically invasive defects such as adsorbed oxygen and hydrogen atoms (see Section 7.4). The latter allow stronger tunability of disorder strength, functionalities (such as intrinsic magnetism) and transport regimes in graphene materials. General analytical results are derived either using the semiclassical Boltzmann transport equation or its quantum generalization (Kubo–Greenwood method). The main transport characteristics in various forms of disordered graphene-related systems are ascertained, and the limitations of the semiclassical transport description are given and illustrated. We also discuss how weak antilocalization phenomena develop and what governs the crossover from weak to strong localization, using a comprehensive disorder model. This is followed by an analysis of the impact of structural defects such as monovacancies and divacancies, as well as grain boundaries in polycrystalline graphene which fix an intrinsic limit for charge mobilities in CVD-grown graphene. Finally, quantum transport in strongly damaged graphene is discussed (Chapter 4), with a focus on defects such as monatomic oxygen and hydrogen adsorbed defects which are the focus of many recent experiments.

5.2.2   Metallic versus insulating state and minimum conductivity

The effect of disorder in metallic materials has been a greatly debated issue. A key question has been to understand how a metallic system will change to an insulating state with the disorder increases in the material. Sir Nevill Mott proposed that the metallic state persists as long as the mean free path remains larger than the Fermi wavelength (Mott, 1990). Beyond that scale, the system was proposed to undergo a discontinuous transition to the insulating regime, with vanishing conductivity at zero temperature. Assuming the conductivity as $\sigma = \frac{e^2}{h} k_F \ell_{el}$, this argument leads to a minimum conductivity given by $\sigma_{min} = \frac{e^2}{h} k_F \lambda_F \sim \frac{e^2}{h}$ (known as the Ioffe–Regel criterion), before it abruptly vanishes at the metal–insulator transition.

It is worth noting that the effect of disorder in graphene is a rather delicate issue since the applicability of usual perturbative treatments (Fermi golden rule) and validity of a semiclassical approach (Bloch–Boltzmann) become questionable close to the Dirac point in many regards. First the assumption of weak disorder requires $\lambda_F \ll \ell_{el}$ which becomes problematic close to the Dirac point since $\lambda_F \sim 2\sqrt{\pi/n} \to \infty$ as $n \to 0$ (when approaching the Dirac point). In the presence of disorder, the density of states is, however, increased compared to the clean case and charge density is then not strictly going down to zero. As discussed and illustrated in this section, different types of disorders produce a varying nature of electronic states close to the Dirac point, with inequivalent degree of localization depending on the underlying broken symmetries.

Weak Coulomb scatterers (charges trapped in the oxide) generate real space or charge inhomogeneities which, in the vicinity of the Dirac point, yield the formation of electron–hole puddles on the scale of about 30 nm. The absence of valley mixing then induces maximum effects of the Berry's phase, Klein tunneling and antilocalization phenomena. The observed minimum conductivity in graphene, and absence of a localization regime, has been related to percolation transport conveyed by Klein-tunneling mechanisms (Katsnelson, 2012, Chenaiov et al., 2007). In contrast, sharp defects such as vacancies or other types of structural defect with strongly broken local symmetries promote the predominance of intervalley scattering events and drive the electronic system to the Anderson insulating state.

5.2.3   Boltzmann transport in two-dimensional graphene

Using the expression derived in Section 3.3, the starting point for computing the relaxation time in graphene within the Fermi golden rule is given by

$$\frac{1}{\tau} = \frac{2\pi}{\hbar} \sum_q (1 - \cos\theta_{k+q}) |\langle k+q|\hat{U}|k\rangle|^2 \delta(\varepsilon_{k+q} - \varepsilon_k), \qquad (5.11)$$

introducing a scattering potential $\hat{U}$ which has an arbitrary form for the moment. Note that the relaxation time is also named *transport time*, and differs from the elastic scattering time which is the lifetime of a plane wave state (it is given by the same Eq. (5.11), excluding the $1 - \cos\theta_{k+q}$ term). The ratio between both timescales allows some discussion about the nature of underlying disorder (Monteverde *et al.*, 2010).

Figure 5.8 shows the different angles introduced for transformation of the integral factors. The following equations are straightforward to demonstrate: $q = 2k\sin\theta_{k+q}/2$ (the momentum transfer), $\theta_q = \pi/2 + \theta_{k+q}/2$, and $\cos\theta_q = -q/2k$. The $\sum_q$ sum is converted into an integral via $\sum_q = S/(2\pi)^2 \int d^2q$, so that

$$\frac{1}{\tau} = \frac{2\pi}{\hbar}\frac{S}{(2\pi)^2}\int_0^\infty qdq \int_{-\pi}^{+\pi} d\theta_q(1 - \cos\theta_{k+q})|\langle k + q|\hat{U}|k\rangle|^2\delta(\varepsilon_{k+q} - \varepsilon_k). \quad (5.12)$$

The term $|\langle k + q|\hat{U}|k\rangle|^2$ can be replaced by $n_i|V(q)|^2(1 + \cos\theta_{k+q})/2$ where n_i is the impurity density, $V(q)$ the Fourier transform of the scattering potential, and the cosine term in parenthesis derives from the chirality factor that arises from the projection of the spinor wavefunctions between the incoming and outgoing states (for the case of no valley mixing). Indeed, recalling that the general form of the eigenstates reads

$$\Psi = \frac{1}{\sqrt{2}}\begin{pmatrix} 1 \\ e^{i\theta_k} \end{pmatrix} e^{i\mathbf{k}.\mathbf{r}}, \quad (5.13)$$

$$e^{i\theta_k} = \frac{k_x + ik_y}{|k|}, \quad (5.14)$$

the overlap between two wavevectors is

$$\int \Psi^\dagger(\mathbf{k})\Psi(\mathbf{k} + \mathbf{q})d\mathbf{r} = \frac{1}{2}(1, 1)\cdot\begin{pmatrix} 1 \\ e^{i\theta_{k+q}} \end{pmatrix} = \frac{1}{2}(1 + e^{i\theta_{k+q}}), \quad (5.15)$$

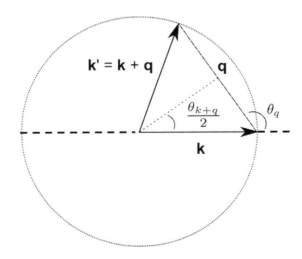

Figure 5.8 The wavevectors used in derivation of the relaxation scattering time. q denotes the total momentum transfer involved in the scattering event.

which yields $| \int \Psi^{\dagger}(\mathbf{k})\Psi(\mathbf{k}+\mathbf{q})d\mathbf{r}|^2 = \frac{1}{2}(1 + \cos\theta_{k+q})$. Equation (5.12) is finally rewritten by changing variables in the integrand using properties of the δ function. Straightforward calculations show that $\delta(\varepsilon_k - \varepsilon_{k+q}) = \frac{1}{\hbar v_F q}\delta(\cos\theta_q + \frac{q}{2k})$, and $1/\tau$ reads

$$
\frac{1}{\tau} = \frac{n_i}{2\pi\hbar^2 v_F} \int_0^{2k} |V(q)|^2 dq \int_{-1}^{+1} \frac{d\cos\theta_q}{\sqrt{1 - \cos^2(\theta_q)}} (1 - \cos^2\theta_{k+q})\delta(\cos\theta_q + \frac{q}{2k})
$$

$$
= \frac{n_i \rho(E_F)}{4\hbar} \int_0^{\pi} |V(q)|^2 (1 - \cos^2\theta_{k+q})d\theta_{k+q}. \tag{5.16}
$$

The first expression of Eq. (5.16) is clearly limited to variations of q from 0 to $2k$, while the contribution of the integrand of Eq. (5.16) cancels at $\theta = \pi$ as a consequence of pseudospin interferences. Depending on the form of $V(q)$, different Boltzmann conductivity behaviors are obtained, including long range Coulomb ($V(q) = 2k_f \sin(\theta/2)) \sim q^{-1}$), Gaussian white noise ($V(q) \sim q_0$) and Gaussian correlated disorder ($V(q) \sim e^{-q^2}$) mimicking a screened potential, as well as resonant scatterers that cause a maximal phase shift of $\pi/2$ between incoming and outgoing wavefunctions (Stauber, Peres & Guinea, 2007). The expression of the Boltzmann conductivity (at zero temperature) can be derived from the Einstein (or Drude) formula $\sigma = 4e^2\rho(E)D(E)$, where $\rho(E) = 2|E|/(\pi \times (\hbar v_F)^2)$ gives the DOS close to the Dirac point, and $D(E) = v_F^2\tau/2$ the diffusion coefficient in the diffusive regime (remember $|E| = \hbar v_F\sqrt{n}$, introducing the charge density n). Then the semiclassical conductivity $\sigma_{sc}(E)$ reads

$$
\sigma_{sc}(E) = \frac{4e^2}{h} \times \frac{|E|\tau}{2\hbar} = \frac{4e^2}{h}\frac{k\ell_{el}(E)}{2}, \tag{5.17}
$$

with the transport time τ given by Eq. (5.16). The final expression of the Boltzmann conductivity depends on the nature of the scattering potential. Short range scatters are defined by fluctuations of the potential profile on the scale of the carbon–carbon spacing (vacancies, adatoms, etc.) and introduce marked valley mixing (coupling between k vectors distant in the Brillouin zone, located at different valley–intervalley scattering). The Anderson model is one example of such disorder, which for sufficiently large potential strength breaks all symmetries of graphene. In contrast, long range scatters are defined by potential profiles which change smoothly at the atomic scale, restricting scattering events to short momentum transfer in reciprocal space. This limits the possibility of scattering events to intravalley transitions, which has profound consequences on all transport properties.

Long range disorder

Coulomb scattering in graphene stems from long range variations in the electrostatic potential caused by the presence of trapped charged impurities in the underlying substrate. This disordered Coulombian potential is further screened by the conduction electrons propagating through graphene, so its local strength becomes density-dependent (greater screening is expected at higher electron density). The long range character of the induced electrostatic interaction could be included in the model of bare Coulomb-type scattering centers such as

$$U_i = \frac{1}{4\pi \epsilon_r \epsilon_0} \sum_{j=0}^{N_{\text{imp}}} \frac{e^2}{|r_i - r_j|}, \tag{5.18}$$

where ϵ_0 and ϵ_r stand, respectively, for the vacuum and relative permittivities and N_{imp} represents the number of scattering centers or impurities. However, the application of the bare Coulomb potential can only be justified for low values of the electron density when the screening effects limiting the range of the potential are negligible (Rycerz, Tworzydo & Beenakker, 2007). To go beyond this, the simplest screened potential would be given by the Thomas–Fermi approximation,

$$U_i = U_{\text{TF}} \sum_{j=0}^{N_{\text{imp}}} \frac{e^{-\xi_{\text{TF}}|r_i - r_j|}}{|r_i - r_j|}, \tag{5.19}$$

where the parameters U_{TF} and ξ_{TF} describe the strength and the range of the scattering centers for the Thomas–Fermi potential. The inclusion of screening makes it possible to achieve both the limits of Coulomb scattering (for low n) and short range scattering (for high n). However, the singularity at $r_i = r_j$ in the Thomas–Fermi potential, Eq. (5.19), causes numerical instabilities. Using a self-consistent calculation of the impurity scattering in the random phase approximation (RPA), the scattering rate is found to be proportional to $\sqrt{n}/n_i$ (assuming a random distribution of charged impurities with density n_i), which leads to a Boltzmann conductivity at high density ($n \gg n_i$) (Adam $et\ al.$, 2007),

$$\sigma_{sc} = \frac{e^2 v_F \tau}{\hbar} \sqrt{\frac{n}{\pi}} = \frac{Ce^2}{h} \frac{n}{n_i}, \tag{5.20}$$

with C a dimensionless parameter related to the scattering strength, and $C \simeq 20$ within the random-phase approximation (taking the dielectric screening from the SiO_2 substrate). Chen $et\ al.$ (2008) experimentally explored the effect of charged impurities on the carrier conductivity by doping graphene with a controlled potassium flux in ultra-high vacuum (UHV). The gate voltage of minimum conductivity was found to become more negative with increasing doping (with reduced mobility), resulting from the electron doping induced by K atoms, which shifts the Fermi level up in energy with respect to the Dirac point. The value of $\sigma(V_g)$ was also found to become more linear with the increasing doping concentration n_i, in agreement with Eq. (5.20) (as seen in Fig. 5.7 main plot).

It is further convenient to use a model for screened potential based on a Gaussian function, to allow a more comprehensive analytical derivation of the semiclassical conductivity (Nomura & MacDonald, 2007, Zhang $et\ al.$, 2009). A realization of the disorder potential is introduced by randomly choosing N_{imp} lattice sites $\mathbf{r}_1, \mathbf{r}_2, \mathbf{r}_3, \dots, \mathbf{r}_{N_{\text{imp}}}$

out of the total number N carbon sites in the disordered sample, and by randomly choosing the potential amplitude U_n at the nth site in the interval $[-W/2, W/2]$ (W given in γ_0 units). We then smooth the potential over a range ξ by convolution with a Gaussian function (see Fig. 5.9 for illustration):

$$U_{\text{imp}}(\mathbf{r}) = \sum_{n=1}^{N_{\text{imp}}} U_n e^{\frac{-|\mathbf{r}-\mathbf{r}_n|^2}{2\xi^2}} . \tag{5.21}$$

Assuming random configurations of different graphene samples with same size, ξ, W, and $n_{\text{imp}} = N_{\text{imp}}/N$, provides a statistical ensemble for a given disorder strength. We fix the impurity effective range $\xi = 3a_{cc} = 0.426$ nm as a typical value for a long range potential, but vary W to describe different screening situations. Such a potential mimics the effect of screened charges trapped in the substrate. The disorder strength is quantified by the dimensionless correlator (Nomura & MacDonald, 2007, Zhang et al., 2009, Rycerz et al., 2007),

$$K_0 = \frac{(L/N_{\text{imp}})^2}{(\hbar v_F)^2} \sum_{i=1}^{N_{\text{imp}}} \sum_{j=1}^{N_{\text{imp}}} \langle U_{\text{imp}}(r_i) U_{\text{imp}}(r_j) \rangle, \tag{5.22}$$

of the random impurity potential (with vanishing average $\langle U_{\text{imp}} \rangle = 0$ over disorder configurations). The impurity potential correlation function equals

$$\langle U_{\text{imp}}(\mathbf{r}) U_{\text{imp}}(\mathbf{r}') \rangle = \frac{K_0 (\hbar v_F)^2}{2\pi \xi^2} e^{-\frac{|r-r'|^2}{2\xi^2}} . \tag{5.23}$$

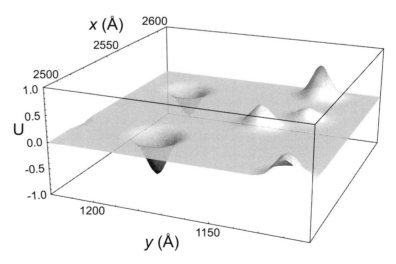

Figure 5.9 The long range disorder potential $U(x, y)$. Courtesy of Frank Ortmann (Ortmann et al., 2011).

Using $|V(q)|^2 = K_0(\hbar v_F)^2 e^{-q^2\xi^2/4}$, the scattering time (Eq. 5.16) can be derived, and after straightfoward calculations the Boltzmann conductivity is given by

$$\sigma_{\text{sc}}(E) = \frac{4e^2}{h} \frac{\pi n\xi^2 e^{\pi n\xi^2}}{K_0 I_1(\pi n\xi^2)} \qquad (5.24)$$

$$= \frac{2\sqrt{\pi}e^2}{K_0 h}\left[(2\pi n\xi^2)^{3/2} + O(n\xi^2)^{1/2}\right], \qquad (5.25)$$

with the carrier density $n = k_F^2/\pi$, and I_1 is the modified Bessel function ($I_1(x) = 1/\pi \int_0^\pi e^{x\cos\theta}\cos\theta\,d\theta + 1/\pi \int_0^\infty e^{(-x\cosh t - t)}dt$). $K_0 = 40.5 n_{\text{imp}}(W/2\gamma_0)^2(\xi/\sqrt{3}a)^4$ where $n_{\text{imp}} = N_{\text{imp}}/N$ denotes the relative concentration (Rycerz *et al.*, 2007, Klos & Zozoulenko, 2010).

The leading term for high density can also be obtained considering the classical diffusion of a particle undergoing small-angle deflections from the random potential U. Defining $z = \pi n\xi^2$, σ_{sc} then exhibits two different limits. For $|z| \ll 1 \to \sigma_{\text{sc}} \sim$ constant while $|z| \gg 1 \to \sigma_{\text{sc}} \sim n^{3/2}$. The situation $|z| \ll 1$ arises when the Fermi wavelength is larger than the effective screening length $\lambda \gg \xi$, which describes a strong quantum scattering, while the opposite condition $|z| \gg 1$ drives to a classical scattering $\lambda \ll \xi$.

Short range disorder: Anderson disorder

A simple approximation for describing a short range potential can be defined as $U(\mathbf{r}) = \sum_j U_j\delta(r - r_j)$, where U_j are onsite energies which are taken as random within a certain energy scale and following a given distribution (uniform, Gaussian, correlated distribution, etc.). By directly applying the Fermi golden rule, one obtains at the simplest approximation level

$$\frac{1}{\tau} = \frac{2\pi}{\hbar} \times \frac{\langle n_{\text{imp}}U_i^2\rangle}{2} \frac{|E|}{2\pi(\hbar v_F)^2}, \qquad (5.26)$$

and defining a dimensionless parameter $W = \frac{\langle n_{\text{imp}}U_i^2\rangle}{4\pi(\hbar v_F)^2}$, we get $\tau = \frac{2\pi}{\hbar}|E|W$ and

$$\sigma_{BB} = \frac{2e^2}{\pi h} \times \frac{1}{W}, \qquad (5.27)$$

which means that the Bloch–Boltzmann conductivity is a constant, independent of the energy close to the Dirac point (illustrated in Fig. 5.13 inset). The Anderson disorder roughly mimics neutral impurities such as structural defects, dislocation lines, or adatoms, although the local geometry and chemical reactivity of defects and impurities actually demand more *ab initio* calculations if aiming at quantitative predictions. An interesting aspect of this model is, however, that numerical simulations can be contrasted to analytical results derived in the self-consistent Born approximation

(SCBA) (Shon & Ando, 1998, Ostrovsky, Gornyi & Mirlin, 2006). First the density of states is obtained as

$$\rho(E) = -\frac{1}{\pi L^2} \sum_{\alpha} \Im m \langle G_{\alpha,\alpha}(E + i\eta) \rangle = 4\Gamma(E)/(\pi n_i u^2), \qquad (5.28)$$

with $\Gamma(E) = \Im m \Sigma(E + i\eta)$ derived from the self-energy, which satisfies the recurrent equation (which is solved numerically)

$$\Sigma(E + i\eta) = \frac{n_i u^2}{2\pi}(E - \Sigma(E + i\eta)) \int_0^{k_c} \frac{k\,dk}{(E - \Sigma(E + i\eta))^2 - (\gamma k)^2}, \qquad (5.29)$$

with n_i the defect density, u^2 the average squared disorder strength. A mapping of this model to the Anderson disorder is possible by adjusting parameters to obtain the same average values and variances, i.e. $n_i(1 - n_i)|u| = W/\sqrt{12}$ and $k_c = \gamma_0/\epsilon_0$, the cutoff where $\epsilon_c = 50\epsilon_0$ in the simulations. ϵ_0 is an arbitrary energy scale assumed to have the same order of magnitude as relevant energies such as E, Δ and Γ in the SCBA, $\Delta(E) = \Re e \Sigma(E + i\eta)$ (Shon & Ando, 1998). Figure 5.10 shows the density of states (DOS), computed with a Lanczos-type method (Section D.1), which is reported as a function of W. The disorder-free DOS (dashed line) shows typical behavior with a linear increase at low energy and the presence of two sharp van Hove singularities at $E = \pm\gamma_0$. As W increases, two different features are observed. At high energies, van Hove singularities are smoothed, whereas close to the charge neutrality point, Anderson disorder enhances the DOS in full agreement with analytical results (Shon & Ando, 1998) (see Fig. 5.10(b) for a close-up).

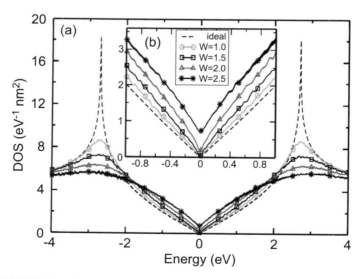

Figure 5.10 (a) DOS of an ideal (dashed lines) and disordered graphene sheets for several values of $W \in \{1; 1.5; 2; 2.5\}$. (b) Zoom in the energy area around the charge neutrality point. (Reproduced with permission from Lherbier *et al.* (2008). Copyright (2008) by the American Physical Society.)

Short range disorder: strong scattering of local impurities

Finally, note that different other types of short range defects such as vacancies and cracks in graphene flakes have also been predicted to produce midgap states in graphene (Stauber *et al.*, 2007). By considering that an incident particle is a massless Dirac particle, we can use 2D scattering theory to access the transport time as well as the conductivity. We consider impurities (distributed randomly in the graphene network with a density n_i) defined by a strong local scattering potential such as $U(r) = V_0 > 0$ if $r < R$, and zero otherwise (deep circular potential well), where V_0 is the potential strength while R is the potential range. We next consider an incident low-energy massless Dirac particle with wavevector k such that $kR \ll 1$ (see Fig. 5.11), so that the scattering amplitude has the form (Katsnelson, 2012)

$$f(\theta) = \frac{e^{2i\delta(k)} - 1}{i\sqrt{2\pi k}}(1 + e^{-i\theta})$$

$$= \frac{-\sqrt{\pi/2k}}{\frac{J_0(\tilde{k}R)}{\tilde{k}RJ_1(\tilde{k}R)} + \ln\left(\frac{2}{\tilde{k}R\gamma_E}\right) + i\pi/2}(1 + e^{-i\theta}),$$

where $\delta(k)$ is the s wave scattering phase shift, $\gamma_E = 1.781$ and J_n are Bessel functions. The wavevector $\tilde{k}$ is defined as $\tilde{k} = |\hbar v_F k - V_0|/\hbar v_F$, $v_F = 10^6$ ms^{-1} in a graphene monolayer. The weak potential limit is given by weak $V_0 \ll \hbar v_F k$ and $\tilde{k} \sim k$, whereas the strong potential limit assumes $V_0 \gg \hbar v_F k$ and $\tilde{k} \sim V_0/\hbar v_F$. Here, we focus on the strong disorder limit, so that $\tilde{k}R \sim V_0 R/\hbar v_F$, which is k-independent. The differential cross-section $dA/d\theta$ and transport times are given by

$$\frac{dA}{d\theta} = |f(\theta)|^2 = \frac{8\sin^2\delta(k)}{\pi k}\frac{1 + \cos\theta}{2},$$

$$\frac{1}{\tau} = n_i v_F \int (1 - \cos\theta)|f(\theta)|^2 d\theta,$$

and the transport cross-section and total cross-section are

$$A_{tr} = \int d\theta(1 - \cos\theta)|f(\theta)|^2 = \frac{4\sin^2\delta}{k}.$$

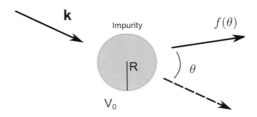

Figure 5.11 A scattering event of incident electron with momentum k, on short range impurity potential, with scattering amplitude $f(\theta)$ in a direction defined by scattering angle θ.

The transport time τ can be determined from A_{tr} using the relation $1/\tau = n_i v_F A_{tr} = 4 n_i v_F \sin^2 \delta(k)/k$, while the conductivity derives from

$$\sigma = \frac{2e^2}{\pi h} v_F k_F \tau(k_F).$$

To derive the k_F dependence of τ and σ, we consider different limits determined by the three terms in the denominator of $f(\theta)$ in the equation for scattering amplitude above. When $J_0(\tilde{k}R) \sim 0$, the logarithmic term predominates. This is possible since $\tilde{k}R \sim V_0 R/\hbar v_F$ can be larger than one even if $kR \ll 1$ (resonant case). In this situation, the phase shift of the scattered wavefunction becomes

$$\delta(k) \sim -\frac{\pi}{2 \ln(kR)} \to 0,$$

which leads to a transport cross-section and transport time

$$A_{tr} \sim \frac{4\delta^2}{k} \sim \frac{\pi^2}{k \ln^2(kR)},$$

$$\tau \sim \frac{k \ln^2(kR)}{n_i v_F \pi^2} \sim k \ln^2(kR). \tag{5.30}$$

This finally gives a conductivity which becomes roughly linear in the charge density as

$$\sigma = \frac{2e^2}{\pi h} \frac{n}{n_i} \ln^2(\sqrt{\pi n} R), \tag{5.31}$$

where n_i is the short range defect density. This equation mimics the one for charged impurities, Eq. (5.20), with a slightly logarithmic dependence of the conductivity on the charge carrier density. One observes the mathematical singularity of Eq. (5.31), which suggests a diverging resistivity $1/\sigma \to \infty$ in the vicinity of the Dirac point, outlining a limit of such derivation. These disorder models are, however, quite complicated to compare with experimental data. This is partly due to the simplification made in the modeling and the various sources of disorder usually present in a real situation. It is, however, very useful to draw basic conclusions with respect to the strength and range of the disorder, which nevertheless encode a certain universality.

5.2.4 Kubo transport: graphene with Anderson disorder

The effect of short range scattering potential (Anderson disorder potential) with the disorder strength (W) is now analyzed using the Kubo conductivity. The mean free path $\ell_{\rm el}(E)$ is deduced from the saturation of the diffusion coefficients (using the numerical method presented in Section 3.4.4 and Section D.2). Figure 5.23(a) in Section 5.2.7 gives several $D(t, E = 0)$ time evolution behaviors at the Dirac point. After the starting quasi-ballistic spreading of the wavepackets, a saturation regime develops and is

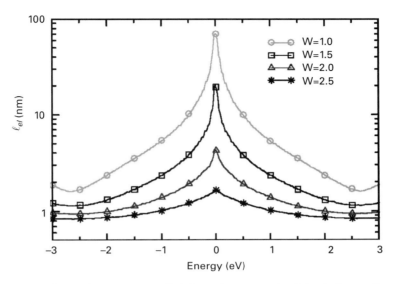

Figure 5.12 Mean free path versus energy (ℓ_{el}) for various strengths of the Anderson disorder potential.

followed by a decay of $D(t, E = 0)$ when quantum interferences are strong enough (see Section 5.2.7). The behavior of $\ell_{el}(E)$ is shown in Fig. 5.12.

The semiclassical conductivity has actually been derived by Shon and Ando (Shon & Ando, 1998) using a perturbation theory and the self-consistent Born approximation (SCBA). The semiclassical part of the conductivity, $\sigma_{xx} \sim \mathrm{Tr}\langle v_x \Im mG(E+i\eta)v_x \Im mG(E+i\eta)\rangle_{\mathrm{conf.}}$, can then be simplified by $\langle G(E)G(E')\rangle \sim \langle G(E)\rangle\langle G(E')\rangle$ (all interference effects driven by Cooperon contributions are neglected), and is then deduced to be (Shon & Ando, 1998)

$$\sigma_{sc}(E) = \frac{1}{2}\frac{e^2}{\pi^2\hbar}\left[\left(\frac{E-\Delta(E)}{\Gamma(E)} + \frac{\Gamma(E)}{E-\Delta(E)}\right)\arctan\left(\frac{E-\Delta(E)}{\Gamma(E)}\right)+1\right]. \quad (5.32)$$

A peculiar energy dependence of $\sigma_{sc}(E)$ is obtained, together with a universal minimum value at the Dirac point, where $\sigma_{sc} = 4e^2/\pi h$ (Fig. 5.13, main plot). Using the Kubo method, σ_{sc} is evaluated for various W (Fig. 5.13), and found to agree very well with the SCBA results. In particular, for $W = 2$, the dimensionless parameter $A = (4\pi\gamma_0)/(n_i u^2)$ is fixed at $A = 21$ to get a convincing fit for energies up to $2\epsilon_0$ with $\epsilon_0 = 0.3$ eV. The SCBA result fails, however, to describe the transport coefficient in the presence of strong disorder, since the neglect of quantum interferences jeopardizes the observation of localization (see Section 3.4.5).

The mobility (μ) versus charge energy is given in Fig. 5.14 for several strengths of the Anderson disorder. Similar energy dependencies are obtained for $\mu(E)$ and $\ell_{el}(E)$. At low energies, a simple Fermi golden rule captures the downscaling of μ with W, although it diverges precisely at the Dirac point. In most semiclassical

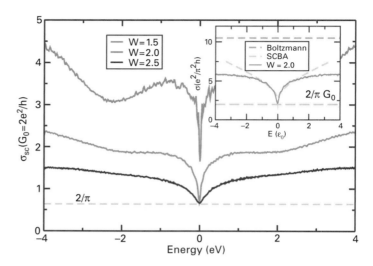

Figure 5.13 Main plot: semiclassical conductivity for Anderson disorder strengths of $W = 1.5, 2.0,$ and 2.5. The dashed line denotes the minimum value of $\sigma_{sc} = 4e^2/\pi h$. Inset: Boltzmann result (horizontal dashed line) and self-consistent Born approximation (dashed line) using Eq. (5.32) with fitting factor $A = 21$ for $W = 2$. (Reprinted from Roche *et al.* (2012). Copyright (2012), with permission from Elsevier.)

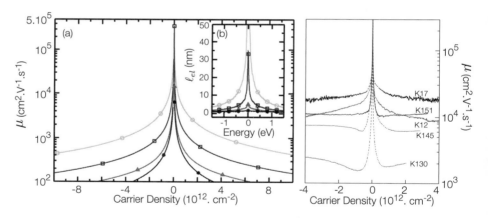

Figure 5.14 Left: charge mobility with carrier density (a) and mean free path (b) for $W = 1, 1.5, 2, 2.5$ (from top to bottom). (Reproduced with permission from Lherbier *et al.* (2008). Copyright (2008) by the American Physical Society.) Right: experimental mobility. (Reproduced with permission from Tan *et al.* (2007), courtesy of Philip Kim. Copyright (2007) by the American Physical Society.)

simulations, the Dirac point charge density becomes vanishingly small, which introduces a nonphysical singularity in the mobility. By numerically integrating the disordered DOS, it is, however, possible to compute a finite charge density at the Dirac point, offering a more meaningful discussion of the Dirac point transport physics (see Section 7.3.1 for a particularly striking example and discussion). Experimental transport measurements of graphene samples deposited onto silicon oxide are also reported in Fig. 5.14 (right), from Tan *et al.* (2007). Different experimental samples with varying quality exhibit a charge density dependence of μ, which is seen to vary with absolute value of μ. The Anderson disorder leads to a theoretical $\mu(n)$ which seems to best reproduce samples with higher mobilities. However, a quantitative analysis is out of reach of a simplified disorder model, and would have to account for many other sources of scattering.

5.2.5 Kubo transport: graphene with Gaussian impurities

In this section, we explore the effect of long range (Gaussian) disorder potential (mathematically defined in Section 5.2.3) on transport features. We first investigate how quantum transmission develops through a finite size system containing a single scatter, following the interesting work of Zhang *et al.* (2009). By diagonalizing the Hamiltonian of a small supercell with N carbon sites (and a single impurity), the band structure $E_k, \psi_k = \sum_{i=1}^{N} a_{ki}|i\rangle$ is obtained as well as the participation ratio $\mathcal{P}_k$ defined by $\mathcal{P}_k = \frac{\left(\sum_{i=1}^{N} a_{ki}^2\right)^2}{N\left(\sum_{i=1}^{N} a_{ki}^4\right)}$. The quantity $\mathcal{P}_k$ is a measure of the localization of electronic states. Localized states are defined by $\mathcal{P}_k \sim 1/N$, whereas extended states require that $\mathcal{P}_k$ remains constant with system size (as shown in Fig. 5.15).

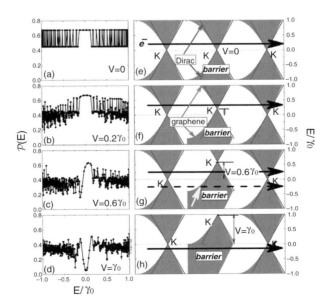

Figure 5.15 (a–d) Participation ratio for a graphene with $N = 70 \times 40$, with a single impurity at the center with different potential barrier height $V \geq 0$, and fixed $\xi = 1.73a$. (e–h) Scattering process corresponding to the left counterparts. (Adapted with permission from Zhang *et al.* (2009). Copyright (2007) by the American Physical Society. Image courtesy of Wu-Ming Liu.)

The results obtained with varying the potential height V ($V > 0$) are shown in Fig. 5.15(a–d). For $V \leq 0.2\gamma_0$, states remain Dirac-like inside and outside the barrier (for incoming-left and outgoing-right electrons), as depicted in Fig. 5.15(e) and (f). The formation of bound states is obtained by further increasing V, as revealed by the small values of $\mathcal{P}$ in the negative energy region near the Dirac point (Fig. 5.15(c)). For positive injected energy (arrow with solid line in Fig. 5.15(g)), the electron is not far from K both inside and outside the barrier, which preserves the Klein tunneling mechanism. In contrast, for negative energy (arrow with dashed line in Fig. 5.15(g)), the presence of a non-Dirac barrier (marked by the arrow/barrier) affects electron tunneling, and induces a localization of the state around the impurity (bound states). When V is increased up to $V \sim \gamma_0$ (Fig. 5.15(h)), the Klein tunneling is totally suppressed owing to the radical change of the electronic structure.

The relative contribution of intervalley versus intravalley scattering has been computed for both the single scatter limit and for an average result of a disordered system (Zhang *et al.*, 2009). The valley-resolved scattering amplitude $\mathcal{A}$ is calculated following Ando (Ando, 1991). The sum of all the scattering amplitudes can be related to the sum over all propagating channels at E_F, which is written as $N_C = \mathcal{A}_{\text{intra}} + \mathcal{A}_{\text{inter}} = \mathcal{A}$ (Fig. 5.16). This interesting analysis allows differentiation of the transport regime depending on the relative strength of intervalley versus intravalley scattering. As seen in Fig. 5.16, the intravalley contribution predominates as long as the onsite impurity potential remains smaller than γ_0. In such a case, Klein tunneling or weak antilocalization phenomena mainly dominate the total conductivity. In the absence of intervalley scattering, one notes that Bardarson and coworkers (Bardarson *et al.*, 2007) have also

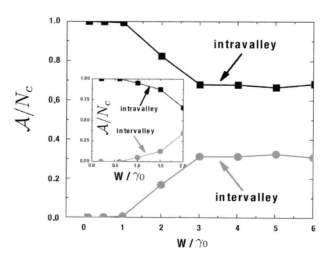

Figure 5.16 Normalized scattering amplitudes of intravalley and intervalley versus W. Each point is an average over 100 samples on a square graphene sheet with 112×64 sites. Inset: same results for the single-impurity case at the Dirac point. (Adapted with permission from Zhang *et al.* (2009). Copyright (2007) by the American Physical Society. Image courtesy of Wu-Ming Liu.)

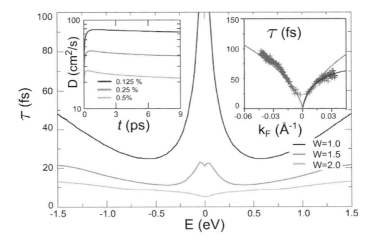

Figure 5.17 Main plot: Transport time τ for several values of the onsite disorder strength W. Left inset: Diffusion coefficient at Dirac point for chosen impurity densities and $W = 2$. Right inset: k_F dependence of the measured transport time for four different graphene samples. (Adapted with permission from Monteverde *et al.* (2010). Copyright (2010) by the American Physical Society.)

numerically studied the one-parameter scaling behavior. Their results show that the Dirac point conductivity scales logarithmically with sample size, but does not reach a scale-invariant limit ($\beta(\sigma) = d\ln\sigma/d\ln L > 0$), confirming that Dirac fermions evade Anderson localization and remain delocalized as long as valleys are not mixed by scattering. In conclusion, the true impact of long range disorder is markedly dependent on the W value.

We can further deepen the analysis of the quantum transport properties using numerical analysis of the quantum Kubo transport. Figure 5.17 (inset) shows the evolution of the wavepacket dynamics (through time-dependent diffusion coefficients) at the Dirac point ($E = 0$) for $W = 2$ and increasing impurity density ($n_i = 0.125\%, 0.25\%, 0.5\%$). The diffusion coefficients reach a saturation regime, typically after 1 ps, indicating a diffusive regime. At much longer timescales, a time-dependent decay of the diffusion coefficient is observed (regardless of the energy and impurity density), indicating the onset of weak localization phenomena. The maximum value of $D(t) = D_{max}$ provides the elastic mean free path $\ell_{el} = D_{max}/2v_F$ and the total transport time $\tau = \ell_{el}/v_F$ where $v_F = 8.7 \times 10^5$ m s^{-1} is the Fermi velocity.

Figure 5.17 (main plot) gives the transport time (deduced from the numerical diffusion coefficient (inset)) for three values of $W = 1.0$, 1.5 and 2.0 in γ_0 units. The behavior of $\tau(E)$ in the vicinity of the Dirac point is found to strongly vary depending on the disorder strength considered. For $W = 2.0$, $\tau(E = 0) \approx 5$ fs, and increases almost linearly at higher energies. Differently for $W = 1.5$ and $W = 1.0$, τ exhibits an upturn close to the Dirac point followed by a saturation value of $\tau(E = 0)$ (about 4 times larger for $W = 1.5$) due to a finite density of states. For the weakest disorder ($W = 1$), $\tau(E) \sim 1/E$ for low energies and $\tau(E) \sim E$ at higher energies.

An experimental estimation of the transport time $\tau(k_F)$ is also shown in Fig. 5.17 (right inset) for four different exfoliated graphene samples with mobilities ranging from 3000 to 5000 cm^{-1} (Monteverde *et al.*, 2010). The wavevector (or energy) dependence is similar to the simulated transport time for $W = 2.0$, which decays as the charge density is reduced. The experimental data are further reasonably fitted using

$$\tau(k_F) = \frac{k_F \ln^2(k_F R)}{\pi^2 n_i v_F}, \tag{5.33}$$

taking $n_i \sim 10^{12}$ cm^{-2} and R in the order of the lattice spacing. This form is consistent with resonant scattering due to short range disorder as introduced in Section 5.2.3, and yields a typical semiclassical behavior of the conductivity given by Eq. (5.31). An opposite energy-dependent behavior of $\tau(k_F)$ is observed experimentally for bilayer graphene, indicating a possible interpretation in terms of larger screening strength (Monteverde *et al.*, 2010).

The behavior of the elastic scattering time can be analyzed using the Fermi golden rule. Together with the energy dependence of the density of states roughly proportional to E, we found an energy-dependent contribution for τ by Fourier transforming the long range potential. For weak disorder the low-energy relaxation time is approximated by $\tau \propto 1/(E(1 - cE^2)) \approx 1/E + cE$, so that a low-energy peak and a linear slope at higher energies separated by a minimum τ at finite E are expected. In the simulations such an expected weak disorder limit is obtained for $W = 1.0$ (Fig. 5.17), while for larger disorder the above estimation has to be revised. For $W = 1.5$, the DOS at the Dirac point is finite and the leading term for $\tau(E)$ at low-energy $1/E$, stemming from the DOS, is strongly reduced. As a result, the minimum for $\tau(E)$ still exists but is relocated to smaller energies ($E \approx 400$ meV). For stronger disorder ($W = 2.0$), the DOS at the Dirac point is large enough that the minimum at finite E disappears.

The Kubo conductivity for the case of Gaussian impurities with density $n_i = 4\%$ and $\xi = 16a$ in the density interval $|z| \leq 35$ (Radchenko, Shylau & Zozoulenko 2012) is shown in Fig. 5.18 (main plot). Therefore for $|z| \gg 1$, σ shows a linear density dependence $\sigma \sim n$, whereas the semiclassical Boltzmann approach (see Eq. (5.25)) predicts a superlinear dependence $\sigma \sim n^{3/2}$, and a vanishing conductivity at the Dirac point. This outlines one limitation of the semiclassical approach in the regime of low densities, where the effects of multiple scattering impact on DOS and semiclassical transport length scales (mean free path) are neglected. A similar discrepancy is obtained for the case $|z| \leq 1$, for which neither the density dependence of σ nor the absolute values is correctly obtained using the Boltzmann equation (Eq. (5.25)) (Radchenko *et al.*, 2012). Another disagreement is also reported in Radchenko *et al.* (2012) for the case of short-range Gaussian impurities. Figure 5.18 (inset) shows the Kubo conductivity (at $E = 0.2\gamma_0$) for the case of $n_i = 2\%$ of Gaussian impurities with onsite impurity potential of $\sim 37\gamma_0$. No logarithmic correction (as predicted by Eq. (5.31)) is obtained with the full Kubo calculation, which instead evidences a plateau of the semiclassical conductivity, similarly to the case of homogeneous Anderson disorder (Section 5.2.4).

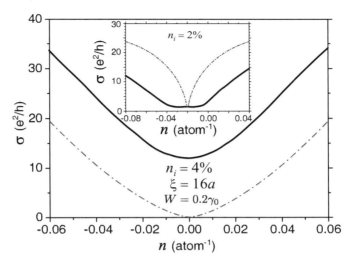

Figure 5.18 Main frame: Conductivity versus electron density n (the number of electrons per C atom) for random Gaussian impurities $n_i = 4\%$ ($\xi = 16a$) using the Kubo formula (bold line) or the Boltzmann result according to Eq. (5.25) (dashed line). Inset: Conductivity versus electron density n for random impurities with density $n_i = 2\%$ for the Kubo conductivity (bold line) and for the Boltzmann conductivity using Eq. (5.31) (dashed line). Curves are shifted to the charge neutrality point at $n \sim 0.02$ (Adapted with permission from Radchenko *et al.* (2012). Copyright (2012) by the American Physical Society. By courtesy of T. M. Radchenko.)

5.2.6   Weak localization phenonema in disordered graphene

Broken symmetries and weak antilocalization: Cooperon contribution

By classifying disorder symmetry classes, a generalized diagrammatic theory of quantum interferences in graphene has been derived (McCann *et al.*, 2006, Kechedzhi *et al.*, 2007, Fal'ko *et al.*, 2007). It describes the contributions of various particle–particle correlation functions (associated to a given class of scattering diagrams), or Cooperons to the total quantum correction of the semiclassical conductivity, whose strength is monitored by several phenomenological parameters (intravalley versus intervalley elastic scattering times). The magnetoconductance $\Delta\sigma(B)$ obtained by this theory is then used for fitting experimental curves. The full derivation of such a theory is a highly technical and tedious exercise which goes beyond the scope of this book. We just summarize the main features concerning the various contributions of Cooperons, which either decrease (weak localization – WL) or increase (weak antilocalization – WAL) the quantum conductivity with respect to its semiclassical value. For such purposes, we recall the effective Hamiltonian describing massless Dirac fermions but including the trigonal warping deformation. Near the corners $K_\pm$ of the hexagonal Brillouin zone, the effective description of the Hamiltonian can be rewritten as (Fal'ko *et al.*, 2007)

$$\hat{\mathcal{H}} = v_F \Pi_z (\sigma_x p_x + \sigma_y p_y) + \mu \Pi_0 [\sigma_y (p_x p_y + p_y p_x) - \sigma_x (p_x^2 - p_y^2)] + \hat{U}_{\text{dis}}, \quad (5.34)$$

where the first term determines the massless Dirac fermions, while the second term accounts for the trigonal warping correction (with a threefold symmetry) to the Dirac cone. $\hat{\mathcal{H}}$ takes into account in-plane nearest neighbor A/B hopping with the first (second) term representing the first (second) order term in an expansion with respect to momentum $\mathbf{p}$ measured from K_+ and K_-. Here $\sigma_0 \equiv \mathbb{I}, \sigma_{x,y,z}$ give the AB lattice space Pauli matrices, whereas the inter-/intra-valley matrices are denoted by $\Pi_0 \equiv \mathbb{I}, \Pi_{x,y,z}$. The latter term in Eq. (5.34) entails the disorder potential. Such an effective Hamiltonian operates in the space of four-component wavefunctions $[\Psi_{K_+}(A), \Psi_{K_+}(B), \Psi_{K_-}(B), \Psi_{K_-}(A)]$ describing the electronic amplitudes on A and B sites and in the valleys $K_\pm$ (see Section 2.2.2). Electrons in the conduction and valence bands differ by the isospin projection onto the momentum direction, so that $\sigma \cdot \mathbf{p}/|\sigma \cdot \mathbf{p}| = 1$ in the conduction band, and $\sigma \cdot \mathbf{p}/|\sigma \cdot \mathbf{p}| = -1$ in the valence band for the K_+ valley. Differently, in the K_- valley, the electron chirality is mirror-reflected, so that $\sigma \cdot \mathbf{p}/|\sigma \cdot \mathbf{p}| = -1$ for the conduction band and $\sigma \cdot \mathbf{p}/|\sigma \cdot \mathbf{p}| = +1$ for the valence band. For an electron in the conduction band, the plane wave states can be written generically as

$$|\Psi_{K_\pm,\mathbf{p}}(\mathbf{r})\rangle = \frac{e^{i\mathbf{p}\cdot\mathbf{r}/\hbar}}{\sqrt{2}} \left(e^{i\theta_p/2} | \downarrow \rangle_{K_\pm,\mathbf{p}} \pm e^{-i\theta_p/2} | \uparrow \rangle_{K_\pm,\mathbf{p}} \right), \tag{5.35}$$

$$|\Psi_{K_\pm,-\mathbf{p}}(\mathbf{r})\rangle = \frac{e^{-i\mathbf{p}\cdot\mathbf{r}/\hbar}}{\sqrt{2}} \left(e^{i\theta_p/2} | \downarrow \rangle_{K_\pm,-\mathbf{p}} \mp e^{-i\theta_p/2} | \uparrow \rangle_{K_\pm,-\mathbf{p}} \right), \tag{5.36}$$

with for instance $| \uparrow \rangle_{K_+,\mathbf{p}} = [1,0,0,0], | \downarrow \rangle_{K_+,\mathbf{p}} = [0,1,0,0]$ and $| \uparrow \rangle_{K_-,\mathbf{p}} = [0,0,1,0]$, $| \uparrow \rangle_{K_-,\mathbf{p}} = [0,0,0,1]$, and with the factors $e^{\pm i\theta_p/2}$ encoding the Berry's phase effects (Section 2.2.2). One defines pseudospin up $| \uparrow \rangle$ (resp. pseudospin down $| \downarrow \rangle$) for restriction of the wavefunction to the A (resp. B) sublattice. The trigonal deformation of the Dirac cone exhibits a $\mathbf{p} \rightarrow -\mathbf{p}$ asymmetry of the electron dispersion inside each valley, as illustrated in Fig. 5.19(a): $E(K_\pm, \mathbf{p}) \neq E(K_\pm, -\mathbf{p})$. Owing to time-reversal symmetry, trigonal warping shows, however, some symmetry between two valleys as $E(K_\pm, \mathbf{p}) = E(K_\mp, -\mathbf{p})$.

The final Cooperon correction to the semiclassical conductivity results from the full (disorder averaged) phase interference of electronic trajectories, which are topologically identical upon time-reversal symmetry (as sketched in Fig. 3.6(a) and Fig. 5.19(b)). Such quantum correction is formally described in terms of the Cooperon contributions (introduced in Section C.3), which are (for spin-1/2 particles) classified as singlets and triplets in terms of isospin (AB lattice space) and pseudospin (inter-/intra-valley) indices (given through Eq. (5.35) and Eq. (5.36)). With regard to the isospin (sublattice) composition of Cooperons, only singlet modes are mathematically found to be relevant (bringing a pole in the corresponding two-particles correlation function for the total backscattering amplitude $\mathbf{p}' = -\mathbf{p}$). This can be seen in the correlator describing two plane waves $\sim \Psi_{K_+,\mathbf{p}} \Psi_{K_-,-\mathbf{p}}$, with $\Psi_{K_+,\mathbf{p}}$ and $\Psi_{K_-,-\mathbf{p}}$, propagating in opposite directions along a ballistic segment of the closed trajectory (as in Fig. 3.6(a)), which displays the generic form (Fal'ko et al., 2007) $| \uparrow \rangle_{\mathbf{K_+,p}} | \downarrow \rangle_{\mathbf{K_-,-p}} - | \downarrow \rangle_{\mathbf{K_+,p}} | \uparrow \rangle_{\mathbf{K_-,-p}} - e^{-i\varphi} | \uparrow \rangle_{\mathbf{K_+,p}} | \uparrow \rangle_{\mathbf{K_-,-p}} + e^{i\varphi} | \downarrow \rangle_{\mathbf{K_+,p}} | \downarrow \rangle_{\mathbf{K_-,-p}}$ containing only sublattice-singlet terms

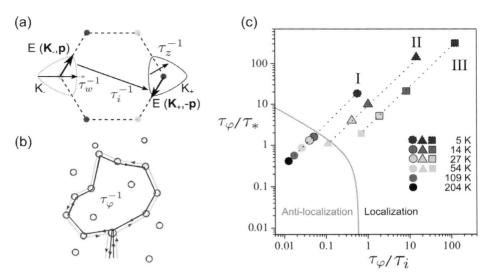

Figure 5.19 (a) Shape of the Fermi surface at finite Fermi energy, illustrating the processes responsible for trigonal warping (τ_w^{-1}), chirality-breaking (τ_z^{-1}) and intervalley scattering (τ_i^{-1}); with $\tau_*^{-1} = \tau_w^{-1} + \tau_z^{-1}$ (Horsell *et al.*, 2008). (b) Real space diagram of the closed trajectories responsible for quantum interferences correction, which are limited in size by the dephasing rate τ_φ^{-1}. (c) The scattering times related to quantum interferences in graphene. The solid curve separates the regions of electron localization and antilocalization. ((b) and (c) are adapted with permission from Tikhonenko *et al.* (2009). Copyright (2009) by the American Physical Society.)

(the first two terms) because triplet terms (the last two terms) disappear after averaging over the momentum direction ($\langle e^{i\varphi} \rangle = 0$).

Diagrammatic calculations show that the total Cooperon can then be mainly divided into *intravalley Cooperons* (interfering trajectories confined to a single valley) and *intervalley Cooperons* (interfering trajectories containing valley mixed trajectories) (McCann *et al.*, 2006). Each Cooperon contribution is confined by two cutoff timescales (see Section 3.4.5 and Section C.3 for details), namely the elastic (temperature-independent) relaxation times related to a given class of elastic scattering events ($\tau_{zz}, \tau_{\perp z}, \tau_{\perp\perp}, \ldots$) and the coherence time τ_φ which dictates the maximum length of interfering trajectories. Denoting $\{x,y\} \equiv \perp$, the total scattering time can actually be split (assuming total x/y symmetry) in a generic form as

$$\frac{1}{\tau} = \frac{1}{\tau_0} + \frac{1}{\tau_{zz}} + \frac{2}{\tau_{\perp z}} + \frac{2}{\tau_{z\perp}} + \frac{4}{\tau_{\perp\perp}}, \qquad (5.37)$$

where disorder effects are included in all three directions of space x, y, z. For instance, in the specific case of Gaussian white noise potential defined by

$$\langle u_{sl}(\mathbf{r})u'_{s'l}(\mathbf{r}')\rangle = u_{sl}\delta_{s,s'}\delta_{l,l'}\delta(\mathbf{r}-\mathbf{r}'),$$

the several elastic scattering times (which are all longer than the momentum relaxation time τ_0) will then be defined as $\hbar\tau_{sl}^{-1} = \pi\rho(E_F)u_{sl}^2$. A nonreconstructed vacancy would contribute to all terms except $u_{\perp z}$ and $u_{z\perp}$, while bond disorder would contribute to all terms except u_{zz}. A more realistic disorder potential will make the analytical calculation of timescales out of reach and numerical simulations are required to go beyond phenomenology, although no direct calculation of those elastic length scales is technically possible (in contrast to $\tau_0 = \tau$, the momentum relaxation time), except through the fitting of the numerical magnetoconductance curves (as discussed in Section 5.2.6). The full diagrammatic calculation of the Cooperon correction to the conductivity reduces to (McCann *et al.*, 2006, Kechedzhi *et al.*, 2007, Fal'ko *et al.*, 2007)

$$\delta\sigma = \frac{2e^2D}{\pi h} \int \frac{d^2q}{(2\pi)^2}(\mathcal{C}_0^x + \mathcal{C}_0^y + \mathcal{C}_0^z - \mathcal{C}_0^0), \qquad (5.38)$$

where the two first Cooperon terms $(\mathcal{C}_0^x + \mathcal{C}_0^y)$ give the intravalley contributions (including trigonal warping), whereas the two others $(\mathcal{C}_0^z - \mathcal{C}_0^0)$ provide the intervalley parts. The last term in Eq. (5.38), $\mathcal{C}_0^0$, is the main Cooperon contribution, which dictates the sign of the quantum correction $\delta\sigma$, when coherence time is sufficiently long $(\tau_\varphi > \tau_i)$. The intervalley component $\mathcal{C}_0^z$ is determined by the intervalley scattering rate (cutoff) $\tau_i^{-1} = 4\tau_{\perp\perp}^{-1} + 2\tau_{z\perp}^{-1}$, while the two intravalley components $\mathcal{C}_0^x$ and $\mathcal{C}_0^y$ are determined by cumulative inter-/intra-valley scattering rates (also including the trigonal warping effect), with $\tau_*^{-1} = \tau_w^{-1} + 2\tau_z^{-1} + \tau_i^{-1}$ and with $\tau_z^{-1} = 2\tau_{\perp z}^{-1} + \tau_{zz}^{-1}$.

Let us consider the effect of the trigonal warping deformation of the Dirac cone on interference phenomena, following Fal'ko *et al.* (2007). In the quantum interference picture, two phases θ_+ and θ_- are accumulated along the time-reversal symmetric paths, and the cumulated π Berry's phase fixes the phase difference of a given trajectory $\delta = \theta_+ - \theta_- = \pi N$ (where N is the winding number of a trajectory), and determines the strength of the WAL correction. Notwithstanding, the asymmetry of the electron dispersion brought about by the trigonal warping (Fig. 5.19(a)) deviates δ from πN. Indeed, since any closed trajectory is a combination of ballistic intervals characterized by momenta $\pm p_j$ (for the two directions) and duration t_j, each segment contribution to the total phase difference is given by $\delta_j = [E(\mathbf{p}_j) - E(-\mathbf{p}_j)]t_j$ (taking the energy versus momentum relation of the second term of the Hamiltonian Eq. (5.34)). Since δ_j are random and uncorrelated, the mean square of the total phase difference simplifies to $\sim\langle(t_j\delta_j)^2\rangle_\varphi t/2\tau_0$. A relaxation rate related to trigonal warping is further deduced from $e^{-\frac{1}{2}\langle\delta^2\rangle_\varphi} = e^{-t/\tau_w}$, which yields $\tau_w^{-1} = 2\tau_0(\mu E_F^2/\hbar v_F^2)^2$, where τ_0 is the momentum relaxation time, $v_F = \sqrt{3}a\gamma_0/2\hbar = 10^6 \text{ m s}^{-1}$, and $\mu = 3\gamma_0 a^2/8\hbar^2$ ($\gamma_0 = 3$ eV, $a = 0.26$ nm). For the typical parameters in measured samples (Tikhonenko *et al.*, 2009), $\tau_w^{-1} = 0.001 \text{ ps}^{-1}$ for the Dirac region ($E_F \simeq 30$ meV, $\tau_0 \sim 0.1$ ps) and $\tau_w^{-1} = 0.3 \text{ ps}^{-1}$ for the highest measured concentration ($E_F \simeq 130$ meV, $\tau_0 \sim 0.05$ ps). Trigonal warping of the Fermi surface is found to have a very weak effect compared to estimated intravalley scattering timescales (Fal'ko *et al.*, 2007). One notes that trigonal warping effects should play an important role whenever τ_w becomes the smallest timescale, suppressing the two intravalley Cooperons together with their corresponding WAL fingerprints.

It is worth noting that two intervalley Cooperons are not affected by trigonal warping due to time-reversal symmetry of the system which requires $E(\mathbf{K}_\pm, \mathbf{p}) = E(\mathbf{K}_\mp, -\mathbf{p})$ (Fig. 5.19(a)). These two Cooperons cancel each other in the case of weak intervalley scattering, thus giving $\delta\sigma \sim 0$. However, intervalley scattering, with a rate τ_i^{-1} larger than the decoherence τ_φ^{-1} breaks the exact cancellation of the two intervalley Cooperons and partially restores weak localization. Following McCann *et al.* (2006), Kechedzhi *et al.* (2007), and Fal'ko *et al.* (2007), the global effect of Cooperons on the transport observable is finally recast (using Eq. (5.38)) as

$$\Delta\sigma(B) = e^2/\pi h \left\{ \mathcal{F}\left(\frac{\tau_B^{-1}}{\tau_\varphi^{-1}}\right) - \mathcal{F}\left(\frac{\tau_B^{-1}}{\tau_\varphi^{-1} + 2\tau_i^{-1}}\right) - 2\mathcal{F}\left(\frac{\tau_B^{-1}}{\tau_\varphi^{-1} + \tau_i^{-1} + \tau_*^{-1}}\right) \right\},$$

$$(5.39)$$

with τ_i, τ_ω, τ_s, and $\tau_B = \hbar/2eDB$ scattering times, while τ_φ denotes the coherence time. The above formula shows that depending on the relative strength of intravalley versus intervalley scattering, negative (weak localization) or positive magnetoresistance (weak antilocalization) will be obtained. The function $\mathcal{F}(x)$ can be approximated by $\mathcal{F}(x) \simeq x^2/24$ for $x \ll 1$, which has different consequences depending on the quantum coherence time.

In the high temperatures regime, τ_φ becomes small enough such that τ_i^{-1} and τ_*^{-1} (which are temperature independent) can be neglected next to τ_φ^{-1} and one easily reduces Eq. (5.39) to

$$\Delta\sigma(B) \simeq -\frac{2e^2}{\pi h} \mathcal{F}\left(\frac{\tau_B^{-1}}{\tau_\varphi^{-1}}\right) < 0. \qquad (5.40)$$

So when the magnetic field increases, $\Delta\sigma(B)$ decreases, and a weak antilocalization effect develops, instead of the usual weak localization correction. *In the low temperatures regime*, τ_φ becomes eventually very long, so that the second and third terms of Eq. (5.39) become negligible compared to the first one, so that

$$\Delta\sigma(B) \simeq \frac{e^2}{\pi h} \mathcal{F}\left(\frac{\tau_B^{-1}}{\tau_\varphi^{-1}}\right) < 0. \qquad (5.41)$$

We observe again the opposite phenomenon to that described at low temperatures, that is weak localization prevails. Finally, *in the intermediate temperatures regime*, when τ_φ^{-1} is in the order of the intervalley and intravalley scattering rates τ_i^{-1} and τ_*^{-1}, the final correction depends on the ratio τ_i/τ_*. When intravalley scattering dominates, then $\tau_* \ll \tau_i$, and weak antilocalization drives the magnetoconductance profile, whereas weak localization takes place if $\tau_i \ll \tau_*$.

This theory has been verified by a series of fine-tuned experimental measurements confirming the existence of weak antilocalization effects, whose origin thus stems from the pseudospin-driven Berry phase interference term (Tikhonenko *et al.*, 2008, 2009,

Horsell *et al.*, 2008). In particular, Tikhonenko and coworkers have experimentally confirmed such a complex phase diagram (see Fig. 5.19, right frame).

It is worth mentioning that the Cooperon theory, presented derived in the perturbative regime, remains at some point phenomenological, owing to the introduced elastic timescales beyond the momentum relaxation time (τ_0), which cannot be derived analytically for a general model of (realistic) disorder. Additionally, if τ_0 is directly connected to the diffusion coefficient in the diffusive regime, the other timescales are not directly accessible from zero-field transport coefficients, and their effect is only revealed through the positive or negative contribution of the total Cooperon to the semiclassical conductivity. Therefore, estimations of these temperature-independent timescales demand particular care and further scrutiny. Actually, numerical simulations give a further access to the connection (and crossovers) between these localization phenomena and the underlying nature of microscopic disorder. In the following, a single additional scattering time is found to be enough to measure the ratio between intravalley versus intervalley scattering, which dictates the dominating localization phenomenon (weak localization or weak antilocalization).

Crossover between weak localization and weak antilocalization: numerical analysis

Let us investigate localization effects under a magnetic field for disordered graphene with Gaussian impurities. We introduce a magnetic field through the Peierls phase (Ortmann *et al.*, 2011), with a flux per hexagon given by $\phi = \oint \mathbf{A} \cdot d\mathbf{l} = h/e \sum_{\text{hexagon}} \varphi_{\alpha\beta}$. A gauge is implemented where $\sum_{\text{hexagon}} \varphi_{\alpha\beta}$ can take integer multiples of $1/(N_x N_y)$ with N_x and $N_y = N_x + 1$ defining the sample size (Ortmann *et al.*, 2011).

The application of an external magnetic field is seen to induce some significant changes in the time dependence of the diffusion coefficients (Fig. 5.20). First for $W = 2$ (referred to as a strong disorder case), the time dependence of $D(t, B)$ exhibits a clear suppression of quantum interferences (Fig. 5.20(a)) with increasing magnetic strength, as expected in the weak localization regime. On reducing W to $W = 1.5$, this trend starts to deviate at $B = 0.137$ T since, after initial suppression of quantum interferences, the diffusion coefficient decreases with increasing magnetic field, although it remains constant at long times (Fig. 5.20(b)). Such a fingerprint of WAL indicates a stronger contribution of intravalley processes (sketched in Fig. 5.20, bottom pictures). This tendency becomes even more pronounced for $W = 1$ (Fig. 5.20(c)), but care in interpretation has to be taken once all intervalley scattering events have been suppressed (as suggested by Zhang *et al.*, 2009, and shown in Fig. 5.15), since activation of the Klein tunneling mechanism might then totally dominate scattering phenomena, jeopardizing the establishment of the diffusive regime, and WAL. This is further discussed below by analyzing the magnetoconductance patterns.

Several (theoretical) magnetoconductance $\Delta\sigma(B) = \sigma(B) - \sigma(B = 0)$ profiles are shown in the left panels of Fig. 5.21 and Fig. 5.22 for two different disorder strengths. For $W = 2$ (Fig. 5.21), the sign of $\Delta\sigma(B)$ (positive magnetoconductance) manifests a weak-localization behavior, regardless of the Fermi energy. Figure 5.21 (right) shows the experimental data obtained by Tikhonenko and coworkers which exhibit similar

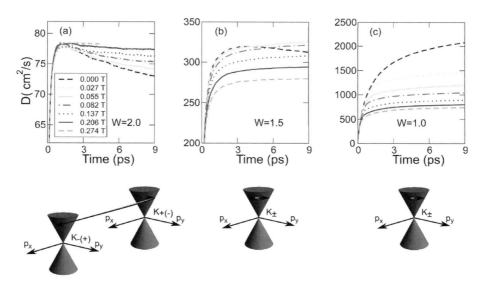

Figure 5.20 Magnetic field dependent Dirac point diffusion coefficient for several values of W ($W = 2$ (a), $W = 1.5$ (b) and $W = 1$ (c)), with fixed $n_i = 0.125\%$ and correlation length $\xi = 0.426$ nm. Adapted from Ortmann *et al.* (2011).

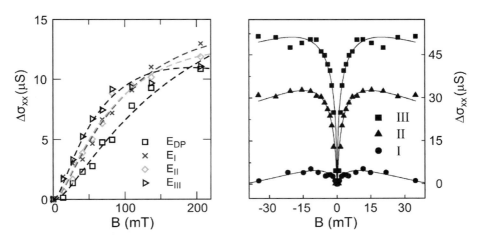

Figure 5.21 Left panel: $\Delta\sigma(B)$ ($n_i = 0.125\%$, $\xi = 0.426$ nm) for four different Fermi level positions ($E_{DP} = 0$, $E_I = 0.049$ eV, $E_{II} = 0.097$ eV and $E_{III} = 0.146$ eV). Averages over 32 different configurations are performed. Dashed lines are fits from analytical curves. Right panel: Experimental data extracted from Tikhonenko *et al.* (2009) obtained at $T = 5$ K. Here I, II and III refer to charge densities at about 2×10^{11} cm^{-2}, 10^{12} cm^{-2} and 2.3×10^{12} cm^{-2} respectively. (Right panel reproduced with permission from Tikhonenko *et al.* (2009). Copyright (2009) by the American Physical Society. Figure reprinted from Roche *et al.* (2012). Copyright (2012), with permission from Elsevier.)

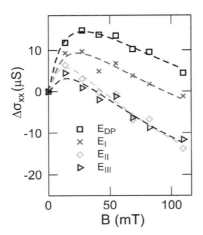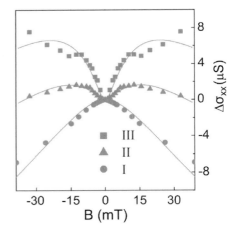

Figure 5.22 Left panel: $\Delta\sigma(B)$ for four different Fermi level positions ($E_{DP} = 0$, $E_I = 0.049$ eV, $E_{II} = 0.097$ eV and $E_{III} = 0.146$ eV) after Ortmann *et al.* (2011). 64 configurations have been averaged. Dashed lines are fits to analytical curves. Right panel: Experimental magneto-conductance curves measured at $T = 27$ K extracted from Tikhonenko *et al.* (2009). (Right panel reproduced with permission from Tikhonenko *et al.* (2009). Copyright (2009) by the American Physical Society. Figure reprinted from Roche *et al.* (2012), Copyright (2012), with permission from Elsevier.)

trends (Tikhonenko *et al.*, 2008, 2009). All experimental curves measured at low temperature ($T = 5$ K) also exhibit positive magnetoconductance whatever the induced charge densities (Fig. 5.21, right panel), which thus agrees with Eq. (5.41).

For smaller disorder ($W = 1.5$), the sign of $\Delta\sigma(B)$ is seen to change at sufficiently high magnetic field, suggesting a crossover from weak localization to weak antilocalization (Fig. 5.22, left panel). The estimation of the elastic mean free path ($\ell_{el} \sim 9-20$ nm) remains much smaller than the simulated sample size, warranting the occurrence of the diffusive regime and quantum interferences. Figure 5.22 (right) shows the experimental data at higher temperature, $T = 27$ K, for the same charge densities as in Fig. 5.21 (Tikhonenko *et al.*, 2008, 2009). Here, the onset of weak antilocalization at a higher temperature agrees with Eq. (5.40) derived earlier.

It is instructive to observe that in Zhang *et al.* (2009) (and Fig. 5.15) the strength of valley mixing is steadily enhanced when increasing disorder from $W = 1$ to $W = 2$ (keeping $\xi = 0.426$ nm). The strong contribution of intervalley scattering for the case $W = 2$ supports the computed positive magnetoconductance. Besides, by decreasing the disorder strength (from $W = 2$ to $W = 1.5$), WAL emerges in conjunction with the reduction of intervalley processes. We can also contrast numerical simulations with a phenomenological law that solely includes an extra single elastic scattering time τ_* (dictating the relative strength of valley mixing). Then for $W \geq 1$ both intravalley and intervalley processes jointly contribute, whereas intervalley scattering is suppressed when $W < 1$. The relevant expression for $\Delta\sigma(B)$ can be rewritten as

$$\Delta\sigma(B) = e^2/\pi h \left\{ \mathcal{F}(\tau_B^{-1}/\tau_\varphi^{-1}) - 3\mathcal{F}(\tau_B^{-1}/(\tau_\varphi^{-1} + 2\tau_*^{-1})) \right\}, \qquad (5.42)$$

where $\mathcal{F}(z) = \ln z + \psi(1/2 + z^{-1})$, $\psi(x)$ is the digamma function and $\tau_B^{-1} = 4eDB/\hbar$ (McCann et al., 2006). Least-squares fits (dashed lines) are superimposed on the simulated $\Delta\sigma(B)$ (symbols), taking $\tau_\varphi = 9$ ps (the maximum computed time). For $W = 2$ and $W = 1.5$, τ_* ranges within $[1.1 - 2.3]$ ps and $[1.5 - 6.3]$ ps, respectively (increasing values with increasing energy), thus confirming the weak localization regime for lowest B, which is fully consistent with McCann et al. (2006) ($\tau_i < \tau_\varphi$).

It is thus demonstrated that pseudospin effects are tunable by adjusting a single disorder parameter W, which denotes the depth of an impurity-driven local Coulomb potential. When $W \geq 1$ (unit of γ_0), local energetics between nearest neighbors A and B sites fluctuate enough to increase intervalley scattering, which progressively predominates over the intravalley contribution (Zhang et al., 2009). The comparison with experimental data, however, deserves an additional comment. Indeed, in the simulation the crossover from weak localization to weak antilocalization occurs as the Fermi level is moved from high energies to lower energies. The experimental data obtained at finite temperatures (Fig. 5.22) show an opposite trend. First, one should mention that the disorder model used here is kept constant with changing charge density. A self-consistent screening calculation of the scattering potential should be more suitable, but screening is expected to be less efficient for lower densities (screening will essentially decrease the value of W). Thus the crossover from WL to WAL is not expected to change by using energy-dependent (and differently screened) scattering potential. Second, the simulations are performed at zero temperature and should therefore be ultimately compared with the lowest temperature measurements. In Fig. 5.21, one sees the absence of WAL, which agrees with the simulation, considering a sufficiently deep onsite impurity potential.

5.2.7 Strong localization in disordered graphene

The simulation using the Kubo formula and Anderson disorder (Section 5.2.4) shows that the semiclassical conductivity remains high and always larger than or equal to the SCBA limit value $\sigma_{sc} = 4e^2/\pi h$. This value is first observed at the Dirac point. However, the conductivity would not be sensitive to localization effects except in the presence of some decoherence mechanisms such as electron–electron scattering or electron–phonon coupling (Lee & Ramakrishnan, 1985). Here at zero temperature, the time dependence of the diffusion coefficient clearly evidences the contribution of localization effects that develop beyond the diffusive regime. Figure 5.23(a) shows $D(E, t)$ at the Dirac point for different Anderson disorder strengths W. A saturation of $D(E, t)$ pinpoints the diffusive regime, while its further decay at longer times (which is enhanced for increasing W) indicates a larger contribution of the quantum interferences correction.

The 2D localization length ξ can be estimated using the fact that the transition to the insulating state occurs when the quantum correction, which scales as $\Delta\sigma(L) = (G_0/\pi) \ln(L/\ell_{el})$ (L is the length scale associated with the propagation time), is of the order of the semiclassical conductivity σ_{sc}. Given that localization occurs when

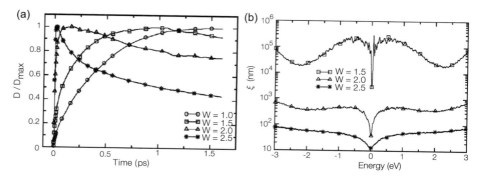

Figure 5.23 (a) Diffusion coefficient $D(E, t)$ as a function of time for various disorder strengths and Fermi energies. $D(E, t)$ has been normalized with respect to its maximum value $D_{max}(E)$ to allow an easier comparison between the different curves. (b) $\xi(E)$ for three disorder strengths (same legend as in Fig. 5.10).

$\Delta\sigma(L = \xi) \simeq \sigma_{sc}$, we thus obtain

$$\xi = \ell_{el}\exp(\pi\sigma_{sc}/G_0). \tag{5.43}$$

Some numerical results for the Anderson disorder, using Eq. (5.43), are shown in Fig. 5.23(b) for several disorder strengths (average over several tens of configurations has been performed). It is clear that the general shape of $\xi(E)$ is mainly dominated by the behavior of $\sigma_{sc}(E)$ (see Fig. 5.13), which is expected because of the exponential dependence of $\xi(E)$ on $\sigma_{sc}(E)$. As a result, although ℓ_{el} diverges when approaching the Dirac point (Fig. 5.12), $\xi(E)$ shows an opposite trend, with a minimum value at lowest energies. It is also worth observing that even for disorder strengths as large as $W \sim 6$ eV, $\xi(E = 0) \geq 10$ nm, while it quickly increases with energy, reaching several microns. For more realistic values of W in the order of 1 eV, $\xi(E) \sim 10\mu$m. Localization of electronic states in graphene with Anderson disorder is therefore very inefficient, although shorter localization lengths are found at the Dirac point. This feature is actually quite general for all types of short range disorder (see Section 7.3.1 for complementary analysis on oxygen-damaged graphene).

5.3   Quantum Hall effect in graphene

The quantum Hall effect, first discovered by Klaus von Klitzing (Klitzing, Dorda & Pepper, 1980), is a quantum-mechanical version of the Hall effect, observed in two-dimensional electron systems in the low-temperatures regime and strong magnetic fields, in which the Hall conductivity σ_{xy} becomes quantized,

$$\sigma_{xy} = 2Ne^2/h,$$

and the longitudinal conductivity becomes vanishingly small, $\sigma_{xx} \sim 0$. The prefactor N in σ_{xy} is the filling factor, and is either an integer ($N = 1, 2, 3, \ldots$) or a fractional number ($N = 1/3, 2/5, 3/7, 2/3, 3/5, \ldots$). The integer quantum Hall effect (IQHE)

(Klitzing *et al.*, 1980) is explained in terms of single-particle orbitals of an electron in a magnetic field and is related to the Landau quantization. Differently, the fractional quantum Hall effect (FQHE) fundamentally relies on strong electron–electron interactions, and the existence of so-called charge–flux composites known as composite fermions (Tsui, Stormer & Gossard, 1982, Laughlin, 1983).

5.3.1 Hall quantization in graphene

Owing to the peculiar nature of the Landau levels spectrum (see Sec. 2.7) with energy spacing given by $E_n = \text{sgn}(n)\sqrt{2\hbar v_F^2 eB|n|}$, the well-known integer quantum Hall effect (IQHE) (Klitzing *et al.*, 1980) observed in conventional two-dimensional electron systems transforms to a relativistic half-integer (anomalous) QHE in graphene whose quantized Hall conductivity becomes (Novoselov *et al.*, 2005b, Zhang *et al.*, 2005, Goerbig, 2011)

$$\sigma_{xy} = 4e^2/h \times (N + 1/2).$$

Such an anomalous QHE was simultaneously reported in the groups of Manchester University (Novoselov *et al.*, 2005b) and Columbia University (Zhang *et al.*, 2005). Figure 5.24 shows both the charge density dependence of the longitudinal resistivity (ρ_{xx}) and Hall conductivity (σ_{xy}) at 14 tesla and 4 K (Novoselov *et al.*, 2005b). Quantized plateaus of the Hall conductivity have also been reported at room temperature and low magnetic fields (Novoselov *et al.*, 2007).

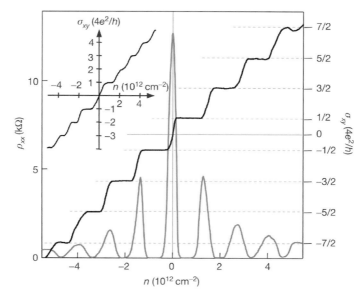

Figure 5.24 Longitudinal resistivity (ρ_{xx}) and Hall conductivity (σ_{xy}) as a function of charge density in monolayer graphene at 14 tesla and 4 K. The inset shows the case of bilayer graphene. (Reprinted by permission from Macmillan Publishers Ltd: *Nature*, Novoselov *et al.* (2005b), copyright (2005). Courtesy of Andre Geim.)

5.3.2   The mystery of the zero-energy Landau level splitting

One striking fingerprint of the graphene Landau level spectrum is the formation of a fourfold degenerate zero-energy Landau level (twofold valley and spin degeneracies) where electrons and holes coexist. The anomalous QHE is actually tightly interwoven with the π Berry's phase and pseudospin degree of freedom, which requires decoupled K_+ and K_- valleys (Ostrovsky, Gornyi & Mirlin, 2008). In the presence of disorder-induced sublattice symmetry breaking and strong valley mixing, QHE is predicted not to differ for any other two-dimensional system (Ostrovsky *et al.*, 2008, Aleiner & Efetov, 2006, Altland, 2006).

Several experiments performed in high-mobility samples have confirmed the presence of additional quantized Hall plateaus at $\sigma_{xy} = 0$, associated with splitting of the zero-energy LL. The level splitting may have different origins, such as spin and/or sublattice degeneracies, which could be driven by a Zeeman interaction or a disorder-induced symmetry-breaking effect (Li, Luican & Andrei, 2009); or electron–electron interactions (Zhang *et al.*, 2006, Jiang *et al.*, 2007, Nomura & MacDonald, 2006), although the issue remains complicated and probably material quality dependent. The nature of the $\sigma_{xy} = 0$ quantized plateau has also been envisioned as a possible manifestation of an unconventional dissipative QHE, which would assume a finite σ_{xx} value at the Dirac point (in between split Landau levels) (Abanin, Lee & Levitov, 2006, Abanin *et al.*, 2007, Zhang *et al.*, 2010, Jia, Goswami & Chakravarty, 2008, Checkelsky, Li & Ong, 2008).

In Abanin, Lee & Levitov (2006), Abanin *et al.* (2007), and Zhang *et al.* (2010), a dissipative QHE (with finite conductivity $\sigma_{xx} \sim 1 - 2e^2/h$) is predicted to be conveyed by the formation of counterpropagating (gapless) edge states carrying opposite spin. Finite σ_{xx} ($\simeq e^2/(\pi h)$) at the Dirac point has also been obtained numerically in some disordered graphene models (introducing bond disorder in a *tight-binding* model), and tentatively related to the formation of extended states centered at zero energy, but without evidence of a fully quantized σ_{xy} (Jia *et al.*, 2008).

The anomalous dissipative nature of the QHE remains, however, in total contradiction with many other experiments reporting divergent Dirac point resistivity (Checkelsky *et al.*, 2008), and a conventional nondissipative QHE regime. The measurement of a temperature-dependent activated behavior of $\sigma_{xx}(T)$ further supports the nondissipative nature of the plateau $\sigma_{xy} = 0$, in the presence of spin-splitting gap opening (Giesbers *et al.*, 2009, Kurganova *et al.*, 2011, Zhao *et al.*, 2012). Thus much effort is still required on both theoretical and experimental sides, to develop a comprehensive picture of QHE in disordered graphene. One can also expect that the physics of QHE in disordered graphene is much richer, with possible unconventional and defect-specific fingerprints of the magnetotransport features and localization/delocalization mechanisms. The possibility to structurally and chemically vary the quality and properties of graphene by chemical substitutions, functionalization, or the formation of hybrid materals (for instance graphene/boron-nitride samples) offers interesting challenges for further exploration of QHE.

5.3.3   Universal longitudinal conductivity at the Dirac point

We briefly analyze here the behavior of the dissipative conductivity (σ_{xx}) of the zero-energy Landau level in the presence of Anderson disorder, and also with or without additional A/B sublattice symmetry-breaking potentials. This illustrates some basic and general features of disorder-induced localization effects on QHE, and the modifications induced by energy-level splitting. The Kubo longitudinal σ_{xx} and Hall conductivities are computed in the presence of disorder and external magnetic fields with varying strengths (within the experimentally accessible range of parameters, from a few tesla to several tens of tesla). $\sigma_{xy}(E)$ is computed using a recently developed order N approach derived by Frank Ortmann (Ortmann & Roche, 2013). The longitudinal conductivity has already been discussed, whereas the novel algorithm for the Hall conductivity $\sigma_{xy}(E)$ is essentially implemented from

$$\sigma_{xy}(E) = -\frac{2}{V} \int_0^\infty dt e^{-\eta t/\hbar} \int_{-\infty}^\infty dE' f(E' - E)$$
$$\times \mathrm{Re}\left[\langle \varphi_{\mathrm{RP}} | \delta(E' - \hat{H}) \hat{j}_y \frac{1}{E' - \hat{H} + i\eta} \hat{j}_x(t) | \varphi_{\mathrm{RP}} \rangle \right], \qquad (5.44)$$

with $\hat{j}_x = \frac{ie_0}{\hbar}[\hat{H}, \hat{X}]$, the current operator ($\hat{X}$ the position operator), while $\eta \to 0$ is a small parameter required for achieving numerical convergence.

Figure 5.25 shows the conductivity $\sigma_{xx}(E, B, W)$ of the zero-energy LL for a graphene with Anderson disorder and a perpendicular magnetic field. A clear absence (or suppression) of localization is seen for disorder up to $W = 2$. The values of $\sigma_{xx}(E = 0, B, W \leq 2)$

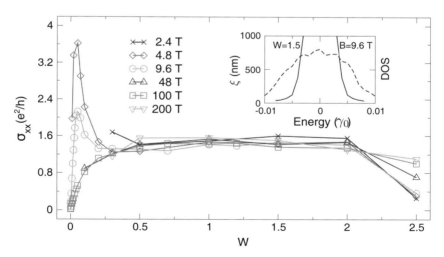

Figure 5.25 Main plot: Dirac point Kubo conductivity with varying Anderson disorder potential W. Localization length $\xi(E)$ (inset) at the center of the zero-energy Landau level (solid line) and density of states (dashed line, and arbitrary units) for $W = 1.5$ and $B = 9.6$ T. (Reproduced with permission from Ortmann & Roche (2013). Copyright (2013) by the American Physical Society.)

are shown in Fig. 5.25 (maximum computational time of the wavepacket spreading is 12 ps). At high enough energies, localization effects come into play in relation to the formation of mobility edges, as manifested by a finite localization length (ξ) at the tails of the zero-energy LL (see $W = 1.5$ in Fig. 5.25). When the disorder strength exceeds $W = 2.5$, all states including the zero-energy LL eventually localize, prohibiting the QHE state (Sheng, Sheng & Weng, 2006, Ortmann & Roche, 2013).

Actually, as seen in Fig. 5.25, three different transport regimes for σ_{xx} are identified depending on the disorder strength and applied magnetic field (Ortmann & Roche, 2013). In the absence of disorder, all states are localized by the magnetic field ($\sigma_{xx} \to 0$), whereas small disorder induces delocalization, as seen in the enhancement of σ_{xx} with W. For disorder such that $W \simeq 0.1$, σ_{xx} increases roughly linearly with W, regardless of the magnetic field (tuned from 4.8 T to 100 T). The value of $d\sigma_{xx}/dW$ depends on the magnetic field, and is seen to be larger for lower B. This actually agrees with the scaling of the magnetic length $l_B \propto B^{-1/2}$, which reduces disorder-induced delocalization effects as B is increased (and magnetic length is shortened).

Strikingly, for a large range of disorder values $W \in [0.3, 2]$ and for magnetic strengths varying between 2.4 T and 200 T, the dissipative conductivity σ_{xx} saturates to a constant value $\simeq 1.4e^2/h$. Finally, in the limit of strong disorder ($W > 2.5$), all states become localized and the system is driven to the insulating state (for moderate B). Differently from the low-W limit, the conductivity is higher for stronger fields indicating reversed roles of disorder and B field compared to the zero disorder limit, i.e. disorder localizes states while the magnetic field tends to suppress localization effects.

By introducing a sublattice symmetry-breaking disorder, pseudospin-split states are found to convey different critical bulk conductivities $\sigma_{xx} \simeq e^2/h$, regardless of the splitting and superimposed Anderson disorder strengths, dictating the width of the $\sigma_{xy} = 0$ plateau (Ortmann & Roche, 2013). The pseudospin-splitting is included through a heuristic model which shifts all onsite energies of A (and B) lattice sites by a constant quantity V_A (and V_B). When all A and B sites are differentiated in energy according to $V_A = -V_B$, a gap of $V_A - V_B = 2V_A$ is naturally formed. The superposition of both potentials mimics some weak imbalance in the adsorption site in the sense of a slightly preferred sublattice (Anderson disorder is such that $|V_A| \ll W$). In between pseudospin-split critical states, σ_{xx} eventually vanishes owing to intervalley-induced localization effects. Interestingly, it is found that by keeping the product pV_A constant, the same splitting strengths and split-gap (p the density of impurities breaking the sublattice symmetry) are obtained.

The analysis of $\sigma_{xx}(L)$ for various energies makes it possible to discriminate localized states from critical states, which remain delocalized as seen by a length-independent σ_{xx}. A typical result (σ_{xx}, σ_{xy}) is shown in Fig. 5.26 (maximum propagation time of wavepackets is $t = 6$ ps while disorder is set to $W = 0.2$). The maximum value of the doubly peaked σ_{xx} is found to be field-independent. Two peaks of σ_{xx} are actually related to disorder-induced pseudospin-splitting. Remarkably, the peak maxima σ_{xx} are not half of the maximum obtained in the unsplit case but reduced by a factor of $\simeq 0.7$. Figure 5.26 finally shows that $\sigma_{xx}(E = 0) \to 0$ while the double-peak height of $\simeq e^2/h$ is robust for different magnetic fields and disorder strength pV_A (Ortmann & Roche, 2013).

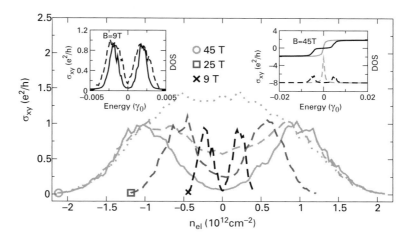

Figure 5.26 Longitudinal and Hall conductivities with and without sublattice impurity potential. Main plot: $\sigma_{xx}(t = 6 \text{ ps})$ for $W = 0.2$ using $pV_A = 0.0005$ (dotted line), $pV_A = 0.001$ (dashed lines), and $pV_A = 0.002$ (solid line). Magnetic fields as indicated. Left inset: $\sigma_{xx}(t = 6 \text{ ps})$ (solid line), DOS (dashed line) for $W = 0.2$, $pV_A = 0.004$. Right inset: Hall conductivity ($W = 0$) σ_{xy} for $pV_A = 0$ and $pV_A = 0.005$, $p = 2.5\%$, together with the corresponding DOS (dashed lines) at 45 T and $W = 0.002$. (Reproduced with permission from Ortmann & Roche (2013). Copyright (2013) by the American Physical Society.)

The Hall conductivity ($W = 0$) σ_{xy} for $pV_A = 0$ (grey) and $pV_A = 0.005$, $p = 2.5\%$ (black) and corresponding DOS (dashed lines) at 45 T and $W = 0.002$ are shown in Fig. 5.26 (right inset).

To date, available experimental data do not provide any universality concerning the critical values of σ_{xx}, which often show electron–hole asymmetry and variability (Giesbers *et al.*, 2009, Kurganova *et al.*, 2011, Zhao *et al.*, 2012). It is expected that the study of QHE in chemically modified graphene (for instance upon atomic deposition, hydrogen or oxygen adsorption, doping, strain, etc.) will reveal a plethora of additional interesting features of transport coefficients, allowing for an in-depth scrutiny of the connection between microscopic localization phenomena and the resulting phase diagram (σ_{xx}, σ_{xy} versus magnetic field and temperature) of the QHE.

5.4   Graphene with monovacancies

This section focuses on the effect of monovacancies (or missing carbon atoms) on quantum transport in graphene. Figure 5.27 shows a scanning tunnel microscope (STM) picture of a single monovacancy on graphite previously irradiated with Ar^+ ions (Ugeda *et al.*, 2010). Although various types of vacancies have been observed in disordered graphene, monovacancies are particularly interesting since they do not break electron–hole (chiral) symmetry. By breaking A/B sublattice symmetry, vacancies offer the possibility (similarly to adsorbed hydrogen atoms) to manipulate the formation of local

Figure 5.27 STM picture of a single vacancy in graphite. (Reproduced with permission from Ugeda *et al.* (2010). Copyright (2010) by the American Physical Society. Courtesy of J. M. Gomez-Rodriguez and Ivan Brihuega.)

magnetic moments and long range magnetic ordering, although the issue is currently the source of intense theoretical and experimental debate.

Monovacancies particularly affect the electronic structure of graphene at the Dirac point, introducing zero-energy modes whose impact on the Dirac point transport physics could be fundamental for explaining experiments in ultraclean graphene (Ponomarenko *et al.*, 2011). Also, transport properties have been predicted to be very sensitive to the way the sublattice symmetry is broken (Ostrovsky *et al.*, 2006). For equally distributed vacancies among the two sublattices, a saturation (or even increase) of the conductivity at (from) $\sigma_0 = 4e^2/\pi h$ with increasing density of vacancies has been reported in the tunneling regime of a very short graphene channel (Ostrovsky *et al.*, 2010). Other calculations have found a saturation or even increase of the conductivity with density of defects (Yuan, De Raedt & Katsnelson, 2010, Zhu *et al.*, 2012), in contradiction with localization theory. We explore here transport in graphene with vacancies using the Kubo method, which makes it possible to provide a more comprehensive explanation of effects of monovacancies on quantum transport properties in disordered graphene.

A vacancy in the honeycomb lattice leaves three dangling covalent bonds, which might eventually reassemble into one double bond and one dangling bond, but here we restrict the study to nonreconstructed vacancies. The vacancies are then theoretically modeled by suppressing hopping terms between the orbital at a vacancy site and its nearest neighbors. Another possibility of describing the vacancy at a given site is to add a very large onsite potential (e.g. in the order of $U_v = -6\gamma_0 = +17.4\,eV$ for the simulation presented later). The Hamiltonian is given by

$$\mathcal{H} = -\sum_{<i,j>} \left(\gamma_0 |i\rangle \langle j| + h.c. \right) - \sum_{<i_v,j>} \left(\gamma_0 |i_v\rangle \langle j| + h.c. \right) + \sum_{i_v} U_v |i_v\rangle \langle i_v|, \qquad (5.45)$$

where i_v indexes the ith vacancy site and the brackets in the summation indices indicate that the corresponding sum is over nearest neighbors. An important issue is the distribution statistics of the vacancies among the two A and B sublattices. Indeed, using the rank-nullity theorem, an imbalance of vacancies between the two sublattices A and B has been shown to create zero-energy modes (Pereira, dos Santos & Neto, 2008).

This demonstration requires taking only first-neighbor interactions into account, and goes as follows. Consider a system with m vacancies on lattice B and $(m+1)$ vacancies on lattice A. Both lattices have the same number of sites: $N_A = N_B = N_{sites}/2$. Using the AB representation of the Hamiltonian, Eq. (5.45), and reordering the real-space basis of subspaces A and B so that sites corresponding to vacancies come last, we get for the interaction matrix $\mathcal{H}_{BA}$ operating from subspace B onto subspace A:

$$\mathcal{H}_{BA} = \begin{pmatrix} \tilde{\mathcal{H}}_{BA} & \mathcal{O}_{m,N_A-m} \\ \mathcal{O}_{N_B-m} & \mathcal{O}_m \end{pmatrix}, \tag{5.46}$$

where $\mathcal{O}_{n,p}$ is the null matrix of dimension $n \times p$. Since there are $m+1$ vacancies on sublattice A, the $N_A - (m+1)$th basis vector in subspace A represents a vacancy. It verifies: $\mathcal{H}_{BA}|N_A - (m+1)\rangle = 0$. Consequently the last line of matrix $\tilde{\mathcal{H}}_{BA}$ is null and this operator has at least one eigenvector associated with the eigenvalue 0. From this vector we can obtain a vector in B space by completing with zeros. We call this new vector φ. Also, we have

$$\mathcal{H}_{BB} = \begin{pmatrix} \epsilon_B \times \mathbb{I}_{N_B-m} & \mathcal{O}_{N_B-m,m} \\ \mathcal{O}_{m,N_B-m} & (\epsilon_B + U_v) \times \mathbb{I}_m \end{pmatrix}, \tag{5.47}$$

where $\mathbb{I}_n$ is the identity matrix of size n, and ϵ_B the onsite energy on the B sublattice in the clean case. By construction, φ verifies: $\mathcal{H}_{BB}(\varphi) = \epsilon_B \varphi$. Finally, applying the total Hamiltonian (5.45) to vector $(0, \varphi)$ yields

$$\mathcal{H} \begin{pmatrix} 0 \\ \varphi \end{pmatrix} = \begin{pmatrix} 0 \\ \mathcal{H}_{BA}(\varphi) + \mathcal{H}_{BB}(\varphi) \end{pmatrix} = \epsilon_B \begin{pmatrix} 0 \\ \varphi \end{pmatrix}. \tag{5.48}$$

Therefore, the additional vacancy on one sublattice creates an electronic state at the onsite energy and located on the other sublattice. Since for graphene $\epsilon_B = \epsilon_A = 0$, these modes will be created at the Dirac point. This creation of modes at zero energy mainly affects the spectrum near the Dirac point. The formation of an energy gap Δ_g has been derived in Pereira et al. (2008), and for a vacancy concentration n, it is predicted that

$$\Delta_g \simeq \frac{\hbar v_F}{d}, \tag{5.49}$$

with $d = n^{-1/2}$ the average distance between vacancies. Note that when vacancies are equally distributed over the two sublattices, zero modes should not appear if $U_v \neq 0$. Also, even if the distribution is numerically enforced to be random over both sublattices, some zero-energy modes appear due to statistical error, as might be the case in the numerical study (Zhu et al., 2012), in which unexplained zero modes have been reported. In the following, we enforce a strict distribution of vacancies among both sublattices, with a focus on the two cases of main interest, namely the situation for which vacancies are equally distributed among different A and B sublattices, and the situation for which all vacancies belong to the same sublattice, say the A sublattice (AA).

5.4.1   Electronic structure of graphene with monovacancies

Electronic structure calculations are performed using the Lanczos recursion method on a sample of 10^6 atoms with periodic boundary conditions (Section D.1). This sample size allows for a randomization of the distribution of vacancies. The chosen parameters for the Lanczos calculations are $N = 1500$ recursion steps with energy resolution of $\eta = 0.005|\gamma_0| = 0.015$ eV. We investigate several concentrations of vacancies n up to 1%, for the two cases where the vacancies are either equally distributed among the two sublattices (AB) or restricted to a single sublattice (AA).

The numerical results for the DOS in the AB case are plotted in Fig. 5.28 for vacancy densities varying from 0.1% to 1%. We restrict the analysis to the energy region around the Dirac point where most of the modifications occur. Note that we also observe a softening of the van Hove singularities at higher energies (not shown here), but hole–particle (chiral) symmetry is conserved.

The vacancies induce an increase of the spectral density around the Dirac point, with a flattening with increasing density of vacancies. Although the DOS seems to increase close to the Dirac point, as in Pereira *et al.* (2008) the chosen numerical resolution to capture the zero-energy physics must be very high. DOS in the AA case are plotted in Fig. 5.29, for the same concentrations. Here again, the system remains particle–hole symmetric. As expected the breaking of A–B symmetry (see Fig. 5.29, insets) generates a peak at zero energy, although it has been softened numerically.

The peak height increases with vacancy concentration and induces a decay of the DOS on each side of the Dirac point, which is actually related to the formation of gaps. For more clarity, we define an energy ϵ up to which the DOS can be considered as negligible and plot, in Fig. 5.30, Δ_g against $n^{1/2}$, which is found to scale linearly, agreeing with Eq. (5.49).

In both AB and AA cases, vacancies preserve hole–particle symmetry and affect the electronic structure around the Fermi energy, although in a different manner. In the

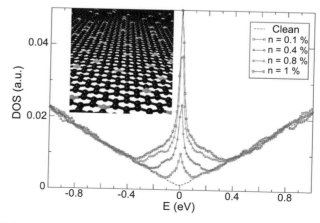

Figure 5.28 DOS for vacancy concentrations $n = 0.1, 0.4, 0.8, 1\%$ distributed equally over the two sublattices, and for the clean case. Inset: Ball-and-stick views of graphene with AB vacancy distribution. Vacant sites and bonds are colored in grey.

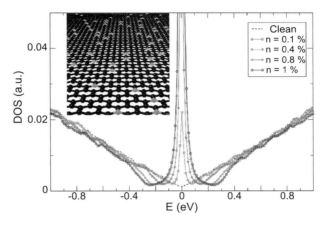

Figure 5.29 DOS for same concentrations of vacancies as in Fig. 5.28 distributed on one sublattice only. Inset: Ball-and-stick views of graphene with AA vacancy distribution. Vacant sites and bonds are colored in grey.

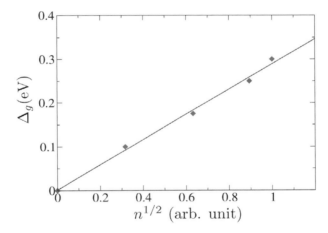

Figure 5.30 Width of the gap Δ_g estimated from Fig. 5.29 as a function of $\sim n_i^{1/2}$, with n_i the density of vacancies. The continuous line is a linear fit of the simulation.

first case the DOS increases and tends to flatten, while for the AA distribution there is a depletion of the DOS around the Fermi energy with a finite concentration of zero modes in the middle.

5.4.2   Transport features of graphene with monovacancies

We calculate the conductivity in the semiclassical and quantum regimes using the Kubo method and analyzing the dynamics of electronic wavepackets (Section D.1) with maximum propagation times of 1000 steps with respective time steps $T_1 = \hbar/\gamma_0 \simeq 1.43$ fs and $T_2 = 10\,T_1$. The random phase states (RPS) evolve during a little more than 15 ps, and the total system is a rectangular sheet of 212×122 Å^2, chosen large enough to limit finite-size effects (Cresti *et al.*, 2013). For each concentration and distribution of vacancies, the maximum of the diffusion coefficient $D^{max}(E)$ is first estimated, together with the mean free path $\ell_{el}(E)$ and the Fermi velocity $v(E)$. From these values, the semiclassical conductivity in the diffusive regime is deduced, σ_{sc}. We also follow the time-evolution of the Kubo conductivity to quantify the contribution of quantum interferences, using the approximation

$$\sigma = \frac{e^2}{2}\rho(E)\frac{D(E,t)}{t}. \qquad (5.50)$$

Numerical results for the fully balanced AB configuration are summarized in Fig. 5.31. The semiclassical conductivity is plotted as a function of energy (main plot, full line) with a density of vacancies of 0.8%. Away from the Dirac point, σ_{sc} increases with energy, while it exhibits a plateau above the value $\sigma_{sc} = 4e^2/\pi h$ (black, dotted line). This result confirms earlier theoretical predictions (Ostrovsky *et al.*, 2010, Zhu *et al.*, 2012). The scale-dependent conductivity for the same concentration of vacancies is also shown after a time $t = 15$ ps (dashed and dotted line), and seen to decay below σ_0, pinpointing the onset of localization phenomena, which are stronger for larger vacancy concentration (as illustrated by the dashed curve which corresponds to the conductivity at timescale $t = 15$ ps for a concentration $n = 1\%$). To confirm the presence of localization effects, the evolution of diffusion coefficients $D(E,t)$ at $n = 0.8\%$ is shown for several energies in the region of the plateau. Values of $D(E,t)$ are seen to reach

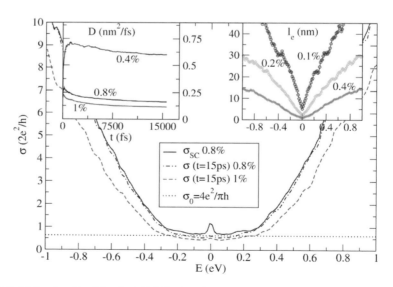

Figure 5.31 Main plot: Semiclassical (full line) and Kubo conductivity (dashed line, dash and points) versus energy for different concentrations of vacancies, equally distributed among sublattices. Left inset: Evolution of diffusion coefficients calculated at $E = 0.15$ eV, for different concentrations of vacancies. Right inset: Elastic mean free path against energy for different vacancy concentrations.

a saturation regime after a few femtoseconds, and then start to decline sublinearly, as expected in a weak localization regime (see Fig. 5.31, left inset).

Figure 5.31 (right inset) further displays $\ell_{el}(E)$ for various concentrations: 0.1, 0.2 and 0.4% (from top to bottom). At the Fermi energy, the mean free path is below 10 nm for a concentration of 0.1%, while for 0.4%, $\ell_{el}(E)$ is just a few interatomic distances, suggesting even shorter localization lengths. Only at sufficiently higher energies do we observe that ℓ_{el} roughly scales as $1/n$. An important feature revealed by these simulations is that the energy scale on which transport is affected by vacancies matches the one found previously when studying the electronic structure (see Fig. 5.28). All these results concerning the AB configuration are consistent with previous studies on weakly hydrogenated graphene (Soriano et al., 2011, Leconte et al., 2011). It is also interesting to note that the Dirac point conductivity scaling follows a power law, which could be a fingerprint of zero-energy modes, see Cresti et al. (2013) for details.

Figure 5.32(a) (main plot) presents the semiclassical conductivity in the AA case for $n = 0.4\%$, for two different energy resolutions: $\eta = 0.005\gamma_0 \simeq 0.015$ eV (dashed line) and $\eta = 0.001\gamma_0 \simeq 0.003$ eV (full line). Both curves exhibit a behavior similar to that found in the AB case (see Fig. 5.31), except for a peak lying in the middle of the plateau, at the Fermi energy. The first curve, for $\eta = 0.015$ eV, saturates above σ_0, as was observed by Leconte et al. (2011). However with increasing degree of resolution in energy (for the same density of vacancies), the conductivity drops dramatically and goes below the classical limit; only the midgap peak is increased. The inset reveals the underlying phenomenon: here are plotted corresponding DOS for the two energy resolutions (the better the resolution, the higher the DOS obtained). Figure 5.32(b) (main plot) exhibits a similar evolution of the $\sigma_{sc}(E)$ when increasing the density of vacancies to $n = 0.8\%$. For both energy resolutions, σ_{sc} decays under σ_0 around the Dirac point, with a marked peak exactly at the Fermi energy. For $\eta = 0.003$ eV and $n = 0.8\%$, the DOS in the region close to the Dirac point ($E < 0.25$ eV) strongly resembles a gap within numerical resolution. This evolution towards a gap when decreasing the resolution confirms the results presented in Pereira et al. (2008), which suggest that in the zero temperature limit ($\eta \to 0$) the system is an Anderson insulator at the Dirac point.

It is interesting to note that calculation of the diffusion coefficients $D(t)$ for states lying inside the gap yields a misleading result. An explanation is given in Fig. 5.32(b) (inset), for the two resolutions $\eta = 0.015$ eV (full lines) and $\eta = 0.003$ eV (dashed lines). These coefficients quickly reach a saturation regime but then surprisingly remain constant at long elapsed times, suggesting a diffusive regime in which quantum effects would be suppressed. This aspect is indeed very peculiar: looking only at $D(t)$, one could conclude that localization effects have been suppressed when enforcing the vacancy distribution on one sublattice only. Nevertheless, the observed formation of a gap in the density of states clearly suggests that $D(t)$ should be driven to zero in the limit $\eta \to 0$, which is confirmed by increasing the energy resolution from $\eta = 0.015$ to 0.003 eV. However, the spectral weight of these states is negligible on average, and should not bring any relevant contribution to the temperature-dependent conductivity.

Let us now look at the transport features of the system for energies out of the gap. To that end, one scrutinizes diffusion coefficients for several energies away from the Fermi

(a)

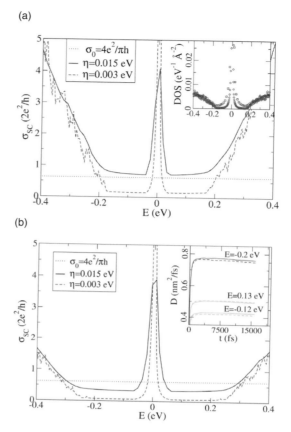

(b)

Figure 5.32 (a) Main plot: $\sigma_{sc}(E)$ for $n = 0.4\%$ of vacancies on one sublattice, for two different energy resolutions $\eta = 0.003$ (dashed line), 0.015 eV (full line). Inset: Corresponding DOS (squares: $\eta = 0.015$ eV, diamonds: $\eta = 0.003$ eV). (b) Main plot: $\sigma_{sc}(E)$ for $n = 0.8\%$ of vacancies on one sublattice, for two different energy resolutions $\eta = 0.003$ eV (dashed line), 0.015 eV (full line). Inset: Evolution of diffusion coefficients at chosen energies.

energy, compared with the situation exactly at the Dirac point, for the AA and AB case with 0.4% of vacancies (Fig. 5.33)

Focusing on the zero-energy modes, the peak observed in σ_{sc} (see Fig. 5.32) suggests a strong contribution of zero-mode states to transport. Here the diffusion coefficients are actually seen to be extremely small (rescaled by a factor 10). After reaching the saturation regime, $D(t)$ strongly decays and saturates to values one order of magnitude below the values for the other coefficients. This zero-energy diffusion coefficient, in the fully balanced case, decays after saturation, thus confirming the contribution of strong localization effects (as shown in Fig. 5.31). Besides, far from the gap region of the AA case, one observes no significant difference between the wavepacket dynamics for the two distributions of vacancies (see the diffusion coefficients at chosen energies $E = -0.8$ and 0.5 eV).

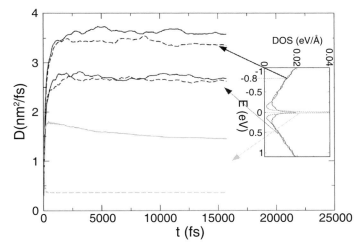

Figure 5.33 Main plot: Diffusion coefficients plotted for different densities and vacancy distributions. $D(E = 0, t)$ has been rescaled by a factor 10. Inset: Corresponding DOS. Arrows show the selected energies for calculating $D(t)$.

5.5 Polycrystalline graphene

5.5.1 Motivation and structural models

Chemical vapor deposition (CVD) enables the growth of very large films of relatively good quality graphene. However, these films are polycrystalline and incorporate a large number of interconnected domains (of different sizes), through grain boundaries which are formed roughly as lines of structural defects and differ randomly in crystalline orientations one from another. The formation of extended topological defects is observed at the interface between grains: the grain boundaries (GBs).

Figure 5.34 (in UHV and low temperature) shows a scanning tunneling microscope picture of a grain boundary in graphene, where the different orientations of the two adjacent grains are clearly visible on the corresponding Moiré pattern. These interface regions contain a large amount of structural defects, especially odd-numbered carbon rings, that will act as strong scattering centers, limiting charge mobilities. Moreover, a grain boundary breaks the translational symmetry of the system so that wavefunctions from the two coalescent grains interfere in a destructive manner along the GB, which generate interface states. A theoretical model (Ferreira *et al.*, 2011) proposes charge accumulation along grain boundaries, creating an electrostatic potential that would cause GBs to act as extended electron scatterers.

We note that certain highly symmetric line defects (as shown in Fig. 2.42(c), Section 2.8.2) can actually give rise to an interesting valley-filtering effect (Gunlycke & White, 2011). For instance, in the case depicted in Fig. 5.35(a), the incoming low-energy

Figure 5.34 Scanning tunneling microscope picture of a grain boundary. Two different Moiré patterns for individual grains are clearly visible. Courtesy of J.M. Gomez-Rodriguez.

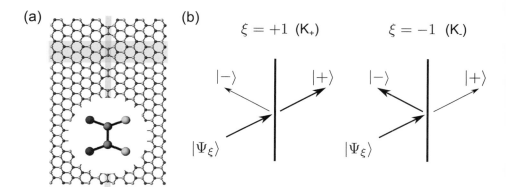

Figure 5.35 (a) A line defect in graphene. (b) Valley states scattering off the line defect. The sublattice symmetric $|+\rangle$ and antisymmetric $|-\rangle$ components of the incident state $|\Psi_\xi\rangle$ are transmitted and reflected, respectively. At the two valleys, K_+ and K_-, an incident quasiparticle state is defined by the valley index ($\xi = \pm 1$) and wavevector k, where the latter points in the direction k given by the angle of incidence θ_k. The scattering is shown to be valley-dependent (Gunlycke & White, 2011). ((a) is reproduced with permission from Gunlycke & White (2011). Copyright (2011) by the American Physical Society.)

electron states are given by Eq. (2.27) that we repeat here:

$$|\Psi_{\xi,s}\rangle = \frac{1}{\sqrt{2}} \begin{pmatrix} 1 \\ se^{+i\xi\theta_k} \end{pmatrix}, \tag{5.51}$$

where ξ is the valley index ($\xi = \pm 1$ gives the wavefunction in the $K_\pm$ valley) and $s = \pm 1$ is the band index (electron or hole band). The angle $\theta_k = \arctan(k_y/k_x)$ defines the angle of the wavevector measured from the center of the corresponding valley. The first (second) component of the vector on the right-hand side of the previous equation gives the probability amplitude in the A (B) sublattice.

Following Gunlycke & White (2011), let us consider waves incident from the left and write $\mathbf{k} = k(\cos\theta_k, \sin\theta_k)$ with $k = |\mathbf{k}|$. Then, using these relations one can write the right-moving solutions as:

$$|\Psi_\xi\rangle = \frac{1}{\sqrt{2}} \begin{pmatrix} 1 \\ ie^{-i\xi\theta_k} \end{pmatrix}. \tag{5.52}$$

In the limit $k \to 0$, the reflection operator and the graphene translation operator perpendicular to the line defect commute. This allows construction of symmetry-adapted states $|\pm\rangle$ that are eigenstates of both operators (do not confuse the $\pm$ with the values of ξ). The reflection operator maps A onto B sites, and vice versa, and therefore it can be represented by the operator σ_x acting on the two sublattices. From the eigenstates of σ_x, we get

$$|\pm\rangle = \frac{1}{\sqrt{2}} = \begin{pmatrix} 1 \\ \pm 1 \end{pmatrix}. \tag{5.53}$$

The graphene states expressed in the symmetry-adapted basis become

$$|\Psi_\xi\rangle = \frac{1 + ie^{-i\xi\theta_k}}{2}|+\rangle + \frac{1 - ie^{-i\xi\theta_k}}{2}|-\rangle. \tag{5.54}$$

Gunlycke & White (2011) found that there exist two symmetric states at the Fermi level without a node on the line defect, carrying quasiparticles across the line defect without scattering. The calculation of the transmission probability across the line defect is

$$T_\xi = |\langle+|\Psi_\xi\rangle|^2 = \frac{1}{2}(1 \pm \sin\theta_k),$$

which immediately means that the line defect is semi-transparent and its transparency depends on the valley ($T_{K_+} + T_{K_-} = 1$). At a high angle of incidence, there can be almost full transmission or reflection, depending on the valley index ($K_\pm$).

The impact on transport of a single boundary has also been measured experimentally (Yu *et al.*, 2011). Figure 5.36, (a) and (b), shows respectively the experimental setup and current–voltage curves measured within two adjacent CVD-grown grains (curves for $V_{7,8}$ and $V_{9,10}$) and at the boundary (curve for $V_{8,9}$). All curves exhibit a linear behavior from which the conductance can be derived (e.g. $G_{8,9} = \frac{dI}{dV}|_{V_{8,9}}$). From this measurement, it is clear that the current is reduced (together with corresponding conductance) when electrons have to cross a grain boundary. A complementary study by Tsen and coworkers (Tsen *et al.*, 2012) combining electron transmission microscopy with electrical measurements has also revealed links between electrical and geometrical properties of GBs. However, these studies have generally been carried out on few samples and for a better examination of the potential of CVD-graphene for applications (such as transparent electrodes), a fundamental connection between the morphologies

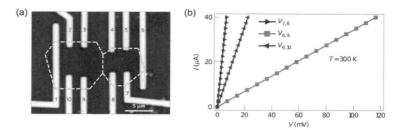

Figure 5.36 (a) An experimental device with multiple electrodes forming contact over two coalesced grains (dashed lines). (b) Current versus bias voltage, measured within each grain and across the grain boundary. The legend shows the corresponding electrode pairs for each curve. (Adapted with permission from Macmillan Publishers Ltd: *Nature Materials*, Yu *et al.* (2011), copyright (2011).)

of polycrystalline graphene and their transport features would be highly desirable (see Section 8.2 for complementary discussion).

To date, however, such a comprehensive description of the correlation between mobility and average grain size, for instance, is crucially missing (Li *et al.*, 2010). Additionally, although CVD grown films can be produced at large scales, their corresponding electrical performances remain a bit disappointing when compared with those of exfoliated graphene. A better understanding of the impact of GBs at the macroscopic scale would be desirable to guide experimentalists for further optimization of the growth processes.

At present, based on atomic-resolution images (Huang *et al.*, 2011, An *et al.*, 2011, Kim *et al.*, 2011, Kurasch *et al.*, 2012) and theoretical models (Yazyev & Louie, 2010*a*), it is a relatively well-established fact that different grains in polycrystalline graphene preferably stitch together predominantly via pentagon–heptagon pairs. Additionally, diffraction-filtered imaging has provided some mapping of the location, orientation and shape of hundreds of grains and boundaries, revealing an unexpectedly small and intricate patchwork of grains interconnected by tilt boundaries (Kim *et al.*, 2011). However, a typical grain boundary is covered by adsorbates, which indicates that the local configuration of atoms in these areas can differ from the ideal pentagon–heptagon chains.

In this section, the electronic and transport properties of polycrystalline graphene models introduced in Kotakoski & Meyer (2012) are presented. The polycrystalline graphene structures are created by initially fixing only the number of grains and the size of the sample and then randomly growing each of the grains with similarly random orientation.

The dynamical growth of the structure stops when all the grains have reached neighboring grains in all directions. The equilibrium structure is then achieved using molecular dynamics simulations (Kotakoski & Meyer, 2012). This procedure leads into corrugated structures with realistic misorientation angle distribution and ring statistics (see Fig. 5.37). Such structures have been additionally flattened while ensuring that the local atomic density at any point in the structure remained reasonable. For the rest of the

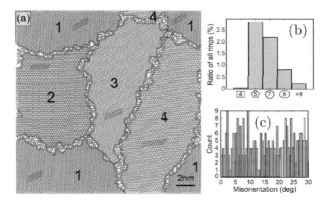

Figure 5.37 Model structures for polycrystalline graphene. (a) Top view of a periodic 20 nm × 20 nm graphene sheet with four grains, as marked by the numbered shaded areas. The lines indicate orientations of the graphene lattice within each grain. (b) Distribution of misorientation angles for the bicrystalline sample structures used in this study. (c) Relative probabilities for nonhexagonal carbon rings in the same structures. (Reproduced with permission from Kotakoski & Meyer (2012). Copyright (2007) by the American Physical Society.)

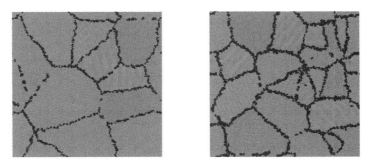

Figure 5.38 Left and right panels are ball-and-stick models for samples S1 and S2 respectively, with grain boundaries outlined in dark tint. Courtesy of Jani Kotakoski.

study, let us focus on the electronic and transport study of two different samples (from here designated S1 and S2; see Fig. 5.38) in which the average grain sizes are 18 nm and 13 nm respectively.

5.5.2   Electronic properties of polycrystalline graphene

The electronic and transport properties of these disordered lattices were investigated using a π–π* orthogonal *tight-binding* (TB) model, described by a single p_z orbital per carbon site, with nearest neighbors hopping γ_0, and zero onsite energies. The study was carried out on about 600×600 Å^2 sheets including $N_{\text{sites}} = 138\,292$ carbon atoms for sample S1 and $N_{\text{sites}} = 137\,985$ carbon atoms for sample S2 (shown in Fig. 5.38).

Periodic boundary conditions were applied to the structures to minimize finite-size effects. A criterion to search the first nearest neighbors was set empirically to $1.15 \times a_{CC}$. The local fluctuations of bond length are small enough to reasonably keep a constant value of γ_0 for the transfer integral. The density of states is then computed using the Lanczos recursion method with $N = 1000$ recursion steps and an energy resolution $\eta = 0.01\gamma_0 \simeq 0.03$ eV (Tuan *et al.*, 2013).

The DOS of S1 and S2 show little difference from that of pristine graphene (Fig. 5.39 main plot). This suggests that grain boundaries correspond to weak disorder preserving electron–hole symmetry. Only the presence of some enhanced density of zero energy modes and a slight smoothing of van Hove singularities at $E = \pm\gamma_0$ reveal the presence of the disorder. Figure 5.39 (inset) shows that S2 (which is more fragmented than S1) has a larger DOS especially close to the charge neutrality point, reflecting a higher density of midgap states (Stauber *et al.*, 2007).

We next identify grain boundaries by searching for atoms for which the bond length of at least one nearest neighbor differs from the pristine carbon spacing ($a_{CC} = 1.42$ Å) by 0.03 Å or more. Figure 5.39 (main plot) shows the average of local DOS (LDOS) over all boundary sites of S1 (applying the recursion method to a random phase state strictly located on boundary sites). The averaged density shows a very marked contribution of midgap states (Stauber *et al.*, 2007), together with a strong suppression of van Hove singularities at $\pm\gamma_0 = \pm2.9$ eV. The electron–hole symmetry is also considerably broken owing to the presence of many odd-membered rings along the GB.

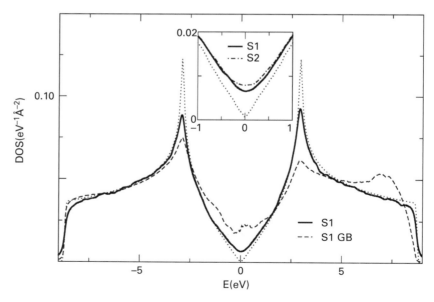

Figure 5.39 Main plot: DOS for pristine graphene (dotted line), polycrystalline graphene S1 (bold line), and the DOS average over grain boundary sites only of S1 (dashed line). Inset: Zoom in the DOS of S1 and S2 (dot-dashed) together with pristine graphene (dashed lines).

5.5.3   Mean free path, conductivity and charge mobility

Figure 5.40 (inset) shows the time dependence of $D(t)$ at the Dirac point for both samples S1 and S2. From the maximum values, $\ell_{el}(E)$ (and σ_{sc}) are deduced (Fig. 5.40, main plot). Despite some genuine electron–hole asymmetry for energies higher than 3 eV (far from the experimentally relevant energy window), ℓ_{el} are found to be weakly changing over an energy window around the charge neutrality point, with $\ell_{el}(E, \text{S1}) \in [6, 10]$ nm for S1 and $\ell_{el}(E, \text{S2}) \in [4, 7]$ nm for S2. In Fig. 5.40, we also show a rescaled energy-dependent value given by $\sqrt{2} \times \ell_{el}(E, \text{S2})$ which is surprisingly close to the behavior of $\ell_{el}(E, \text{S1})$ for S1. It turns out that the corresponding grain size perfectly matches with such a rescaling factor; that is, the grain size for S2 is about $\sqrt{2}$ smaller than the typical grain size of S1.

We have therefore identified a remarkably simple scaling law linking the grain size morphologies with transport length scales. We observe in Fig. 5.41 (inset) the corresponding energy dependence of $\sigma_{sc}(E)$, which manifests energy-dependent variation similar to the mean free path (as well as some linear dependence with charge density in the low-energy limit). An interesting feature is that $\sigma_{sc}(E)$ remains much larger than the minimum value $4e^2/\pi h$ (horizontal dashed line), which fixes the theoretical limit in the diffusive regime, as derived within the self-consistent Born approximation and valid for any type of disorder. This indicates that polycrystalline graphene remains a good conductor.

The charge mobility, $\mu(E) = \sigma_{sc}(E)/en(E)$, with $n(E)$ being the carrier density, is found to vary within $100 \text{ cm}^2\text{V}^{-1}\text{s}^{-1}$ to $\sim 2 \times 10^4 \text{ cm}^2\text{V}^{-1}\text{s}^{-1}$ for $n = 10^{12} - 10^{13}\text{cm}^{-2}$, in very satisfactory agreement with the typically reported values for polycrystalline graphene in recent literature (Tsen *et al.*, 2012). We stress that the computed values of $\mu(E)$ are valid down to the charge neutrality point (that is to the smallest charge density

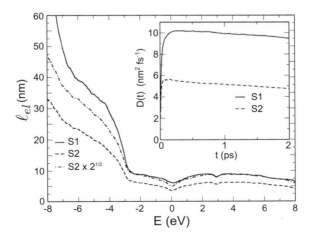

Figure 5.40 Main plot: Mean free path for S1 ($\ell_{el}(E, \text{S1})$, solid curve) and S2($\ell_{el}(E, \text{S2})$, dashed curve), together with a rescaled value $\sqrt{2} \times \ell_{el}(E, \text{S2})$ (dot-dashed curve). Inset: Time evolution of D at the Dirac point for sampler $S1$ and $S2$.

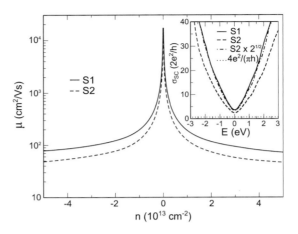

Figure 5.41 Charge mobility $\mu(E)$ for both samples (main plot) along with the semiclassical conductivity $\sigma_{sc}(E)$ (inset). Horizontal dashed line in the inset gives $4e^2/\pi h$. (Adapted with permission from Tuan *et al.* (2013). Copyright (2013) American Chemical Society.)

$n(E)$), since we account for the disorder-induced finite DOS, which yields nonzero charge density (and thus no singularity as $1/n(E)$).

We finally note the weak time-dependent decay of $D(t)$ after the saturation value, indicating a small correction due to quantum interferences which will eventually drive the system to a weak localization regime, as observed in transport measurements (Yu *et al.*, 2011). The values obtained for ℓ_{el} and $\sigma_{sc}(E)$ allow an estimation of the localization length ($\xi(E)$) of electronic states. Using the scaling analysis ($\xi(E) = \ell_{el}(E)\exp(\pi h\sigma_{sc}(E)/2e^2)$ (Lee & Ramakrishnan, 1985)) localization lengths in the order of about $\xi \simeq 10\mu m$ are obtained over a large energy window around the charge neutrality point. This contrasts with values in order of $\xi \simeq 10$ *nm* usually obtained with typically 1% of structural defects or covalently bonded adatoms (Lherbier *et al.*, 2011).

5.6   Amorphous graphene

5.6.1   Structural models

We finally study transport properties in strongly amorphous models of graphene, prepared using the Wooten–Winer–Weaire (WWW) method (Wooten, Winer & Weaire, 1985), introducing Stone–Wales defects (Stone & Wales, 1986) into the perfect honeycomb lattice. The structures are constructed by imposing periodic boundary conditions, while the entire network is relaxed with the Keating-like potential (Keating, 1966, Kapko, Drabold & Thorpe, 2010). Parts of two different networks are shown in Fig. 5.42(a),(b). The samples respectively contain 10 032 and 101 640 atoms, each of them with three-fold coordination as the honeycomb lattice, but topologically distinct. Samples 1 and 2 are characterized by a number of parameters, as given in Table 5.1.

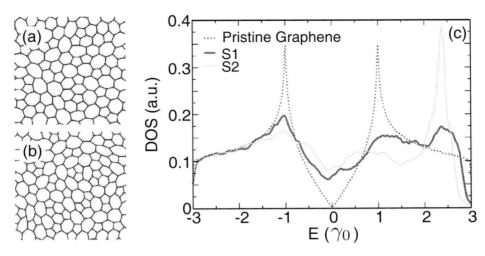

Figure 5.42 Structures of two generated amorphous graphene samples S1 (a) and S2 (b). (c) DOS. The pristine crystalline graphene case (dashed line) is also shown for comparison.

Table 5.1 Comparison of sample specifications.

	S1	S2
Number of atoms	10032	101640
Percent. of n-membered rings ($n = 5/6/7$)	24/52/24	44/12/44
$\langle n^2 \rangle - \langle n \rangle^2$	0.47	0.88
RMS deviation of bond angles	11.02°	18.09°
RMS deviation of bond lengths	0.044 Å	0.060Å
Fermi energy (γ_0)	0.03	0.05

For sample 1 (S1), 24% of the elementary rings are pentagons, 52% hexagons and 24% heptagons, while sample 2 (S2) has a larger share of odd-membered rings.

Comparing samples 1 and 2 the root mean square angular deviation from the mean of 120 deg is 11.02 deg for sample 1 and 18.09 deg for sample 2. The second moment of the ring statistics $\langle n^2 \rangle - \langle n \rangle^2$ is 0.47 for sample 1 and 0.88 for sample 2. The Fermi energy is $0.03\gamma_0$ for sample 1 and $0.05\gamma_0$ for sample 2. Thus we see that sample 2 is furthest from the pristine honeycomb, but it is likely that sample 1 is closer to physical reality as it is less strained. Nevertheless, having an extreme sample like S2 to study is useful to provide the limit of maximum amorphousness.

5.6.2 Electronic properties of amorphous graphene

Electronic and transport properties of such amorphous graphene models have been investigated using a π–π^* orthogonal *tight-binding* (TB) model with nearest neighbors hopping γ_0, and zero onsite energies. No variation of the hopping elements with disorder is included in the model as bond-length variation does not exceed a few percent (see Table 5.1).

All dependence on disorder stems from the ring statistics, which is the dominating effect (Dinh *et al.*, 2013). Figure 5.42(c) shows the density of states (DOS) of the two disordered samples considered, together with the pristine case (dashed line). Sample 1 (S1) exhibits 52% of hexagonal rings and displays several striking electronic characteristics. The zero-energy DOS is strongly enhanced, while the electron–hole symmetry of the band structure is broken (Lherbier *et al.*, 2011). The hole part of the spectrum still resembles the pristine graphene DOS, with a smooth van Hove singularity, whereas the electron part is more strongly modified with the occurrence of a large peak in the vicinity of the upper conduction band edge. By reducing the ratio of even- versus odd-membered rings (S2), the second maximum is found to be enhanced further at $E = 2.5\gamma_0$ while the spectral weight at $E = 3\gamma_0$ is suppressed. The redistribution of DOS at the upper conduction band edge reveals the massive presence of odd-membered rings and its strength with increasing number of such rings relates to the statistical distribution of rings with DOS features.

5.6.3   Mean free path, conductivity and localization

Figure 5.44 shows the time dependence of the normalized diffusion coefficient $D(t)/D_{\max}$ for two chosen energies and for both samples. For energy $E = -2\gamma_0$, the short-time ballistic regime is followed by saturation of the diffusion coefficient, typically after 0.1 ps. From the saturation values, $\ell_{\mathrm{el}}(E)$ and σ_{sc} are deduced and reported in Fig. 5.43. A striking feature is the very low value of the mean free path ℓ_{el} below 0.5 nm for the energy window around the Fermi level, in which the DOS departs from that of the pristine graphene structure. For negative energies (holes) far from the charge neutrality point, a considerable increase of more than one order of magnitude in the mean free paths is observed. The increase occurs for smaller binding energies for S1 than for

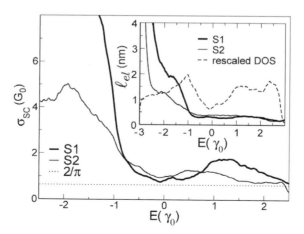

Figure 5.43 Semiclassical conductivity for both amorphous samples. The horizontal dotted line gives the value $4e^2/\pi h$. Inset: Elastic mean free path versus energy for the two samples.

S2, in good correlation with the changes observed in the DOS (which, around the van Hove singularity, deviates from the pristine graphene one more strongly for S2).

The value of σ_{sc} (Fig. 5.43) shows a minimum about $\sigma_{sc}^{\min} = 4e^2/\pi h$, as already obtained for other types of disorder (Section 5.2.4). However, in contrast to prior studies, the conductivity minimum is obtained over an energy range of several eV around the Dirac point. This indicates that transport is strongly degraded in the amorphous network compared to pristine graphene, for which the conductivity increases rapidly when the Fermi level is shifted away from the Dirac point. The charge mobility, $\mu(E) = \sigma_{sc}(E)/en(E)$, with $n(E)$ being the carrier density, is found to be about 10 cm^{2}V^{-1}s^{-1} for $n = 10^{11} - 10^{12}$cm^{-2}, orders of magnitudes lower than those usually measured in graphene samples (Tan *et al.*, 2007). Such low conductivity and mobility values should be measured at room temperature, where the semiclassical approximation is expected to hold.

The short ℓ_{el} obtained (Fig. 5.43, inset) and minimum $\sigma_{sc}(E)$ suggest the marked contribution of quantum interferences, and the formation of a strong insulating regime. Quantum interferences are evidenced by the time-dependent decay of the diffusivity $D(t)/D_{\max}$ (see Fig. 5.44). For $E = -2\gamma_0$, such decay is weak but clearly more pronounced for the S2 sample (which is more disordered than S1). In contrast, localization effects are strongly apparent at the Dirac point and develop at much shorter timescale (few nanoseconds). These differences are reflected in the corresponding localization lengths (ξ).

Based on the scaling theory of localization, an estimate of the localization length of electronic states can be inferred from $\xi(E) = \ell_{el}(E)\exp(\pi h\sigma_{sc}(E)/2e^2)$ (Lee & Ramakrishnan, 1985, Evers & Mirlin, 2008). The results are shown in Fig. 5.44. The amorphous samples are confirmed to be extremely poor conductors, with localization lengths as low as $\xi \sim 5-10$ nm over a large energy window around the charge neutrality

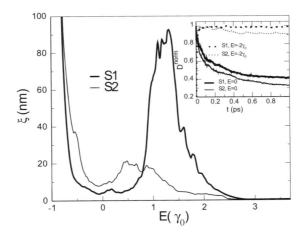

Figure 5.44 Main plot: Localization lengths as a function of the carrier energy. Inset: Normalized time-dependent diffusion coefficients for two selected energies for samples S1 and S2.

point. One also observes that ξ can vary by more than one order of magnitude depending on the disordered topology of the sample and rings statistics.

5.7   Phonon transport in graphene-related materials

The study of phonon transport in graphene-based materials is of particular relevance for fundamental understanding of coherent transport in low-dimensional materials (for instance to explore possible Anderson localization of phonons), as well as in view of engineering novel materials with a high thermoelectric figure of merit. The thermal conductivity of suspended and supported single graphene layers has been found to be anomalously high (Balandin *et al.*, 2008, Seol, Jo & Moore, 2010) with micrometer-long phonon mean free paths.

The possibility of Anderson localization of acoustic phonons in disordered materials remains an intriguing phenomenon, which is expected to occur in situations of strong interference between multiple scattering paths (Anderson, 1958, Lagendijk, van Tiggelen & Wiersma, 2009). Disordered SWNT-based bundles exhibit a certain tendency towards a thermal insulating regime (Che, Çagin & Goddard, 2000, Prasher, Hu & Chalopin, 2009). For instance, isotope disorder strongly scatters high-energy phonon modes, resulting in very low mean free path, although ballistic conduction is robust for low-energy phonon modes (Savić, Mingo & Stewart, 2008). The contribution of localization effects of high-energy modes is, however, not easily demonstrated in the scaling behavior of the thermal conductance of the disordered material, so that onset of Anderson localization is elusive. Graphene nanoribbons can be strongly disordered by mixing isotope impurities (Mingo *et al.*, 2010), and edge disorder (Li *et al.*, 2010). Following the spirit of the real space Kubo formalism, a time-dependent phonon wavepacket formalism is derived here for calculation of the thermal conductance.

5.7.1   Computational phonon propagation methodology

The Kubo formalism has been used for investigating thermal transport in disordered binary alloys or nanocrystralline silicon (Alam & Mookerjee, 2005, Allen & Feldman, 1989), but with a computational method of limited capability. Here the real space implementation of the Kubo formula for phonon propagation is given, establishing a direct computational bridge between phonon dynamics and the thermal conductance (Li *et al.*, 2010, Li *et al.*, 2011). The starting Hamiltonian which describes the phonon spectrum, taking only the harmonic interactions into account, is given by

$$\mathcal{H} = \sum_i \frac{\hat{p}_i^2}{2M_i} + \sum_{ij} \Phi_{ij}\hat{u}_i\hat{u}_j, \qquad (5.55)$$

where $\hat{u}_i$ and $\hat{p}_i$ are the displacement and momentum operators for the ith degree of freedom, M_i is the corresponding mass, and Φ is the force constant tensor. Based on the linear response theory, the phonon conductivity σ can be defined as $\Omega T^{-1} \int_0^\beta d\lambda \int_0^\infty dt \langle \hat{\mathcal{J}}^x(-i\hbar\lambda)\hat{\mathcal{J}}^x(t)\rangle$ (Allen & Feldman, 1989). $\hat{\mathcal{J}}^x$ is the x component of the energy flux operator $\hat{\mathbf{J}}$, and it can be expressed as $\hat{J}^x = 1/2\Omega \sum_{ij}(X_i - X_j)\Phi_{ij}\hat{u}_i\hat{v}_j$, where $\hat{v}_j$ is the velocity operator and X_i is the equilibrium position of the atom to which the ith degree of freedom belongs and $\beta = 1/(K_B T)$. After some algebra, σ becomes

$$\sigma = -\frac{\pi}{\Omega}\int_0^\infty d\omega \, \hbar\omega \frac{\partial f_B}{\partial T} \text{Tr}\{[\hat{X}, D]\delta(\omega^2 - D)[\hat{X}, D]\delta(\omega^2 - D)\}, \tag{5.56}$$

where f_B is the Bose distribution function, and $D_{ij} = \Phi_{ij}/\sqrt{M_i M_j}$ is the mass normalized dynamical matrix. The dynamical matrix can be obtained given the atomistic configuration of the system and it defines an equation of motion which is of second order in time. Using the fact that the operators D and H have the same energy spectrum ($\hbar^2 D = H^2$), and defining $\hat{V}_x = [\hat{X}, H]/i\hbar$, one can write the thermal conductance of a one-dimensional system as

$$\kappa = \frac{\pi\hbar^2}{L^2}\int_0^\infty d\omega \, \hbar\omega \frac{\partial f_B}{\partial T} \text{Tr}\{\hat{V}_x\delta(\hbar\omega - H)\hat{V}_x\delta(\hbar\omega - H)\}. \tag{5.57}$$

The thermal conductance can also be derived from the Landauer formalism (Mingo, 2006) or the nonequilibrium Green function approach (Yamamoto & Watanabe, 2006) as

$$\kappa = \frac{1}{2\pi}\int_0^\infty d\omega \, \hbar\omega \frac{\partial f_B}{\partial T} T(\omega), \tag{5.58}$$

with $T(\omega)$ being the phonon transmission function. Comparing the two formulas, the transmission function $T(\omega)$ is defined as

$$T(\omega) = \frac{2\pi^2\hbar^2}{L^2} \text{Tr}\{\hat{V}_x\delta(\hbar\omega - H)\hat{V}_x\delta(\hbar\omega - H)\}, \tag{5.59}$$

which has the same form as the electron transmission function ($T_{el}(E)$) derived from the Kubo–Greenwood formula (Roche & Mayou, 1997),

$$T_{el}(E) = \frac{2\pi^2\hbar^2}{L^2} \text{Tr}\{\hat{V}_x\delta(E - H_{el})\hat{V}_x\delta(E - H_{el})\}. \tag{5.60}$$

This equivalence allows us to implement the order N algorithm related to the Lanczos method. The key quantity of the method is the transmission function expressed in Eq. (5.60). One rewrites $T(\omega)$ in terms of the diffusion coefficient,

$$T(\omega) = \frac{2\omega\pi}{L^2}\text{Tr}\left[\delta(\omega^2 - D)\right]\lim_{t\to\infty}\frac{d}{dt}(t\mathcal{D}(\omega, t)), \tag{5.61}$$

where the diffusion coefficient $\mathcal{D}$ is given by

$$\mathcal{D}(\omega, t) = \frac{1}{t}\frac{\text{Tr}\left[(\hat{X}(t) - \hat{X}(0))^2\delta(\omega^2 - D)\right]}{\text{Tr}\left[\delta(\omega^2 - D)\right]}, \tag{5.62}$$

where $\hat{X}(t)$ is the position operator in the Heisenberg picture. In the diffusive regime $\mathcal{D}(\omega, t) = \mathcal{D}_{max}(\omega)$, so that the transmission function reduces to

$$T(\omega) = \frac{2\omega\pi}{L^2}\text{Tr}\left[\delta(\omega^2 - D)\right]\mathcal{D}_{max}(\omega), \tag{5.63}$$

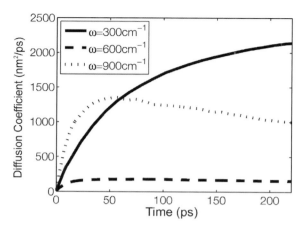

Figure 5.45 Time-dependent diffusion coefficients $D(\omega, t)$ for the system of SWNT with 10.7% of ^{14}C isotope disorder at three different frequencies. From Li *et al.* (2011).

where in the ballistic regime $\mathcal{D}(\omega, t) = v^2(\omega)t$, with $v(\omega)$ being the average group velocity over states with frequency ω. Since the number of channels is $N_{ch}(\omega) \approx 2\omega\pi \operatorname{Tr}\left[\delta(\omega^2 - D)\right] v(\omega)/L$, the phonon mean free path can be approximated by $\ell(\omega) = \mathcal{D}_{max}(\omega)/v(\omega)$. By computing $\mathcal{D}(\omega, t)$, one can thus deduce $\mathcal{D}_{max}(\omega)$, $v(\omega)$, and $\ell(\omega)$. Note that the numerator in Eq. (5.62) can be rewritten as $\operatorname{Tr}\{[X, U(t)]^\dagger\delta(\omega^2 - D)[X, U(t)]\}$, and approximated by $\operatorname{Tr}\{[X, \mathcal{U}(\tau)]^\dagger\delta(\omega^2 - D)[X, \mathcal{U}(\tau)]\}$, with $\mathcal{U}(\tau) = e^{-iD\tau}$, and $\tau = t/2\omega$. The trace can be efficiently calculated through an average over a few random phase states of atomic displacements as $N\langle\psi|[X, \mathcal{U}(\tau)]^\dagger\delta(\omega^2 - D)[X, \mathcal{U}(\tau)]|\psi\rangle$, N being the number of degrees of freedom, and the braket corresponds to the local density of states (LDOS) associated with the vector $[X, \mathcal{U}(\tau)]|\psi\rangle$, which is calculated by using a Chebyshev expansion of $\mathcal{U}(\tau)$. LDOS is obtained by using the Lanczos method, and $\operatorname{Tr}\left[\delta(\omega^2 - D)\right]$ is also calculated by averaging the LDOS of $|\psi\rangle$ through the Lanczos method.

5.7.2   Disordered carbon nanotubes with isotope impurities

As a matter of test and illustration of the above method, we first consider an SWNT(7, 0) with 10.7% of ^{14}C impurities, which has been also studied using the Green functions method (Savić *et al.*, 2008). An analytical expression for one-dimensional systems with isotopic disorder is also used for comparison:

$$\ell(\omega) = (12aN_{uc}N_{ch}(\omega))/\left(\pi^2 f\left|\frac{\Delta M}{M}\right|^2\rho_{uc}^2(\omega)\omega^2\right),$$

where a is the length of the lattice vector in the translational direction, N_{uc} is the number of atoms in each unit cell, ρ_{uc} is the density of states per unit cell, f is the percentage of isotopic impurities having mass difference ΔM, and M is the average mass of the atoms.

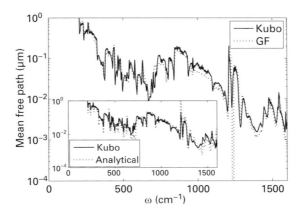

Figure 5.46 Frequency-dependent MFP in SWNT with isotopic disorder (10%) calculated by the GF method, the Kubo method, and the analytical formula. From Li *et al.* (2011).

The evolution of the wavepacket dynamics for different frequencies can be evaluated (see Fig. 5.45). At low enough frequency, for instance $\omega = 300$ cm^{-1}, a quasi-linear increase of $D(\omega, t)$ at short time indicates ballistic motion, whereas a further saturation at longer times to some maximum value characterizes diffusive transport (cf. Eq. 5.63). Diffusion coefficients for phonons with $\omega = 600$ cm^{-1} exhibit such saturation after a few tens of ps. At high enough frequency, $D(\omega, t)$ will decay at long times, as illustrated for phonons with $\omega = 900$ cm^{-1}. From maximum values of $D(\omega, t)$, the frequency-dependent mean free paths of coherent phonons can be evaluated. The results obtained compare very well with those obtained with the Green functions approach (Savić *et al.*, 2008), as well as with the analytical expression (Fig. 5.46).

5.7.3   Disordered graphene nanoribbons with edge disorder

Next, we consider zigzag graphene nanoribbons (zGNR) of different widths with edge disorder. We use the fourth nearest neighbor force constants for building the dynamical matrices (Zimmermann, Pavone & Cuniberti, 2008). The ribbon widths are defined by the number of zigzag chains $N_z = 20$, 40, and 80, and the relative amount of edge defects (removed carbon atoms at the edges) is chosen to be 10%, and additionally 15% for $N_z = 80$ (Fig. 5.47).

Figure 5.48 shows how $\ell(\omega)$ is governed by the ribbon width and disorder strength. At fixed disorder, $\ell(\omega)$ increases with ribbon width. One notes that, for low-frequency modes, the MFP are several hundreds of nanometers, and due to large values of N_{ch} the possibility of accessing the Anderson localization regime is clearly jeopardized in the thermal conductance, as also discussed for small diameter disordered carbon nanotubes (Savić *et al.*, 2008). We next focus on the zGNR(80), and obtain the transmission spectra according to $\mathcal{T}(\omega) = N_{ch}/(1 + L/\ell(\omega))$.

The resulting frequency-dependent transmission function $\mathcal{T}(\omega)$ for different ribbon lengths is plotted in Fig. 5.49, while the thermal conductance κ using Eq. (5.58) is

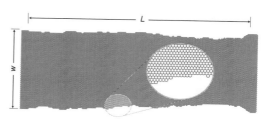

Figure 5.47 A short portion of an edge disordered GNR with width $w = 17.04$ nm, length $L \sim 50$ nm and disorder density of 10%. From Li *et al.* (2010).

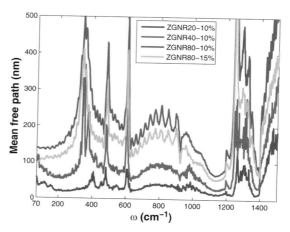

Figure 5.48 Phonon mean free path for zGNR of widths $N_z = 20, 40,$ and 80 (4.26, 8.52, and 17.04 nm, respectively) with disorder density of 10%, and also the $N_z = 80$ and 15% disorder for comparison. Adapted from Li *et al.* (2010).

shown in Fig. 5.50. The downscaling of $\mathcal{T}(\omega)$ directly impacts on κ, which is reduced by one order of magnitude for 2 μm (at room temperature). For low-frequency phonons, ($\omega < 70$ cm^{-1}) the computation time to reach $D_{\max}$ is much longer than the rest of the spectrum.

To evaluate the contribution of these modes to κ, a linear extrapolation for the transmission is used and the results are compared with those obtained from the GF method. Our analysis shows that this approximation causes an error of less than 1.5% for the thermal conductance at room temperature. The conductivity of edge disordered GNRs is obtained using $\sigma = \kappa L/A$, A being the cross-sectional area of the ribbons, which is taken as the interplane distance of graphite layers. At room temperature, $\sigma = 1006$ W m^{-1}K^{-1} and $\sigma = 671$ W m^{-1}K^{-1} with edge disorder 10% and 15%, respectively, for $L = 2$ μm.

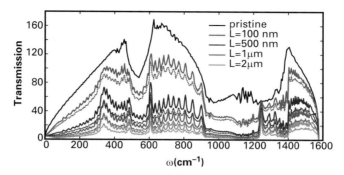

Figure 5.49 Transmission spectra for different lengths of zGNR(80) with edge disorder of 10% (solid) and 15% (dashed). From Li *et al.* (2010).

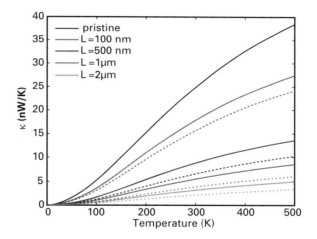

Figure 5.50 Temperature and length-dependent behaviors of the zGNR(80) thermal conductance (10% disorder (solid lines), 15% (dashed lines)). From Li *et al.* (2010).

5.8   Graphene quantum dots

Electron–electron interactions in graphene systems are generally expected to have an important role, especially: (i) close to the Dirac point, where the low carrier concentration strongly reduces the screening effects of the Coulomb interactions; (ii) in flat bands where the quenching of the kinetic energy may lead to enhanced interactions; (iii) when confinement is strong as in graphene quantum dots. Other effects such as spin–orbit coupling may also play an important role and this is addressed in Section 2.6.

Quantum dots (Kastner, 1992, Kouwenhoven, Marcus & McEuen, 1997, Alhassid, 2000) are obtained by confining the electrons in a sample in all three spatial dimensions, much as in an atom but in a length scale orders of magnitude larger. Furthermore, the possibility of contacting them using electrodes allows us to reveal their electronic properties via transport measurements, reaching a much higher magnetic flux than would be possible in atoms. Notwithstanding, it must be noted that as a result of the larger size the quantization energies of these artificial atoms (of about 1 meV) are smaller than for atoms and therefore low temperatures are required to resolve them.

Quantum dots made of nanometer-thick graphite layers were demonstrated in early work (Bunch *et al.*, 2005) and a few years later graphene quantum dots were achieved (Geim & Novoselov, 2007, Stampfer, Guttinger & Molitor, 2008, Ponomarenko, Schedin & Katsnelson, 2008) attracting much attention (Güttinger & Molitor, 2012). But why graphene instead of GaAs or other materials? In the following, a brief overview on graphene quantum dots tries to shed light on this and other questions. But first let us review some basics on the Coulomb blockade.

5.8.1   Generalities on Coulomb blockade

The phenomenon known as Coulomb blockade (Kastner, 1992, Kouwenhoven *et al.*, 1997) takes place when the quantum dot is weakly coupled to the leads as represented in Fig. 5.51(a). The conductance then falls below e^2/h and the charge inside the island gets quantized. The relevant energy scales are: the level spacing $\delta\varepsilon$ between the single

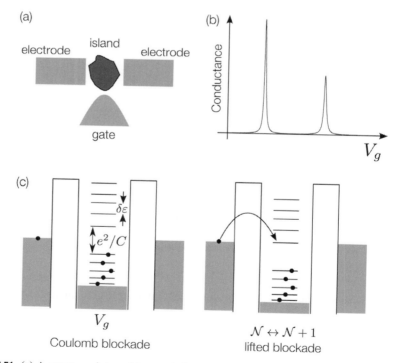

Figure 5.51 (a) A quantum dot weakly coupled to electrodes and in the presence of a gate voltage. (b) Two consecutive conductance peaks vs. gate voltage showing a blockade region in between. (c) Representation of the energy levels for two situations: when blocking is active (left) and at a conductance peak (right).

particle levels of the dot, the intrinsic level widths $\Gamma_{L(R)}$ of a given energy level due to the left and right electrodes, the charging energy E_c, and the thermal energy $k_B T$.

The first experiments on the Coulomb blockade were done in metallic grains (Giaever & Zeller, 1968) where the mean level spacing $\Delta = \langle \delta \varepsilon \rangle$ is much smaller than $k_B T$. In this situation we have $\Delta \ll k_B T \ll E_c$ and the grain energy spectrum can be considered as a continuum. The electrostatic energy of $\mathcal{N}$ electrons in the quantum dot is given by (Kastner, 1992)

$$U(\mathcal{N}) = (\mathcal{N}e)^2/(2C) - \mathcal{N}e\eta V_g, \qquad (5.64)$$

where C is the effective capacitance of the quantum dot, and ηV_g is the energy shift of the states in the quantum dot due to the gate voltage V_g. By rewriting the previous equation as $U(\mathcal{N}) = (Q - Q_0)^2/(2C) + constant$ with $Q = \mathcal{N}e$ and $Q_0 = C\eta V_g$, one sees that $U(\mathcal{N})$ is a parabola with minimum at $Q = Q_0$. Now suppose that V_g is chosen such that a given value of $\mathcal{N}$ minimizes U; then the energy needed to add or take an electron out of the quantum dot is $e^2/(2C)$ and there is an energy gap of $E_c = e^2/C$ for excitations. For low enough temperatures ($k_B T < e^2/(2C)$) this leads to a blockade in the electron and hole flow to the quantum dot.

However, when $Q_0 = (\mathcal{N} + 1/2)e$ one has $U(\mathcal{N}) = U(\mathcal{N}+1)$, i.e. the configurations with $\mathcal{N}$ and $\mathcal{N} + 1$ electrons are degenerate and the tunneling of charge to and out of the dot is allowed, leading to a conductance peak. This is observed as a series of conductance peaks separated by regions of vanishing conductance as the gate voltage is changed.

When the quantum dots are small enough such that the level spacing is larger than $k_B T$ one is in the regime of *quantum* Coulomb blockade: $k_B T < \Delta \ll E_c$. A term due to the filling of the discrete levels E_j must be added to $U(\mathcal{N})$:

$$U(\mathcal{N}) = Q^2/(2C) + \sum_{j=1}^{\mathcal{N}} \left(E_j - e\eta V_g \right). \qquad (5.65)$$

The separation between the $\mathcal{N}$th and the $\mathcal{N}+1$th conductance peaks can be obtained from

$$\left(e\eta V_g \right)_{\mathcal{N}+1} - \left(e\eta V_g \right)_{\mathcal{N}} = e^2/C + (E_{\mathcal{N}+1} - E_{\mathcal{N}}) = E_c + \delta\varepsilon. \qquad (5.66)$$

Therefore, the separation between the conductance peaks is controlled by the charging energy plus a smaller term due to the spacing of the dot levels. A scheme summarizing this is presented in Fig. 5.51(b), (c).

Besides providing information on the quantum dot level spacing, the fluctuations in the intensity of the conductance peaks in the quantum Coulomb blockade regime offer complementary information on the wavefunctions in the quantum dot. Indeed, using the standard (sequential) theory (Beenakker, 1991) one obtains for the conductance peak $G_{\max}$ in resonance with the jth dot level

$$G_{\max} = \frac{2e^2}{h} \frac{\pi}{2k_B T} \frac{\Gamma_{j,L}\Gamma_{j,R}}{(\Gamma_{j,L} + \Gamma_{j,R})}. \qquad (5.67)$$

This expression is obtained for a single nondegenerate level E_j by assuming that the intrinsic level widths are the smallest energy scale in the problem ($\Gamma_j = \Gamma_{j,L} + \Gamma_{j,R} \ll \delta\varepsilon, k_B T$) and a *constant interaction model* (Alhassid, 2000) (E_c independent of the number of electrons in the dot). Since Γ_j can be related to the overlap between the electrodes and the dot wavefunctions, the conductance peaks provide information on the latter (Jalabert, Stone & Alhassid, 1992).

Studies of a large number of Coulomb blockade conductance peaks and the associated level spacings reveal reproducible universal statistical features, i.e. features that are independent of the particular device; see Alhassid (2000) and references therein. They can be explained by thinking of these islands as nano/meso-scale electron billiards which are classically chaotic. This fact leads to a peculiar level spacing statistics exhibiting repulsion among neighboring levels described by the Wigner–Dyson distribution (Alhassid, 2000). The theoretical level spacing statistics can be derived within random matrix theory and follows one of the Gaussian random ensembles: Gaussian orthogonal (GOE) or Gaussian unitary (GUE) depending on whether the time-reversal symmetry is broken or not (Alhassid, 2000). Conductance peaks also obey specific statistics which can be derived by similar methods (Jalabert *et al.*, 1992, Alhassid, 2000).

Experimentally, achieving good statistics for the level spacings and the conductance peaks is a difficult task, since as the gate voltage is changed the charging energy and the tunneling into the leads may be modified as well. For the case of the conductance peaks, many independent realizations can be obtained by applying a variable magnetic field thereby improving the statistics (Chang *et al.*, 1996; Patel *et al.*, 1998).

5.8.2   Confining charges in graphene devices

In quantum dots formed of semiconducting materials like GaAs, gating carefully chosen regions is enough to confine the charges. However, this is ineffective for graphene devices due to Klein tunneling (see Section 4.1). The absence of a true bandgap in bulk graphene makes controlling the electron flow a challenging task. Notwithstanding, tunable confinement has been demonstrated by etching graphene into nanoribbons which serve as the contacts between a larger graphene sample and the electrodes (Geim & Novoselov, 2007, Stampfer *et al.*, 2008, Ponomarenko *et al.*, 2008) or in graphene nanoribbons *pn* junctions (Liu *et al.*, 2009).

One must be aware, however, that a disorder potential in the nanoribbons may induce electron-hole puddles producing charged islands along the ribbon direction (Todd *et al.*, 2009, Stampfer *et al.*, 2009, Liu *et al.*, 2009). This introduces new energy scales such as the charging energy of the quantum dots that are formed, thereby making it a more difficult problem than one would have naively expected. Figure 5.52 (reproduced from Stampfer *et al.*, 2009) illustrates this issue.

Tunability of a graphene quantum dot can be achieved by using a back gate as well as side gates, which allow pinching off the constrictions independently. An example of a tunable Coulomb blockade device is shown in Fig. 5.53 (a) and (b). A trace of the source-drain current as a function of the gate voltage for this device is shown in Fig. 5.53 (c). The conductance oscillations were shown to be regular over more than

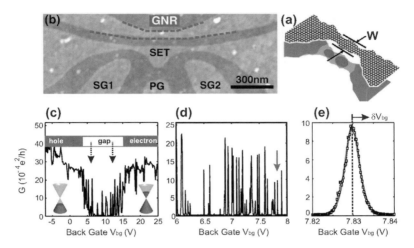

Figure 5.52 (a) An etched graphene nanoribbon (GNR) of width W and the underlying charge puddles. (b) Scanning force microscope image showing an etched graphene nanoribbon together with a single electron transistor (SET) used to detect single charging events in the ribbon. SG1, PG, and SG2 are lateral gates used for operation of the SET. (c) Low bias conductance of the nanoribbon versus back gate showing a gap between the regimes of hole and electron transport. (d) and (e) are zooms of (c); (e) shows a detail of a single sharp resonance within the gap. (Reproduced with permission from Stampfer *et al.* (2009). Copyright (2009) by the American Physical Society. Courtesy of C. Stampfer.)

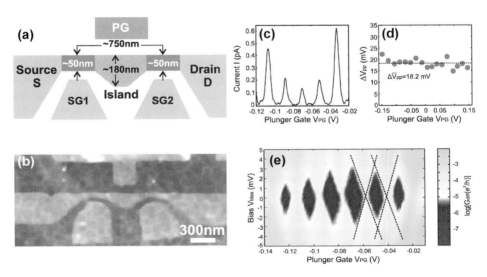

Figure 5.53 (a) and (b) An all-graphene quantum-dot device. (c) The oscillation in the source-drain current as a function of the plunger gate pg as indicated in (a). The peak spacing for 18 consecutive current peaks is plotted in (d). The differential conductance as a function of the bias and plunger gate voltages is shown as a density plot in (e). (Reprinted with permission from Stampfer *et al.* (2008). Copyright (2008), American Institute of Physics. Courtesy of C. Stampfer.)

ten conductance peaks with a period of 18.2 mV, as shown in Fig. 5.53(d). Figure 5.53(e) shows a density plot of the differential conductance as a function of the bias and plunger gate voltages.

From the width of the Coulomb diamonds along the vertical direction one can infer a charging energy $E_c \sim 3.5$ meV. An independent estimation of E_c can be obtained by modeling the island as a disk with diameter d and calculating its self-capacitance (Kouwenhoven et al., 1997, Güttinger et al., 2012) $C_{\text{disk}} = 4\epsilon_0 \epsilon d$, where ϵ can be estimated as the average of the dielectric constant of the oxide underneath and the vacuum ϵ, $\epsilon = (\epsilon_{\text{ox}} + 1)/2 \simeq 2.5$. Since this simple model underestimates the capacitance of the island, which is increased by the capacitive coupling of the island to the gates and the leads, the resulting charging energy $E_c^{\text{disk}} = e^2/C_{\text{disk}}$ can be used as an upper bound for the charging energy of the island E_c ($E_c \sim 2.5 \times E_c^{\text{disk}}$) (Güttinger et al., 2012).

From Fig. 5.53(d) one can see that there are no important fluctuations in the separation between Coulomb peaks, from which we can conclude that the island spacing is much smaller than E_c. This is confirmed by Fig. 5.53(e) which shows no additional lines due to excited states or cotunneling events.

Smaller dots ($d < 100$ nm) may reveal important information about the level statistics of the graphene island (Ponomarenko et al., 2008). But is there any difference between graphene quantum dots and those made of conventional semiconductors? This is an interesting and debated question (Ponomarenko et al., 2008, Libisch, Stampfer & Burgdörfer, 2009, Huang, Lai & Grebogi, 2010). Whereas in the absence of a magnetic field for a chaotic dot made of a usual semiconductor material one would expect level statistics given by the GOE, the level spacing of a chaotic Dirac or neutrino billiard was predicted to follow the GUE (Berry & Mondragon, 1987) (a time-reversal symmetry breaking due to chirality). Although experimental evidence in this direction has been presented (Ponomarenko et al., 2008), time-reversal symmetry at zero magnetic field is restored when both valleys are taken into account (see for example Beenakker, 2008) and other authors have shown numerical evidence for GOE statistics in both clean chaotic billiards (Libisch et al., 2009) and disordered dots with defects (Huang et al., 2010).

So far we have seen that tunable confinement is possible in single-layer graphene devices by connecting the graphene island with the electrodes through graphene constrictions. There are two factors that complicate the controllability of these confining constrictions: (a) the formation of electron-hole puddles as mentioned before, and (b) edge roughness. Electron-hole puddles may be lessened by suspending the device, thereby isolating it from the substrate. An alternative would be to use a substrate such as hexagonal boron nitride (Dean et al., 2010) where disorder effects are reduced. Edge roughness, on the other hand, changes the ribbon properties as predicted (Querlioz et al., 2008, Evaldsson et al., 2008, Mucciolo et al., 2009), increasing the differences between different geometries (as observed in Han et al., 2007, Chen et al., 2007), introducing localization and inducing energy gaps. Improving the edge sharpness can be achieved, for example, by optical annealing (Begliarbekov et al., 2011) or by bottom-up fabrication of GNRs (Cai et al., 2010).

A different strategy was presented in Allen, Martin & Yacoby (2012), where quantum confinement has been demonstrated in suspended bilayer graphene with external electric fields used to open a bandgap. Given that a suspended 2D sample is used, the experiments are clean from edge disorder and substrates effects.

Other interesting phenomena not mentioned before include controlling the spin degree of freedom in graphene quantum dots and exploring elastic and inelastic cotunneling phenomena. The interested reader may find more in specialized reviews (Güttinger *et al.*, 2012).

5.9 Further reading and problems

- For a general presentation on disorder effects in carbon nanotubes, see Roche *et al.* (2006).
- The reader interested in transport through edge states and the influence of disorder (not covered here) may follow the presentation in Wimmer (2009). A clever way to do interferometry with these states is discussed in Usaj (2009).
- For a very recent and detailed review on graphene quantum dots we recommend Güttinger *et al.* (2012).

Problems

5.1 *Conductance through a SWNT with a single vacancy.*
As another example of application of Landauer's theory to carbon-based devices, let us consider the case of transport through a metallic carbon nanotube with a single vacancy. The simplest forms of introducing the vacancy within a π-orbitals model are: (i) disconnecting one of the orbitals, or (ii) introducing a very large onsite energy at one lattice point. Both models give equivalent results for the Landauer conductance.
(a) Calculate the conductance in the presence of this defect.
(b) You are encouraged to rationalize the conductance decrease due to the vacancy using the mode-decomposition introduced in Section 4.2.2. (For a complementary approach that uses an effective Hamiltonian instead of the *tight-binding* model used here we refer to Matsumura & Ando (2001).)

5.2 *Conductance in the presence of a single Stone–Wales defect.*
Consider a nanotube with a single Stone–Wales defect.
(a) By using a simple *tight-binding* model, where only the topology of the hoppings is changed, compute the density of states and the conductance.
(b) Repeat your calculation for a graphene nanoribbon. Do the results depend on the position of the defect?

5.3 *Transmission through a line defect and valley filtering.*
Consider the line defect in Fig. 5.35(a).
(a) Consider electrons that are incident on the line defect from the left. Express the solutions for the wavefunctions on each valley (in bulk graphene) for these left-moving electrons.
(b) Following Gunlycke & White (2011), note that in the $k \to 0$ limit the reflection operator and the graphene translation operator perpendicular to the line defect

commute. Then obtain symmetry adapted states $|\pm\rangle$ that are eigenstates of both operators.

(c) Compute the transmission probabilities for states on each valley and show that this line defect acts a valley filter.

5.4 *Elastic mean free path of carbon nanotubes and graphene nanoribbons.*

(a) Follow up our discussion in Section 5.1 based on White & Todorov (1998) and derive Eq. 5.7 ($\ell_{el} = 18\sqrt{3}a_{cc}(\gamma_0/W)^2 N$) for the elastic mean free path in carbon nanotubes. Discuss the dependence of ℓ_{el} on the tube diameter.

(b) Discuss the dependence of ℓ_{el} on the width for the case of a graphene nanoribbon. (*Hint*: For this last question you may follow Areshkin *et al.* (2007).)

5.5 *Level spacings of GaAs and graphene quantum dots compared.*

(a) Compare the typical level spacing for a quantum dot made of GaAs and one made of graphene with the same diameter d. In which one are confinement effects going to be more important?

(b) Look for experimental data in the literature to confirm your assertion and point out the main differences between the conductance oscillations in small and large diameter dots.

 ** Additional exercises and solutions available at our website.

6   Quantum transport beyond DC

In this chapter we give a flavor of quantum transport beyond DC conditions, when time-dependent potentials are applied to a device. Our main focus is on Floquet theory, one of the most useful approaches for driven systems. Section 6.4 is devoted to an overview of some of the most recent advances on driven transport in graphene-related materials, while Section 6.5 presents an illustrative application to laser-illuminated graphene.

6.1   Introduction: why AC fields?

Though less explored, quantum transport beyond the DC conditions considered in previous sections also offers fascinating opportunities. Alternating current (AC) fields such as alternating gate voltages, alternating bias voltages or illumination with a laser can be used to achieve *control* of the electrical response (current and noise), thereby providing a novel road for *applications*. Furthermore, there are many *novel phenomena* unique to the presence of AC fields such as quantum charge pumping (Thouless, 1983, Altshuler & Glazman, 1999, Büttiker & Moskalets, 2006, Switkes *et al.*, 1999), i.e. the generation of a DC current even in the absence of a bias voltage due to quantum interference,[1] coherent destruction of tunneling (Grossmann *et al.*, 1991) or laser-induced topological insulators (Lindner, Refael & Galitski, 2011, Kitagawa *et al.*, 2011).

The activity in this area has grown rapidly in the arena of nanoscale systems (Platero & Aguado, 2004, Kohler, Lehmann & Hänggi, 2005). Notwithstanding, it was not until the last few years that advances in the applications to graphene-related systems started to flourish (see the overview in Section 6.4).

Let us consider the two paradigmatic situations schematized in Fig. 6.1, namely a graphene device with two time-dependent gate voltages applied to different parts of the sample (a), or subjected to laser illumination (b). In both situations the electronic excitations can be modeled through a time-dependent Hamiltonian of the general form:

$$\hat{\mathcal{H}}(t) = \hat{\mathcal{H}}_0 + \hat{\mathcal{H}}_1(t). \tag{6.1}$$

In the first case, a simple model of the alternating gates would be to add a harmonic modulation of the site energies in a *tight-binding* Hamiltonian (Orellana & Pacheco,

---

[1] For a closed system, the generation of a circulating current is called "quantum stirring" (Sela & Cohen, 2008).

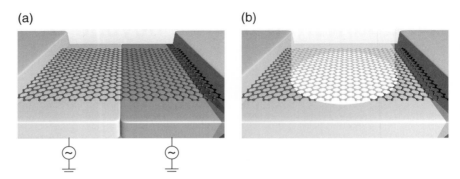

Figure 6.1 Two paradigmatic situations of transport beyond the DC limit: a graphene sample contacted between two electrodes in the presence of (a) two AC gates in the sample region (represented by shaded areas), and (b) laser illumination applied perpendicularly to the graphene plane.

2007, Foa Torres & Cuniberti, 2009), whereas in the second case laser illumination can be included through a time-dependent vector potential (Syzranov, Fistul & Efetov, 2008, Oka & Aoki, 2009, Calvo *et al.*, 2011).

This time dependence renders the usual separation of space and time variables inapplicable, thereby making the calculation of the transport response more cumbersome. For the particular case of time-periodic Hamiltonians, one can exploit the time periodicity to achieve elegant and simple solutions, as shown in Section 6.3. Furthermore, if the time variation is sufficiently slow one may use the appealing adiabatic theory mentioned in Section 6.2.

6.2   Adiabatic approximation

To obtain meaningful approximations, two relevant timescales need to be weighted: the electronic traversal time through the sample τ_D, and the period of the time-periodic potential T. If τ_D is much shorter than T, the Hamiltonian is effectively static during the electrons' trip through the device. The corrections due to the slow time-dependence of the potential can be calculated through the "adiabatic" theory (Brouwer, 1998, Büttiker, Thomas & Pretre, 1994, Entin-Wohlman, Aharony & Levinson, 2002, Kashcheyevs, Aharony & Entin-Wohlman, 2004).

Within this framework, the current which flows in response to a cyclic variation of a set X_j of device-control parameters is expressed in terms of the scattering matrix $S(X_j)$ of the system. (Brouwer, 1998, Büttiker *et al.*, 1994). Since the scattering matrix $S(X_j)$ is calculated for a stationary situation where the potentials are *frozen* at time t, the response of the system is assumed to depend on time only through the parametric time-dependence $X_j(t)$. In general, the current will contain a component due to the bias voltage as well as a component which survives even for vanishing bias voltage (Brouwer, 1998). This last component is called the *pumped current*, while the first one is a generalization of the Landauer current (Entin-Wohlman *et al.*, 2002).

This approximation works reasonably well for alternating gates where the achievable frequencies determine an energy scale $\hbar\Omega$ which is much smaller than all the other energy scales in the problem. Some illustrations of its use in graphene-related materials can be found in Prada, San-Jose & Schomerus (2009) and Zhu & Chen (2009) and references therein. In situations where the leading-order (adiabatic) correction vanishes, going beyond the adiabatic approximation becomes crucial (Foa Torres *et al.*, 2011, San-Jose *et al.*, 2011, Zhou & Wu, 2012).

6.3   Floquet theory

Floquet theory (Shirley, 1965, Sambe, 1973, Kohler *et al.*, 2005) is based on the application of a theorem due to Gaston Floquet to obtain the solutions of the Schrödinger equation with a time-periodic potential.

Formulated long before Bloch's theorem, Floquet's theorem (Floquet, 1883) can be thought of as its analog for *time*-periodic (instead of space-periodic) Hamiltonians. Given a Hamiltonian with a time period T, there is a complete set of solutions of the form

$$\psi_\alpha(r, t) = \exp(-i\varepsilon_\alpha t/\hbar)\phi_\alpha(r, t), \tag{6.2}$$

where ε_α are the quasi-energies and $\phi_\alpha(r, t + T) = \phi_\alpha(r, t)$ are the Floquet states. It can be shown that Floquet states corresponding to quasi-energies differing in an integer multiple of $\hbar\Omega$ are physically equivalent, i.e. linearly dependent, and therefore only states within $-\hbar\Omega/2 \leq \varepsilon_\alpha < \hbar\Omega/2$ need to be considered. This is the analog of the first Brillouin zone when using Bloch's theorem. Furthermore, one notices that the states given by Eq. (6.2) separate the slow dynamics (contained in the exponential factor) from the fast dynamics (given by Floquet states $\phi_\alpha(r, t)$).

Replacing the solutions given by Eq. (6.2) into the time-dependent Schrödinger equation (TDSE) gives

$$\left[\hat{\mathcal{H}}(r, t) - i\hbar\frac{\partial}{\partial t}\right]\phi_\alpha(r, t) = \varepsilon_\alpha\phi_\alpha(r, t). \tag{6.3}$$

This equation has the same form as the usual time-independent Schrödinger equation but with two main differences:

(1) Equation (6.3) is an eigenvalue problem in Floquet space $\mathcal{R} \otimes \mathcal{T}$, where $\mathcal{R}$ is the usual Hilbert space and $\mathcal{T}$ is the space of periodic functions with period $T = 2\pi/\Omega$. The Fourier space $\mathcal{T}$ is spanned by the set of orthonormal vectors $\langle t |n\rangle \equiv \exp(in\Omega t)$, where n is an integer. If $|i\rangle$ is a basis for $\mathcal{R}$, a suitable basis for Floquet space (Sambe, 1973) is $\{|i, n\rangle \equiv |i\rangle \otimes |n\rangle\}$. The index n can be assimilated to the number of "photon" excitations (or modulation quanta) in the system as noted by Shirley (1965).
(2) The role of the Hamiltonian is played by the Floquet Hamiltonian

$$\hat{\mathcal{H}}_F = \hat{\mathcal{H}} - i\hbar\frac{\partial}{\partial t}, \tag{6.4}$$

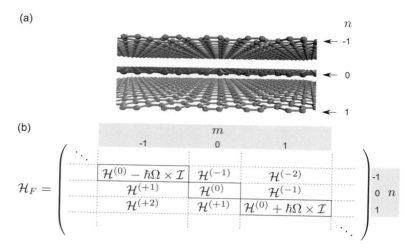

Figure 6.2 (a) By using Floquet theory, the TDSE is mapped to an equivalent time-independent problem in a higher dimensional space. This scheme shows a monochromatically driven graphene sample along with the replicas corresponding to up to ±1 photon excitations. Though formally the number of replicas is infinite, in practice the ($\mathcal{T}$) space is truncated until convergence of the desired response function is achieved. (b) Floquet Hamiltonian in matrix form with blocks defined according to the number of "photon" excitations n, m. $\mathcal{H}^{(n)} = \frac{1}{T} \int_0^T dt \mathcal{H} e^{-in\Omega t}$ is the nth Fourier component of the Hamiltonian.

whose matrix elements in Floquet space are given by

$$\langle i, m| \hat{\mathcal{H}}_F |j, n\rangle = \frac{1}{T} \int_0^T dt \, \langle i| \hat{\mathcal{H}} |j\rangle \, e^{-i(n-m)\Omega t} + n\hbar\Omega\delta_{n,m}. \qquad (6.5)$$

Therefore, the time-dependent problem has been mapped to a time-independent one in a higher dimensional space (as represented in Fig. 6.2(a)). The diagonal elements of the Hamiltonian take into account the number of "photons" in the system while the matrix elements connecting the different states are determined by the Fourier components of the time-dependent potential. The Floquet Hamiltonian can be cast in the block-matrix form shown schematically in Fig. 6.2(b).

6.3.1 Average current and density of states

To obtain the current through the device one may assume: (i) non-interacting electrons, (ii) a harmonic time-dependent field limited to a finite region of the sample, and (iii) a sample connected to semi-infinite electrodes where thermalization takes place. Under these assumptions, a coherent calculation gives an appealing expression for the time-averaged current (Kohler *et al.*, 2005):

$$\bar{I} = \frac{1}{T} \int_0^T dt \, I(t) = \frac{2e}{h} \sum_n \int \left[T_{R,L}^{(n)}(\varepsilon) f_L(\varepsilon) - T_{L,R}^{(n)}(\varepsilon) f_R(\varepsilon) \right] d\varepsilon, \qquad (6.6)$$

where $T_{R,L}^{(n)}(\varepsilon)$ is the probability for an electron on the left (L) with energy ε to be transmitted to the right (R) reservoir while exchanging n photons. These probabilities are weighted by the usual Fermi–Dirac distribution functions $f_{R(L)}$ for each electrode.

To compute the transmission probabilities between the different inelastic channels, as needed to obtain the total current, Eq. (6.6), we may relate them to the Floquet–Green function (see also Martinez 2003, Foa Torres 2005), the Green's functions for the Floquet Hamiltonian, $G_F = (\varepsilon\mathcal{I} - \mathcal{H}_F)^{-1}$, in a way similar to the usual scattering theory (Kohler *et al.*, 2005, Stefanucci *et al.*, 2008):

$$T_{R,L}^{(n)}(\varepsilon) = \text{Tr}\left[\Gamma_{R,n}(\varepsilon)G_{(R,n),(L,0)}(\varepsilon)\Gamma_{L,0}(\varepsilon)G_{(R,n),(L,0)}^{\dagger}(\varepsilon)\right], \tag{6.7}$$

where $G_{(R,n),(L,0)}(\varepsilon)$ is the block matrix for the Floquet–Green function connecting the left and right electrodes with the exchange of n photons. Note that the subindex F was omitted to simplify the notation. The escape rates to the electrodes now take into the account the different possible elastic and inelastic channels through the additional subindex n:

$$\Gamma_{\alpha,n}(\varepsilon) = \text{i}\left[\Sigma_\alpha(\varepsilon + n\hbar\Omega) - \Sigma_\alpha^{\dagger}(\varepsilon + n\hbar\Omega)\right], \tag{6.8}$$

where $\Sigma_\alpha(\varepsilon)$ is the usual retarded self-energy correction due to electrode $\alpha = L, R$.[2] For an expression of the noise as characterized by the zeroth moment of the current–current correlation function we refer to Kohler *et al.* (2005). For non-interacting systems, this formalism gives results equivalent to those of the nonequilibrium Green's functions (Keldysh) formalism (Arrachea & Moskalets, 2006).

In the presence of an AC field, one may compute the DC component of the spectral function. This time-averaged density of states is given by Oka & Aoki (2009), Zhou & Wu (2011)

$$DOS(\varepsilon) = -\frac{1}{\pi}\lim_{\eta\to 0^+}\text{Im}\left[\sum_i \langle i, 0| G_F(\varepsilon + i\eta) |i, 0\rangle\right], \tag{6.9}$$

which is the trace of the Floquet–Green function restricted to the block corresponding to zero photons.

It is important to note that a truncation of Floquet space is needed to compute these probabilities. Indeed, one considers only the Floquet states $|j, n\rangle$ within some range for n, i.e. $|n| \le N_{\max}$. This range can be successively expanded until the answer converges, giving thus a variational (*non-perturbative*) method. The case of an AC bias may seem difficult to handle within this scheme, since it introduces a modulation not restricted to the sample. However, one can use a suitable gauge transformation to map an AC bias voltage into a time-dependent field acting on the borders of the sample region (Kohler *et al.*, 2005).

---

[2] A derivation of Eq. (6.7) in this multichannel case can be found in the appendix of Stefanucci *et al.* (2008).

6.3.2   Homogeneous driving and the Tien–Gordon model

If the time-dependent field is applied homogeneously to the sample, i.e.

$$\hat{\mathcal{H}}_{\text{sample}}(r) = \hat{\mathcal{H}}^0_{\text{sample}}(r,t) + eV_{ac}\cos(\Omega t)\hat{\mathcal{I}}, \qquad (6.10)$$

a separation of the space and time variables in the time-dependent Schrödinger equation is still possible (Platero & Aguado, 2004):

$$\psi(r,t) = \psi^0(r,t)\exp(-ie(V_{ac}/\hbar\Omega)\sin(\Omega t)), \qquad (6.11)$$

where $\psi^0(r,t)$ are the solutions in the absence of the time-dependent potential. By expanding the complex exponential in Fourier series,

$$e^{-i(eV_{ac}/\hbar\Omega)\sin(\Omega t)} = \sum_n J_n\left(\frac{eV_{ac}}{\hbar\Omega}\right) e^{in\Omega t}, \qquad (6.12)$$

one already has the solutions in the Floquet form (Kohler *et al.*, 2005). After some algebra one gets

$$\bar{I} = \frac{2e}{h}\sum_n \left|J_n\left(\frac{eV_{ac}}{\hbar\Omega}\right)\right|^2 \int T(\varepsilon)(f_L(\varepsilon) - f_R(\varepsilon))d\varepsilon. \qquad (6.13)$$

Here $T = T_{R,L} = T_{L,R}$ is the transmission in the absence of driving force. This has the same form as the solution proposed by Tien & Gordon (1963) to describe photon-assisted processes. Note that if inversion symmetry is broken, $(x \to -x)$ $T_{R,L} \neq T_{L,R}$ and therefore there could be a pumped charge even with only one time-dependent parameter. This is called *single-parameter* pumping (or monoparametric pumping) and requires going beyond the adiabatic approximation (which gives a vanishing response in such a case).[3]

6.3.3   Time-evolution operator

In the general case of a time-dependent Hamiltonian, the time-evolution operator $\hat{\mathcal{U}}(t,t_0)$ (defined such that $\hat{\mathcal{U}}(t,0)\psi(t_0 = 0) = \psi(t)$) can be written as

$$\hat{\mathcal{U}}(t,0) = \mathsf{T}\exp\left(-\frac{i}{\hbar}\int_0^t dt\,\hat{\mathcal{H}}(t)\right), \qquad (6.14)$$

where T is the time-ordering operator. When the Hamiltonian is time-periodic, $\hat{\mathcal{H}}(t + nT) = \hat{\mathcal{H}}(t)$ (n an integer), the time-evolution operator shares the same time periodicity. Therefore, knowing the time-evolution operator at one driving period allows building a

---

[3] See for example Moskalets & Buttiker 2002, Foa Torres 2005, Kaestner *et al.* 2008.

map into the evolved states $\psi(t = nT)$ at *any* integer multiple of T, however large is n. Indeed,

$$\hat{\mathcal{U}}(nT,0) = \mathsf{T} \prod_{0}^{n} \exp\left(-\frac{i}{\hbar} \int_{0}^{T} dt\, \hat{\mathcal{H}}(t)\right) \tag{6.15}$$

$$= \prod_{0}^{n} \mathsf{T} \exp\left(-\frac{i}{\hbar} \int_{0}^{T} dt\, \hat{\mathcal{H}}(t)\right) \tag{6.16}$$

$$= \left[\hat{\mathcal{U}}(T,0)\right]^{n}, \tag{6.17}$$

where in the second line one uses the fact that since the Hamiltonian is T-periodic, the terms in the product are equal and therefore commute, allowing the time-ordering operator to be brought inside the product.

Furthermore, it can be shown that $\hat{\mathcal{U}}(T,0)$ is related in a striking way to the Floquet Hamiltonian:

$$\hat{\mathcal{U}}(T,0) = \exp\left(-\frac{i}{\hbar} \hat{\mathcal{H}}_F T\right), \tag{6.18}$$

i.e. a Floquet state with quasi-energy ε_α acquires a phase $\exp(-i\epsilon_\alpha T/\hbar)$ after one period T. Hence, if one wants to compute the stroboscopic evolution of a system with a T-periodic Hamiltonian at times $t = nT$, it can be done just as for time-independent Hamiltonians but the role of the stationary eigenfunctions is now played by the Floquet states and instead of the Hamiltonian one needs to use the Floquet Hamiltonian.

6.4   Overview of AC transport in carbon-based devices

Interest in AC transport in carbon-based devices has been inspired by diverse sources. From developing new methods able to cope with more atoms or give the full time-dependent response (in contrast with time-averaged values) (Stefanucci *et al.*, 2008, Perfetto, Stefanucci & Cini, 2010), to proposing different setups that exploit the peculiarities of these materials for enhancing quantum pumping both in the adiabatic (Prada *et al.*, 2009, Zhu & Chen, 2009, Grichuk & Manykin, 2010, Alos-Palop & Blaauboer, 2011) and non-adiabatic regimes (Foa Torres *et al.*, 2011, San-Jose *et al.*, 2011, Zhou & Wu, 2012),[4] to finding new ways of generating backscattering in spite of Klein tunneling effects in graphene (Savelev, Häusler & Hänggi, 2012),[5] or even the generation and control of laser-induced topological states in a monolayer (Kitagawa *et al.*, 2011, Gu *et al.*, 2011) and bilayer graphene (Suárez Morell & Foa Torres, 2012). Although most of the work in these issues has been theoretical, there are already several experiments available. Quantum pumping for example has been studied in carbon nanotubes (Leek *et al.*, 2005) and more rencently also in graphene (Conolly *et al.*, 2013) where

---

[4] The issue of quantum pumping in graphene nanomechanical resonators is also discussed in Low *et al.* (2012).

[5] Savelev *et al.* (2012) offer a solution for the problem of an *arbitrary* spacetime-dependent scalar potential.

it promises to close the metrological triangle by allowing us to redefine the ampere. A related ratchet effect probing inversion symmetry breaking due to substrate or adatoms has also been probed recently (Drexler *et al.*, 2013). The role of defects in AC transport through graphene-based materials is an important issue that requires further progress. In the context of adiabatic quantum pumping, the role of topological defects was examined in Ingaramo & Foa Torres (2013).

A strong momentum to this area is also expected to come from the experimental community working on laser-induced effects and nanophotonics. For recent reviews see Bonaccorso *et al.* (2010) and Glazov & Ganichev (2013).

Floquet theory, in particular, has been applied to a variety of carbon-based devices including Fabry-Pérot interferometers with AC gating (Foa Torres & Cuniberti, 2009, Rocha *et al.*, 2010) and non-adiabatic quantum pumps (Foa Torres *et al.*, 2011, San-Jose *et al.*, 2011, Zhou & Wu, 2012). Another example is developed in more detail in Section 6.5 where the captivating possibility of opening a bandgap in graphene through illumination with a laser field (Syzranov *et al.*, 2008, Oka & Aoki, 2009, Calvo *et al.*, 2011) is discussed.

Further theoretical studies propose new ways of achieving states akin to those of a topological insulator through laser illumination, a *Floquet topological insulator* (Lindner *et al.*, 2011, Kitagawa *et al.*, 2011).[6] A topological insulator (Hasan & Kane, 2010) exhibits a bulk bandgap as for a usual insulator, but has surface or edge states which are gapless and protected by time-reversal symmetry. These states turn out to be insensitive to smooth changes in the potential or disorder. To date, the materials exhibiting topological properties are scarce and the ability to control their transport features is very challenging and limited (Hasan & Kane, 2010). Being able to change the topological properties in the *same* material by using a laser field could open fascinating doors for novel ways of controlling electronic states of matter (Lindner *et al.*, 2011), particularly in graphene devices (both monolayer (Kitagawa *et al.*, 2011, Gu *et al.*, 2011) and bilayer graphene (Suárez Morell & Foa Torres, 2012)). The description of the topological properties of a system with a time-dependent Hamiltonian, however, remains a challenging technical task (Rudner *et al.*, 2013). One of the proposals (Kitagawa *et al.*, 2011) suggests building an effective time-independent Hamiltonian chosen so that its dynamics mimics the one of the time-dependent system at $t = n \times T$ (T being the period of the field). Using this stroboscopic picture, the topological properties result from applying the usual classification for time-independent systems to the effective Hamiltonian. Many more questions, however, remain open: Is it possible to produce laser-induced protected states in graphene in an experimentally relevant regime? What is the nature of those states and how robust are their conduction properties?

Perez-Piskunow *et al.* (2013) advance in this direction by presenting the first analytical solution and a proposal for unveiling "laser-induced chiral edge states" in graphene. The interplay between spin–orbit interaction and a laser in bulk graphene was explored in Scholz *et al.* (2013).

---

[6] For an overview this topic we refer to Cayssol *et al.* (2013).

6.5 AC transport and laser-induced effects on the electronic properties of graphene

Among the many promising areas sparked by graphene research, graphene photonics (and optoelectronics) is one of the brightest.[7] Since the very beginning light has been one of the best tools for non-invasive characterization of carbon-based materials (Jorio *et al.*, 2011). But light can also be used for achieving useful functions that take advantage of the extraordinary properties of these materials: from improved energy harvesting[8] and novel plasmonic applications,[9] to graphene photodetectors. (Konstantatos *et al.*, 2012)

Recently, the captivating possibility of controlling the electronic properties of graphene through simple illumination with a laser field (Syzranov *et al.*, 2008, Oka & Aoki, 2009, Kibis, 2010) has been examined through atomistic calculations (Calvo *et al.*, 2011, 2012), calculations of the optical response (Zhou & Wu, 2011, Busl, Platero & Jauho, 2012), among other interesting issues (Abergel & Chakraborty, 2009, Savelev & Alexandrov, 2011, San-Jose *et al.*, 2012). The basic idea is that laser illumination may couple states on each side of the Dirac point, inducing a bandgap at energies $\pm\hbar\Omega/2$, if the field intensity and frequency Ω are appropriately tuned.

Here we analyze this in more detail following recent studies (Oka & Aoki, 2009, Calvo *et al.*, 2011, Calvo *et al.*, 2012). We start by considering an electromagnetic field modeled in a semiclassical approximation: a monochromatic plane wave of frequency Ω traveling along the z axis, perpendicular to the plane defined by the graphene sheet (see Fig. 6.3(a)).

By using a Weyl's gauge, the electromagnetic field is represented through a vector potential $\mathbf{A}(t) = (A_x \cos(\Omega t)\,\mathbf{x} + A_y \cos(\Omega t + \phi)\,\mathbf{y})$ (the electric field being directly related to the time derivative of $\mathbf{A}$), where A_x, A_y, and the phase ϕ can be set to go from linear to circular polarization. The interaction with the laser field is modeled through the Hamiltonian

$$\mathcal{H}(t) = v_F \hat{\sigma} \cdot \left[\mathbf{p} - e\mathbf{A}(t)\right], \tag{6.19}$$

where $v_F \simeq 10^6$ m/s denotes the Fermi velocity as usual and $\hat{\sigma} = (\hat{\sigma}_x, \hat{\sigma}_y)$, the Pauli matrices describing the pseudospin degree of freedom.

As made clear below, a correct description of our problem crucially requires a solution valid beyond the adiabatic approximation. Here we can exploit the Floquet theory introduced in Section 6.3, which is an appropriate approach for such electron–photon scattering processes. Once the Hamiltonian is defined one can proceed by computing the matrix elements of the Floquet Hamiltonian leading to the replica picture discussed in Section 6.3. This is the basis for the calculation of the transport properties and the time-averaged DOS. We skip the details here, which are left for Problem 6.3 (more can be found in Calvo *et al.*, 2013), and discuss some results.

---

[7] See for example Bonaccorso *et al.*, 2010, Xia *et al.*, 2009, Karch *et al.*, 2011.

[8] See for example Gabor *et al.* (2011) and the more recent results in Tielrooij *et al.* (2013).

[9] Graphene plasmonics is a blooming field; see for example Koppens, Chang & Garcia de Abajo (2011), Chen *et al.* (2012).

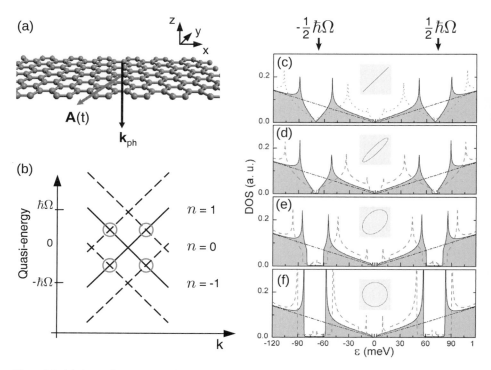

Figure 6.3 (a) A graphene sample illuminated by a laser field perpendicular to the graphene plane. (b) The quasi-energy spectra along a particular k direction. The crossing points including the first two Floquet replicas are marked with circles. (c–f) Average density of states for (c) linear, (d) $\varphi = 0.125\pi$, (e) $\varphi = 0.375\pi$, and (f) circular polarizations taking $\hbar\Omega = 140$ meV. The black solid line is for $I = 32$ (mW/μm^2), while the gray dashed line corresponds to $I = 130$ (mW/μm^2). The case in the absence of irradiation is shown with a dash-dotted gray line for comparison. (Adapted with permission from Calvo *et al.* (2011). Copyright 2011, American Institute of Physics.)

Figure 6.3(c–f) shows how the time-averaged density of states (DOS) for bulk graphene changes as the polarization goes from linear to circular. For linear polarization, one observes a strong depletion for energies close to $\pm\hbar\Omega/2$. These depletions evolve into gaps, called *dynamical gaps* (Syzranov *et al.*, 2008, Oka & Aoki, 2009), for circular polarization. Furthermore, close to the Dirac point a mini-gap opens for circular and elliptic polarizations. A closer scrutiny of these figures shows that the gaps mentioned are areas with a small negligible DOS in the bulk limit and for the parameter range explored here.

To rationalize this behavior, one can take advantage of the Floquet picture explained before. A scheme with the quasi-energy spectra close to the Dirac point including $n = 0, \pm1$ photons is shown in Fig. 6.3(b). The dispersion relation for the states $\{|\mathbf{k}, n\rangle_{\pm}\}$ for $n = 0$ is represented with a solid line, while those for $n = \pm1$ are shown with dashed lines. The effects of the AC field are expected to be stronger at the crossing points marked with circles, leading to the opening of energy gaps at those points provided that the Hamiltonian has a nonvanishing matrix element. From geometrical

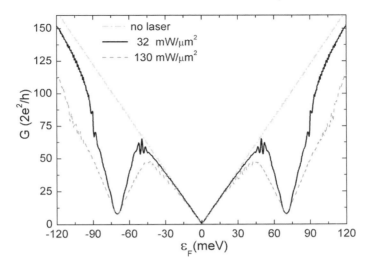

Figure 6.4 DC component of the conductance calculated for a graphene stripe of $1\,\mu$m times $1\,\mu$m in the presence of a linearly polarized laser as a function of the Fermi energy. The solid line is for a laser power of $32\ \mathrm{mW}/\mu\mathrm{m}^2$ while the dashed line corresponds to $130\ \mathrm{mW}/\mu\mathrm{m}^2$. (Reprinted with permission from Calvo *et al.* (2011). Copyright 2011, American Institute of Physics.)

considerations one can see that the crossing of the states differing in one photon lies exactly at $\pm\hbar\Omega/2$. These degeneracies are increased by the AC field, leading to the gaps observed in Fig. 6.3(c–f). Analytical expressions for the dependencies of these gaps on the field parameters can be obtained in a direct way (Calvo *et al.*, 2011). While the dynamical gaps depend linearly on the amplitude of the vector potential, this dependence is quadratic at the mini-gap around zero (which is produced by a virtual photon emission and re-absorption process).

But, are there any consequences on the conductance? To answer this question one can compute the DC component of the conductance within a *tight-binding* model and following a transport calculation, such as the one introduced in Section 6.3. The results are shown in Fig. 6.4 for the case of linearly polarized light along the *x* direction. In this case, the calculation for an armchair ribbon can be done in an efficient way by using the mode decomposition introduced in Section 4.2.2. One can observe that the conductance also shows an important depletion around the dynamical gaps.

Here, we have chosen the laser wavelength within the mid-infrared region, i.e. $\lambda \simeq 9\mu$m. A careful analysis (Calvo *et al.*, 2011) shows that in this frequency region the effects should be maximized while keeping reasonable power levels. Furthermore, gating graphene to reach dynamical gaps located at ~ 70 meV is also within experimental reach.

Further work along these lines hints that lateral confinement in laser-illuminated graphene nanoribbons may have an important role, especially in small ribbons of a few nanometers wide (Calvo *et al.*, 2012). Bilayer graphene also shows similar features, though trigonal warping effects may introduce important changes in the low-energy spectra (Suárez Morell & Foa Torres, 2012). Moreover, the emergence of chiral edge states bridging the dynamical gaps (Perez-Piskunow *et al.*, 2013) offers a playground connecting the physics of topological insulators, graphene, and driven systems.

We close this section by pointing out that this physics may have an impact on other areas such as condensates (Crespi *et al.*, 2013) and photonic crystals, where experiments are already available (Rechtsman *et al.*, 2013). For the case of graphene this mechanism may compete with others (photothermal for example) and therefore experimental data would be crucial. We expect that future experimental work in this area may help to unveil this interesting physics.

6.6   Further reading and problems

- For a review on AC transport in nanostructures see Kohler *et al.* (2005).
- The reader interested in Floquet theory may also enjoy the foundational papers (Shirley, 1965, Sambe, 1973).
- For a very nice introduction to quantum pumping we suggest Büttiker & Moskalets (2006) and references therein.

Problems

6.1   *Quantum pumping: generating a current at zero DC-bias.* In this exercise we try to throw light on the mechanism behind quantum charge pumping. To this end, we follow Büttiker & Moskalets (2006) and consider a one-dimensional system consisting of two regions with alternating potentials, $V_1 = V_0 \cos(\Omega t)$ and $V_2 = V_0 \cos(\Omega t + \phi)$, separated by a distance D.

(a)   Analyze the transmitted amplitudes (specially the inelastic ones) and show that the phase difference ϕ may induce a directional asymmetry in the transmission probability, so that $T_\leftarrow \neq T_\rightarrow$. ($V_0$ can be considered to be small to simplify the analysis. Then only channels with plus or minus one photon need be included.)

(b)   Interpret the asymmetry found before in terms of an interference effect in Floquet space.

(c)   Obtain the dependence of $T_\leftarrow - T_\rightarrow$ on the phase difference ϕ.

6.2   *AC gated Fabry-Pérot interferometers and the quantum wagon-wheel effect.* Reconsider the simple model for a Fabry-Pérot resonator made of an infinite CNT of Problem 4.7. This time we follow Foa Torres & Cuniberti (2009) and Rocha *et al.* (2010), and add a homogeneous alternating potential to the central part of the system, which is modeled by adding an onsite term $eV_{AC} \cos(\Omega t)$. V_{AC} and Ω are the oscillation amplitude and frequency respectively.

(a)   Compute the changes in the conductance (dI/dV_{bias}) as a function of a bias voltage V_{bias} (which can modeled through a symmetric shift of the leads' onsite energies by $\pm V_{bias}$) and a stationary gate voltage V_{gate} applied to the central part of the system. Check the contour plots of the conductance as a function of these two variables for different values of the driving frequencies. Show that the conductance recovers the pattern found for the stationary case ($V_{AC} = 0$) when $\hbar\Omega$ is commensurate with the mean level spacing in the central region Δ (*wagon-wheel effect*). Compare this with the results obtained for other values of the frequency.

(b) Compute the zero frequency noise (see Kohler *et al.* (2005) for explicit formulas) as a function of the driving parameters V_{AC} and $\hbar\Omega$. Show that this time the stationary values are recovered when $\hbar\Omega$ is commensurate with *twice* the mean level spacing (named *quantum* wagon-wheel effect because it relies on the phase shift of the carriers (Foa Torres & Cuniberti, 2009)). Interpret your results.

6.3 *Laser-illuminated graphene.* Consider the case of laser-illuminated graphene as presented in Section 6.5.

(a) Starting from the Hamiltonian given by Eq. (6.19), write its Fourier components and give an expression for the matrix elements of the Floquet Hamiltonian following Eq. (6.5).

(b) Obtain analytical expressions for the gaps within a small intensity approximation.

(c) Design the steps needed for evaluation of the average DOS (Eq. (6.9)).

(d) Design a numerical code for numerical evaluation of the DOS. You may help yourself by using the codes already available through our website.

** Additional exercises and solutions available at our website.

7   *Ab initio* and multiscale quantum transport in graphene-based materials

This chapter illustrates the several possible computational approaches that can be used towards a more realistic modeling of disorder effects on electronic and transport properties of carbon-based nanostructures. Multiscale approaches are first presented, combining *ab initio* calculations on small supercells with *tight-binding* models developed from either a fitting of *ab initio* band structures, or a matching between conductance profiles with a single defect/impurity. Chemical doping with boron and nitrogen of carbon nanotubes and graphene nanoribbons is discussed in detail, as well as adsorbed oxygen and hydrogen impurities for two-dimensional graphene, being both of current fundamental interest. Finally, fully *ab initio* transport calculations (within the Landauer–Büttiker conductance framework) are discussed for nanotubes and graphene nanoribbons, allowing for even more realism, albeit with limited system sizes, in description of complex forms of edge disorder, cluster functionalization or nanotube interconnection.

7.1   Introduction

In the following sections, disordered and chemically doped carbon nanotubes and graphene nanoribbons are explored. The main scientific goal consists in illustrating how defects and impurities introduce resonant quasi-localized states at the origin of electron–hole transport asymmetry fingerprints, with the possibility of engineering *transport (or mobility) gaps*. Several multiscale approaches are described to develop various *tight-binding* models from first-principles calculations. A first technical strategy (illustrated on boron-doped nanotubes, Section 7.2.2) consists in designing a *tight-binding* model by fitting the *ab initio* band structures. Such an approach is used to describe doped metallic nanotubes, but actually ceases to be accurate for graphene nanoribbons, owing to complex screening effects introduced by edges.

A second approach consists in adjusting the *tight-binding* parameters by searching for a good match with first-principles transport calculations for a short nanotube or ribbon with a single dopant. This is illustrated in Sections 7.2.3 and 7.5.11 for nitrogen-doped nanotubes and boron (and nitrogen)-doped graphene nanoribbons. Using the re-parameterized *tight-binding* models, mesoscopic transport can then be investigated in micrometer-long and disordered semiconducting nanotubes and graphene ribbons with random distribution of chemical impurities. By studying the statistics of transmission coefficients, transport length scales such as the mean free paths (ℓ_{el}) and localization lengths (ξ) are predicted, together with crossovers between transport regimes. The scaling between ℓ_{el} and ξ is demonstrated to be in full agreement with the generalized form of the Thouless relationship, thus offering the first quantitative test for fundamental theories of mesoscopic physics (Section 7.2.3).

Finally, full *ab initio* transport calculations in both disordered graphene ribbons and defective carbon nanotubes are presented in Section 7.5, for various types of defects such as edge defects, or grafted molecules and randomly distributed clusters. Although these calculations are much more computationally demanding, they prove to be essential in a situation of enhanced chemical complexity at the nanoscale, while they offer more possibilities for quantitative comparison with experimental data.

7.2   Chemically doped nanotubes

7.2.1   Tight-binding Hamiltonian of the pristine carbon nanotube

The electronic properties of an armchair (n, n) CNT are first described using a *tight-binding* Hamiltonian with a single p_z orbital per site and only nearest neighbors hopping integrals. Within this approach, the Hamiltonian only depends on the network connectivity. Such an assumption is valid and accurate enough for describing the energy bands near the charge neutrality point. Note that the numerical studies are restricted to weak disorder, meaning that the elastic mean free path $\ell_e \gg \lambda_F$ ($\lambda_F = \frac{3}{2}\sqrt{3}a_{cc} \approx 3.7$ Å).

7.2.2   Boron-doped metallic carbon nanotubes

In order to investigate the transport properties of boron-doped (metallic) carbon nanotubes, the first step consists in a calculation of the electronic structure of armchair $(10, 10)$ nanotubes containing 2500 cells using periodic boundary conditions (10^5 atoms) and the zone-folding approximation (ZFA) (Latil *et al.*, 2004). A conventional *tight-binding* Hamiltonian can be defined as follows:

$$\hat{\mathcal{H}} = \sum_{\alpha=1}^{N} \varepsilon_\alpha |\alpha\rangle\langle\alpha| + \sum_{\langle\alpha,\beta\rangle} \left[\gamma_{\alpha\beta}|\alpha\rangle\langle\beta| + h.c.\right], \tag{7.1}$$

where the first sum is achieved over all the $p_\perp$ orbitals while the second is limited to first neighbors of the α site. In Eq. (7.1), the matrix elements ε_α denote the onsite energies, while $\gamma_{\alpha\beta}$ are the hopping integrals.

The ZFA technique applied to boron-doped carbon nanotubes consists in deriving the local electronic properties in the vicinity of an atomic substitution. Unfortunately, the *tight-binding* methods are usually not suitable to account for charge transfer between different atomic species. However, by adding a corrective electrostatic potential to the onsite energies, the effects of electric fields, charge transfer, or electric dipole moments can be taken into account. The corresponding corrections (added as extra terms to the onsite energies) are usually calculated self-consistently (solving a Schrödinger–Poisson equation). The onsite energies ε_α in Eq. (7.1) are labeled ε_C, ε_B for carbon and boron atoms, respectively. The γ_{CC} and γ_{BC} describe the inequivalent hopping integrals. Practically, these parameters are obtained by fitting the ZFA band structure to the *ab initio* calculations (Latil *et al.*, 2004).

In order to capture the electronic fingerprints of carbon orbitals in the vicinity of the B impurity, the electronic structure of a supercell containing 31 carbon atoms and a single B atom is first examined using a DFT approach within the LDA approximation. The electronic density for the last (half) occupied band is found to be distributed only on the $p_{\perp}$ orbitals for atoms located close to the impurity, up to the third-nearest neighbors of the B impurity. This localization of the HOMO-LUMO band allows us to restrict the correction to carbon atoms only up to this level of accuracy. Additionally, such a result suggests that the hopping integrals between sites are not affected by the charge transfer. Moreover, the boron atom is supposed to be "carbon-like," i.e. $\gamma_{CC} = \gamma_{BC} = \gamma$. Consequently, only six parameters need to be adjusted: the single hopping integral γ, the carbon and boron onsite energies ε_C and ε_B, and the renormalized carbon onsite energies ε_3, ε_2 and ε_1 (resp. third-, second- and first-nearest neighbors of the boron impurity).

These parameterizations are performed using a least square energy minimization scheme between DFT-LDA and ZFA band structures. At first, the *ab initio* electronic band structure of an isolated graphene sheet is used to fit the hopping terms. As a low boron density is considered, the chemical potentials (Fermi energies) of the two subsystems are equal, resulting in $\varepsilon_C = E_{F,\text{ supercell}} = E_{F,\text{CNT}}$. The band structure obtained with the optimal parameters can thus be favorably compared to the DFT-LDA band structure (not shown here) (Latil *et al.*, 2004). The best fit for the hopping integral gives $\gamma = 2.72$ eV, while onsite energies are $\varepsilon_B = +2.77$ eV, $\varepsilon_1 = -0.16$ eV, $\varepsilon_2 = +0.21$ eV, $\varepsilon_3 = +0.39$ eV and $\varepsilon_C = -1.56$ eV. Finally, the spectrum is shifted to fix the charge neutrality point (E_F) to 0 eV.

As illustrated in Fig. 7.1(a), the DOS of a 0.1% B-doped (10, 10) CNT exhibits the typical acceptor peak (E_1), as confirmed by *ab initio* calculations (Choi *et al.*, 2000). Depending on the energy, three different transport regimes are obtained, as illustrated in Fig. 7.1(b). At energy E_2 above the Fermi level (far from the impurity resonance level),

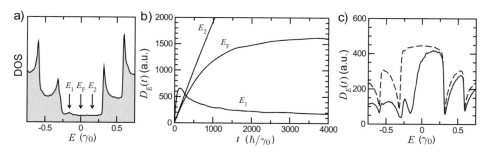

Figure 7.1 (a) Density of states of a 0.1% B-doped (10, 10) CNT. (b) $D_E(t)$ for three different energies (indicated by arrows in (a)). $D_E(t)$ for energy E_1 is ten times magnified. (c) $D_E(t)$ at an elapsed time $t = 200\hbar/\gamma_0$, for the same B-doped CNT (solid line) and a pristine CNT (dashed line). Courtesy of S. Latil.

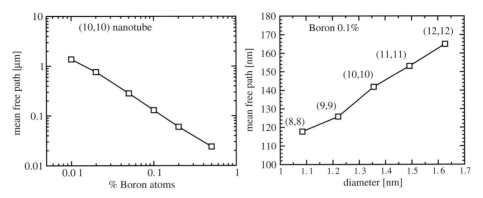

Figure 7.2 Mean free path ℓ_e estimated at the Fermi energy for B-doped (n, n) nanotubes. Left: Evolution of ℓ_e for a $(10, 10)$ nanotube with varying boron densities. Right: Evolution of ℓ_e vs. tube diameter for a fixed concentration of B atoms (0.1%).

the diffusion coefficient scales almost linearly with time, indicating a quasi-ballistic motion and a weak sensitivity to the presence of boron impurities.

Differently, at the Fermi energy, the saturation of $D(E_F, t) \rightarrow D_0 \sim \ell_e v_F$ denotes a diffusive regime. The extracted ℓ_e is found to decay linearly with dopant concentration, following Fermi's golden rule (Fig. 7.2, left). However, the mean free path is also predicted to increase linearly with the nanotube diameter (Fig. 7.2, right). This upscaling with diameter for a fixed disorder is a unique character of metallic nanotubes, as derived in Section 5.1. Interestingly, the typical values obtained numerically for ℓ_e turn out to agree reasonably well with experimental data (Liu *et al.*, 2001) ($\ell_e \sim$ 175–275 nm for boron-doped nanotubes with diameters in the range 17–27 nm, and 1.0% of boron impurities).

Finally, at the resonant energy of the quasi-bounded states (E_1), the diffusivity exhibits a $\sim 1/t$ behavior, typical signature of a localization phenomenon. In Sections 7.2.3 and 7.5.11, a more extensive analysis of quantum interferences and localization phenomena is achieved, including extraction of the localization lengths, directly from the scaling analysis of transmission coefficients (for 1D systems) or using the predicted logarithmic law to describe weak localization quantum corrections (for two-dimensional disordered graphene).

It is worth mentioning that the chemical disorder induces strong electron–hole conduction asymmetry (see Fig. 7.1(c)). This is further observed in the length dependence of conductance. Indeed, by increasing the (effective) channel length of the device denoted L_{device} from ∼10 nm to ∼1 μm (Fig. 7.3), the contribution of quantum interferences comes into play and further amplifies the electron–hole conductance asymmetry. For instance, the hole conductance has been totally suppressed, whereas electron conductance remains close to its maximum value ($2G_0$) for a nanotube length of about 1.2 μm. Such a phenomenon of resonant scattering will actually be suggested to open interesting perspectives for the design of unipolar graphene-based field effect transistors with improved ON/OFF current ratio (see Section 7.5.11).

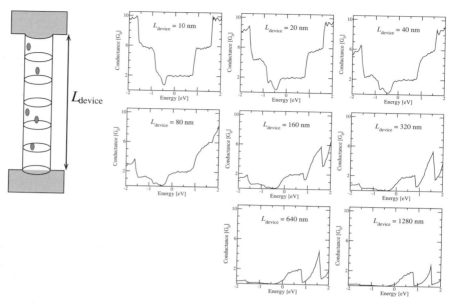

Figure 7.3 Quantum conductance for a device (with varying channel length) made of a single (10, 10) nanotube containing 0.1% of boron impurities. Courtesy of Sylvain Latil.

7.2.3   Nitrogen-doped metallic carbon nanotubes

The single nitrogen impurity case

An alternative approach to capture the effect of chemical impurities and resonant scattering is presented below. Using first-principles calculations, the energy-dependent conductance profile is first computed for a finite length armchair (n, n) CNT, where a single carbon atom (labeled δ) has been substituted by a nitrogen impurity. Next, the evolution of scattering potential around an impurity is computed self-consistently using an *ab initio* code (SIESTA (Soler *et al.*, 2002)). Effective onsite and hopping matrix elements are then directly extracted from the *ab initio* simulations (thus including structural optimization of the system), and a simplified *tight-binding* Hamiltonian with a single orbital per site is developed using the following form:

$$\hat{\mathcal{H}}_\delta = \sum_\alpha V_{\delta,\alpha} \, |\alpha\rangle \, \langle\alpha| - \sum_{<\alpha,\beta>} \left(\gamma_{\alpha\beta} \, |\alpha\rangle \, \langle\beta| + h.c. \right). \tag{7.2}$$

The main effect of a single nitrogen impurity can be captured by a proper renormalization of onsite energetics through $V_{\delta,\alpha}$, conserving the hopping integrals $\gamma_{\alpha\beta} \approx \gamma_0 = 2.9$ eV. The evolution of the *ab initio* onsite energies for π orbitals as a function of the distance to the impurity can be fitted by a Gaussian-like function over a range of 10 Å (Adessi *et al.*, 2006), as depicted in Fig. 7.4. In the present simulation, nitrogen atoms in substitution are compared to physisorbed potassium atoms. The potential well created by a N impurity in substitution is clearly much deeper than the one associated

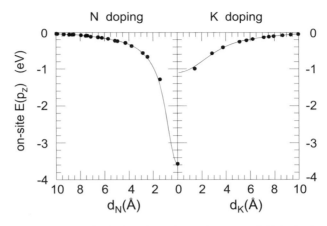

Figure 7.4 Onsite p_z orbital energies in doped graphene sheet upon (left) N doping and (right) K doping. Filled circles correspond to the two p_z orbitals in the *ab initio* calculation (solid line gives the Gaussian fit). The abscissa indicates the distance (in Angstroms) from a carbon atom to the N impurity (in the substitution case), and to the hexagon center on which the K impurity projects (in the physisorption case). (Reproduced with permission from Adessi *et al.* (2006). Copyright (2006) by the American Physical Society. Courtesy of Ch. Adessi.)

with the partially screened K$^+$ ion (Fig. 7.4). In particular, the ability of adsorbed K ions to trap electrons is significantly reduced as compared to N impurities. We note that even though screening of Coulomb potential is known to be much weaker in low dimension, the K-induced potential does not seem to be much longer range than the one generated by the nitrogen impurity. However, some arbitrariness still remains in the choice of the *ab initio* atomic-like basis (spatial extent, completeness, etc.), which could significantly alter the impurity potential with fluctuations as large as ~ 1 eV. One reasonable approximation consists in adjusting the onsite π potential impurity around the obtained *ab initio* value, and optimizing the agreement between *tight-binding* and *ab initio* conductance profiles (see below).

Such an impurity potential breaks the reflection symmetry plane of the nanotube, generating two resonant quasibound states in the conductance profile, shown in Fig. 7.5. The first resonance (located at low energy in the π^* band) is of even parity (s-wave) and broad in energy. In contrast, the second sharper resonance (located at higher energy) has odd parity (p-wave). Both resonances suppress one conduction channel of given parity. The energy position of the s-wave resonance can be finally adjusted with respect to the first van Hove singularity (less than a tenth of an eV) by tuning the onsite nitrogen potential, therefore improving the agreement with the *ab initio* calculation (Choi *et al.*, 2000).

Doping with a random distribution of nitrogen atoms

Using such an effective impurity potential $V_{\delta\alpha}$, it is now possible to explore the mesoscopic transport properties of chemically doped nanotubes with nitrogen impurities (for a fixed doping density n_{dop}). The total Hamiltonian of the disordered nanotube is

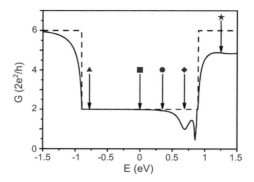

Figure 7.5 Conductance curve versus energy for both the pristine (dashed line) and doped with a single nitrogen impurity (solid line) (10, 10) armchair nanotube. (Adapted with permission from Avriller *et al.* (2006). Copyright (2006) by the American Physical Society.)

written as

$$\hat{\mathcal{H}}(\Omega) = \sum_{\delta \in \Omega} \sum_{\alpha} V_{\delta,\alpha} |\alpha\rangle \langle\alpha| - \gamma_0 \sum_{<\alpha,\beta>} \left(|\alpha\rangle \langle\beta| + |\beta\rangle \langle\alpha| \right), \qquad (7.3)$$

where Ω denotes an ensemble of impurity distributions, satisfying the chosen doping level. All disorder configurations have the same probability ($P(\Omega)$) of occurring, and the resulting conductance average is computed from $\overline{G} = \sum_{\Omega} P(\Omega) G(\Omega)$. The mean distance between impurities $\ell_{\text{imp}} = a/2nn_{\text{dop}}$ becomes a new length scale of the problem. For a (10, 10) CNT, $\ell_{\text{imp}} = 0.12/n_{\text{dop}}$ Å, and for a doping rate $n_{\text{dop}} = 0.1\%$, the mean distance between impurities is $\ell_{\text{imp}} \approx 12$ nm. Consequently, the nanotube lengths must satisfy $L \geq l_{\text{imp}}$.

The normalized conductance $T(\Omega) = G(\Omega)/G_0$ depends on the distribution Ω of impurities, hence becoming a random variable, statistically defined by its mean value $\overline{T} = \frac{1}{\text{card}(\{\Omega\})} \sum_{\Omega} T(\Omega)$ and its root mean squared (RMS) fluctuation $\Delta T = \sqrt{\overline{T^2} - \overline{T}^2}$. Analysis of the dependence of the conductance probability distribution $P(T; E, L)$ on energy E and length L gives access to all transport length scales (ℓ_{el} and ξ). For instance, the elastic mean free path is extracted from the curve $\overline{T}(L)$, using the interpolation given by Eq. (7.4). The values obtained for ℓ_{el} (Fig. 7.6) are in very good agreement with estimations based on the Kubo method (Avriller *et al.*, 2007). The whole range of transport regimes is discussed below for the (10, 10) nanotube with 0.1% nitrogen impurities (Avriller, 2008).

Conductance statistics
Conductance profiles in the quasi-ballistic regime
All results on conductance statistics are presented for a selected energy ($E = 0.35$ eV, indicated by an arrow and filled circle in Fig. 7.5). The dispersion of conductance values (by varying disorder configuration) for $L = 20$ nm $< \ell_{\text{el}} \sim 122$ nm is given in Fig. 7.7. $P(T)$ is found to be very narrow with a maximum close to its ballistic value ($T \approx 2$). The

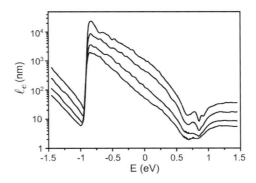

Figure 7.6 Electronic mean free paths in a $(10, 10)$ nanotube doped with nitrogen impurities with $n_{\text{dop}} = 0.05\%, 0.1\%, 0.2\%, 0.3\%$ (from top to bottom) estimated using the Kubo method presented in Section 3.4.4. (Adapted with permission from Avriller *et al.* (2006). Copyright (2006) by the American Physical Society.)

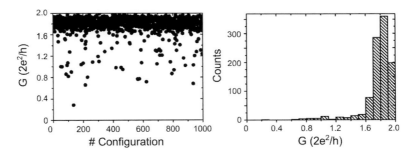

Figure 7.7 Left: Conductance values (at $E = 0.35$ eV) as a function of the disorder configuration ($L = 20$nm $\leq \ell_{\text{el}}$). Right: Corresponding conductance histogram. (Reproduced with permission from Avriller *et al.* (2007). Copyright (2007) World Scientific.)

first two moments of the transmission distribution exhibit well-defined values $\bar{T} = 1.8$ and $\Delta T = 0.2$, pinpointing a crossover from a ballistic to a diffusive regime.

The length scaling of the conductance obtained can be captured by adding a resistance quantum per channel to the diffusive contribution, i.e. $R \approx R_0/N_\perp + R_0/N_\perp (L/\ell_{\text{el}})$, giving

$$\bar{T} = N_\perp \frac{1}{1 + \dfrac{L}{\ell_{\text{el}}}}, \qquad (7.4)$$

which yields to the expected asymptotic limits $\bar{T} \approx N_\perp$ for $L \ll \ell_{\text{el}}$ and $\bar{T} \approx N_\perp \ell_{\text{el}}/L$ for $L \gg \ell_{\text{el}}$, but remains approximative at the transition ($L \sim \ell_{\text{el}}$). The transition regime occurs when the tube length is similar to the mean free path. Figure 7.8 depicts the dispersion of T in the ballistic regime, that is for $L \sim \ell_{\text{el}} = 122$ nm. Here, the distribution $P(T)$ becomes Gaussian-like and fully symmetric, with first moments given by $\bar{T} = 1.0$ and $\Delta T = 0.3$.

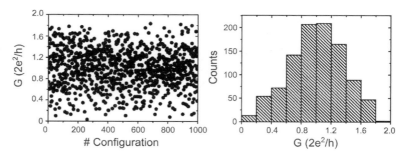

Figure 7.8 Left: Conductance values (at $E = 0.35$ eV) as a function of the disorder configuration ($L = 122nm \sim \ell_{el}$). Right: Corresponding conductance histogram. (Reproduced with permission from Avriller *et al.* (2007). Copyright (2007) World Scientific.)

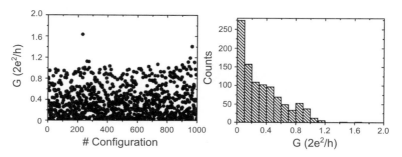

Figure 7.9 Left: Conductance values (at $E = 0.35$ eV) as a function of the disorder configuration ($L = 400$ nm $> \xi$). Right: Corresponding conductance histogram. (Reproduced with permission from Avriller *et al.* (2007). Copyright (2007) World Scientific.)

Conductance profiles in the localized regime

The localized regime is reached when the nanotube length L becomes longer than the localization length ξ, with $\exp\{\overline{\ln T}\}$ decreasing as $\overline{\ln T} = -L/\xi$ (Kostyrko, Bartkowiak & Mahan, 1999, Hjort & Stafstrom, 2001, Gómez-Navarro *et al.*, 2005). For example, the corresponding dispersion values for T in such a regime ($L = 400$ nm $> \xi = 170$ nm) are reported in Fig. 7.9. The formation of the localized regime is evidenced by a strongly asymmetric distribution $P(T)$, a peak near zero transmission and a long tail toward higher values of T.

The first moments $\bar{T} \simeq \Delta T = 0.3$ do not reflect the shape of $P(T)$. In order to characterize such a regime, the most suitable statistical indicator is actually given by $\overline{\ln T}$ (Anderson *et al.*, 1980, Abrahams *et al.*, 1979). Figure 7.10 shows the dispersion of $\ln T$ for $L = 2000$ nm $\gg \xi = 170$ nm. The distribution $P(\ln T)$ becomes more symmetric with first moments $\overline{\ln T} = -11.7$ and $\Delta \ln T = 4.8$.

Transport regimes and crossovers

By scrutinizing the length dependences of $\Delta T/\bar{T}$ and $|\Delta \ln T/\overline{\ln T}|$ at $E = 0.35$ eV, the crossover between transport regimes can be identified (Avriller *et al.*, 2007). Indeed, $\Delta T/\bar{T}$ is found to increase as a function of L, whereas $|\Delta \ln T/\overline{\ln T}|$ peaks near $L = \ell_{el}$

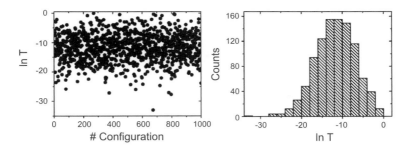

Figure 7.10 Left: Conductance dispersion as a function of the sample number for the case $L = 400$ nm $> \xi$. Right: Corresponding conductance histogram. (Reproduced with permission from Avriller *et al.* (2007). Copyright (2007) World Scientific.)

and further decays with increasing length. The crossing point of the two curves occurs at $L = L_c \approx 385$ nm (in our case) where the relative fluctuations of $\bar{T}$ and $\overline{\ln T}$ become similar. The regime remains thus quasi-ballistic as long as $\Delta T/\bar{T} < 1$, whereas the localization regime forms when $\Delta T/\bar{T} > 1$ and $|\Delta \ln T/\overline{\ln T}| < 1$. At $L = L_c$, the resistance of the nanotube is of the order of the resistance quantum, which thus confirms the onset of the localization regime, following an argument given by Thouless in 1977 (Thouless, 1977). Indeed, Thouless proposed viewing the normalized conductance as the ratio between two characteristic energy scales, namely $G/G_0 \approx E_{th}/\Delta$ where $E_{th} = \hbar D/L^2$ and Δ, the mean level spacing. Localization takes place when $E_{th} \approx \Delta$, that is when the conductance becomes smaller than the quantum of conductance (Thouless, 1977).

Localization length (ξ) and Thouless relationship
The localization length at $E = 0.35$ eV is extracted from the curve $\overline{\ln T}(L)$. Taking into account points for which $L > L_c$, the curve is fitted with the scaling law $\overline{\ln T} = -L/\xi$. Random matrix theory (RMT) allows some connection between both transport length scales by studying statistical properties of eigenvalues of the $\hat{t}_{LR}\hat{t}_{LR}^\dagger$ matrix. In disordered wires, the statistics of the joint distribution of transmission coefficients $P(T_1, \dots, T_{N_\perp}; L)$ has been studied in-depth (Mello, Pereyra & Kumar, 1988). In the asymptotic metallic (localized) regime (when $\ell_{el} \ll L \ll \xi$ ($\xi \ll L$)), the joint probability distribution can actually be written analytically. The ratio ξ/ℓ_{el} is found to be driven by the symmetry class of the Hamiltonian and not the microscopic nature of the underlying disorder model.

The general relation between mean free path and localization lengths was derived by Beenakker for multimode wires, $\xi/\ell_{el} = \frac{1}{2}\{\beta(N_\perp - 1) + 2\}$ (Beenakker, 1997). The β coefficient depends on the symmetry class of the Hamiltonian under time-reversal transformation. When the system is invariant under time reversal (belonging to the orthogonal class) $\beta = 1$, whereas $\beta = 2$ when time-reversal symmetry is broken (the system then belonging to the unitary class). The first case is obtained in a metallic system without spin orbit coupling or magnetic field, whereas the second arises in the presence of a magnetic field. This relationship is a generalization of the Thouless relationship for

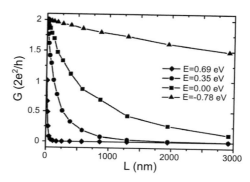

Figure 7.11 Length dependence of the Landauer conductance for a disordered $(10, 10)$ nitrogen-doped nanotube at several energies and fixed doping $n_{dop} = 0.1\%$. Averages are performed over 200 configurations of disorder. (Reproduced with permission from Avriller *et al.* (2007). Copyright (2007) World Scientific.)

1D disordered systems where $\xi = 2\ell_{el}$ (Thouless, 1973) and for disordered wires in the limit of a high number of channels $\xi \approx N_\perp \ell_{el}$.

In the case of $(10, 10)$ nanotubes doped randomly with nitrogen impurities, the weak disorder approximation and geometric restriction of the wire are satisfied. It is, however, difficult to establish the extent the *tight-binding* model presented for doped nanotubes matches with the scattering matrix hypothesis, that is if the universal mean values of RMT are equivalent to our model mean values. The difficulty comes from the fact that the CNT's Hamiltonian is the sum of a given initial periodic Hamiltonian combined with a random potential profile driven by the chemical impurities, whereas RMT is based on an entropy ansatz, regardless the underlying energetics of the problem. Accordingly, the energy-dependent behavior of transport scaling properties is not within the reach of the RMT. Only in specific regions of the spectrum can the fluctuating part dominate and drive to universal behavior.

For $(10, 10)$ nanotubes doped with 0.1% of nitrogen atoms, the scaling properties of the averaged conductance $\bar{T}$ are strongly energy-dependent. Figure 7.11 illustrates the weak energy-dependence of conductance $G(L)$, for L varying from 10 to 3000 nm. At $E = -0.78$ eV, the conduction remains quasi-ballistic ($\bar{T} \approx 2$), whereas at the quasibound state resonance energy $E = 0.69$ eV, strong localization develops. Such a possibility of tailoring the conduction regime (and conductance value) from a ballistic motion to an Anderson insulator is unique to graphene-related systems and proves to be extremely useful for device applications in Section 7.5.11

7.3 Two-dimensional disordered graphene with adatoms defects

7.3.1 Monatomic oxygen defects

In this section, we focus on the impact on quantum transport of monatomic oxygen atoms adsorbed on graphene. Atomic oxygen has been observed experimentally by means of scanning tunneling measurements (Hossain *et al.*, 2012) as illustrated in

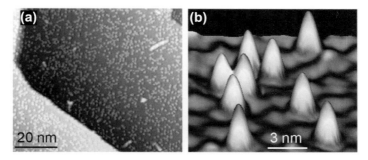

Figure 7.12 (a) STM image of epitaxial graphene after exposure to monatomic oxygen. (b) High-resolution STM picture of several chemisorbed oxygen atoms on graphene. (Adapted by permission from Macmillan Publishers Ltd: *Nature Chem*. Hossain *et al.* (2012), copyright (2012). Courtesy of Mark Hersam.)

Fig. 7.12. A weak ozone treatment (O_3) of the graphene sample actually generates such epoxy defects, as discussed experimentally by Moser and coworkers (Moser *et al.*, 2010). The influence of physisorbed O_2 molecules on the electronic properties of graphene turns out to be negligible in contrast to these epoxy groups.

An important observation is that oxygen atoms find their equilibrium position by bridging two first-neighbor carbon atoms, so that locally the A/B sublattice symmetry is not fully broken (consistent with the absence of local magnetism). The bridge between the oxygen and its carbon neighbors slightly displaces locations of both carbon atoms but does not form a covalent bond. The resulting epoxy defects are found to be stable at room temperature.

In order to analyze transport properties of oxygen-functionalized graphene, *ab initio* calculations are mandatory to develop a suitable *tight-binding* Hamiltonian. Indeed, a standard π electron orthogonal TB model based on first nearest-neighbor interactions of the p_x and the p_z orbitals of oxygen with carbon can be easily derived from DFT simulations (Leconte *et al.*, 2011). The combined contribution of the s and p_z orbitals of carbon binding with oxygen is reduced to a single orbital. Practically, an *ab initio* band structure calculation is performed on a supercell containing a single epoxy defect. The bands near the Fermi energy are then fitted using TB parameters, as detailed in Leconte *et al.* (2010, 2011).

The effect of epoxy defects on transport allows us to discuss a longstanding debated issue related to the validity of the semiclassical Boltzmann approach in disordered graphene. Consequently, the corresponding assumptions and the current debate are summarized first.

Does Boltzmann conductivity capture Dirac point physics?

At low energy, the presence of electron-hole puddles generates transport percolation precluding Anderson localization (Das Sarma *et al.*, 2011). In first transport experiments, the origin of the absence of Anderson localization and the reported minimum conductivity were attributed to these puddles (see Section 5.2.1 for a complete overview).

In a puzzling experiment on clean graphene sandwiched between two boron-nitride layers, the suppression of electron-hole puddles was observed to result in a large increase of the Dirac point resistivity, suggesting a transition to the Anderson localization regime (Ponomarenko *et al.*, 2011). However, the role of quantum interferences has been fiercely questioned, with an alternative scenario being argued based on Boltzmann transport (that is, absence of localization phenomena) and ascribing the divergence of the resistivity to a vanishingly small density of states (Das Sarma *et al.*, 2011).

The use of the Boltzmann approach relies on approximating the conductivity as

$$\sigma^*_{\text{Drude}}(E) = (4e^2/h) \times k\ell_{\text{el}}(E)/2, \qquad (7.5)$$

using $\rho(E) = 2|E|/(\pi \times (\hbar v_F)^2)$. From Eq. (7.5), the conventional downscaling with defect densities can be deduced as $\sigma^*_{\text{Drude}}(V_g) \simeq 1/n_i$. But the conductivity decay is stronger close to the Dirac point, suggesting a different interpretation of the data (Ponomarenko *et al.*, 2011), as argued in Das Sarma *et al.* (2012). However, this interpretation has still to be studied with care since, within the self-consistent Born approximation, the semiclassical conductivity should reach a limit value given by $4e^2/\pi h$. Figure 7.13 illustrates this minimal conductivity: σ^*_{Drude} (left) and the exact result σ_{Kubo} (right). The approximation performed in Eq. (7.5) actually drives the semiclassical conductivity to zero, especially in the vicinity of the Dirac point, in contradiction with the exact result, which remains larger or equal to $\frac{4e^2}{\pi h}$. Both, the Kubo conductivity and experimental data (Moser *et al.*, 2010) confirm that, especially close to the Dirac point, localization effects are significant up to about 100 K, driving the system to an Anderson insulating regime.

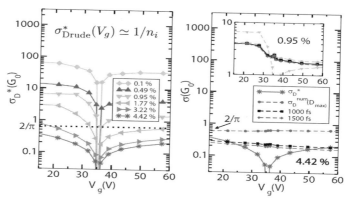

Figure 7.13 Left: Boltzmann conductivity versus gate voltage (or energy) for graphene with several densities of epoxide defects. Right: Kubo conductivity and semiclassical conductivity for two defect densities 0.95% (inset) and 4.42% (main plot). Two other curves depicting conductivity in the localization regimes at longer times are also drawn. Adapted from Leconte *et al.* (2010).

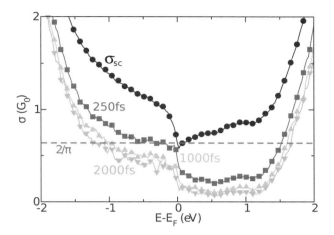

Figure 7.14 Kubo conductivity (at different times of wavepacket evolution) in a graphene sample containing 4.42% of impurities randomly distributed in the plane. The minimum value of $\sigma_{sc} = 4e^2/\pi h$ is indicated by the horizontal dashed line. (Reprinted from Roche *et al.* (2012). Copyright (2012), with permission from Elsevier.)

Localization effects in oxygen-damaged graphene

Figure 7.14 presents the semiclassical conductivity (σ_{sc}) together with the quantum Kubo conductivity computed at different timescales (beyond the diffusive regime) for 4.42% epoxy defects randomly distributed in graphene. These calculations are achieved by computing $D_x(E, t)$ and by evaluating $\sigma(E, t) = \frac{1}{4}e^2\rho(E)D(E, t)$. Oxygen defects produce quasibound states at some resonant energies, thus breaking the symmetry between electron and holes transport. Such asymmetry has already been seen in the local density of states but further develops in the energy profiles of $\ell_e(E)$, $\sigma_{sc}(E)$, and localization contributions (Leconte *et al.*, 2011). For long enough timescales, $\sigma(E, t) \ll 4e^2/h\pi$, indicating the strong contribution of quantum interferences and localization effects.

The time evolution of the Kubo conductivity clearly shows quantum interference effects, which can easily be understood from the scaling theory of localization. In a two-dimensional disordered system (such as the one pictured in Fig. 7.15 (left)), two different scaling behaviors are predicted depending on the strength of quantum interferences, namely the weak localization regime defined by (Leconte *et al.*, 2011, Lherbier *et al.*, 2012),

$$\sigma(L) - \sigma\big|_{D^{\max}} = -\frac{e^2}{\hbar\pi^2}\ln\left(\frac{L}{\sqrt{2}\ell_{el}}\right), \tag{7.6}$$

and the strong localization regime driven by $\sigma(L) \sim \exp\left(-\frac{L(t)}{\xi}\right)$, where ξ is the localization length, while $L(t) = 2\sqrt{2\Delta X^2(t)}$ is the average length scale probed by the wavepacket. The transition from the weak to the strong regime occurs at $k_F\ell_{el} \sim 1$. For a defect density of 4.42%, $\ell_{el} \leq 3\text{Å}$ satisfies this criterion for all energies between

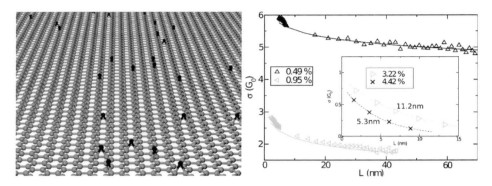

Figure 7.15 Left: Ball-and-stick model of epoxy defects adsorbed on graphene. Right: Weak localization corrections to σ_{sc} for different impurity densities at energy 0.8 eV for weak defect (main) and strong defect density (inset). Adapted from Roche *et al.* (2012).

0.5 and 1 eV. The value of ξ can thus be estimated, either using $\xi(E) = \ell_{el} \exp(\pi \sigma_{sc}/G_0)$ (Lee & Ramakrishnan, 1985), or directly extracted from the exponential scaling decay of conductance with length.

Figure 7.15 (right) illustrates that in the weak disorder limit (up to $\sim 1\%$ of defects) both the numerical $\sigma(L)$ (symbols) and the analytical $\sigma_{D^{\max}} - e^2/\hbar\pi^2 \ln\left(\frac{L}{\sqrt{2}\ell_{el}}\right)$ (solid lines) are in reasonable agreement. The fitting loses its quality for larger defect density owing to the transition to strong localization. Assuming an exponential decay of the conductivity, ξ values are estimated to be of the order of 11.2 and 5.3 nm for defect densities of 3.22% and 4.42% respectively (at $E - E_F = 0.8$ eV), making quantitative comparison with experimental data possible (Moser *et al.*, 2010). Using the Landauer–Büttiker method, the effect of epoxy defects on quantum transport in graphene nanoribbons has been found to generate mobility gaps and larger electron–hole transport asymmetry (Cresti *et al.*, 2011).

7.3.2   Atomic hydrogen defects

Adsorption of hydrogen atoms on graphene introduces sp^3 defects, thus breaking the AB symmetry and turning the material into a large bandgap insulator (*graphane*) in the large density limit.[1] In the present section, we study the low hydrogen density limit and contrast the transport results depending on the underlying A/B sublattice symmetry breaking. This issue is of genuine concern, since the Lieb theorem predicts that any imbalance between A and B sites generates ferromagnetic ordering in the groundstate, with total magnetic moment related to $S = 1/2|n_A - n_B|$ with $n_{A,B}$ the functionalized sites of type A (or B) (Lieb, 1989). Such a magnetism is taken into account through a self-consistent (spin-dependent) Hubbard Hamiltonian, in which the Coulomb interaction is accounted

---

[1] *Graphane* was theoretically described in Sluiter & Kawazoe (2003) and Sofo, Chaudhari & Barber (2007), and experimental evidence was presented in Elias *et al.* (2009).

for by means of the Hubbard model in its mean-field approximation (Soriano *et al.*, 2011), and defined as follows:

$$\mathcal{H} = \gamma_0 \sum_{<i,j>,\sigma} \left(c_{i,\sigma}^\dagger c_{j,\sigma} + h.c. \right) + U \sum_i \left(n_{i,\uparrow} \langle n_{i,\downarrow} \rangle + n_{i,\downarrow} \langle n_{i,\uparrow} \rangle \right), \tag{7.7}$$

where $c_{i,\sigma}^\dagger$ ($c_{j,\sigma}$) is the creation (annihilation) operator in the lattice site i (j) with spin σ, U is the onsite Coulomb repulsion, and $n_{i,\downarrow}$, $n_{i,\uparrow}$ are the self-consistent occupation numbers for spin-down and spin-up electrons, respectively. The ratio U/t is adequate to reproduce spin density obtained from first-principles calculations. To compute $\langle \hat{n}_{i\uparrow} \rangle = \int dE f(E_F - E) \rho_{i\uparrow}(E)$, a self-consistent procedure is used: $\langle \hat{n}_{i\sigma} \rangle_0 \Rightarrow \mathcal{H} \Rightarrow \rho_{i\sigma} \Rightarrow \langle \hat{n}_{i\sigma} \rangle$. Once the convergence is achieved, two different sets of spin-dependent onsite energies $\varepsilon_{i\uparrow} = U \langle \hat{n}_{i\uparrow} \rangle (1 - \langle \hat{n}_{i\downarrow} \rangle)$ and $\varepsilon_{i\downarrow} = U \langle \hat{n}_{i\downarrow} \rangle (1 - \langle \hat{n}_{i\uparrow} \rangle)$ are obtained, allowing estimation of the magnetization $\mathcal{M}_i = \langle \hat{n}_{i\uparrow} \rangle - \langle \hat{n}_{i\downarrow} \rangle / 2$. Assuming weak spin–orbit coupling, two different spin-dependent Kubo conductivities can be predicted:

$$\sigma_{\uparrow,\downarrow}(E,t) = (e^2/2) \text{Tr}[\delta_{\uparrow,\downarrow}(E - \hat{H})] D_{\uparrow,\downarrow}(E,t), \tag{7.8}$$

with $\text{Tr}[\delta_{\uparrow,\downarrow}(E - \hat{H})/S]$ and $D_{\uparrow,\downarrow}(E,t)$ respectively the spin-dependent density of states per surface unit at Fermi energy E and the diffusion coefficients.

Figure 7.16 (inset) shows $D_\uparrow(E,t)$ at three selected energies for $n_x = 0.8\%$, assuming the hydrogen defects are randomly distributed in the graphene matrix but equally on each A/B sublattice. The diffusion coefficients reach a saturation regime after a few hundreds of femtoseconds, and then exhibit a logarithmic decay (fingerprint of weak localization). The corresponding semiclassical and Kubo conductivities evaluated at long times are illustrated in Fig. 7.16 (main plot). The Drude conductivity σ_{sc} is seen

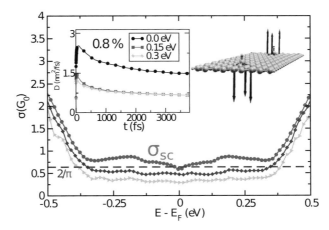

Figure 7.16 Main plot: Kubo conductivities for the nonmagnetic state with $n_x = 0.8\%$ hydrogen impurities: semiclassical value $4e^2/h\pi$ (horizontal dashed line), quantum conductivity at 250 fs (rhombus symbols) and 2000 fs (triangle symbols). Inset: Diffusion coefficients (spin-up channel) for selected energies. Antiferromagnetic spin polarization (opposite arrows) on H defects is also pictured in the ball-and-stick model. (Reprinted from Roche *et al.* (2012). Copyright (2012), with permission from Elsevier.)

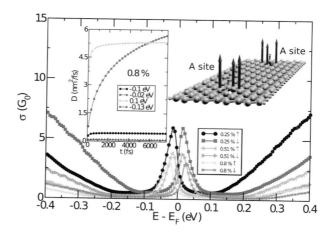

Figure 7.17 Spin-dependent Kubo conductivities (at elapsed time $t = 7600$ fs) for different hydrogen defect densities, enforcing a local ferromagnetic ordering (AA). Diffusion coefficients (spin-up channel) for selected energies (inset). Ball-and-stick model illustrates two hydrogen defects with ferromagnetic spin polarization (arrows). (Reprinted from Roche *et al.* (2012). Copyright (2012), with permission from Elsevier.)

to remain larger than $4e^2/\pi h$, whereas quantum interferences yield $\sigma_{\uparrow,\downarrow}(E, t) \leq \sigma_{\text{sc}}$, as reported for two elapsed times $t = 250$ fs and $t = 2000$ fs (Fig. 7.16 (main plot)).

If hydrogen defects are solely occupying one of the two sublattices, an unconventional transport regime develops. Figure 7.17 (inset) gives the corresponding diffusion coefficients which are found to saturate at long enough times, but without further decay, in contrast with Fig. 7.16. The corresponding saturation of the Kubo conductivity to its semiclassical value indicates a puzzling absence of localization effects (Fig. 7.17, main plot), which presently lacks theoretical understanding. Indeed, although such a choice of defect functionalization preserves one sublattice free from sp^3 contamination, the related suppression of quantum interferences remains difficult to capture using analytical arguments. One prediction is, however, that the existence of local ferromagnetic ordering could be reflected in an anomalously robust metallic state, a phenomenon which could be further discussed in relation to the concept of a *"supermetallic state"* introduced by Ostrovsky, Gornyi and Mirlin for monovacancies (Ostrovsky, Gornyi & Mirlin, 2006; Cresti *et al.*, 2013). Recent experiments on hydrogenated graphene have shown some modulations of pure spin currents, tentatively related to the interaction between propagating spins and hydrogen-induced local magnetic moments (McCreary *et al.*, 2012).

7.3.3   Scattering times

An interesting observation lies in the relation between the energy dependence of the transport times and the nature of underlying disorder and symmetry-breaking mechanisms. This is actually of great interest for determining the relation between local symmetry-breaking effects and resulting transport features at the mesoscopic scale. To illustrate this point, τ computed with the Kubo approach is shown for oxygen adatom

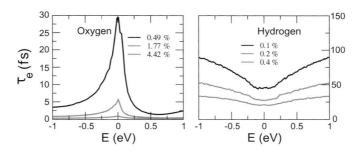

Figure 7.18 Scattering time τ for various densities of adsorbed monatomic oxygen (left) or hydrogen defects (right) (Roche *et al.*, 2012).

(left) and hydrogen adatom (right) densities in Fig. 7.18. The behavior of τ for epoxide defects actually looks very similar to that of the long range Coulomb impurities with onsite potential depth ($W = 1.5$, Fig. 5.17, main plot), whereas hydrogen defects look closer to the short range disorder case (Fig. 5.17 for $W = 2$).

The fingerprint of $\tau(E)$ for hydrogen defects looks very similar to the Anderson disorder with $W = 2$. As mentioned above, adsorbed hydrogen atoms locally break the sp^2 symmetry and A/B degeneracy, in contrast to epoxide defects. Interestingly, the absolute values of scattering times are only weakly sensitive to the defect density, and $\tau(E = 0) \sim 25 - 30$ fs for $n_i \simeq 0.4\%$, but their energy dependence reflects local symmetry-breaking mechanisms. These features are fingerprints for specific defects, thus allowing possible discussion about potential sources of disorder in experiments (Monteverde *et al.*, 2010).

7.4 Structural point defects embedded in graphene

As already described in detail in Section 2.8, structural defects (such as vacancies) in graphene can be intentionally introduced by ion or electron beam irradiation. Using Ar^+ irradiation of carbon nanotubes for instance, an Anderson localization was induced and experimentally observed using STM techniques (Gómez-Navarro *et al.*, 2005). Usually, it is believed that single vacancies (also called monovacancies) freely migrate and recombine to easily form divacancy defects (Lee *et al.*, 2005). The conductivity of irradiated two-dimensional graphene has however been found to saturate at the Dirac point above e^2/h even down to cryogenic temperatures, remaining a puzzling unexplained feature (Chen *et al.*, 2009). Recent STM analysis of irradiated graphene has also revealed the signature of resonant states produced by divacancies (Ugeda *et al.*, 2010). In the present section, we discuss how a strong (Anderson) localized regime can be tuned by varying the density of such types of structural defects (Banhart, Kotakoski & Krasheninnikov, 2011, Kotakoski *et al.*, 2011, Cockayne *et al.*, 2011), which can be intentionally introduced in the graphene substrate.

Here, we focus on the three structural imperfections (already observed experimentally) described in Section 2.8, namely the Stone–Wales (SW) defect and two types of

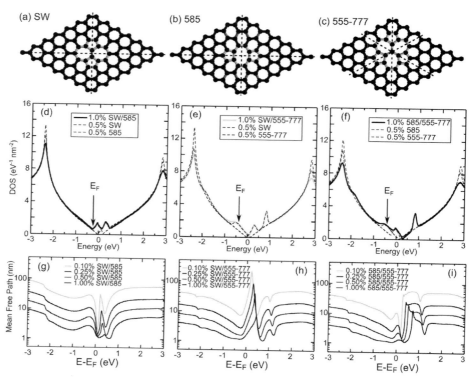

Figure 7.19 Three structural point defects: (a) Stone–Wales, (b) 585, and (c) 555-777 divacancies. The different symmetry axes are outlined in dashed lines. TB densities of states are shown for a large graphene plane containing a defect concentration of 1% (thick solid lines) of (d) SW/585, (e) SW/555-777, and (f) 585/555-777. These DOS are compared with DOS obtained with 0.5% of each defect separately (dashed lines). The position of the Fermi energy is indicated by a vertical arrow. Mean-free paths (ℓ_{el}) for concentrations ranging from 0.1% to 1.0% of (g) SW/585, (h) SW/555-777, and (i) 585/555-777 structural defects. (Adapted from Lherbier *et al.* (2012). Copyright (2012) by the American Physical Society. Courtesy of Aurélien Lherbier.)

divacancies (missing two carbon atoms). Differently to monovacancies, those defects are all nonmagnetic. SW defects are commonly observed in sp^2 carbon-based materials (Lee & Stone, 1985), and can be seen as generated by a 90° rotation of a carbon–carbon bond. This topological transformation produces two heptagons connected with two pentagons (Fig. 7.19(a)). The first divacancy reconstruction yields to the formation of two pentagons and one octagon (named 585, Fig. 7.19(b)), while the second relates to the formation of three pentagons and three heptagons (named 555-777, Fig. 7.19(c)). *Ab initio* calculations (Lherbier *et al.*, 2012) suggest that the formation energy of the 555-777 divacancy is smaller than that of the 585 divacancy by about 0.9 eV in graphene, which differs from nanotubes where curvature stabilizes the former reconstruction (Lee *et al.*, 2005).

As described in Section 2.8, *tight-binding* models for pristine graphene and for defects are derived by extracting the suitable TB parameters directly from the SIESTA Hamiltonian used to calculate the *ab initio* band structures (Lherbier *et al.*, 2012). The effects of these topological defects on the electronic properties of graphene have also been investigated in Chapter 2. TB densities of states of randomly distributed structural defects in the honeycomb lattice, as depicted in Fig. 2.41, confirm that the defect signatures, inducing energy resonances, are preserved in the DOS of random disordered system.

In order to model even more realistic systems, the case of a mixture of defects is now explored by considering graphene planes containing 50%–50% of SW/585, SW/555-777, and finally 585/555-777. The corresponding DOS estimated for a defect concentration of 1% in total (half of one type and half of the other type) are represented in Fig. 7.19(d–f)). These DOS containing a mixture of two types of defects are compared also with the DOS of graphene planes containing a single type of defect separately (defect concentration: 0.5%). The features observed in the DOS of the mixed systems are roughly the sum of the individual features of each defect type. The particular case of SW/585 is interesting in the sense that the resonance peaks of these two defects are almost symmetric with respect to the Dirac point (Fig. 7.19(d)), tending to overlap and leading to an increase of the DOS at the Dirac point. In this special situation, there is no more a clear minimum of DOS associated with the Dirac point. By adding other types of defects, a large increase of the DOS at the Dirac point can be foreseen (Fig. 7.19). This is actually what is observed in Haeckelite planes (Terrones *et al.*, 2000) and in highly defective or amorphous graphene membranes (Holmström *et al.*, 2011, Lherbier *et al.*, 2013). However, this increase of DOS comes from resonant states mainly localized around the defects, which will therefore not participate in the transport of charge carriers, but will rather degrade it (Lherbier *et al.*, 2013).

The corresponding elastic mean free paths for graphene structures containing varying densities of SW/585 defects, SW/555-777 defects, or 585/555-777 defects (ranging from 0.1% to 1.0%) are illustrated in Fig. 7.19(g–i). The energy dependence of ℓ_{el} exhibits dips associated with the defect resonance energies (or equivalently bumps in the DOS). The mean free path changes by no more than one order of magnitude, regardless of the energy and defect nature. For a defect concentration of $\sim$0.1%, mean free path ranges within 60–200 nm for the longest values and 2–10 nm for the shortest. For a defect concentration $\geq$ 1%, ℓ_{el} can eventually be shorter than 10 nm for the whole spectrum.

7.5   *Ab initio* quantum transport in 1D carbon nanostructures

7.5.1   Introduction

Coherent quantum transport in mesoscopic and low-dimensional systems can be rigorously investigated either with the Kubo–Greenwood (Kubo, 1966) or the Landauer–Büttiker formalisms (Büttiker *et al.*, 1985). The first approach, that has been explicitly illustrated in the previous sections, derives from the fluctuation–dissipation theorem. This technique allows evaluation of the intrinsic conduction regimes within the linear response, and gives a direct access to the fundamental transport length scales, such as the elastic mean free path (ℓ_{el}) and the localization length (ξ). While ℓ_{el} results from the elastic backscattering driven by static perturbations (defects, impurities) of an otherwise clean crystalline structure, ξ denotes the scale beyond which quantum conductance decays exponentially with the system length (L), owing to the accumulation of quantum interference effects that progressively drive the electronic system from weak to strong localization.

The coherence length L_ϕ gives the scale beyond which localization effects are fully suppressed owing to decoherence mechanisms, such as electron–phonon (e–ph) or electron–electron (e–e) couplings, treated as perturbations on the otherwise noninteracting electronic gas (weak localization regime). When ℓ_{el} becomes longer than the length of the nanotube in between voltage probes, the carriers propagate ballistically, and contact effects prevail.

In such a situation, the Landauer–Büttiker formalism becomes more appropriate, since it rigorously treats transmission properties for open systems and arbitrary interface geometries. Besides, its formal extensions (nonequilibrium Green's functions (NEGF) and Keldysh formalism, see Appendix C) further enable us to investigate quantum transport in situations far from the equilibrium, of relevance for high-bias regimes or situations with a dominating contribution of Coulomb interactions (Di Ventra, 2008, Datta, 1995).

Interestingly, to investigate coherent quantum transport in a graphene ribbon or a nanotube of length L with reflectionless contacts (ideal contact) to external reservoirs, both transport formalisms are formally fully equivalent. In the following, the Landauer–Büttiker formalism is used and the corresponding conductance $G(E) = 2e^2/h \times T(E)$ is evaluated from the transmission coefficient $T(E) = \text{Tr}\{\hat{\Gamma}_L(E)\hat{G}_S^{(r)}(E)\hat{\Gamma}_R(E)\hat{G}_S^{(a)}(E)\}$, given as a function of the retarded Green function $\hat{G}^{(r)}(E) = \{E\mathbb{I} - \hat{H}_S - \hat{\Sigma}_L(E) - \hat{\Sigma}_R(E)\}^{-1}$, and $\hat{\Sigma}_R(\hat{\Sigma}_L)$ the self-energy accounting for the coupling with the right (left) electrode (Di Ventra, 2008, Datta, 1995, Pastawski & Medina, 2001). The Landauer–Büttiker formula can be implemented with effective models, such as a *tight-binding* Hamiltonian fitted on first-principles calculations (as depicted previously), or a fully *ab initio* Hamiltonian. The *ab initio* electronic transport calculations presented in the following sections are performed within the nonequilibrium Green's functions formalism and using the one-particle Hamiltonian obtained from the DFT calculations as implemented in the TRANSIESTA (Brandbyge *et al.*, 2002) or SMEAGOL codes (Rocha *et al.*, 2006).

When defects in a nanostructure are introduced, the supercells containing the perturbation are connected to perfect nanotube- or ribbon-based leads. To simulate open boundary conditions, the self-energies associated with the leads are included within the self-consistent calculation of the potential. Finally, the electronic transmission functions are evaluated using the Fisher–Lee relation (Fisher & Lee, 1981) (see Appendix C). Note that, for the computation of the transmission functions, a convergence study of the supercell size has to be considered in order to obtain a good screening of the perturbed

Hartree potential due to the defect. It is then possible to experimentally investigate the energy dependence of the conductance by modulating the density of charge using a capacitive coupling between the nanostructure channel and an external gate.

7.5.2   Carbon nanotubes

As a reminder, for a carbon nanotube of length L between metallic contact reservoirs, the transport regime is ballistic if the measured conductance is L-independent, and only given by the energy-dependent number of available quantum channels $N(E)$ times the conductance quantum $G_0 = 2e^2/h$, that is $G(E) = 2e^2/h \times N(E)$, including spin degeneracy. Such an ideal situation occurs only in the case of perfect (reflectionless) or ohmic contacts between the CNT and metallic voltage probes. In this regime, the expected energy-dependent conductance spectrum is easily deduced, from band structure calculations, by counting the number of channels at a given energy. For instance, metallic armchair nanotubes present two quantum channels at the Fermi energy $E_F = 0$, or charge neutrality point, resulting in $G(E_F) = 2G_0$. At higher energies, the conductance increases as more channels become available to conduction. For illustration, the electronic bands and conductance of a $(5, 5)$ metallic tube are displayed in Fig. 7.20 within the symmetric $\pi-\pi^*$ _tight-binding_ model.

This quantum conductance of armchair carbon nanotubes within a nearest-neighbor π orbital _tight-binding_ Hamiltonian is in good agreement (Fig. 7.21(b)) with _ab initio_ calculations (Fig. 7.21(a)). Indeed, the $(5, 5)$ armchair carbon nanotube is found to be a metallic nanowire with two linear electronic energy bands which cross at the Fermi level and contribute two conductance quanta $(=4e^2/h)$ to the conductance when the tube is defectless. These two quantum channels at the charge neutrality point also lead to a plateau of conductance over a quite important interval of energies ($\sim$2.5 eV).

In Fig. 7.21(b) and Fig. 7.21(c), the _ab initio_ electronic properties and the corresponding conductance are presented for $(9, 0)$ and $(10, 0)$ zigzag nanotubes, respectively.

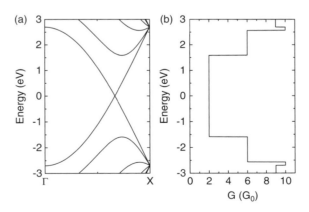

Figure 7.20 Band structure (a) and conductance (b) for $(5, 5)$ armchair nanotube calculated within the nearest neighbor π-orbitals tight-binding model.

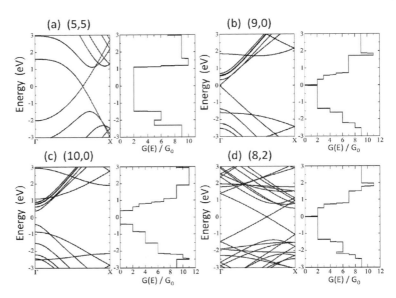

Figure 7.21 *Ab initio* band structures and quantum conductances for: (a) $(5, 5)$ armchair; (b) $(9, 0)$; (c) $(10, 0)$ zigzag; and (d) $(8, 2)$ chiral nanotubes. (Reproduced from Dubois *et al.* (2009), with kind permission from Springer Science and Business Media.)

The first-principles calculations confirm the general features of the electronic structure obtained in the *tight-binding* approach. Indeed, as mentioned in Chapter 2, the opening of a secondary gap (pseudogap) at the Fermi energy produced by the curvature of the graphitic walls in the $(9, 0)$ nanotube can be observed. In the $(10, 0)$ case, predicted to exhibit a semiconducting behavior, a primary gap of 0.8 eV is obtained, leading to a zero transmission for that specific energy window. Finally, in order to be as exhaustive as possible, the *ab initio* electronic properties and the conductance of $(8, 2)$ nanotubes are illustrated in Fig. 7.21(d). Although the single-band model would have proposed a metallic tube, the first-principles calculations predict a semiconducting system with a very small pseudogap related to the curvature of the nanotube, analogous to the $(9, 0)$ nanotube.

Note that all these *ab initio* conductance values are the uppermost theoretical limits that would be experimentally measured. In practical situations, lower values are observed since reflectionless transmission at the interface between the voltage probes (metallic leads) and the nanotubes is fundamentally limited by interface symmetry mismatch, inducing Bragg-type backscattering. Additionally, topological and chemical disorders, as well as intershell coupling, introduce intrinsic backscattering along the tube, which also reduce its transmission capability. To account for both effects, one generally introduces $T_n(E) \leq 1$, the transmission amplitude for a given channel, at energy E, so that $G(E) = G_0 \sum_{n=1,N_\perp} T_n(E)$ (Datta, 1995) (see also the discussion in Section 4.2.2), as illustrated in the following for the case of topological defects, doping, and chemical functionalization in carbon nanotubes.

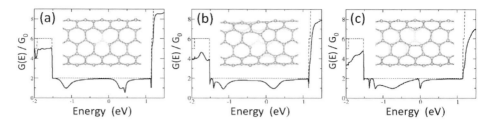

Figure 7.22 *Ab initio* quantum conductance of a $(5, 5)$ armchair nanotube containing a bare D_{3h} monovacancy (a); and a reconstructed C_s vacancy in two different positions: (b) tilted, or (c) perpendicular to the axis of the tube. Atomic structures of the corresponding defects are shown in insets. (Reproduced from Dubois *et al.* (2009), with kind permission from Springer Science and Business Media.)

7.5.3   Defective carbon nanotubes

The effects of impurities and local structural defects on the conductance of metallic carbon nanotubes have been calculated using *ab initio* techniques within the Landauer–Büttiker formalism (Choi *et al.*, 2000, Dubois *et al.*, 2009). For example, with a point (single-atom) vacancy, the conductance presents one broad dip (valence region) and two narrower dips (conduction region), as illustrated in Fig. 7.22(a). The reduction of conductance at the broad dip is $1G_0$ with approximately half reflection of both π and π^* bands. Because a single impurity breaks the mirror symmetry planes containing the tube axis, an eigen-channel is a mixture of the π and π^* bands. Consequently, an electron in an eigen-channel is either completely reflected or completely transmitted. The location of the broad dip (-1.2 eV with respect to the charge neutrality point) is quite different from the results obtained using a single-band *tight-binding* model which predicts a single dip exactly at the Fermi level (Chico *et al.*, 1996).

Actually the electron–hole symmetry is no longer valid in a more realistic *ab initio* calculation, and the dip position moves. Moreover, two other narrower dips are observed closer to E_F and originate from resonant scattering by quasibound states derived from the broken σ bonds around the vacancy (Fig. 7.22(a)). The σ bonds between the removed atom and its neighbors are broken, and dangling bonds are produced which are mainly composed of π orbitals parallel to the tube surface. Since σ bond states are orthogonal to the π valence band states, a very weak coupling is present between them. Among three quasibound states derived from three dangling bonds, one is an s-like bonding state, which lies well below the first lower subband (outside the scope of the figure). The other two states are orthogonal to it (i.e. partially anti-bonding) and give rise to the two narrower dips in the conduction region, as shown in Fig. 7.22(a).

The interaction among the dangling σ bonds actually causes substantial atomic relaxations, at the origin of reconstruction of the bare monovacancy (D_{3h} symmetry) into a more stable vacancy structure exhibiting C_s symmetry (Fig. 7.22(b–c), Amara *et al.*, 2007). More specifically, the D_{3h} vacancy undergoes a Jahn–Teller distortion upon relaxation, where two of the atoms near the vacancy move closer, forming a pentagon-like structure while the third atom is slightly displaced out of the plane. In addition, the vacancy can adopt two different positions related to the hexagonal network of the nanotube: a tilted position (Fig. 7.22(b)) or a perpendicular position (Fig. 7.22(c)) regarding the axis of the nanotube. The tilted vacancy is found to be the most stable configuration with a $\Delta E = 1.34$ eV energy difference compared to the perpendicular case (Zanolli & Charlier, 2010). Both vacancies have a significant influence on the electronic structure of the tube, at the origin of important backscattering to incoming electrons at resonant energies. In fact, the accurate positions of the vacancy-related quasibound state levels depend on various factors such as atomic configuration, orientation versus the axis of the tube, and the nanotube diameter (Choi *et al.*, 2000, Zanolli & Charlier, 2010).

Additionally, since the localized orbitals of unsaturated carbon atoms are expected to behave as magnetic impurities (Shibayama *et al.*, 2000, Lehtinen *et al.*, 2004a, 2004b), a monovacancy (but not a divacancy) is expected to hold a net magnetic moment. It is worth mentioning that periodic-boundary-condition calculations can lead to results in contradiction with these theoretical predictions. Interestingly, the total magnetic moment of CNTs containing the tilted monovacancy oscillates with the length nd_0 of the $1 \times 1 \times n$ supercell (d_0 being the length of the (5, 5) unit cell), and finally goes to zero as illustrated in Fig. 7.23 (square symbols). It is worth noting that these oscillations are not due to a poor k-point sampling of the Brillouin zone, since accurate convergence studies have been performed (Zanolli & Charlier, 2010).

In fact, these oscillations of the total magnetic moment are due to a long range interaction between the periodic images of the magnetic moments mediated by the conduction electrons of the metallic tube, also called indirect exchange coupling (Kirwan *et al.*, 2008). This indirect coupling is defined as the energy required to rotate the magnetic moments from the ferromagnetic to the antiferromagnetic configuration, that is $J = E_{\uparrow\uparrow} - E_{\uparrow\downarrow}$, where $E_{\uparrow\uparrow}$ and $E_{\uparrow\downarrow}$ are the total energies of the ferromagnetic and

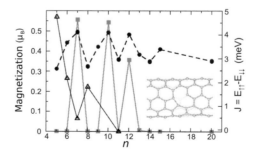

Figure 7.23 Total magnetization (μ_B) for a $1 \times 1 \times n$ supercell containing one monovacancy calculated within the periodic-boundary-condition scheme (grey curve ■—■) and for the open system with $1 \times 1 \times n$ *core* region (black dotted curve ●—●). Indirect exchange coupling (J, meV) for the $1 \times 1 \times 2n$ supercell containing two monovacancies at distance nd_0 (grey curve △—△). All the calculations are performed for a (5, 5) CNT containing a monovacancy reconstructed in the tilted configuration, as illustrated in the inset. (Adapted with permission from Zanolli & Charlier (2010). Copyright (2010) by the American Physical Society.)

antiferromagnetic configurations, respectively. The interaction results in an oscillatory behavior of J, whose amplitude decreases as a power law of the distance between the magnetic impurities. Such a power law strongly depends on the nature of the impurity and on the dimensionality of the system.

When the supercell contains a single defect, periodic boundary conditions either allow a non-spin-polarized configuration or a ferromagnetic coupling between the periodically repeated magnetic impurities. Hence, when their distance is such that the antiferromagnetic coupling is the lowest-energy configuration (i.e. $J > 0$), the periodic system is forced into either a non-spin-polarized or a ferromagnetically coupled state, depending on the relative energetic ordering of the three possible magnetic configurations (ferromagnetic, antiferromagnetic and non-spin-polarized) (Venezuela *et al.*, 2009).

To check this assumption, J has been calculated for double supercells of various length, $2nd_0$, containing two monovacancies at nd_0 distance, retrieving the expected damped oscillatory behavior (Fig. 7.23, triangles). Indeed, the single-monovacancy supercells having zero-magnetization (zero values in the "square symbols" grey curve of Fig. 7.23) correspond to a two-monovacancy system where the antiferromagnetic coupling is favored ($J > 0$ in the "triangle symbols" grey curve of Fig. 7.23).

Even though the indirect coupling is quite weak (J is of the order of a few meV), it clearly affects the periodic-boundary-condition computation of the magnetic properties of defected CNTs. Consequently, to accurately describe the local magnetic properties of isolated vacancies, an open system with a single magnetic impurity has to be considered. The defected CNTs are thus modeled as two semi-infinite sections of nanotube and a central region (or *core*) containing a single defect site, as described in Zanolli & Charlier (2009). Using this approach, the magnetization of the tilted monovacancy has always been found to be finite and converges for a core region consisting of $1 \times 1 \times 15$ cells (see Fig. 7.23, black dashed curve).

It can further be noted that both the open system and the periodic-boundary-condition approaches can be considered as first approximations to "real" CNTs with low or high densities of defects, respectively. On the one hand, the open system scheme can approximate the experimental case of low densities of vacancies only when the separation between defects is larger than the interaction range of the indirect exchange coupling. This interaction is found to depend on the location of the magnetic impurity on the hexagonal network of the host CNT (Kirwan *et al.*, 2008) and can vary substantially between 3 to 10 nm. Consequently, the open system approach can model a "real" defected CNT where vacancies are more than $\sim$3–10 nm apart.

On the other hand, the periodic-boundary-condition approach presents some intrinsic limitations in describing the opposite limit of a high density of vacancies in CNTs, as recently illustrated for magnetic impurities in graphene (Venezuela *et al.*, 2009). Indeed, a periodic-boundary-condition calculation with one defect per unit cell always forces the periodically repeated magnetic impurities to be in the same spin state (either non-spin-polarized or ferromagnetically coupled).

The antiferromagnetic or non-collinear spin-polarization of the magnetic impurities cannot be described within such an approach. Besides, the periodicity of a PBC

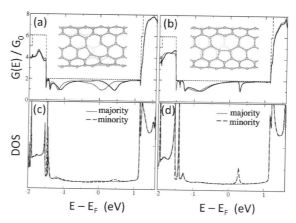

Figure 7.24 *Ab initio* quantum conductance and DOS calculated for a (5, 5) CNT containing a monovacancy reconstructed in its most stable (a, c) and next stable (b, d) configurations, respectively. The DOS allows one to identify the degree of localization of the dips in the transmission curve to facilitate the comparison. (Adapted with permission from Zanolli & Charlier (2010). Copyright (2010) by the American Physical Society.)

calculation is an ideal model which imposes spurious interactions that are not present in the real system, where the distribution of defects is random. Consequently, the PBC technique cannot model "real" CNTs with a high density of defects, unless the defects are equally and ideally spaced and present either no coupling (no spin-polarization) or a ferromagnetic coupling.

After clarifying how to study the magnetic properties of a single defect site in CNTs, the open system scheme is used to predict both the magnetization and the spin-polarized conductance of defected tubes (see Fig. 7.24). From these calculations, it can be seen that the open system scheme makes it possible to recover the expected magnetic behavior of monovacancies. In addition, these results definitely show that the magnetization of defected CNTs can be ascribed to the presence of unsaturated carbon atoms, in agreement with a naive picture of a dangling bond.

The analysis of the conductance curves (Fig. 7.24) reveals that a carbon nanotube containing a monovacancy may act as a spin filter within some specific energy windows. For instance, the conductance at E_F of the *tilted* reconstruction of the mono-vacancy (Fig. 7.24(a)) drops from $2G_0$ to G_0 for one spin channel (*majority* spin carriers) while the *minority* spin conductance is almost unaffected. Consequently, at E_F, half of the electrons with majority spin orientation will be filtered out while minority electrons will be almost fully transmitted. The situation is reversed at 0.4 eV: majority electrons are almost fully transmitted while half the minority electrons are reflected. A similar behavior is predicted for the *parallel* monovacancy (Fig. 7.24(b)).

The conductance dips correspond to states which are quasi-localized (wide dips, Fig. 7.24(a)) or strongly localized (sharp dips, Fig. 7.24(b)) on the under-coordinated carbon, as can be seen from the height of the peaks of the density of states (DOS, Fig. 7.24(c) and Fig. 7.24(d), respectively) computed from the Green's function of the open system.

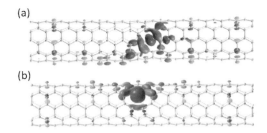

Figure 7.25 Wavefunctions illustrating the degree of localization of the electronic states in a (5,5) CNT containing a monovacancy reconstructed in its most stable (a) and next stable (b) state. The tilted vacancy induces a more extended state (a) than the parallel one (b). (Adapted with permission from Zanolli & Charlier (2010). Copyright (2010) by the American Physical Society.)

The information on the degree of localization of the states on the defect site obtained from the conductance and the DOS computed within the open system scheme helps in better understanding why the range of the exchange coupling is so long and affects the magnetization computed when using periodic boundary conditions. As an example for the monovacancy in the *parallel* configuration, electronic states are seen to be more localized on the defect site (Fig. 7.25(b)) and, hence, little coupling between adjacent cells is found within the periodic-boundary-condition scheme. On the other hand, for the monovacancy in the *tilted* configuration, the electronic states localized on the defect sites are clearly extended over the whole cell (Fig. 7.25(a)), resulting in a strong coupling of the magnetic impurities. The spatial extension of the quasi-localized states is inversely proportional to the tube radius R and the indirect coupling will have the same $1/R$ dependence (Kirwan *et al.*, 2008). For this reason, oscillations in the magnetic moment are less pronounced in the *parallel* case and in the graphene case.

7.5.4 Doped carbon nanotubes

Analogously to point defects, boron or nitrogen impurities, substituting carbon atoms, produce quasibound impurity states of a definite parity and reduce the conductance by a quantum unit $2e^2/h$ via resonant backscattering (Choi *et al.*, 2000), as presented in Fig. 7.26. The conductance of the doped tube is found to be virtually unchanged at E_F, meaning that the impurity potential does not scatter incoming electrons at this energy. On the other hand, two pronounced dips are observed in the conductance below E_F for the boron impurity since it acts as an acceptor dopant. The amount of the conductance reduction at these dips is $1 G_0$ (Fig. 7.26(a)). The upper dip is caused by an approximate half reflection from states of both π and π^* bands. Because a single impurity breaks the mirror symmetry planes containing the tube axis, an eigen-channel is a mixture of the π and π^* bands. Consequently, an electron in such an eigen-channel is either completely reflected or completely transmitted. Associated with the two conductance dips, the density of states (DOS) around the boron impurity exhibits two peaks arising from the presence of quasibound states (Fig. 7.26(a)). The lower peak is too close (~ 1 meV) to be seen separately from the peak originating from the van Hove singularity of the lower subbands. A nitrogen substitutional impurity has similar effects on the conductance (Choi *et al.*, 2000), but on the opposite site of the charge neutrality point since it acts as a donor dopant (Fig. 7.26(b)). In summary, a substitutional boron or nitrogen impurity produces quasibound states of definite parity (resonant states) made of π orbitals perpendicular to the tube surface below or above E_F, in close analogy to the acceptor or donor levels in semiconductors, and the conductance is reduced at the corresponding quasibound state energies.

Experimentally, hetero-doped carbon nanotubes are quite easily synthesized by CVD techniques (Cruz-Silva *et al.*, 2008) using for example benzylamine and triphenylphosphine as nitrogen and phosphorus sources, respectively. These P–N-doped nanotubes are thermodynamically stable, as predicted theoretically when scrutinizing the defect formation energies (Cruz-Silva *et al.*, 2009). Analysis of the relaxed structures confirms that phosphorus maintains an sp^3 hybridization, and bonds to the carbon atoms with tetrahedral orbitals, inducing structural strain in the carbon network in order to accommodate the longer P–C bonds and the larger sized P ion (Fig. 7.27(a)). Total energy calculations also confirm that curvature helps to reduce the structural strain caused by the phosphorus, and that the P–N defect is energetically more stable than the phosphorus impurity alone.

The electronic band structure shows the presence of "semi-localized" states around the P–N doping atoms (Cruz-Silva *et al.*, 2009). In contrast to nitrogen, these states do not modify the intrinsic nanotube metallicity. Electronic transport calculations on pristine and P–N-doped nanotubes clarify the different effects of the dopants on their conductance (Fig. 7.27(b, c)). The calculation of the quantum conductance shows that zigzag phosphorus-doped nanotubes do not modify the intrinsic semiconducting behavior (Fig. 7.27(b)), in contrast to what is observed for N-doped nanotubes (Choi *et al.*,

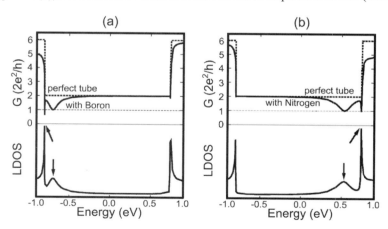

Figure 7.26 Effects of a boron (a) and a nitrogen (b) impurity on the conductance of a (10, 10) carbon nanotube. The conductance as a function of the incident energy exhibits two dips. The local density of states (LDOS) around the impurity presents two peaks (indicated by arrows) and rapid changes, respectively, associated with the dips in the conductance. (Reproduced from Choi *et al.* (2000). Copyright (2000) by the American Physical Society. Courtesy of Hyoung Joon Choi.)

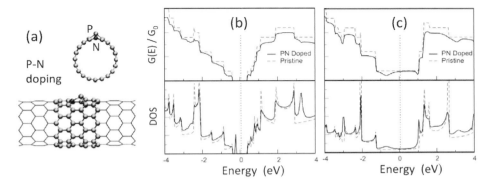

Figure 7.27 (a) Atomic structures of phosphorus–nitrogen-doped (10, 0) nanotube. The phosphorus atom protrudes from the nanotube wall due to the longer P–C bonds. (b, c) Quantum conductance and density of states plots for P–N-doped (b) zigzag (10, 0) and (c) armchair (6, 6) carbon nanotubes (solid line), compared with pristine nanotube values (dashed line). The conductance of the semiconducting tube is almost insensitive to the presence of the dopant. Localized states cause scattering and hence a reduction of the conductance close to the corresponding energies for the metallic tube. (Adapted from Cruz-Silva *et al.* (2009).)

2000). Phosphorus–nitrogen doping in a (10, 0) nanotube only creates bound and quasi-bound states around the phosphorus atom, that are dispersionless giving sharp peaks in the density of states (Fig. 7.27(c)). These states are normal to the nanotube surface and do not contribute to the electronic transport in semiconducting nanotubes, while in the case of a metallic nanotube, these states behave as scatterers, creating small dips in the conductance at specific energies. These electronic properties are also very useful for fast response and ultra-sensitive sensors operating at the molecular level. Such molecular selectivity has been predicted in CO, NH_3, NO_2, and SO_2 adsorbed on P–N-doped nanotubes (Cruz-Silva *et al.*, 2011). In fact, the adsorption of different chemical species onto the doped nanotubes modifies the dopant-induced localized states, which subsequently alter the electronic conductance. Although SO_2 and CO adsorptions cause minor shifts in electronic conductance, NH_3 and NO_2 adsorptions induce the suppression of a conductance dip. Conversely, the adsorption of NO_2 on P–N-doped nanotubes is accompanied by the appearance of an additional dip in conductance, correlated with a shift of the existing values. Overall these changes in electric conductance provide an efficient way to detect selectively the presence of specific molecules (Cruz-Silva *et al.*, 2011).

7.5.5   Functionalized carbon nanotubes

The transport properties of carbon nanotubes can be tailored by molecular functionalization (Star *et al.*, 2003, Collins *et al.*, 1998, Balasubramanian *et al.*, 2008). Indeed, functionalized carbon nanotubes display tunable structural and electronic properties, and promise to become for nanoelectronics a match for what DNA is for the life sciences. For instance, by grafting photoactive molecules onto the nanotube, the resulting nanotube-based devices could be controlled optically (Campidelli *et al.*, 2008, Simmons *et al.*, 2007). The recent development of synthetic methods to attach ligand molecules has been a major breakthrough, and opens the possibility to use molecular self-assembly and nanolithography techniques to arrange nanotubes in a device. An accurate control of the physical properties can also be made possible by target molecular functionalizations that tune the electronic response or the structural conformations.

In order to graft molecules at the nanotube surface, two types of chemical functionalization are usually considered, namely physisorption (non-covalent functionalization) and chemisorption (covalent functionalization). Both methods provide effective pathways for modifying the intrinsic properties of electric transport along CNTs, but a trade-off has to be found to add a new functionality to the tube without excessively damaging its electronic and transport features (Balasubramanian *et al.*, 2008).

Non-covalent functionalization

In the case of physisorption, the non-covalent adsorption of molecules has the advantage of enabling CNT functionalization while preserving their electronic structure, since the original sp^2 hybridized bonds and conjugation remain unaltered (Tournus *et al.*, 2005). Consequently, the scattering efficiency resulting from molecule deposition remains negligible owing to weak bonding (see for instance Latil *et al.* (2005)) and vanishing charge transfer. Nevertheless, physisorption effects on electronic conduction have been suggested to critically depend on the nature of molecular species and their HOMO-LUMO gap positioning with respect to the Fermi level of the host tube (Latil *et al.*, 2005). For example, benzene molecules yield vanishing modulations of the intrinsic conductance, whereas azulene molecules (with a HOMO-LUMO gap of about $\sim 2eV$) produce substantial elastic backscattering in the nanotube, resulting in mean free paths of the order of a few micrometers for large coverage. Such a possibility of creating/removing a reversible elastic disorder by a simple adsorption/desorption of molecules covering the nanotube surface opens interesting perspectives for experimental studies and potential applications in nanotechnology.

Covalent functionalization: a few examples

In the case of chemisorption, covalent functionalization of CNT involves the formation of saturated sp^3 bonds which markedly breaks the nanotube π conjugation (Zhao *et al.*, 2004). The diazonium addition is a commonly used technique (Cabana & Martel, 2007, Lee *et al.*, 2005), but can result in dramatic loss in tube transport capability if too many addends are chemisorbed. The impact of covalent functionalization on tube conductance can, however, be significantly reduced by a suitable choice of the addends. To circumvent such a problem, [2 + 1] cycloaddition reactions have been theoretically proposed (Lee *et al.*, 2005, Lee & Marzari, 2006). Such functionalization is driven by grafted carbene (or nitrene) groups that induce bond cleaving between adjacent sidewall carbon atoms, maintaining the sp^2 hybridization and providing sites for further attachment of more complex molecules and related functionalities. A transport study based on a nonorthogonal *tight-binding* Hamiltonian has first reported strong differences between monovalent and divalent additions in short length nanotubes (Park, Zhao & Lu, 2006).

Using *ab initio* calculations, Lee and Marzari (Lee *et al.*, 2005, Lee & Marzari, 2006) further demonstrated that cycloaddition reactions induce the grafting of dichlorocarbene groups (CCl_2) which preserve most of the conductance in (5, 5) CNT metallic nanotubes, in contrast to phenyl-type functionalization. However, these first-principles calculations have been limited to ultrashort nanotube segments with length below 50 nanometers. Below, various processes of chemical functionalization are detailed, and their impact on electronic and transport properties in micrometer-long and disordered nanotubes is analyzed following (Lopez-Bezanilla, 2009).

Diazonium salts are a group of organic compounds sharing a common functional group with the characteristic structure of $R - N_2^+ X^-$ where R can be any organic ligand such alkyl or aryl and X is an anion such as a halogen. Phenyl group (denoted with the formula C_6H_5 and sometimes abbreviated as Φ) is the aryl component of diazonium salts which is widely used in chemistry to functionalize carbon nanotubes to form nanotube composites. Diazonium salts provide a selective chemical reaction that favors the covalent attachment to metallic CNTs. Such a selectivity characteristic is used for sorting nanotubes of different chiralities (Strano *et al.*, 2003). In the grafting process, the aryl diazonium cation gets one electron from the substrate and subsequently becomes an aryl radical by losing a N_2 molecule (see Fig. 7.28). The attachment of a phenyl group onto the nanotube sidewall makes an anchorage point for further grafting of more complex molecules with specific functionalities (Campidelli *et al.*, 2008). Due to the sp^3 rehybridization induced by the grafting of the phenyl group, this new chemical function strongly affects local features of the carbon-based systems.

Charge transport in metallic single-walled CNTs with random distribution of phenyl and carbene functional groups bonded to the tube sidewalls has been investigated within a first-principles approach (López-Bezanilla *et al.*, 2009*b*, López-Bezanilla *et al.*, 2009*a*). The disorder introduced by the grafted groups breaks both translational and rotational symmetries. The conductance modulations and conduction regimes (from quasi-ballistic to diffusive) have been investigated as a function of both incident electron energy and functional groups coverage density on long nanotubes from a few hundreds of nanometers to the micron scale. First-principles mesoscopic transport study demonstrates that carbene cycloaddition preserves ballistic conduction up

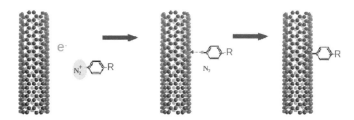

Figure 7.28 Grafting of phenyl groups onto the CNT sidewall. First, an electron is extracted from the nanotube by means of the reaction with diazonium reagents. A N_2 molecule is formed and the aryl group gets chemically attached to the nanotube, resulting in the creation of a stable C–C covalent bond and a radical which is further passivated by another aryl group in a similar process.

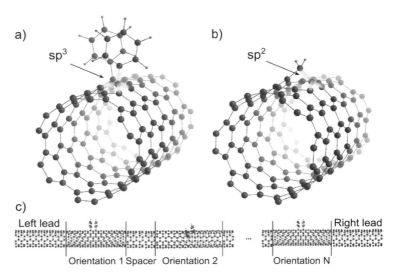

Figure 7.29 sp^2-like and sp^3-like functionalization. Top panel: Atomic structures of the building blocks assembled further in longer structures with both kinds of covalent grafting: (a) sp^3-like functionalization with phenyl groups; and (b) sp^2-like functionalization through divalent addition by carbene groups. (c) CNT functionalized with phenyl groups constructed by assembling individual sections. (Adapted with permission from López-Bezanilla *et al.* (2009a). Copyright (2009) American Chemical Society.)

the micron scale, whereas the grafting of phenyl groups yields a mean free path in the nanometer scale, leading to a strong localization regime for commonly studied nanotube lengths.

The computational approach is based on a large set of *ab initio* calculations which are first performed to get the Hamiltonian and overlap matrix associated with small tube sections functionalized by single groups (see Fig. 7.29(a) and (b)). Such a set of building blocks Hamiltonian is further used to reconstruct a micrometer long tube Hamiltonian formed by a random mixture of functionalized and pristine tube portions to introduce both rotational and translational disorder (see Fig. 7.29(c)). Upon building the small block Hamiltonian with one defect, periodic boundary conditions are used. The length of the building block is chosen such that geometric and energetic perturbations induced by functional groups vanish as the edges are reached. The renormalization procedure used here takes advantage of the locality of the orbital basis set used in the *ab initio* simulation (SIESTA code), allowing us to consider the system as formed by nearest-neighbors' interacting sections. As depicted in Fig. 7.29 (bottom panel), an armchair CNT is divided in segments, so that H is partitioned in onsite energy diagonal blocks and nearest-neighbor coupling blocks. By coupling in a random way functionalized and pristine building blocks, CNTs as long as desired can easily be built up (López-Bezanilla *et al.*, 2009b, 2009a, López-Bezanilla, Blase & Roche, 2010).

The case of a single functional group is first investigated. The effect of a pair of phenyl rings or a carbene group in the electronic structure of an armchair CNT is directly

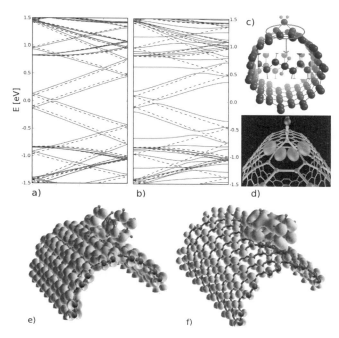

Figure 7.30 Energy bands of a functionalized armchair CNT with a carbene group (a), and with a pair of phenyl rings (b). (c) Ball-and-stick image of a carbene group grafted onto the semiconducting nanotube in a skewed orientation. (d) Carbene functionalization when the C–C bond is perpendicular to nanotube axis (Courtesy of N. Marzari). (e) and (f) The LDOS of the hybrid CNT-pair of phenyl at two energies: in (e) the density of states has been projected over an interval of energy close to the Fermi energy; in (f) the projection is over the localized state at -0.6 eV.

observable in the band energy diagram. Figure 7.30(a) illustrates the unaltered band structure of a (10, 10) CNT upon functionalization with a carbene group. The bond between neighboring circumferential carbon atoms is broken and new bonds between carbene carbon atom and two nanotube carbon atoms are formed. Figure 7.30(b) and Fig. 7.29(c) depict a CNT with a carbene group covalently attached in an orientation that favors a stable configuration. The system CNT-CH_2 reaches its configuration energy minimum by displacing the carbon atoms that serve as an anchorage site to the carbene, which entails a rupture of the original C–C bond of nanotube atoms. If one considers only a nearest neighbor scheme, every C atom is found to be bonded to three C atoms, preserving the original π orbitals network. In Fig. 7.30(d), the preserved π orbitals centered in the altered nanotube carbon atoms are shown in a representation based on Wannier functions approach (Marzari & Vanderbilt, 1997). As seen below, carbene bond orientation in the nanotube plays a key role in electronic transport properties. The orientations depicted in Fig. 7.30(c) and Fig. 7.30(d) are the most stable configurations for zigzag and armchair CNT, respectively.

As can be seen in Fig. 7.30(b), two flat bands at energy values that coincide with the conductance dips in Fig. 7.31(a) show up in the energy band diagram. These

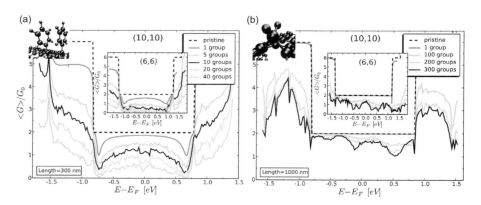

Figure 7.31 Conductance profiles (in units of G_0) for phenyl and carbene functionalization (averaged over 40 different random configurations): (a) Conductance for various molecular coverages of a 300 nm long (10, 10) CNT functionalized with phenyl groups. (b) Same as in (a), but for a 1000 nm long nanotube functionalized with divalent addition of carbene groups. The insets give the same information but for the (6, 6) nanotubes. (Adapted with permission from López-Bezanilla *et al.* (2009*b*). Copyright (2009) American Chemical Society.)

non-dispersive bands indicate the presence of molecular states which have localized the system wavefunction around the phenyl groups, as observed in Fig. 7.30(f) where the local density of states (LDOS) is plotted for the state at energy −0.6 eV. At this energy, the largest contribution to the DOS comes from the orbitals associated with the functional groups, i.e system wavefunction is localized over the molecules, unlike Fig. 7.30(e) where the wavefunction is homogeneously spread along the system.

The diameter of such a (10, 10) tube is close to the limit separating the area of stability for closed carbene configurations (larger tubes) and opened geometries (smaller tubes) (Lee & Marzari, 2006). Closed configurations introduce significant backscattering and the advantage of cycloaddition is therefore lost for larger diameter tubes. Further, on the basis of activation energy calculations for desorption, carbene is found not to be thermally stable on large tube diameters and graphene. Due to the planar-like geometry of large CNTs, carbene groups do not induce displacement of C atoms from the original structure (as in the case of small diameter armchair nanotubes), and thus spontaneously desorb at room temperature (Margine, Bocquet & Blase, 2008). Paired configurations in the 1,4-geometry (para – where two phenyls are grafted as third-nearest neighbors) are investigated keeping intact the conjugation properties of the CNT. Several arguments suggest that such a configuration is the most likely to occur in nanotubes: (a) the grafting of a first radical is known to enhance the reactivity of a carbon atom at an odd number of bonds away from it, (b) the 1,4-configuration is slightly more stable than the 1,2-configuration (ortho), and (c) isolated phenyls have been shown to spontaneously diffuse or desorb at room temperature on standard diameter tubes (Margine, *et al.*, 2008).

Increasing the number of grafted phenyl groups has a strong impact on the conductivity of both 300 nanometers long (6, 6) and (10, 10) nanotubes, as shown in Fig. 7.31(a). The sp^3 bond between the phenyl and the tube surface reduces the conductance for all energies but with marked suppression of one conduction channel at two symmetric

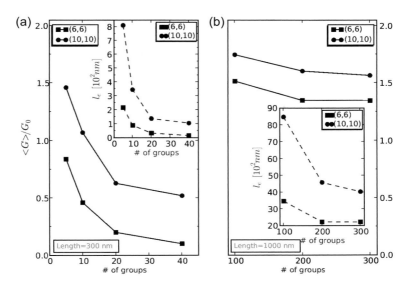

Figure 7.32 (a) Disorder average conductance at the charge neutrality point (main plot) and estimated elastic mean free path (inset) as a function of grafted phenyl groups density for a 300 nm long nanotube. (b) The same for grafted carbene groups for a nanotube length of 1μm. (Adapted with permission from López-Bezanilla *et al.* (2009*a*). Copyright (2009) American Chemical Society.)

energy resonances for a single grafted group. By increasing the coverage density, a stronger damping of the conductance pattern is observed. For a fixed molecule density, this effect is enhanced for smaller nanotube diameter. The conductance is found to roughly decay inversely proportional to the coverage density (Fig. 7.32). Using a conventional phenomenological law, the disorder average transmission coefficient can be related to the elastic mean free path as $\bar{T} = <G>/G_0 = N_\perp(1 + \frac{L}{\ell_{el}})^{-1}$. This expression allows prediction of some approximated range and scaling behavior of the mean free path (Fig. 7.32).

With simpler disorder models (such as Anderson disorder), analytical forms and scaling behavior of elastic mean free paths have been derived for both carbon nanotubes (White & Todorov, 1998) or graphene nanoribbons (Areshkin & White, 2007). In particular, ℓ_{el} was demonstrated to upscale linearly with tube diameter for a fixed disorder strength. Here, for the 300 nm long (6, 6) nanotube, ℓ_{el} quickly decays with coverage density to reach $\ell_{el} \sim 15$ nm when 40 groups are attached to the sidewalls. The same number of functional groups on a larger diameter nanotube (10, 10) will, however, yield $\ell_{el} \sim 100$ nm. The calculated mean free path also presents some upscaling with nanotube diameter, although the scaling behavior cannot be extracted in detail. As evidenced by these results, sp^3 bonds are clearly not favorable for good conduction efficiency of hybrid nanotubes, and thus not suitable for applications.

In contrast, the cycloaddition of carbene groups induces only a small downscaling of the conductance in the first plateau, with increasing the coverage density (see Fig. 7.32). In contrast to the phenyl groups, an asymmetry of the conductance decay is already observed for the single molecule case. The much weaker change of conductance indicates a quasi-ballistic regime. The values of ℓ_{el} are given in the inset to Fig. 7.32(b), and are in the range of $\sim 2 - 9 \ \mu$m, depending on coverage density and nanotube diameter. In this case a rough linear scaling of ℓ_{el} is observed with tube diameter (López-Bezanilla *et al.*, 2009*b*). Note that the coverage density in the case of carbene groups is larger than for the phenyl case, demonstrating the weak effect of such functionalization on transport properties of pristine nanotubes, which is crucial for further envisioning the use of long hybrid nanotubes.

Finally, transport regimes can also be characterized with the help of histograms. For a given distribution δ of grafting groups in the nanotube, the normalized conductance $T(\delta) = G(\delta)/G_0$ will depend on the particular chosen distribution. The conductance then becomes a random variable of which first moments are its mean value $\bar{T} = \frac{1}{\delta} \sum_\delta T(\delta)$ and its mean square deviation $\Delta T = \sqrt{\overline{T^2} - \bar{T}^2}$, where equiprobability for each configuration of disorder is assumed. The histograms presented in Fig. 7.33 corroborate the differences between the two types of functionalization when analyzing

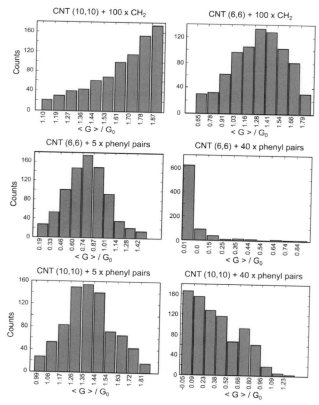

Figure 7.33 Conductance histograms for functionalized metallic armchair CNT (lengths of 0.3 μm (phenyl) and 1 μm (carbene)), determining the spreading of the conductance values obtained at the Fermi level. In the two upper panels, a predominance of large conductance values, indicating a quasi-ballistic regime, is observed. In the two middle panels, the conductance behavior of small diameter armchair nanotubes for both low and high grafting rates is illustrated.

transport properties. Carbene-based functionalization hardly affects the properties of the pristine CNTs and high conductance values are obtained in the vicinity of $G = 1.8G_0$, revealing quasi-ballistic transport. For nanotubes with phenyl functional groups, nearly zero conductance values are obtained when 40 groups are grafted in a 300 nm long (6, 6) tube, with low dispersion ΔT associated with the conductance. In this case, the system is found in the localized regime. In a (10, 10) tube, for only five phenyl grafted groups a similar behavior of the conductance is observed as in the former case. However, for a higher grafting rate a decreasing with conductance the inverse of group number is observed, with a trend to saturation around $\langle G \rangle = 0$. Similar studies on semiconducting carbon nanotubes also find important differences in transport properties depending on the type of chemical bonding (López-Bezanilla, Blase & Roche, 2010), results which have been confirmed experimentally (Bouilly, Cabana & Martel, 2012). Graphene nanoribbons are also strongly affected by chemical functionalization, and become largely insulating even for small defect densities (López-Bezanilla *et al.*, 2009*a*)

In summary, by using a fully *ab initio* transport approach, transport regimes in chemically functionalized long carbon nanotubes can be explored, comparing two different and important types of chemical bonding. The results provide evidence of good conduction ability in the case of carbene cycloadditions, whereas paired phenyl addends are found to yield a strongly diffusive regime, with estimated mean free path ranging from a few tens of nanometers to 1 nm, depending on the coverage density and incident electron energy. Functionalizing is thus an interesting tool to develop novel devices based on modified carbon nanotubes that have specific organic molecules covalently linked to their walls, thus combining the exceptional structural stability and electronic properties of microscopic wires with the diversity and tunability of material properties that come from the molecular attachments.

7.5.6   Carbon nanotubes decorated with metal clusters

Due to their high surface to volume ratio, CNTs are promising candidates as active elements for extremely sensitive gas-sensing devices, since their conductance can be easily perturbed by interaction with gas molecules (Zanolli & Charlier, 2009, Charlier *et al.*, 2009, Goldoni *et al.*, 2010). However, the response of pristine CNTs to gases is weak and scarcely selective since the ideal carbon hexagonal network is held together by strong sp^2 bonds characterized by a low chemical reactivity with the molecular environment. Consequently, functionalization of the CNT sidewalls is mandatory to improve both the sensitivity and the selectivity of CNT-based gas sensors (Peng & Cho, 2003). In particular, functionalization with metal nanoparticles (NPs) can lead to highly sensitive and selective gas sensors, thanks to the extraordinary catalytic properties of metal NPs (Charlier *et al.*, 2009).

Although the sensing ability of CNTs decorated with metal NPs relies on the huge chemical reactivity of the cluster surface, the whole CNT–NP system acts as the detection unit of the device. Indeed, the interaction with gas molecules results in an electronic charge transfer between the molecule and the CNT–NP sensor, which affects the position of the Fermi energy and, hence, the conductivity of the detection unit. Such a conductivity modification can, for instance, be measured by embedding mats of metal-decorated carbon nanotubes in a standard electronic device (Charlier *et al.*, 2009). Since these mats usually behave as p-doped semiconductors, the adsorption of an extra electron coming from molecules exhibiting a donor character will induce a resistance increase. Analogously, the interaction with molecules exhibiting an acceptor character will lead to a reduced resistance. In addition, functionalization of CNTs with metal nanoparticles can be exploited to improve the sensor selectivity, since different metals will present different reactivities toward different molecules. Hence, a gas sensor device can be fabricated by depositing on a microsensor array several sets of CNT mats, each decorated with different types of metallic NPs (Leghrib *et al.*, 2010).

First-principles modeling can be used to investigate sensing responses of CNTs decorated with gold NPs in order to understand accurately the detection ability of these nanosystems for specific gas species (Zanolli *et al.*, 2011). These *ab initio* simulations allow discussion of the problem from a microscopic point of view, giving information on the electronic charge transfer, the interaction strength (such as binding energies and bond lengths), and the quantum electron conductance. Gold nanoparticles are chosen as the sensing metal since the Au cluster interacts weakly with the carbon sp^2 network, and hence such a functionalization perturbs the electronic and transport properties of the host nanotube only slightly. To illustrate different sensing responses, molecules that cause a decrease (NO_2, acceptor character) or an increase (C_6H_6, donor character) or leave unaffected (CO) the Fermi energy of the Au–CNT system are considered.

As mentioned in the previous sections, CNTs can be either metallic or semiconducting depending on their diameter and chirality. Mats of CNTs, such as those employed experimentally, consist of a mixture of metallic and semiconducting tubes. The conductance of the mat can thus be described as a percolation process through the contact points of metallic nanotubes, since a semiconducting tube would not allow electron conductance around the Fermi energy. Hence, a change in resistance of the mat would depend only on the metallic tubes. The tube diameter should not play a crucial role since the sensing mechanism is based on the catalytic properties of the metal NP rather than on the tube curvature. Consequently, a (5, 5) single-walled CNT decorated with a Au_{13} nanoparticle is used as a detection model system to sense various gas molecules (i.e. NO_2, CO, C_6H_6). A 13-atom gold cluster is chosen because Au_{13} is close to the smallest cluster size for which the most stable structure is three-dimensional (Gruber *et al.*, 2008). Decoration of (5, 5) SWNTs with Au_{13} nanoclusters having initially cuboctahedral and icosahedral symmetries has been considered in Zanolli *et al.* (2011). Even though the total energy of cuboctahedral Au_{13} is $\sim$1 eV lower than the icosahedral, the relaxed SWNT-icosahedral system presents a total energy 1.15 eV lower and binding energy 0.6 eV higher than the relaxed SWNT-cuboctahedral. Hence, the latter configuration, a (5, 5) SWNT decorated with icosahedral Au_{13} NP, is considered in the following. After relaxation, the atomic structure of the adsorbed gold NP is quite distorted and the original symmetry of Au_{13} is lost, as shown in Fig. 7.34(a).

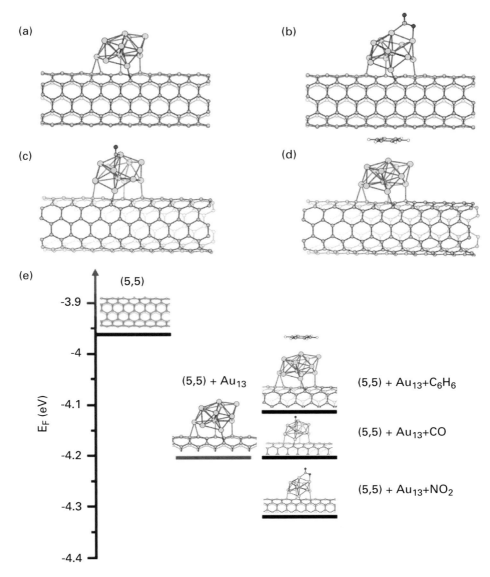

Figure 7.34 (a–d) Ball-and-stick models illustrating fully *ab initio* optimized atomic structures of a (5, 5) SWNT decorated with a Au_{13} nanocluster (a) and with various adsorbed molecules: NO_2 (b), CO (c), and C_6H_6 (d). (e) Computed shift of the Fermi energy (ΔE_F) of a pristine (5, 5) SWNT after Au_{13} decoration and after interaction with the above mentioned molecules. (Adapted with permission from Zanolli *et al.* (2011). Copyright (2011) American Chemical Society.)

The gold NP binds weakly to the SWNT (binding energy ~ -2.4 eV) since the electronic *d* shell of gold is closed and, hence, the Au–C bond is not favored (Park *et al.*, 2009). Consequently, the adsorption of Au_{13} slightly perturbs the band structure of the SWNT (not shown here) and causes only a small shift of the Fermi energy toward lower energies, as schematically represented in Figure 7.34(e). This situation is equivalent to a p-doping of the tube, consistent with the small electronic charge transfer (~ 0.06 e$^-$) from the tube to the Au–NP.

In order to model gas adsorption, first-principles structural optimization techniques are used starting from molecules approaching the relaxed SWNT–NP system from several different initial configurations. The resulting relaxed ground-state geometries of Au$_{13}$–SWNT with and without adsorbed molecules are depicted in Fig. 7.34. Considerations based on basic chemistry/physics indicate that the relaxed structures can be considered as reliable. For instance, NO$_2$ is a polar molecule with positive charge localized on the nitrogen and negative charge on one of the oxygen atoms. Electron interaction with the gold NP will repel the negatively charged oxygen and attract the positively charged nitrogen, as shown in Fig. 7.34(b). Due to the high electronegativity of oxygen, the CO molecule would preferentially bind to gold through the carbon atom (Fig. 7.34(c)), as also found in other work dedicated to the adsorption of CO on various Au nanoclusters (Mpourmpakis, Andriotis & Vlachos, 2010). Lastly, the benzene–Au–CNT relaxed structure (Fig. 7.34(d)) can be explained by considering that benzene would preferentially bind thanks to the delocalized π electron ring, i.e. with the benzene ring parallel to a gold surface. Gas adsorption is found not to alter the Au–SWNT interaction, as indicated by the almost unchanged Au–SWNT bond lengths. Computed binding energies and bond lengths show that NO$_2$ is the most strongly interacting molecule, while benzene is the interacting one. NO$_2$ is predicted to accept a significant fraction of electronic charge (~ 0.5 e$^-$) from the Au–SWNT system, consistent with the quite well-known electron acceptor character of that molecule. Indeed, the LUMO of NO$_2$ is located below the Fermi energy of the Au$_{13}$–(5, 5) system, hence allowing a large charge transfer toward the molecule. The electron charge acquired by NO$_2$ is provided by both the Au-NP (which donates 0.289 e$^-$) and the (5, 5) tube (which donates 0.218 e$^-$). The nature of the charge transfer between NO$_2$ and Au–SWNT indicates that the adsorbed molecule injects majority carriers (*holes*) in the metal-decorated SWNT, hence strengthening its p-type semiconducting behavior and lowering further the Fermi energy of the system (Fig. 7.34(e)). The predicted injection of holes after NO$_2$ adsorption explains the decrease of resistance as measured experimentally in the Au–CNT samples (Zanolli *et al.*, 2011).

Due to the weak bonding between gold and carbon, it follows that the interaction between benzene, which consists of six carbon and six hydrogen atoms, and Au–SWNT is extremely weak and results only in an increase of the Fermi energy (Fig. 7.34(e)). However, the p-type semiconducting behavior of the overall system is maintained. It is worth noting that the shift of the Fermi energy with respect to a reference structure (in the present case, the Au–SWNT) indicates only the donor (increase of E_F) or acceptor (decrease of E_F) nature of the molecule and does not provide information on the actual occurrence of the charge transfer. The latter depends on the interaction strength between the molecule and Au–SWNT, i.e. binding energy and bond length. The benzene–Au–SWNT interaction possesses all the characteristics of a physisorption, as indicated by the relatively large bond length (~ 3.9 Å) and a low binding energy (-0.193 eV). The interaction is so weak that no significant charge transfer could be computed, suggesting that Au-decorated CNTs are not suitable for C_6H_6 detection, consistent with the experimental results using Au–CNT sensors (Zanolli *et al.*, 2011). Metals that exhibit a stronger interaction with carbon should be more suitable for benzene detection, since the sensing mechanism relies on the binding of the gas molecule to the nanocluster, as discussed later. However, since the DFT formalism does not accurately model weak long-range interactions such as van der Waals forces, the real interaction between benzene and Au–SWNT could be stronger than the one predicted within the present *ab initio* approach. The inaccuracy in modeling van der Waals interactions will affect the description of the detection of NO_2 and CO less since these molecules bind ion-covalently to the Au nanocluster. The adsorption of a single CO molecule on Au–CNT results in a small fraction of electron charge (0.164 e$^-$) transferred toward CO, but no shift of the Fermi energy is detectable with respect to the Au–CNT system (Fig. 7.34(e)). However, the interaction between CO and Au nanoclusters is quite complex (Mpourmpakis *et al.*, 2010), since it depends strongly on the overlap of the 5σ and $2\pi^*$ orbitals of the CO with the d orbitals of the gold NP, resulting in a mechanism of electron donation/back-donation. Indeed, CO donates electrons from its σ orbitals to the d_σ orbitals of gold, and the filled d_π orbitals of Au donate electrons back into the empty π orbitals of CO.

In order to check the sensing ability of CNTs, *ab initio* quantum transport calculations are performed by modeling the gas–Au–SWNT structure via an open system, consisting of a central scattering region and left and right semi-infinite electrodes. The ballistic conductance of the (5, 5) nanotube at energy E is proportional to the number of conducting channels, that is, two energy bands crossing at the Fermi energy (E_F), resulting in a conductance of $2G_0$ in the corresponding energy window (Fig. 7.35(a), dashed lines). Any perturbation of the ideal SWNT structure, such as metal decoration and/or gas adsorption, introduces new scattering centers, thus affecting the conductance curve, as illustrated in Fig. 7.35. Au-functionalization of SWNTs induces several dips in the conductance curve of the pristine (5, 5) tube (Fig. 7.35(a)), and, in particular, the conductance at the Fermi energy $G(E_F)$ drops to $1G_0$, indicating the complete suppression of one transmission channel.

The adsorption of a single NO_2 molecule at the surface of the Au-NP results in a different distribution of the $G(E)$ dips (Fig. 7.35(b)), with a remarkable increase of the conductance at the Fermi energy ($G(E_F)$) from $1G_0$ to $1.92G_0$ (+92%). Such a predicted enhancement of the conductance is consistent with the decrease of the resistance measured experimentally on the Au-decorated CNT mats in the presence of NO_2 gas (Zanolli *et al.*, 2011). In addition, this result indicates that gold functionalization of CNTs improves the detection of NO_2 with respect to oxygenated-defected SWNTs, where adsorption of NO_2 was predicted (Zanolli & Charlier, 2009) to cause a small change of $G(E_F)$ (4.6%), and integration of the conductance curve over a properly selected energy interval around E_F was needed to reveal NO_2 adsorption.

The CO adsorption is also found to affect the conductance of Au–SWNTs (Fig. 7.35(c)), but not as strongly as NO_2. In particular, the percentage change of $G(E_F)$ after CO adsorption corresponds to an increase of $\sim$10%, hence predicting a small decrease in the resistance, in good agreement with experimental results (Zanolli *et al.*,

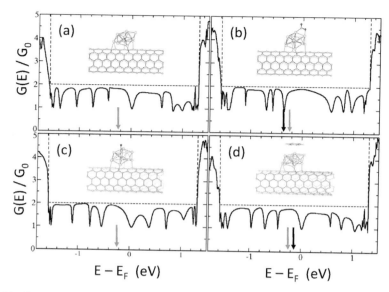

Figure 7.35 *Ab initio* quantum conductance of a (5, 5) SWNT decorated with a Au_{13} nanocluster (a) and with various adsorbed molecules: NO_2 (b), CO (c), and C_6H_6 (d) at the surface of this gold cluster. The energy zero is aligned with the Fermi energy of a pristine (5, 5) tube, the light-grey arrow at −0.25 eV indicates the Fermi energy of the Au_{13}–(5, 5) system, while the dark-grey arrow in panels (b) and (d) indicates the Fermi energy in the presence of NO_2 (b) and C_6H_6 (d). The position of the Fermi level is unchanged after CO adsorption. (Adapted with permission from Zanolli *et al.* (2011). Copyright (2011) American Chemical Society.)

2011). However, the predicted responsiveness of Au–SWNT detectors is similar to oxygenated-defected SWNTs (Zanolli & Charlier, 2009), while experimental measurements suggest that Au functionalization improves the detection of CO (Zanolli *et al.*, 2011). Such a discrepancy between experiments and simulations could be explained by the multiple adsorption of CO molecules on the same Au-NP that could happen experimentally, while the adsorption of a single CO molecule has been modeled. Indeed, the adsorption of more CO molecules further affects the computed conductance (Kauffman *et al.*, 2010), hence resulting in a stronger effect on the measured resistance. It is worth noting that charge transfer considerations alone are not sufficient to describe the electrical behavior of CNTs, especially in the case of CO, whose interaction with gold is quite complex, as it involves a charge transfer in both directions: to and from the CO molecule. A more powerful tool, such as *ab initio* quantum conductance calculations corroborated by a careful analysis of the $G(E)$ curve, is thus mandatory to make accurate predictions.

Lastly, the adsorption of C_6H_6 leaves the overall shape of the conductance curve almost unaffected (Fig. 7.35(d)). The percentage change of $G(E_F)$ is less than 5%, suggesting that it would be quite difficult to achieve benzene detection via a measurement of resistance in Au-decorated CNTs. This theoretical prediction is also consistent with experiment (Zanolli *et al.*, 2011). Indeed, neither the O_2–CNT nor Au–CNT sensor was responsive to benzene for any tested concentration. Finally, it should be noted that a quantitative description of realistic gas sensors should take into account the effect of configurational disorder. The decrease (or increase) of conductance after gas adsorption remains actually unchanged even when $\sim$300 random distributions of such configuration are considered (Rocha *et al.*, 2008). Hence, the study of individual configurations can reasonably be used to understand and analyze experimental results.

In summary, *ab initio* simulations predict that gold decoration of CNTs improves the detection of both NO_2 and CO with respect to oxygenated-defected CNTs, while benzene remains undetected. In particular, in the case of NO_2 detection, the computed shift of the Fermi energy and the nature of the charge transfer are directly correlated to the resistance decrease measured experimentally (Zanolli *et al.*, 2011), suggesting electronic mechanisms at the microscopic level that can be extrapolated at the macroscopic level to explain the Au–CNT behavior in the presence of specific gas molecules. Charge transfer considerations alone are not sufficient to describe how carbon oxide adsorption affects the resistance of the Au–CNTs (Zanolli *et al.*, 2011). Indeed, in the latter case, quantum electron transport simulations have proven to be essential to predict the decrease of the resistance, in agreement with the experimental data. Finally, the lack of benzene sensitivity is explained in terms of the weak interaction strength that arises between the molecule and the Au-NP, but also between that particle and the nanotube. In the following discussion, spin-polarized quantum transport in metallic carbon nanotubes decorated with transition metal (Ni_{13}) magnetic nanoclusters is used to detect physisorbed molecules such as benzene via a magnetic process (Zanolli & Charlier, 2012).

CNTs decorated with transition metal magnetic nanoparticles have been suggested to be good candidates for spin-dependent transport applications. Indeed, spin-polarized currents can propagate throughout CNTs on extremely long distances (spin diffusion length $\sim$1.5 μm (Tsukagoshi, Alphenaar & Ago, 1999, Hueso *et al.*, 2007)) thanks to the weak spin-orbit and hyperfine interactions in carbon-based materials, as well as to the large carrier velocity in CNTs. In this section, a new approach to detecting individual gas molecules based on local magnetic moment measurement is proposed. By analogy with gold clusters, a metallic (5, 5) single-wall CNT decorated with a Ni_{13} nanocluster has been chosen as an ideal model for the CNT–NP system. Indeed, a cluster of 13 Ni atoms corresponds to the minimal size to present a spherical shape (closed atomic shell), and exhibits extraordinary energetic stability (Baletto & Ferrando, 2005). The icosahedron is known to be the lowest energy structure for Ni_{13}, and is also favored for larger cluster sizes (Parks *et al.*, 1994). In addition, the magnetic properties of transition metal nanoparticles are strongly size-dependent: the smaller the cluster, the higher the magnetic moment per atom. Experimental measurements have found a magnetic moment of $\sim$0.96 μ_B/atom in Ni_{13} (Apsel *et al.*, 1996). Finally, benzene has been selected as the target molecule to detect, since it is a highly toxic gas and only a few studies have reported benzene detection in the ppb range (Leghrib *et al.*, 2010) using conventional techniques.

Firstly, the atomic structures of an isolated Ni_{13} NP, with an icosahedral symmetry, have been fully *ab initio* optimized, finding a Ni–Ni bond length and spin magnetization in excellent agreement with the literature: $d_{Ni-Ni} \sim 2.25$–2.29 Å, and M $\sim 8\mu_B$ (Calleja *et al.*, 1999). Secondly, the equilibrium atomic structures and electronic properties of CNT–NP systems (before and after C_6H_6 adsorption) have been computed *ab initio*. After relaxation, the atomic structure of the Ni_{13} NP keeps its original icosahedral shape, as depicted in Fig. 7.36(a). The Ni_{13} cluster interacts strongly with the CNT, as indicated by large binding energies ($E_B = -4.5$ eV) and by the relatively short inter-distance ($d_{NP} \sim 2$Å). Consequently, this strong interaction between the attached NP and the tube greatly perturbs the pristine CNT electronic structure and, hence, its quantum conductance around E_F, as described later. Also, Ni_{13} NP is found to exhibit a donor character, with a large electronic charge transfer from the cluster towards the tube (~ 0.636 e$^-$). Thirdly, the adsorption of a single benzene molecule on the CNT–NP nanosensor has also been investigated theoretically. In order to model gas adsorption, first-principles conjugate gradient minimization is performed starting from a benzene ring lying on the top of the NP and parallel to the CNT axis. Indeed, benzene essentially binds thanks to its ring of delocalized π orbitals, i.e. with the C_6H_6 ring parallel to the binding surface. The resulting *ab initio* fully relaxed atomic structures are illustrated in Fig. 7.36(b). Benzene adsorption is found to slightly alter the electronic and structural properties of the CNT–NP hybrid system, as indicated by the almost unchanged CNT–NP distances ($d_{NP} \sim 2.2$ Å) and binding energies ($E_B = -4.4$ eV). Benzene is found to accept a significant fraction of electronic charge (~ 0.316 e$^-$) from the CNT–Ni_{13} hybrid system. Both the CNT–NP and the NP–C_6H_6 interactions exhibit the conventional features of a chemisorption, i.e. high binding energies and relatively short bond lengths.

Magnetic properties and spin-polarized conductances have been calculated using a nonperiodic open system consisting of a central scattering region and two semi-infinite pristine (5, 5) CNTs as left and right electrodes. The scattering region includes the relaxed CNT–NP hybrid system, either with or without adsorbed benzene. The use of long pristine electrodes is mandatory in order to ensure a good screening of the Hartree

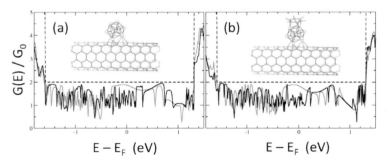

Figure 7.36 Conductances in G_0 units of Ni_{13}–CNT (a) and Ni_{13}–CNT–C_6H_6 (b) hybrid systems. Ball-and-stick models illustrating fully *ab initio* relaxed atomic structures of a (5, 5) CNT decorated with a Ni_{13} nanoparticle before (a) and after (b) the adsorption of a single benzene molecule are shown inset. Grey and black solid lines describe the majority and minority spin channels. Conductances per spin channel of the pristine (5, 5) CNT are depicted with dashed lines. The zero energy corresponds to the Fermi level of the pristine (5, 5) CNT. (Adapted with permission from Zanolli & Charlier (2012). Copyright (2012) American Chemical Society.)

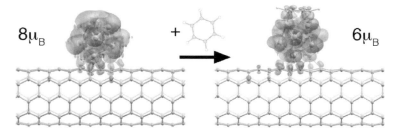

Figure 7.37 Difference between spin-up and spin-down densities ($\rho_\uparrow - \rho_\downarrow$) on the CNT–Ni$_{13}$ hybrid system before (a) and after (b) C$_6$H$_6$ adsorption. Light and dark grey isocurves correspond to positive and negative isodensities of $\rho_\uparrow - \rho_\downarrow$, respectively. The local magnetic moment is found to decrease from $8\mu_B$ to $6\mu_B$ when benzene is adsorbed on the Ni$_{13}$ magnetic cluster. (Adapted with permission from Zanolli & Charlier (2012). Copyright (2012) American Chemical Society.)

potential (Zanolli & Charlier, 2009). The magnetization is computed *ab initio* on the CNT–NP (+C$_6$H$_6$) hybrid systems. The net spin polarization of the CNT–Ni$_{13}$ system corresponds exactly to the isolated Ni$_{13}$ NP computed by first principles (M $\sim$ $8\mu_B$). After benzene adsorption, the electronic charge is redistributed and new bonds are created leading to a decrease of the magnetization to $\sim 6\mu_B$. The difference between spin-up and spin-down densities ($\rho_\uparrow - \rho_\downarrow$) on the CNT–Ni$_{13}$ hybrid system after C$_6$H$_6$ adsorption (Fig. 7.37) reveals that the spin magnetization resides mostly on the Ni$_{13}$–C$_6$H$_6$ part. However, spin polarization is also induced in the CNT due to the strong coupling between the carbon π states and the Ni d orbitals.

In the present CNT–Ni$_{13}$ hybrid system, the very strong interaction between the NP d orbitals and the carbon π electrons heavily alters the pristine CNT electronic structure, inducing quasi-nondispersive energy bands close to EF coming from the NP and distorting the π electron orbitals in the proximity of NP (Zanolli & Charlier, 2012). Consequently, the attached transition metal nanoparticle introduces several scattering centers, thus leading to the many closely spaced dips observed in the conductance curve of Ni-decorated CNTs before (Fig. 7.36(a)) and after (Fig. 7.36(b)) benzene adsorption.

Despite the overall strong electron scattering, the conductance of the CNT–Ni$_{13}$ hybrid system is characterized by an energy window of about ~ 0.5 eV near E_F where the majority spin current (light grey curve in Fig. 7.36(a)) is almost completely transmitted ($G \sim 2G_0$) and the minority spin current (dark curve in Fig. 7.36(a)) is highly backscattered leading to a decrease of the minority conductance to $\sim 1G_0$, i.e. to a drastic suppression of one transmission spin channel. This spin-dependent conducting behavior can be easily explained by analyzing the CNT–Ni$_{13}$ band structure: the majority spin bands differ slightly from the bands of the pristine (5, 5) CNT, while the minority spin bands are highly perturbed due to the CNT–NP interaction (Zanolli & Charlier, 2012). Consequently, CNT–NP hybrid systems could be used to induce spin polarization in unpolarized charge carriers injected via low-resistance nonmagnetic electrodes. Near E_F, the spin-up electrons will not be affected at all by the presence of the NP, while half of the spin-down electrons will be filtered.

In conclusion, the present first-principles approach based on nonequilibrium Green's functions techniques has been proved crucial to provide microscopic insight concerning the behavior of gas nanosensors based on metal-decorated CNTs. Indeed, for CNTs randomly decorated with conventional metal (like Au), the computed shift of the Fermi energy and the nature of the charge transfer have been directly correlated to the resistance decrease measured experimentally, suggesting that electronic mechanisms at the microscopic level can be extrapolated at the macroscopic level to explain the chemisorption of specific gas molecules at the surface of metal–CNT hybrid systems. When the interaction between the NP and the tube increases, as is the case for CNTs randomly decorated with magnetic clusters (like Ni), physisorption can be investigated at the microscopic level. In particular, benzene adsorption on the CNT–Ni_{13} hybrid system results in a drastic change of the net magnetic moment. This new magneto-sensing property, suggested theoretically, could be exploited as a highly sensitive novel gas detection technique based on the measurement of local magnetic moment in these hybrid CNT–NP systems using various experimental techniques such as magnetic AFM or SQUID magnetometer. Finally, this theoretical research also illustrates how first-principles techniques can be used as powerful predictive tools to understand, improve, and design the next generation of gas sensors.

7.5.7   Graphene nanoribbons

In contrast to carbon nanotubes, quantum transport properties of GNRs are expected to strongly depend on whether their edges exhibit the armchair or the zigzag configuration. Indeed, in the previous sections, the electronic properties of GNRs have been predicted using the single-band model and *ab initio* calculations, revealing a dependence on the edge topology (Nakada *et al.*, 1996). The armchair GNRs are semiconductors with energy gaps which decrease as a function of increasing ribbon width. As mentioned earlier, the gaps of the N-aGNRs depend on the N value, separating the ribbons into three different categories (all exhibiting direct band gaps at Γ). The band structures and the corresponding quantum conductances of two armchair GNRs are illustrated in Fig. 7.38(a, b). The 16-aGNR is a ~ 0.8 eV gap semiconductor, thus inducing a quite large energy interval where no transmission is allowed (Fig. 7.38(a)), while the gap of the 17-aGNR is reduced to less than 0.2 eV (Fig. 7.38(b)). The region of zero conductance is also reduced accordingly, and a very small external electric field would induce an electronic transmission through one channel ($1G_0$ for the conductance).

Identically, nanoribbons with zigzag-shaped edges also exhibit direct band gaps which decrease with increasing width. However, in zGNRs, quantum transport is dominated by edge states which are expected to be spin-polarized owing to their high degeneracy. Indeed, for topological reasons, zigzag-shaped edges give rise to peculiar extended electronic states which decay exponentially inside the graphene sheet (Nakada *et al.*, 1996). These edge states, which are not reported along the armchair-shaped edges, come with a twofold degenerate band at the Fermi energy over one third of the Brillouin zone.

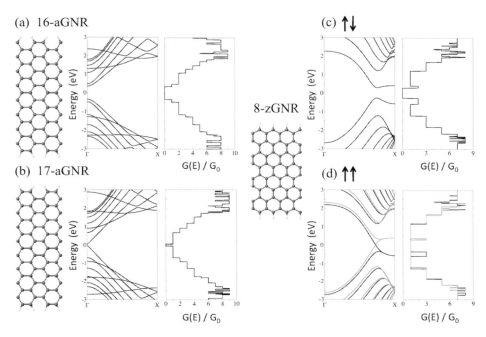

Figure 7.38 Atomic structures, electronic band structures, and quantum conductances (electronic transmission) of various graphene nanoribbons: (a) 16-armchair GNR; (b) 17-armchair GNR. (c, d) 8-zigzag GNR with (c) antiparallel (↑↓) or (d) parallel (↑↑) spin orientations between the two magnetic edges. The spin-dependent transport is evaluated for both magnetic configurations of the 8-zGNR but is only visible for the parallel (↑↑) spin orientations (ferromagnetic). In such a case, one spin orientation is labeled α-spin (in black solid line) while the other is labeled β-spin (in grey solid line). (Reproduced from Dubois *et al.* (2009), with kind permission from Springer Science and Business Media.)

The ground state of zGNRs with hydrogen passivated zigzag edges presents finite magnetic moments on each edge with negligible change in atomic structure, thus suggesting zGNRs to be attractive for spintronics (Son, Cohen & Louie, 2006a). Indeed, upon inclusion of the spin degrees of freedom within *ab initio* calculations (LSDA), the zGNR are predicted to exhibit a magnetic insulating ground state with ferromagnetic ordering at each zigzag edge and antiparallel spin orientation between the two edges (Son, Cohen & Louie, 2006a). The total energy difference between ferromagnetic (↑↑) and antiferromagnetic (↑↓) couplings between the edges is of the order of ~20 meV per edge atom for an 8-zGNR, but this decreases as the width of the ribbon increases and eventually becomes negligible if this width is significantly larger than the decay length of the spin-polarized edge states (Lee *et al.*, 2005). Because the interaction between spins on opposite edges increases with decreasing width, the total energy of an N-zGNR with antiferromagnetic arrangement across opposite edges is always lower than that of a ferromagnetic arrangement for low values of N ($N \leq 30$).

The band structures and the spin-dependent quantum conductances of an 8-zGNR are illustrated in Fig. 7.38(c, d) in the two respective magnetic configurations (↑↓ and ↑↑)

of the ribbon edges. The $\uparrow\downarrow$ spin configuration of the 8-zGNR conserves the semiconducting behavior of the GNR family, and its electronic transmission function displays a gap of 0.5 eV around the Fermi energy (Fig. 7.38(c)). On the contrary, in the $\uparrow\uparrow$ spin configuration, the 8-zGNR becomes metallic, inducing a nonzero electronic transmission function at the Fermi energy (Fig. 7.38(d)). In addition, the spin-dependent conductance calculation also reveals that the transmission of π electrons with one type of spin orientation (α-spin) is favored for an energy region around -0.5 eV below the charge neutrality point. On the contrary, π^* electrons with the other orientation (β-spin) are more easily transmitted around $+0.3$ eV above the Fermi energy.

7.5.8   Graphene nanoribbons with point defects

However, ideal zigzag GNRs are not efficient spin injectors due to the symmetry between the edges with opposite magnetization. In order to obtain net spin injection, this symmetry must be broken (Wimmer *et al.*, 2008). Incorporating defects (such as vacancies or adatoms) in the GNR or imperfections at the edge, which usually cannot be avoided experimentally, breaks the symmetry between the edges and could thus influence the spin conductance of the GNR (Fig. 7.39(a)). In addition, the introduction of magnetic point defects in zGNRs favors a specific spin configuration of the edges. As an example, the $\uparrow\uparrow$ spin configuration is favored when vacancies or adatoms are introduced around the ribbon axis. Consequently, point defects are also expected to play a key role in the transport properties of zGNRs (Dubois *et al.*, 2009). *Ab initio* calculations of the electronic transmission functions have been performed within the Landauer approach using a supercell containing the point defect connected to two leads consisting of a few unit cells of ideal 8-zGNR. Both the ($\uparrow\downarrow$) semiconducting and the ($\uparrow\uparrow$) metallic spin configurations of the ribbon are considered (Fig. 7.39).

The main impact of the magnetic point defects on the transport properties is a global reduction of the transmission associated with the π and π^* electrons. This is related to a decrease of the transmission probability of some $\pi-\pi^*$ conduction eigenchannels compared to the pristine 8-zGNR. Within the ($\uparrow\downarrow$) semiconducting configuration (Fig. 7.39(b, c), bottom), the presence of defects essentially reduces the conductance for energies ranging from -0.80 to -0.3 eV (π channels) and from 0.3 to 0.5 eV (π^* channels), inducing a slight breaking of the spin degeneracy. Within the ($\uparrow\uparrow$) metallic configuration, a similar reduction of the conductance is observed (Fig. 7.39(b, c), top). However, the defects also induce sharp drops in the transmission function around the Fermi level. At these energies, the electronic states localized on the defect are spin-polarized and can only mix with one of the two spin conduction channels. Consequently, the spin degeneracy of the electronic transmission function is raised just around the Fermi energy. In summary, when adatoms and vacancies are introduced, the parallel spin orientation may be preferred and the local magnetic moment of the defect adds to the contributions of the edges. Furthermore, a spin-polarized transmission is observed at the Fermi energy, suggesting a use of defect-doped graphene nanoribbons as a spin-valve device (or spin-filter) in future spin-based electronics.

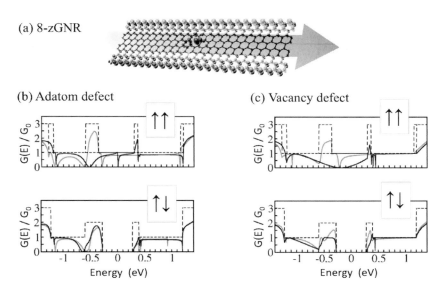

Figure 7.39 Effect of a point defect on the *ab initio* quantum conductance of an 8-zGNR (a). Electronic transmission functions are estimated in the presence of an isolated carbon adatom (b) or an isolated vacancy (c) localized around the ribbon axis, in either the ↑↑ configuration (top panel) or the ↑↓ configuration (bottom panel) of the ribbon edges. The α-spin and β-spin components are shown in dark and light grey, respectively, while the dotted lines indicate the number of π channels in the pristine nanoribbon. (Adapted from Dubois (2009).)

7.5.9   Graphene nanoribbons with edge reconstruction

In contrast to carbon nanotubes, GNRs exhibit a high degree of edge chemical reactivity, which, for instance, prevents the existence of truly metallic nanoribbons (Cresti *et al.*, 2008, Barone, Hod & Scuseria, 2006, White *et al.*, 2007). Additionally, the discrepancy between the theoretical electronic confinement gap and the experimentally measured transport gap has been attributed to localized states induced by edge disorder (Querlioz *et al.*, 2008, Evaldsson *et al.*, 2008, Mucciolo, Castro Neto & Lewenkopf, 2009). Several experimental studies have also reported the characterization of individual edge defects. To date, several defect topologies of edge disorder (reconstruction and chemistry) have been proposed for GNRs, and *ab initio* calculations have shown the stability of certain types of geometries such as the Stone–Wales reconstruction (Wassmann *et al.*, 2008, Koskinen, Malola & Häkkinen, 2008, Huang *et al.*, 2009). Indeed, at room temperature, the zigzag edge is found to be metastable and a planar reconstruction implying pentagons and heptagons (zz57) spontaneously takes place (Fig. 7.40(a)). Such a zz57-reconstruction self-passivates the edge with respect to adsorption of atomic hydrogen from a molecular atmosphere. Indeed, the formation of triple bonds with alternating single bonds is suggested by the nearly isolated dimers at the ribbon edge, thus removing the dangling bond bands (due to the absence of hydrogen) away from the Fermi level by lifting the degeneracy by almost 5 eV (Koskinen, Malola & Häkkinen, 2008). Because the dangling bond bands shift to elusive energies, the corresponding chemical reactivity

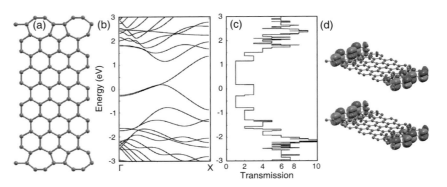

Figure 7.40 Atomic structure model (a), electronic band structure (b), and quantum conductance (c) of an 8-zGNR with a reconstructed edge containing pentagons and heptagons (zz57). The π electronic states surrounding the Fermi energy are represented in (d) and are found to be mainly localized on the zz57-reconstructed edge. (Reproduced from Dubois *et al.* (2009), with kind permission from Springer Science and Business Media.)

is also reduced, stabilizing the zz57 edge. This zz57-reconstruction also modifies the electronic structure of the ribbon. The presence of the edge states around the Fermi level (Fig. 7.40(b)) makes this reconstruction ideal for conductance measurements, in contrast to armchair ribbons where the edge state is absent. Indeed, *ab initio* quantum conductance of a zz57-reconstructed 8-zGNR has been calculated (Dubois *et al.*, 2009) and is presented in Fig. 7.40(c). The electronic transmission is predicted to be quite high at the charge neutrality point ($G = 3G_0$), compared to the conventional conductance ($G = 1G_0$) predicted for pristine zGNRs at the Fermi energy (Fig. 7.38(d)). In the zz57-edge reconstruction, π electrons are easily transmitted and these electronic channels are localized at the zz57-reconstructed edge (Fig. 7.38(d)). Consequently, this novel thermodynamically and chemically stable reconstruction could play a key role in the formation of angular joints in nanoribbons (Li *et al.*, 2008). A knowledge of the atomic structure and the stability of the possible ribbon edges is a crucial issue to control the experimental conditions of the formation of graphene nanoribbons of desired properties for future nanoelectronics.

7.5.10   Graphene nanoribbons with edge disorder

Most of the transport studies of edge-disordered GNRs have assumed simplified defect topologies (Querlioz *et al.*, 2008, Evaldsson *et al.*, 2008, Mucciolo *et al.*, 2009). However, a few *ab initio* calculations have also analyzed a much larger complexity of edge reconstruction and edge chemistry, given the reported stability of certain types of geometries such as the Stone–Wales reconstruction (Wassmann *et al.*, 2008, Koskinen, Malola & Häkkinen, 2008, Huang *et al.*, 2009). Several experimental studies have also reported the characterization of individual edge defects either by means of Raman, scanning tunneling or transmission electron microscopy (Cançado *et al.*, 2004, Liu *et al.*, 2009, Girit *et al.*, 2009). Consequently, it is necessary to investigate the impact of realistic edge defect topology on the electronic transport properties of long and disordered GNRs.

The electronic and transport properties of aGNRs have been predicted to strongly depend on the geometry of the edge reconstruction (Dubois *et al.*, 2010). Indeed, the transport signature due to a single defect at low energy turns out to range from a full suppression of either hole or electron conduction, to a vanishingly small contribution of backscattering. Besides, hydrogenation of the chemically active defects is found to globally restore electron and hole conduction as described below.

Figure 7.41 illustrates various types of possible defect along the ribbon edge using Clar's sextet representation (Clar, 1964, 1972). Among all the defects, the reconstructed geometries that preserve the benzenoid structure of pristine aGNRs turn out to weakly affect the electronic transmission. The most striking example is the conductance profile of the D_4 defect (Fig. 7.41(e)). The conductance remains very close to its maximum quantized value as found in the pristine case, with weak backscattering mainly observed in higher subbands. Slightly differently, the double heptagon and pentagon defect (D_3) exhibits two conductance suppression dips (Fig. 7.41(d)), symmetric with respect to the

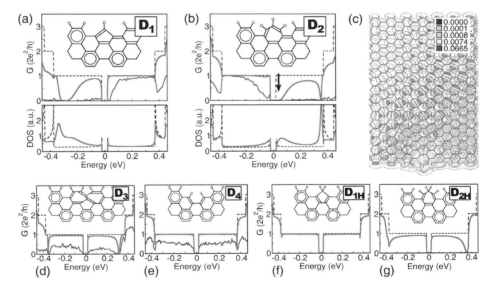

Figure 7.41 Conductance profiles of 35-aGNR for six different edge geometries: pristine ribbon (dashed black lines), single defect (light grey thick solid lines), and average conductance for 500 nm long aGNRs containing 30 defects (dark grey thin noisy lines). TDOS are given for (a) and (b). (Top) Edge defect topologies containing odd-membered rings: (a) a single pentagon defect – D_1, (b) A single heptagon – D_2. (c) Spatial representation of the transmission eigenchannel at the energy marked by an arrow in panel (b). (Bottom) Edge reconstruction involving only benzenoid defects: (d) 2 heptagons and 1 pentagon – D_3, (e) Small hole due to a dimer extraction – D_4. (f) D_{1H} and (g) D_{2H} denote the dihydrogenated pentagon and heptagon, respectively. Insets: Clar's sextet representation for each edge geometry. (Adapted with permission from Dubois *et al.* (2010). Copyright (2010) American Chemical Society.)

charge neutrality point, recalling the signature of an sp^3-type defect (López-Bezanilla *et al.*, 2009*b*).

Other topologies preserving the benzenoid structure also yield very similar results to those of the D_3 and D_4 defects (not shown here). In marked contrast, edge defects containing mono-hydrogenated odd-membered rings convey much stronger backscattering efficiency. Conductance fingerprints for single reconstructed pentagon and heptagon defects are illustrated in Fig. 7.41(a) and Fig. 7.41(b), respectively. Interestingly, a marked acceptor (donor) character develops for the pentagon (heptagon) defect, as evidenced by the strong electron–hole conductance asymmetry. Such an effect, already observed in nanotube junctions (Charlier, Ebbesen & Lambin, 1996), is due to the charge transfer taking place in the $\pi-\pi^*$ bands when odd-membered rings are embedded in a perfect hexagonal network. According to the Mulliken decomposition of the electronic density, a slight excess of π electrons (+0.152) is found on the pentagonal ring and a small deficit of π electrons (−0.135) is reported on the heptagon. Five-membered rings (D_1) have thus an acceptor character, whereas seven-membered rings (D_2) exhibit a donor character, and even-membered rings (D_3, D_4) are predicted to be neutral in a planar hexagonal network (Tamura & Tsukada, 1994).

Total and local DOS have been computed for D_1 and D_2 (Fig. 7.41(a) and (b)) and reveal the energy position of quasibound states, which are responsible for the conductance drops (Choi *et al.*, 2000). The charge density contour plot (Fig. 7.41(c)) represents the electronic state incident from the left and totally reflected by the quasibound state at the energy indicated by an arrow in Fig. 7.41(b). The backscattering associated with quasibound states is a general mechanism. The acceptor, donor, and neutral characteristics reported respectively for the D_1, D_2, and $D_{3/4}$ topologies are actually very robust regardless of the ribbon width. However, the broadening of the conductance dips is expected to decrease for larger ribbon widths. The intensity of the observed electron–hole conductance asymmetry is thus likely to depend on the actual ribbon geometry.

The high chemical reactivity of defects such as D_1 and D_2 is further explored by assuming a dehydrogenation of the carbon atom sitting at the edge. Resulting conductance profiles for the new defects D_{1H} (Fig. 7.41(f)) and D_{2H} (Fig. 7.41(g)) strongly differ from the monohydrogenated cases (D_1 and D_2). Indeed, in the presence of additional passivation, the conductance is fully restored for D_{1H}, whereas the signature for D_{2H} becomes similar to that of D_3. In both cases, the initial strong reduction of electron or hole conductance is markedly suppressed, suggesting a possibility to tune the transport properties from a metallic to a truly insulating state (or vice versa) upon varying the coverage of monatomic hydrogen, as discussed experimentally for 2D graphene (Bostwick *et al.*, 2009).

The Clar's sextet representation (Clar, 1964, 1972) (Fig. 7.41 (insets)) provides a pictorial scheme to understand the impact of edge defects (Baldoni, Sgamellotti & Mercuri, 2008) on transport properties. Clar's theory proposes a simplified description of the π electronic structure of hydrocarbons on the basis of the resonance patterns that maximize the number of benzenoid sextets drawn for the system. The benzenoid sextets, depicted as plain circles in the insets of Fig. 7.41, are defined as the carbon hexagons that stem from the resonance between two Kekulé structures with alternating single and double bonds. These are associated with a benzene-like delocalization of the π electrons over the carbon ring. As a consequence, Clar's representation gives direct insights into the aromaticity of the π electronic structure. According to Clar's theory, pristine aGNRs are fully benzenoid (i.e. all π orbitals are involved in a benzenoid sextet). This ideal picture is not preserved in the presence of defects. Upon introduction of the D_1 and D_2 topologies, the bonding of the aGNR can be seen as the superposition of two mirroring Kekulé structures that partially destroy the benzenoid character of the aGNRs (Fig. 7.41(a) and (b)). By increasing the localization of π electrons in carbon–carbon double bonds, such defects destroy the local aromaticity at the ribbon edge and are thus expected to have a large effect on the $\pi - \pi^*$ conduction channels. On the contrary, the dihydrogenation of both defects D_1 and D_2 fully restores the benzenoid character of the ribbon, as illustrated by Clar's sextet representations (Fig. 7.41(f) and (g)).

To further substantiate the effect of these topological defects on the mesoscopic transport properties, the behavior of long disordered aGNRs is explored with random distribution of edge defects. In Fig. 7.41, the conductance of 500 nm long disordered aGNRs containing 30 defects (light grey lines) is superimposed onto the single defect results (dark grey lines). Computed conductances are averaged over 20 different disorder configurations. The original conductance fingerprints of the defects considered are further amplified when longitudinal disorder is introduced, resulting for D_1 (D_2) in an almost fully suppressed hole (electron) conduction for low defect density.

As mentioned earlier, *ab initio* transport calculations are highly computationally demanding, and the development of an accurate *tight-binding* (TB) model is therefore extremely useful for achieving a complete mesoscopic study. Consequently, most studies dedicated to edge disorder in GNRs rely on a simple topological nearest-neighbor TB Hamiltonian. However, when compared to *ab initio* results, this turns out to generally yield a wrong description of transport properties in the case of non-neutral defects.

Figure 7.42 shown the conductance profiles of three defects (D_1, D_2, and D_4) computed either within a topological TB model (black solid lines) or within an adjusted TB model (grey solid lines). Although the agreement between *ab initio* and topological TB results is good for the D_4 defect, strong discrepancies are observed for D_1 and D_2 defects. In the latter cases, the topological model leads to a severe underestimation of the backscattering efficiency. The fitting of the TB parameters from *ab initio* calculations directly by adjusting the conductance profiles for a single defect is definitely more accurate (Avriller *et al.*, 2007).

Within the fitted TB model, the transport properties of disordered aGNRs are investigated by considering lengths varying up to $L = 5 \ \mu$m and random distribution of defects with a density of 6×10^{-2} nm^{-1}. Figure 7.43 shows the conductance of long disordered aGNRs (averaged over 100 different configurations), as a function of energy and ribbon length. Also, the regions of the (L, E) plane where the conductance is experimentally insignificant (i.e. $G(E) \leq 0.01G_0$ (Han *et al.*, 2007)) are delineated by white lines, allowing determination of the length-dependent conduction gaps.

In the presence of charged defects (i.e. pentagon (D_1) and heptagon (D_2) defects), the conductance scaling behavior of long, disordered aGNRs presents a striking

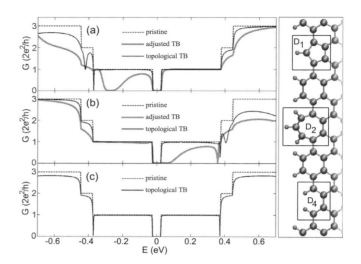

Figure 7.42 Conductance profiles of 35-aGNR for three isolated defects: (a) D_1, (b) D_2, (c) D_4. Ball-and-stick models of the defects are illustrated on the right. Black dashed lines represent the conductance of the pristine ribbon. The *ab initio* computed curves are shown in light grey solid curve. Black and dark grey lines correspond to the topological and fitted TB models, respectively. (Adapted with permission from Dubois *et al.* (2010). Copyright (2010) American Chemical Society.)

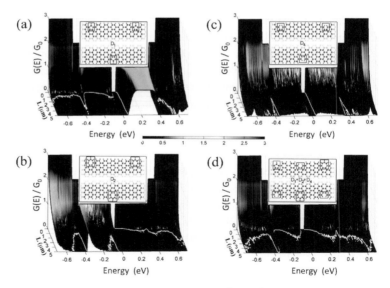

Figure 7.43 Conductance of a 35-aGNR with a 6×10^{-2} nm^{-1} density of (a) D_1 defects, (b) D_2 defects, (c) D_4 defects, and (d) mix of $\{D_1, D_2, D_4\}$ defects. The conductance is given as a function of the carrier energy and the length of the ribbon (L). $G(L, E)$ has been averaged over 100 different defect distributions. The white lines delineate the (L, E) regions for which the conductance value $G \leq 0.01 G_0$ (criterion for the experimental conduction gap). (Adapted with permission from Dubois *et al.* (2010). Copyright (2010) American Chemical Society.)

electron–hole asymmetry. Figure 7.43(a, b) clearly shows that depending on the energy at which carriers are injected, the electronic transport ranges from ballistic to localized regimes. In D_1-defected aGNRs (Fig. 7.43(a)), the propagation of electrons in the first plateau remains quasi-ballistic up to a length $L > 5\,\mu$m. As a consequence of the acceptor character of D_1 defects, the hole conduction, in the same energy window, is almost fully suppressed for length $L < 0.5\,\mu$m. In contrast, in the presence of D_2 defects (Fig. 7.43(b)), holes in the first plateau remain conductive up to length $L > 5\,\mu$m, while the donor character of the D_2 defects suppresses electron conduction in the same energy window. Obviously, such disordered edge defect profiles with a single defect type are rather unlikely, but this example shows however that spectacular fluctuations of transport length scales occur in some specific situations.

The comparison with more realistic defect distributions is rather instructive. Indeed, the conductance scaling in the presence of a random distribution of D_4 defects, and a mix of three types of defects $\{D_1, D_2, D_4\}$ is detailed for the same edge defect density in Fig. 7.43(c) and Fig. 7.43(d), respectively. For both defect distributions, the conductance decay is rather homogeneous within the first electron and hole plateaus. However, while the marked electron–hole asymmetries associated with the D_1 and D_2 defects compensate each other, the presence of odd-membered rings continues to crucially impact the electron and hole localization. This can be seen by looking at the white line in the (L, E) plane. The presence of defects noncompliant with the benzenoid character of aGNRs (i.e. D_1 and D_2) thus appears to strongly affect the ballistic propagation of carriers, even for low defect concentration (Dubois *et al.*, 2010). This has been further emphasized by the estimated localization lengths that are shown in Fig. 7.44 as a function of the carrier energy.

While the distribution of D_4 defects gives rise to a localization length $\xi \sim 1\,\mu$m for low energy carriers, the introduction of charged defects strongly reduces the average value in the $[-0.5, 0.5]$ eV energy window. Note that the fluctuations of ξ in correspondence with the van Hove singularities for D_1 defects are due to the increased scattering induced by the high DOS at these points. In the case of the mixed disorder, these fluctuations are absent or considerably reduced owing to the strong smearing effect that disorder produces in the DOS.

In summary, the electronic and quantum transport properties of edge-disordered graphene nanoribbons have been investigated using both fully *ab initio* techniques and accurately parameterized *tight-binding* models (Dubois *et al.*, 2010). Single topological defects such as pentagons and heptagons have been predicted to induce a strong electron–hole transport asymmetry. Besides, conduction gaps driven by defect-induced localization effects have been found to depend not only on the defect density and ribbon length but also on the geometry and chemical reactivity of edge imperfections. The above analysis is drawn from physical processes that are general in essence. The scope of these results should therefore extend to realistic edge profiles. In particular, similar fluctuations of the conductance have been reported for other GNR topologies whose ground-state π electronic structure is aromatic according to Clar's rule. Note that the GNR topologies that come with partially filled states at the edge are known to break the graphene aromaticity (Wassmann *et al.*, 2008). Therefore, the present conclusions

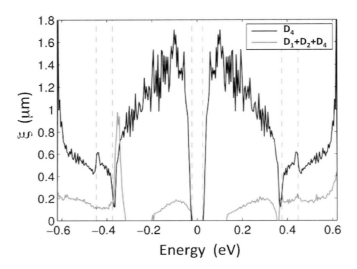

Figure 7.44 Localization length ξ for a 35-aGNR with a 6×10^{-2} nm^{-1} density of D_4 defects (dark grey solid line) and a mix of $\{D_1, D_2, D_4\}$ defects (light grey solid line). The vertical dashed lines correspond to the position of the van Hove singularities. ξ for the D_4 defects is up to almost one order of magnitude longer than in the case of mixed defects. (Adapted with permission from Dubois *et al.* (2010). Copyright (2010) American Chemical Society.)

in terms of Clar's sextet theory are unlikely to apply for such ribbon geometries. In addition, hydrogenation has been identified as a possible new route to tune the robustness of the electronic conductance against edge roughness. As reported experimentally in graphene (Bostwick *et al.*, 2009), the controlled deposition of monatomic hydrogen is a key ingredient in exploring a metal/insulator transition in these 1D materials.

7.5.11   Doped graphene nanoribbons

Finally, doping may also be used to tailor the electronic and transport properties of GNRs. In carbon-based materials, *p*-type (*n*-type) chemical doping can be achieved by boron (nitrogen) atom substitution within the carbon matrix, leading to interesting nanodevices, which are crucial for building logic functions and complex circuits (Derycke *et al.*, 2001). As previously mentioned for metallic carbon nanotubes, boron (B) and nitrogen (N) impurities yield quasibound states that strongly backscatter propagating charge for specific resonance energies (Choi *et al.*, 2000). In contrast with CNTs, doping in GNRs turns out to display even more complex features depending on the dopant position, ribbon width, and edge symmetry. Indeed, the energies of the quasibound states in GNRs are strongly dependent on the position of the impurity with respect to the ribbon edges (Biel *et al.*, 2009b). Binding energies of the bound state associated with the broad drop in conductance are found to increase as the dopant approaches one of the edges of the ribbon. The large variation of resonant energies with dopant position indicates that random distribution of impurities will lead to a rather uniform reduction of conductance over the occupied states part of the first conduction plateau (Biel *et al.*, 2009*b*). These predictions are in sharp contrast to the case of CNTs, where resonant energies do not depend on the position of the dopant around the tube circumference. In addition, doping effects are also found to depend on the ribbon symmetry and width, leading for example to a full suppression of backscattering for symmetry-preserving impurity potentials in armchair ribbons. Finally, chemical doping could be used to enlarge the band gap of a fixed GNR width, resulting in the enhancement of device performances (Biel *et al.*, 2009*a*). All these predictions calculated in chemically doped GNRs are illustrated in the following.

The impact of substitutional (boron and nitrogen) doping and edge disorder can be investigated using first-principles methods. A self-consistent calculation (Biel *et al.*, 2009*a*) provides the profile of the scattering potential around the impurity location, which generally produces quasibound states strongly localized around the defect at a resonance energy. For a boron (nitrogen) impurity, an *ab initio* study can be first performed for the infinite armchair GNR, replacing one of the carbon atoms by the boron (or nitrogen) dopant. In a second step, the onsite and hopping self-consistent Hamiltonian matrix elements (on a localized basis set) are then used to build up the *tight-binding* Hamiltonian.

The *tight-binding* model is developed by adjusting the onsite and hopping self-consistent Hamiltonian matrix elements (on a localized basis set) in order to reproduce the *ab initio* conductance fingerprints of a single impurity. This approach turns out to be much more accurate than a simple fit of band structure. Transport calculations based on the Landauer–Büttiker approach are then performed using these TB Hamiltonians and eventually taking into account a random distribution of impurities and disorder average.

Figure 7.45 shows the quantum conductance for a 35-aGNR as a function of energy for different positions of a single boron impurity along the lateral dimension of the ribbon. A large variation of resonant energies with dopant position is observed indicating an increase in binding energy of the bound state, a feature not previously observed in carbon nanotubes owing to rotation symmetry (Biel *et al.*, 2009*b*). This feature also indicates that a random distribution of impurities will lead to a rather uniform reduction of conductance over the lowest energy window of occupied states.

An astonishing feature is the observation of full suppression of backscattering even in the presence of bound states, when the impurities are located exactly at the center of the ribbon. To understand such a symmetry effect, it is worth noting that GNRs do not always present a well-defined parity associated with mirror reflections with respect to their axis. An ideal odd-index aGNR retains a single mirror symmetry plane (perpendicular to the ribbon plane containing the ribbon axis), and its eigenstates thus present well-defined parity with respect to this symmetry plane. The eigenstates of the doped ribbon keep the same parity with respect to this mirror plane, provided that the potential induced by the dopant preserves this symmetry. For the case of an odd-index aGNR, this can only occur when the dopant is located exactly at the central dimer line, as seen in Fig. 7.46. The wavefunctions Ψ_π^*, at the Γ point associated with the first band below the charge neutrality point (CNP), and Ψ_{QSB} do not mix because of opposite parity, which

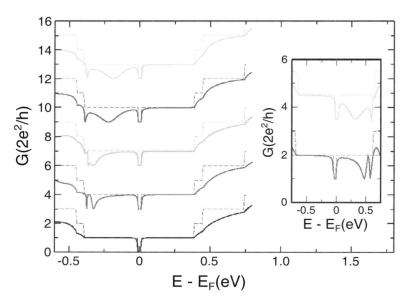

Figure 7.45 Main plot: Conductance for 35-aGNR and a single boron impurity displaced from the center to the ribbon edge (bottom to top). Conductance for the undoped case is given in dashed lines. Except for the bottom curve, all others have been upshifted for clarity. Insets: Same as in main plot for two selected nitrogen dopant positions (at the edge (top), and off-center (bottom) for the 20-aGNR). (Adapted from Biel *et al.* (2009*a*).)

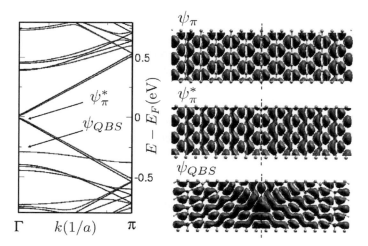

Figure 7.46 Left: Band structure of the 35-aGNR. Black lines correspond to the undoped ribbon; grey lines correspond to the case of B at the center. Right: Real space projections of several eigenstates at Γ-point (dashed line denotes the ribbon axis). Ψ_π denotes the valence band state for pristine ribbon, Ψ_π^* for same energy but with a single B-defect at center of ribbon, and Ψ_{QSB} the projection of the quasibound state. Corresponding locations of the two latter states are indicated by arrows in the left panel. (Adapted from Biel *et al.* (2009*b*).)

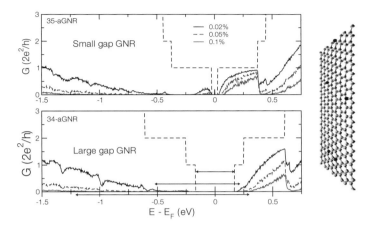

Figure 7.47 Average conductances as a function of energy for the pseudo-metallic 35-aGNR for doping rates ≈ 0.02%, 0.05%, and 0.1% (from top to bottom). The dashed black line corresponds to the ideal (undoped) case. Averages have been performed over ~500 disorder realizations with a ribbon length of ≈ 1 μm. Bottom: Same for the semiconducting 34-aGNR. Right: A randomly doped GNR.

suppress backscattering (Fig. 7.45, bottom curve). For any other position of the dopant, the well-defined parity of the wavefunctions is not preserved, and backscattering develops mainly at the position of the energy resonance of the quasibound state.

Let us examine the impact of different doping rates on ribbons of about 4 nm width, namely the pseudo-metallic 35-aGNR and the semiconducting 34-aGNR. Ribbons with a length up to 1 μm are considered, and impurities are uniformly distributed over the whole ribbon length and width, with a restriction preventing the overlap of the scattering potentials of individual dopants.

Figure 7.47 illustrates the conductance as a function of energy for the (a) 35- and (b) 34-aGNRs, for doping rates between ≈ 0.02% and 0.1%. Here, electrodes are treated as semi-infinite, perfect GNRs, and disorder is included only in the region between the electrodes (channel). As a result of the acceptor-like character of the impurity states induced by the boron dopant, the conductance is affected in a clear asymmetric fashion for energy values below or above the CNP, as evidenced by the opening of a large mobility gap that extends well beyond the first conductance plateau below the small initial electronic bandgap. The mobility gap width reaches almost 1 eV, of the order of the silicon energy gap.

The asymmetry of the electron/hole conduction is also spectacularly evidenced by scrutinizing the energy-dependent transport regimes. Figure 7.48 shows the length dependence of the conductance for several configurations of randomly doped 35-aGNRs, with a doping rate of ≈ 0.05%, at an energy = 0.25 eV below (top panel) and above (bottom panel) the CNP. For energy values lying in the conduction band (bottom panel), the conductance for various random configurations is found to slowly decay with ribbon length with values $G \simeq G_0 = 2e^2/h$, indicating the robustness of a quasi-ballistic regime.

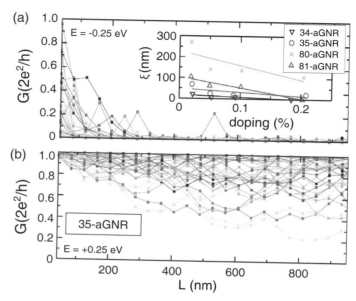

Figure 7.48 Conductances as a function of length for a set of doping realizations of the 35-aGNR at 0.25 eV below (a) and above (b) the CNP for a doping rate of $\approx 0.05\%$. Inset: Localization lengths as a function of doping rate for the 34- and 35-GNRs (at $E = -0.2$ eV), and for 80- and 81-aGNRs (at $E = -0.1$ eV). Solid lines are fits related to calculated values. (Adapted with permission from Biel *et al.* (2009a). Copyright (2009) American Chemical Society.)

In contrast, for energies lying in the valence band (top panel), a strong exponential decrease of the conductance is observed (inset), with large fluctuations associated with different defect positions (main panel). The exponential decay of the conductance with the length of the ribbon is related to the Anderson localization, which has already been observed experimentally at room temperature in defected metallic carbon nanotubes (Gómez-Navarro *et al.*, 2005), due to their long phase coherence lengths.

Figure 7.48 (inset) shows localization lengths (ξ) extracted from $\langle \ln G/G_0 \rangle = -\xi/L$ (with L the ribbon length) for hole transport at different doping rates. A statistical average over about 500 disorder samples is performed. The value of ξ ranges within 10–300 nm, depending on the ribbon width and doping density, and scales as $1/n_i$. The value of ξ is also observed to further increase with the ribbon width but in a nonlinear fashion, owing to the nonuniformity of the disorder potential with the dopant position.

Since the width of these GNRs is reaching the capability limits of standard lithographic techniques (as for instance produced by the IBM group (Chen *et al.*, 2007)), it is interesting to study the case of aGNRs with widths of ≥ 10 nm, as for instance shown in Fig. 7.49 for an 80- and an 81-aGNR. This figure presents the conductance of a 10 nm wide armchair nanoribbon with low boron doping. For a doping density of $\approx 0.2\%$, the ribbon manifests a mobility gap of about ~ 1 eV. When lowering the doping level to 0.05%, the mobility gap reduces to about 0.5 eV and finally becomes less than 0.1 eV for lower density. The 0.2% case is obtained for a fixed nanoribbon width and length, so that optimization would be required upon upscaling either lateral or longitudinal sizes.

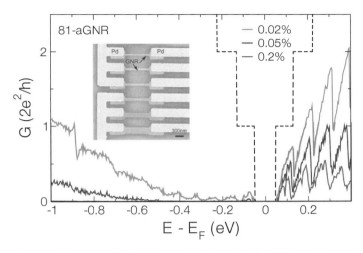

Figure 7.49 Main panel: Same as Fig. 7.48 for semiconducting 81-aGNR and three selected doping rates ($\approx$ 0.02%, 0.05% and 0.2%, from top to bottom). (Adapted with permission from Biel *et al.* (2009*a*). Copyright (2009) American Chemical Society.) Inset: SEM image of various patterned graphene nanoribbons fabricated by lithography. (Reprinted from Chen *et al.* (2007). Copyright (2007) with permission from Elsevier.)

7.5.12   GNR-based networks

The controlled formation of narrow graphene strips with well-defined edges has been an intense topic of research. Significant progress in this direction has been achieved recently, as described in recent reviews (Terrones *et al.*, 2010, Jia *et al.*, 2011). Well-defined crystallographic edges displaying zigzag or armchair morphologies are byproducts of CVD-grown graphene inside a TEM after processing using Joule heating (Jia *et al.*, 2009), by scanning tunneling lithography (Tapaszto *et al.*, 2008), or through catalytic cutting (Ci *et al.*, 2008, Datta *et al.*, 2008). Large scale synthesis of GNRs from the unzipping of carbon nanotubes along their axis has also been demonstrated (Terrones, 2009, Terrones *et al.*, 2010). In particular, plasma etching has been used to produce very narrow nanoribbons from single-wall carbon nanotubes (Ouyang, Dai & Guo, 2010, Jiao *et al.*, 2010, Wang & Dai, 2010). The use of CNTs as starting material to produce GNRs presents the advantage that all the already available technology of CNT production can be exploited. In fact, arrays and networks of CNTs can be used to produce well-aligned arrays and networks of GNRs (Jiao *et al.*, 2010). Even though the edges of such nanoribbons are not atomically smooth, devices constructed with arrays of such nanoribbons display relatively high I_{ON} to I_{OFF} ratio. The nanoribbon cross-points have been suggested for use in applications in logic electronic devices. Based on the bottom-up fabrication of GNRs from the deahalogenation of self-assembled polyphenilenes, a controlled synthesis of a GNRs-based nanonetwork has been confirmed (Cai *et al.*, 2010).

Focusing on systems that can be experimentally realized with existing techniques, both in-plane conductance in interconnected graphene nanoribbons and tunneling conductance in out-of-plane nanoribbon intersections have been studied (Botello-Méndez *et al.*, 2011*a*). Both *ab initio* and semi-empirical simulations confirm the possibility of designing graphene nanoribbon-based networks capable of guiding electrons along desired and predetermined paths. In addition, some of these intersections exhibit different transmission probability for spin-up and spin-down electrons, suggesting the possible applications of such networks as spin filters (Botello-Méndez *et al.*, 2011*a*). Furthermore, the electron transport properties of out-of-plane nanoribbon cross-points of realistic sizes are described using a combination of first-principles and tight-binding approaches. The stacking angle between individual sheets is found to play a central role in dictating the electronic transmission probability within the networks.

As described in detail in the previous sections, the electronic transport properties of isolated aGNRs are sensitive to the shape of the edges and the ribbon width (Yang *et al.*, 2007). These GNRs exhibit a band gap E_g that decreases with the width. Consequently, a region of zero conductance is observed in aGNRs due to the absence of conduction channels in the energy window around E_g. Conversely, the electronic band structure of zGNRs exhibits localized edge states close to the Fermi energy (E_F). The states extend along the edges and correspond to nonzero density of states at E_F and a high electronic conductance at the charge neutrality point. Spin polarized calculations reveal that such edge states carry a finite magnetic moment with a ferromagnetic ordering along the edges but an antiferromagnetic ordering between them (Lee *et al.*, 2005). The interaction between the edges leads to a magnetic insulating ground state. The edge interaction and the band gap decay with the GNR's width. As the width of the zGNR increases, such interaction weakens, and the difference in energy between the parallel and antiparallel spin ordering along the edges disappears. Therefore, the spin polarized electron conductance of a narrow zGNR exhibits vanishing conduction around the charge neutrality point, and two sharp peaks of conductance due to the presence of edge states.

The conductance across continuous or in-plane networks of GNRs has already been investigated using a simple single-orbital nearest-neighbor *tight-binding* method (Jayasekera & Mintmire, 2007, Areshkin & White, 2007). The properties of in-plane cross-points are found to be very sensitive to the geometry of the junction. Notably, under such an approximation, which does not consider spin degrees of freedom, the conductance along a zGNR and across a cross-point exhibits high conductance ($\sim 0.8G_0$) (Jayasekera & Mintmire, 2007). In contrast, *ab initio* simulations (Botello-Méndez *et al.*, 2011*a*) indicate that the electronic transport of GNRs is significantly affected when they are assembled into networks or branches. A notable exception is the cross-point between two zGNR terminals, forming an angle of 60° in which the scattering is minimal, as illustrated in Fig. 7.50 along the (1–4) or (2–3) path, respectively.

First-principles spin-polarized calculations reveal that these 60° intersections have interesting spin transmission behavior. In order to comply with the periodic boundary conditions, the most stable configuration (illustrated in Fig. 7.50(b)) exhibits different spin alignment at the edges of the two 60° turns at the intersection. The existence of the antiferromagnetic state combined with the presence of the junction effectively breaks

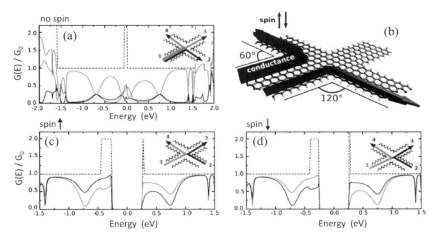

Figure 7.50 (a) *Ab initio* quantum conductance of an in-plane cross-point between two 5-zGNRs for various transmission paths (paramagnetic state – non-spin polarized calculation). The dark grey and light grey arrows in the inset represent the spin configuration corresponding to the curves shown in (c, d). (b) Ball-and-stick representation of the zGNR intersection. Spin-dependent conductances are found to be higher along specific paths of the cross-point. (c, d) Spin-polarized *ab initio* quantum conductance of a zGNR as represented in (b). Note the different spin-up and spin-down transmission probabilities along the same path (e.g. 1–4). While the total transmission (spin up + spin down) is symmetric along various paths, the spin state effectively breaks the structure's symmetry (i.e. vertical C_2 axis), yielding a spin-dependent conductance. The dashed grey line represents the conductance of an isolated 5-zGNR in the corresponding spin state. (Adapted with permission from Botello-Méndez *et al.* (2011a). Copyright (2009) American Chemical Society.)

the left–right symmetry (Fig. 7.50(b)). As a consequence, the transmission probability is different for the two spin channels. For instance, the low-energy spin-up electron conductance along the 1–4 path is significantly higher than the spin-down conductance along this path (compare the light grey lines in Fig. 7.50(c,d)). Opposite behavior is observed for the 2–3 paths. However, note that the symmetric pathways in the total conductance are retrieved (the sum of spin-up and spin-down contribution).

An alternative arrangement for GNR networks is through out-of-plane or bilayer cross-points. In these cases, the interaction between the two GNRs is weak, and the changes in conductance across the GNRs and the conductance between them are driven by tunneling across the GNR cross-point (Botello-Méndez *et al.*, 2011a). As illustrated in Fig. 7.51, this hypothesis is confirmed by the *ab initio* quantum conductance along and across two zGNRs. At the cross-point between two 6-zGNRs, the conductance along the GNR is mainly preserved. Indeed, the loss of transmitted electrons through tunneling at a cross-point is almost nonexistent, regardless of their spin polarizations (Fig. 7.51(b)). However, there is only very little tunneling for the single electron channel across the two 6-zGNRs. Only the electrons localized at the edges tunnel across the GNRs (Fig. 7.51(c)).

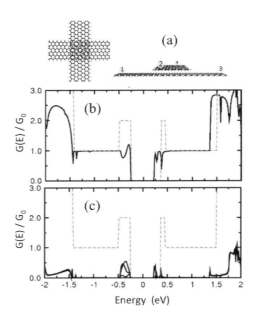

Figure 7.51 (a) Ball-and-stick models of an out-of-plane conducting zGNR network composed of two 6-zGNRs (4 terminals are present). *Ab initio* quantum conductance along (b) and across (c) out-of-plane zGNR networks depicted in (a). The dashed lines represent the conductance of isolated GNRs. Note that the small tunneling current (c) between the zGNRs depends on the type of intersection and the localization of the electrons at a particular energy (e.g. edge states). (Adapted with permission from Botello-Méndez *et al.* (2011*a*). Copyright (2009) American Chemical Society.)

First-principles calculations are frequently limited by computational resources to investigate routinely the transport properties of realistic nanoribbon networks, which include a much larger number of atoms than the systems studied above. For this reason, a single-band *tight-binding* model based on a Slonczewski–Weiss–McClure (SWMC)-like parameterization (Slonczewski & Weiss, 1958) has been proposed to investigate larger systems (Botello-Méndez *et al.*, 2011*a*). In order to describe accurately the properties of graphene and GNRs, interactions up to the third-nearest neighbor (White *et al.*, 2007) have to be considered by setting a cutoff interaction of 3.7 Å, and an exponential decay of the hopping parameter of the form $e^{-\eta(d-d_0)}$, where d is the separation between two carbon atoms, and d_0 is the C–C equilibrium distance (1.42 Å for in-plane interactions, and 3.35 Å for out-of plane interactions). Different hopping parameters (γ_0) have been used for the A and B sites of the carbon hexagonal lattice in graphene. In order to properly describe the band gap of GNRs, corrections to the onsite (ε) and hopping parameters of edge atoms (γ_{edge}) and their interaction with other atoms have to be included (Gunlycke & White, 2008). One TB parameter has also been used for the out-of-plane interactions (γ_1). Starting from such a model, the numerical values of the onsite and hopping parameters can be fine-tuned using an evolutionary algorithm: the fitness function being defined as a weighted error of the calculated band structure with

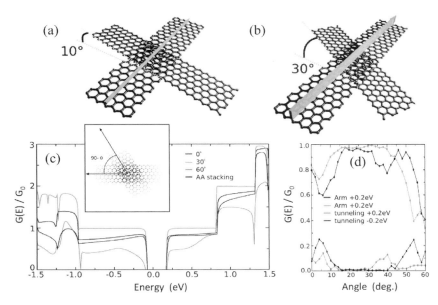

Figure 7.52 Quantum conductance through an aGNR/zGNR out-of-plane cross-point as a function of the stacking angle θ. Atomic models depicting an 11-aGNR and a 6-zGNR intersecting at (a) 10° and (b) 30°. (c) Conductances along the aGNR at the cross-point for various intersection angles compared to AA stacking when $\theta = 90°$ (see inset). (d) Quantum conduction as a function of the stacking angle θ for an aGNR/zGNR cross-point with a ~5 nm width. (Adapted with permission from Botello-Méndez *et al.* (2011*a*). Copyright (2009) American Chemical Society.)

respect to the first-principles DFT-PBE band structure calculations. The weight has to be chosen in order to minimize the error at low energies. The *ab initio* band structures of graphene, GNRs, bilayer graphene, and GNR networks have been used as references, confirming a satisfactory agreement with the TB model, in the energy region close to the charge neutrality point (Botello-Méndez *et al.*, 2011*a*).

In addition, DFT calculations within the LDA or GGA approximation are known to underestimate the weak attraction between graphene layers (Dion *et al.*, 2004). However, the exponential decay of this TB approach can successfully model the difference of interaction between the AA and AB stacking within GNR cross-points, fixing a separation distance between the layers to 3.35 Å (which corresponds to the experimental values of the AB stacking in graphite). Starting from a cross-point, an armchair nanoribbon can be rotated by an angle θ, as shown in Fig. 7.52. The conductance along and across the nanoribbons is found to strongly depend on this specific stacking angle θ.

Figure 7.52(d) presents the conductance at ± 0.5 eV as a function of the stacking angle θ both along the 11-aGNR, and across this aGNR through a 6-zGNR each with a width of ~5 nm. A maximum of conductance along the aGNR is observed for θ values around 30° with almost a ten-fold increase compared to its value at $\theta = 0°$ for -0.5 eV. Such an increase is due to the fact that the interaction between the GNRs is minimal at $\theta = 30°$. Conversely, at $\theta \gg 30°$, the overlap and interaction between the GNRs increase, thus enhancing the tunneling probability and increasing the scattering in the isolated GNR. A similar behavior is observed for different energies for selected stacking angles (Fig. 7.52(c)). A direct comparison between the *AA* and *AB* stacking is also presented in Fig. 7.52(c), for an 11-aGNR/6-zGNR intersection, suggesting that the effect of different stacking orders at $\theta = 0°$ has less impact on the conductance than the stacking angle.

In summary, first-principles and *tight-binding* calculations performed on a number of GNR cross-points (Botello-Méndez *et al.*, 2011*a*) and their quantum transport properties indicate that GNR networks are appealing for potential applications and could play an important role in the development of carbon-based electronics. The quantum transport through in-plane GNR cross-points has been found to be severely scattered at the intersections, except for 60° zGNR terminals, confirming that patterned graphene and GNRs could be used for functional devices and current flux guides (Romo-Herrera *et al.*, 2008). In addition, the transmission probability at these intersections is different for spin-up and spin-down electrons, suggesting the possibility of their use as spin filters. Furthermore, the tunneling transmission at the intersection of bilayer GNR networks is calculated to be very sensitive to the stacking angle between the ribbons. The edge state channels are remarkably robust and could be tuned with an external electric field in order to induce tunneling from an aGNR, thus allowing future development in band-to-band tunneling GNR-based transistors (Schwierz, 2010).

7.6   Conclusion

In conclusion, the present chapter has overviewed some basics of electronic and quantum transport properties in low-dimensional carbon-based materials including 2D graphene, graphene nanoribbons, and carbon nanotubes. Although CNTs and GNRs share similar electronic confinement properties due to their nanoscale lateral sizes, the effects of boundary conditions in the perpendicular direction with respect to the system axis trigger very different electronic and transport properties. For each nanostructure, a simple *tight-binding* approach (single-band model) has been proposed to describe their specific electronic behavior, and, when necessary, *ab initio* calculations have been used to accurately complete the picture. Both 1D systems have also been perturbed using topological defects (vacancies, adatoms), multi-structures and stacking, chemical doping and functionalization to tailor their electronic structure, etc. The effects of these topological, chemical, structural perturbations on the quantum transport of both CNTs and GNRs have been predicted.

To conclude, because of their remarkable electronic properties and structural physical properties, CNTs or GNRs are expected to play an important role in the future of nanoscale electronics. Not only can nanotubes be metallic, but they are mechanically very stable and strong, and their carrier mobility is equivalent to that of good metals, suggesting that they would make ideal interconnects in nanosized devices. Further, the intrinsic semiconducting character of other tubes, as controlled by their topology, allows us to build logic devices at the nanometer scale, as already demonstrated in many laboratories. Similarly the combination of 2D graphene for interconnects together with graphene nanoribbons for active field effect transistor devices could allow completely carbon-made nanoelectronics.

The complete understanding of fundamental electronic and transport concepts in low-dimensional carbon-based nanomaterials definitely needs theoretical modeling and advanced quantum simulation, together with joint studies with experiments. Theory has been very important to initiate, validate, and orientate carbon nanotube science, particularly as far as electronic properties are concerned. Yet, in 1992, one year after the discovery of nanotubes by S. Iijima, several groups theoretically predicted their unique behavior as metals or semiconductors. Similarly, the electronic properties of 2D graphene and graphene ribbons were explored decades before the fabrication of those nanostructures. Since carbon-based nanomaterials have still probably not revealed all their secrets, numerical simulations still have successful days to come in predicting new interesting atomic topologies and their corresponding structural and electronic properties, thus expanding their potential impact in carbon science.

8   Applications

This section presents a brief overview of the most promising graphene applications in information and communication technologies, reflecting current activities of the scientific community and the authors' own views.

8.1   Introduction

The industrial impact of carbon nanotubes is still under debate. Carbon nanotubes exist in two complementary flavors, i.e metallic conductors and semiconductors with tunable band gap (scaled with tube diameter), both exhibiting ballistic transport. This appears ideal at first sight for creating electronic circuits, in which semiconducting nanotubes (with diameter around 1–2 nm) could be used as field effect transistors, whereas metallic single-wall tubes (or large-diameter multiwalled nanotubes), with thermal conductivity similar to diamond and superior current-carrying capacity to copper and gold, would offer ideal interconnects between active devices in microchip (Avouris, Chen & Perebeinos, 2007). Nanotube-based interconnects have been physically studied over almost a decade, with companies such as Samsung, Fujitsu, STMicroelectronics, or Intel acting significantly or encouraging academic research (Coiffic et al., 2007). The current-carrying capability of bundles of multiwalled nanotubes has been practically demonstrated to fulfill the requirements for technology and thus could replace metals (Esconjauregui et al., 2010), although a disruptive technology step remains to be achieved to integrate chemical vapor deposition (CVD) growth at the wafer-scale, a step of no defined timeline.

Concerning active devices (able to switch ON and OFF the current flow), the main bottleneck to fabricating nanotube-based field effect transistors (CNT-FET) is the mixing between metallic and semiconducting tubes generally obtained after growth. Even though chemical sorting has been demonstrated (Liu & Hersam, 2010), this technique remains rather unrealistic for widespread use in mainstream nanoelectronics of nanotubes with well-defined current–voltage characteristics. Besides, the CVD growth of billions of semiconducting nanotubes in a geometrically controlled manner, in spite of much effort, remains *pure science fiction*. Longstanding efforts at IBM Yorktown in the group of Phaedon Avouris (Avouris et al., 2007) have definitely provided a deep understanding of how a nanotube transistor works, but unfortunately use of the fantastic characteristics of ballistic CNT-FET – *with down to zero Schottky-barrier height for some suitable choice of metallic contact (palladium) and tube diameter (in the order of 2 nm)* – seems neither for today nor tomorrow. In early 2000, Samsung reported on some carbon nanotube-based field emission displays (based on the outstanding electron emission capability of nanotubes), but the technology was not mature enough for market delivery, and (as far as we know) has been abandoned.

Other efforts have sought to create new hybrid material systems with superior properties, such as reinforcing fillers in elastomeric matrices. Modified CNTs can enhance the adhesion between CNTs and a polymer matrix. The composite retains the performance advantages from both types of materials, such as mechanical properties and electric properties, etc. The potential applications for such composites are very versatile, for example, circuit boards with conductive coating, conductive coating on color, conductive adhesive, nanocarbon conductive coating on electrical grounding network, nanocapacitor, photoelectric conversion device, conductive membrane, antistatic coating, antistatic fiber product, and conductive coating plated on nonmetallic material, etc. Many efforts have been made, and some products currently containing a fraction of nanotubes for enhancing material performance (especially painted, reinforced, and light composites) can now be found in the market place.

Differently, graphene appears as a much more promising enabling platform for a plethora of applications, in a way that is similar and complementary to present Si technology and plastics. Indeed, in contrast to carbon nanotubes, the advantages of graphene for advancing nanoelectronics lie in its planar structure and demonstrated wafer-scale growth production (either for epitaxial graphene on SiC (Nyakiti *et al.*, 2012) or CVD-graphene (Hwang *et al.*, 2012), which is compatible with state-of-the-art CMOS integration platforms. Combined with their outstanding electronic, thermal and mechanical properties, graphene and related layered materials such as MoS_2 (Wang *et al.*, 2012), or even the recently discovered two-dimensional silicon layer known as silicene (Vogt *et al.*, 2012) bring realistic promise of creating a new, more powerful and versatile, as well as sustainable and economically viable, technology platform for innovation in information and communication technologies (ICT).

Below, we present a very brief overview of some of the most promising foreseen applications of graphene, including more medium-term possibilities such as flexible or high-frequency electronics, or longer term (but potentially revolutionary) technologies such as graphene resonators, plasmonics, and spintronics.

8.2   Flexible electronics

Flexible electronics is the next ubiquitous platform for the electronics industry (Nathan *et al.*, 2012). The realization of electronics with performance equal to that of established technologies based on rigid platforms, but in lightweight, foldable, and flexible formats, would enable many new applications. Flexible electronics is making truly conformal, reliable or even transparent electronic applications possible in the foreseeable future.

It is also essential for rigid ultra-compact devices with tight assembly of components. Flexible electronics could also bring reduced cost and large electronic system integration by using novel mass manufacturing approaches such as printed electronics, roll-to-roll manufacturing or lamination, hitherto unavailable from more traditional brittle material and device platforms.

The transparent nature of graphene is suitably complemented by high thermal and electrical conductivities, which are higher than those of any metal. At the same time, graphene is an elastic thin film, and behaves as an impermeable membrane, while being chemically inert and stable. Thus it seems ideal for the production of next generation transparent conductors, substituting indium-tin oxide (ITO) in the manufacturing of various types of displays and touch screens. Graphene inks are already proven to significantly exceed the performance (e.g. in terms of mobility) of commercial alternatives (Mukhopadhyay & Gupta, 2011). Therefore, as a thin flexible ultra-strong film and an extremely good conductor, graphene is the most natural choice also for flexible and printed (and possibly transparent) electronic systems (Torrisi *et al.*, 2012). Thanks to its intrinsic flexibility, graphene could also be ideal for sensors and devices that would adhere to and interact with the human body, enabling new consumer and medical applications not allowed by present technology.

8.3   High-frequency electronics

The potential of graphene transistors (Avouris, 2010) for high-frequency electronics was recently demonstrated by several groups using exfoliated graphene (Meric *et al.*, 2008, Liao *et al.*, 2010), SiC-based graphene (Moon *et al.*, 2009, Lin *et al.*, 2010), and CVD-based graphene (Wu *et al.*, 2011). The most recent studies show current gain cutoff frequencies (f_T) in the 100–300 GHz range, with room for improvement at both the material and device levels. Yet, field effect transistors (FETs) based on III-V semiconductors have already reached f_T above 640 GHz (Kim & Park, 2010), while for silicon-based MOSFETs, the highest frequency to date is 485 GHz (Lee *et al.*, 2007). It is, however, not yet fully established whether or not graphene will effectively predominate over conventional crystalline semiconductors when approaching the terahertz range, a spectral window which will offer a huge amount of applications (radars, communications, medical devices, etc.). In parallel, graphene is being explored for large scale electronics on flexible substrates via CVD growth of graphene on metal foils associated with transfer methods (Bae *et al.*, 2010). However, the combination of these two properties, namely high speed and flexibility, remains an open challenge. In particular for the viable development of fast and flexible electronic applications in the areas of portable/wearable communicating devices with low power consumption, this combination should be achieved with a source of material adapted to low-cost manufacturing methods such as inkjet printing.

The high potential of graphene for high frequency electronics on rigid substrates was recently demonstrated using solution-based graphene transistors at gigahertz frequencies, ideally combining the required properties to achieve high speed flexible electronics on plastic substrates (Sire *et al.*, 2012). The fabricated graphene flexible transistors have current gain cutoff frequencies of 2.2 GHz and power gain cutoff frequencies of 550 MHz (see Fig. 8.1). Radio frequency measurements directly performed on bent samples also show the remarkable mechanical stability of these devices and demonstrate the advantages of solution-based graphene field effect transistors over other types of flexible transistors based on organic materials.

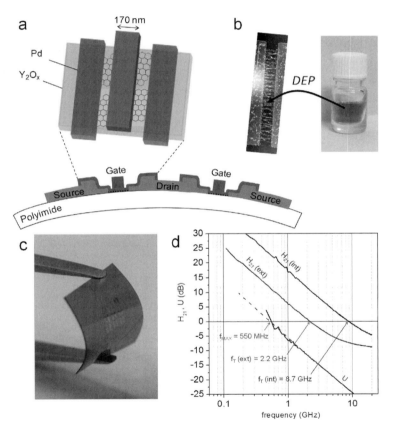

Figure 8.1 (a) Schematic cross-sectional and top views of a top-gate graphene transistor in a ground-signal-ground configuration. (b) AFM image of the graphene flakes deposited from solution by dielectrophoresis (DEP), acquired before the deposition of the source, drain, and top-gate electrodes. (c) Graphene devices fabricated on a polyimide substrate. (d) High frequency performance: evolution of the current gain (H_{21}) and power gain (U) as a function of frequency measured at $V_{GS} = -0.6$ V and $V_{DS} = -0.65$ V ($I_D = -16.8$ mA). $H_{21(ext)}$ is the as-measured (extrinsic) value, $H_{21(int)}$ is the post de-embedding (intrinsic) value. The cutoff frequencies are $f_{T(ext)} = 2.2$ GHz, $f_{T(int)} = 8.7$ GHz and $f_{MAX} = 550$ MHz. The dashed line corresponds to a slope of -20 dB/dec. (Adapted with permission from Sire *et al.* (2012). Copyright (2012) American Chemical Society. Courtesy of V. Derycke.)

8.4   Optoelectronics–photonics–plasmonics

8.4.1   Opacity of graphene and fine structure constant

Graphene's optical properties are particularly striking, with an unexpectedly high opacity (of 2.3%) for an atomic monolayer, uniquely defined by the fine structure constant, $\alpha = e^2/\hbar c \sim 1/137$, the fundamental constant describing in quantum electrodynamics the coupling between light and relativistic electrons (Nair *et al.*, 2008). Figure 8.2(c) shows a photograph of a 50 μm aperture which is partially covered by graphene and its bilayer (the line scan profile gives the intensity of transmitted white light along the dashed line), where the contrast between the air and the transmission through the monolayer shows that only 2.3% of incident light is absorbed (and twice as much for the bilayer).

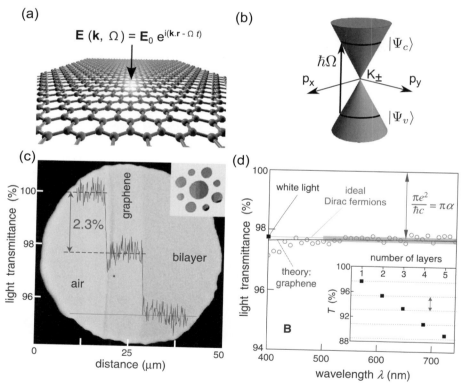

Figure 8.2 (a) Incident light on graphene. (b) Excitation processes responsible for light absorption. Electrons from the valence band are excited into empty states in the conduction band, conserving their momentum and gaining $\hbar\Omega$. (c) A 50 μm aperture partially covered by graphene and its bilayer. (Inset) Sample design: A 20 μm-thickness metal support structure has several apertures of 20, 30, and 50 μm diameter with graphene crystallites placed over them. (d) Transmittance spectrum of single-layer graphene (open circles). (Inset) Transmittance of white light as a function of the number of graphene layers (squares). The dashed lines correspond to an intensity reduction by $\pi\alpha$ with each added layer. ((b,c) from Nair *et al.* (2008). Reprinted with permission from AAAS.)

Such a universal value of graphene's opacity can be easily derived from the calculation of the total light absorption. Assuming an electromagnetic wave (frequency $\Omega/2\pi$) with a perpendicular polarization of the electric field ($\mathbf{E}(\mathbf{k}, \Omega) = \mathbf{E}_0 e^{i(\mathbf{k}.\mathbf{r} - \Omega t)}$) with respect to the graphene plane (Fig. 8.2(a)), the interaction between light and Dirac fermions can be simply captured through

$$\mathcal{H} = v_F \sigma . \left(\mathbf{p} - \frac{e}{c}\mathbf{A}\right),$$

where the second term ($(ev_F/c)\sigma.\mathbf{A} = (ev_F/i\Omega)\mathbf{E}_0$) can be treated as a perturbation inducing electronic transitions (Fig. 8.2(b)). Then, the incident energy flux W_i is given by $(c/4\pi)|\mathbf{E}_0|^2$, whereas the energy absorbed by the graphene per unit time can be computed using the Fermi golden rule as

$$W_a = \frac{2\pi}{\hbar} |\langle \Psi_c|(ev_F/i\Omega)\sigma . \mathbf{E}_0|\Psi_v\rangle|^2 \times \rho(\hbar/2) \times \hbar\Omega, \qquad (8.1)$$

where one must take into account momentum-conserving excitation processes from an initial $|\Psi_v\rangle$ state to final $|\Psi_c\rangle$ states (Fig. 8.2(b)) with energy gain $E = \hbar\Omega$. The quantity $\rho(E)$ is the the density of states given by $\rho(E/2 = \hbar/2) = \hbar\Omega/\pi\hbar^2 v_F^2$. Averaging over all initial and final states and taking into account the valley degeneracy, one gets $W_a = (e^2/4\hbar)|\mathbf{E}_0|^2$, so that the total absorption is given by

$$\mathcal{P} = \frac{W_a}{W_i} = \frac{\pi e^2}{\hbar c} = \pi\alpha,$$

which is independent of the material parameter since v_F cancels out. One notes that the dynamic conductivity (which can be derived using the Kubo–Greenwood approach) is found to be $W_a/|\mathbf{E}_0|^2 = e^2/4\hbar$. Therefore, graphene practically does not reflect light since $(1 - T) = \pi\alpha \sim 0.023$, or equivalently the light transmittance is about 97.7% (as seen in Fig. 8.2(d)).

8.4.2   Harnessing graphene with light: a new dimension of possibilities

The generation of light from electricity (e.g. lasers or light-emitting diodes) or the conversion of photons into electrical signals (e.g. photodetectors and solar cells) is the basis of photonics and optoelectronics. Optoelectronics comprises every process based on the simultaneous manipulation of light and charge. Active optoelectronic devices such as switches and modulators, for example, are of great importance in different technological areas, ranging from ubiquitous displays, biomedical imaging, remote sensing, optical communications and security, to high-tech frequency modulators. Plasmonics is another field that is developing at an increasing pace, paving the way for applications such as nanoscale waveguides, sensing and optical routing. In such a complex and extensive field, a number of challenges must be overcome to further expand the performance of existing devices, or even create entirely new technologies. Graphene offers an all-in-one solution to the challenges of future optoelectronic technologies (Bonaccorso et al., 2010), exploiting its tunable optical properties, broadband absorption (from UV to THz frequencies), and high electrical mobility for ultrafast operation

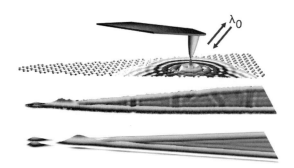

Figure 8.3 Schematic imaging of propagating and localized graphene plasmons by scattering-type SNOM. Top: The experimental configuration used to launch and detect propagating surface waves in graphene (represented as rings). The metalized AFM tip is illuminated by an infrared laser beam. Middle: Near-field amplitude image acquired for a tapered graphene ribbon on top of 6H-SiC. The imaging wavelength is 9.7 μm. Bottom: Image of the calculated local density of optical states (LDOS) at a distance of 60 nm from the graphene surface. (Reprinted by permission from Macmillan Publishers Ltd: *Nature*, Chen *et al.* (2012), copyright (2012). Courtesy of Frank Koppens.)

(Mueller, Xia & Avouris, 2010). In addition, active graphene optoelectronic devices can be integrated monolithically, or as hybrids with silicon platforms, or flexible and stretchable substrates.

Spatial images of graphene surface plasmons have been reported recently (Chen *et al.*, 2012). Such plasmons are surface waves coupled to the charge carrier excitations of the conducting sheet. Due to the unique characteristics of graphene, light can be squeezed into extremely small volumes and thus facilitate strongly enhanced light–matter interactions. In addition, the plasmon wavelength can be tuned and plasmon propagation can be monitored, confined (see Fig. 8.3), and even be switched ON/OFF *in situ*, simply by tuning the carrier density by electrostatic gates. These prospects pave the way towards ultrafast modulation of nanoscale optical fields, resonantly confined in graphene nanostructures or propagating along graphene ribbons (Koppens, Chang & Garcia de Abajo, 2011). In addition, the capability of trapping light in very small volumes could give rise to a new generation of nanosensors with applications in diverse areas such as medicine and bio-molecules, solar cells, and light detectors, as well as quantum information processing.

8.5   Digital logic gates

To develop logic gates based on graphene, graphene-based transistors should be fabricated with high ON/OFF current ratio and low OFF-state current. However, to date, no satisfactory solution has been found to opening a band gap without large reduction of charge mobility. Band-gap engineering of graphene is being explored by patterning graphene nanoribbons (GNRs), but state-of-the-art GNRs exhibit low mobilities (~ 200 cm^2 V^{-1}s^{-1}), as a consequence of carrier scattering mainly occurring at ribbon edges (as discussed in Section 2.8.3). To eliminate unwanted scattering, GNRs should have crystallographically smooth edges (Datta *et al.*, 2008, Ci *et al.*, 2009, Cai *et al.*, 2010) and be deposited on insulating substrates.

This leads to an enormous fabrication challenge as GNR widths ~ 1 nm are required to reach the Si band gap ($\simeq 1$ eV), mandatory for technology-relevant current switching performance. Finally, complementary logic is currently realized through electrostatic doping, which imposes limits on supply voltages in logic gates. In order to lift this restriction, GNRs should be chemically doped (Wang *et al.*, 2009, Farmer *et al.*, 2008, Lohmann, von Klitzing & Smet, 2009), but this doping should not introduce additional scattering centers for maintaining high mobility of crystallographically smooth GNRs, although alternatives have been proposed (see Section 7.5.11).

8.6 Digital nonvolatile graphene memories

Nonvolatile memories are the most complex and advanced semiconductor devices, following Moore's law down to 20 nm feature size. State-of-the-art nonvolatile memories consist of floating-gate flash cells, in which the information is stored by charging/discharging an additional floating gate embedded between the standard control gate and semiconductor channel of a MOSFET. Aggressive scaling of CMOS technology has a negative impact on the reliability of nonvolatile memories. Parasitic capacitances between the adjacent cells increase with scaling, leading to a cross-talk (Park *et al.*, 2008). Diminished lateral area leads to reduced gate coupling and to higher voltages (Shin, 2005), increasing the number of array cells leads to a reduced sensing current and increased access times (Nobunaga *et al.*, 2008). For these reasons, alternative materials and storage concepts have been actively investigated, including implementation of graphene in nonvolatile memories (Hong *et al.*, 2010, Hong *et al.*, 2011, Park *et al.*, 2011, Zhan *et al.*, 2011, Wu *et al.*, 2012) (see Fig. 8.4).

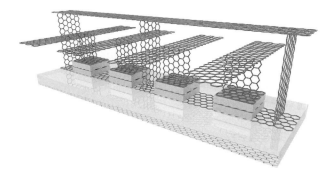

Figure 8.4 Two cells graphene NOR gate flash memory. Graphene is used for conductive FET channels (Stützel *et al.*, 2010, Zhan *et al.*, 2011) and bit lines, control gates (Park *et al.*, 2011) and word lines, and floating gates (Hong *et al.*, 2011). Courtesy of Roman Sordan.

The use of graphene in nonvolatile memories is facing less challenge than in logic gates, because memory operation requires only a large ON/OFF ratio (assuming that the ON current is not too low). In this case the voltage gain of memory GFETs is irrelevant, as reliability of a memory state readout depends only on the sensitivity of the sense amplifiers connected to the bit lines. Graphene could be used in nonvolatile memories as channel (Stützel *et al.*, 2010, Zhan *et al.*, 2011), resistive switch (Wu *et al.*, 2012), and storage layer, i.e. replacement of floating (Hong *et al.*, 2011) or control gates (Park *et al.*, 2011).

8.7   Graphene nanoresonators

Nano-electromechanical (NEM) mass sensing is another viable alternative for highly sensitive devices. NEM resonators can be described as harmonic oscillators. Mass sensing consists of monitoring the shift of the mechanical resonance frequency induced by the adsorption onto the resonator of the particles. The mass of a graphene sheet is ultra-low, so even a tiny amount of deposited atoms makes up a significant fraction of the total mass. Changes in mass due to adsorbed molecules can be sensed by their effect on the resonant frequency of a membrane or cantilever (Bahreyni, 2008).

Mass sensors can be fabricated from graphene membranes and cantilevers (Kim & Park, 2010), with principal frequencies sensitive to changes of mass which was initially reported to be of the order of 10^{-16} fg (Sakhaee-Pour, Ahmadian & Vafai, 2008). Circular drum structures of few-layers graphene were also found to have linear spring constants ranging from 3.24 to 37.4 N m^{-1} and could be actuated to about $18-34\%$ of their thickness before exhibiting nonlinear deflection (Wong *et al.*, 2010).

Measurements using CNT resonators (Lassagne *et al.*, 2008, Chiu *et al.*, 2008, Jensen, Kim & Zettl, 2008, Chen *et al.*, 2009) have been further successfully achieved, with $0.2 - 1$ zg sensitivity (1 zg $= 10^{-21}$ g), and in an impressive *tour de force*, Adrian Bachtold and his team fabricated a nanotube-resonator with such a low noise level that the change of mass due to a single proton could have been detected (Chaste *et al.*, 2012). This device is thus reaching a high enough sensitivity to measure the smallest unit of mass, aka the *yoctogram*, which is just one septillionth of a gram (1 yg $= 10^{-24}$ g). Figure 8.5 shows images of suspended graphene and nanotube resonators, together with the mass sensitivity for the nanotube (right) upon naphthalene adsorption, and the high quality factor for the graphene resonator (left).

For commercial purposes, graphene resonators should be, however, more suitable than CNTs, since large-scale arrays of graphene resonators can be engineered using top-down techniques (van der Zande *et al.*, 2010). As each resonator will have a small capture cross section, large arrays of resonators will be needed. A problem one may foresee is to multiplex the readout of such a large-scale array. An equally important problem is to establish fabrication techniques that will produce resonators with uniform and controlled parameters.

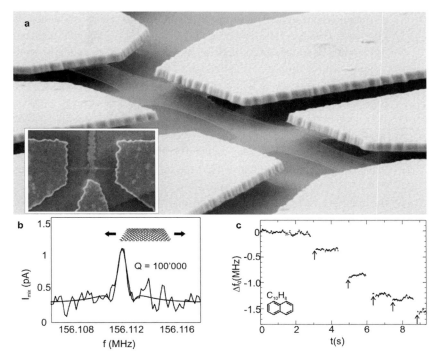

Figure 8.5 (a) A suspended graphene sheet patterned in a Hall bar configuration. The inset shows a nanotube resonator. (b) Resonance lineshape of a mechanical resonator made of graphene at 90 mK. (c) Mass sensing experiment. The shifts in resonance frequency correspond to the adsorption of naphthalene molecules. (Reprinted by permission from Macmillan Publishers Ltd: *Nature Nanotechnology*, Chaste *et al.* (2012), copyright (2012). All images courtesy of Adrian Bachtold.)

8.8 Spintronics

As highlighted in the International Technology Roadmap for Semiconductors (ITRS), devices relying on spintronics (that take advantages of the spin degree of freedom and the intrinsic nonvolatility of magnetization which does not leak with time with the same ease as electric charge) hold unique prospects for ICT. Among potential channels for spintronics, graphene, already acclaimed for its potential for more-than-Moore electronics, is very promising. Indeed, graphene could offer true capability for efficient spin manipulation and for the creation of a full spectrum of spintronic nanodevices beyond CMOS, while being compatible with more-than-Moore CMOS and nonvolatile low-energy MRAM memories (Dery *et al.*, 2012). Ultra-low-energy rewritable microchips, transistors and logic gates, including information storage and processing on a common circuit platform, could be envisioned.

However, while long spin transport in graphene has been demonstrated (Tombros *et al.*, 2007, Dlubak *et al.*, 2012), the reported spin diffusion times remain several orders of magnitude of what is theoretically predicted for clean graphene (Huertas-Hernando, Guinea & Brataas, 2006), whereas the related sources for spin dephasing and scattering remain debated in the literature and strongly material-dependent.

Spin–orbit coupling in graphene is supposed to be weak first because of low atomic number carbon ($Z = 6$, while spin–orbit interaction scales as Z^4). Moreover, the natural occurrence of zero nuclear spin isotope C^{12} is close to 99% and makes hyperfine interaction a vanishingly small decoherence mechanism. Theoretical calculations show that clean graphene exhibits a very low intrinsic (intra-atomic) spin–orbit coupling $\lambda_I \sim 12$ μeV, with a related spin-split gap about 25 μeV (which can be derived using a tight-binding Slater–Koster model (Gmitra et al., 2009, Konschuh, Gmitra & Fabian, 2010)), whereas the application of an external electrical field (perpendicular to the graphene layer) results in gap-closing. Such low spin–orbit coupling should produce relaxation times in the microsecond scale.

In contrast, experiments at room temperature on spin injection in monolayer graphene on SiO_2 substrates (Tombros et al., 2007) report relatively short spin relaxation times (in the order of 1 ns), several orders of magnitude lower than the theoretical prediction. Proposals to explain the unexpectedly short spin relaxation lengths include spin decoherence due to interactions with the underlying substrate, the presence of a random distribution of impurities and the adsorption of molecules, the generation of ripples or corrugations, the presence of strain, topological lattice disorder, graphene edges, etc. (see for instance Han & Kawakami, 2011). The presence of a dielectric oxide, low impedance contacts, or enhanced spin-flip processes, do not seem to affect the spin-relaxation times.

The nature of spin relaxation is actually a fundamental debated issue. Following what is known for metals and semiconductors, two mechanisms have been proposed in graphene, namely the Elliot–Yafet (EY) type and the Dyakonov–Perel (DP) mechanism (Fabian et al., 2007). The Elliott–Yafet mechanism has been derived for spin relaxation in metals, and related the spin dynamics with electron scattering off impurities or phonons. Each scattering event changes the momentum, with a finite spin-flip probability, which is derived by a perturbation theory (assuming weak spin–orbit scattering). This gives rise to weak-antilocalization phenomena in the low-temperature regime, and more interestingly a typical scaling behavior of the spin relaxation time with momentum relaxation as $\tau_s^{EY} \sim \alpha \tau_p$. In the Dyakonov–Perel mechanism (for large band-gap semiconductors without inversion symmetry), electron spins precess along a magnetic field which depends on the momentum. At each scattering event, the direction and frequency of the precession changes randomly. The scaling behavior is opposite to the other mechanism, $\tau_s^{DP} \sim \hbar^2/(\lambda_R^2 \tau_p)$. The most recent theoretical derivation in monolayer graphene (taking into account the Dirac cone physics) reports on some variation of the scaling as $\tau_s \sim \epsilon_F^2 \tau_p/\lambda_R^2$, which is of EY type (Ochoa, Castro Neto & Guinea, 2012). However, such a result is derived assuming absence of intervalley scattering and no Rashba coupling, while the corresponding estimation of spin relaxation times still remains much too long compared to experiments, demanding more generalized and nonperturbative treatments of spin dephasing phenomena in complex and disordered graphene materials.

In Avsar *et al.* (2011), CVD-grown monolayer and bilayer graphene samples were compared, with the surprising result of a dominant EY type of relaxation mechanism in monolayer and a DP type in bilayer graphene. Interestingly, typical transport timescales were found to be $\ell_{el} = v_F \times \tau_p \sim 20-30$ nm, with $\tau_s \sim 175-230$ ps for monolayer, whereas bilayer samples exhibit $\ell_{el} = v_F \times \tau_p \sim 30-50$ nm, $\tau_s \sim 260-340$ ps. Both types of samples show mean free paths of a few tens of nanometers and similar magnitude of spin relaxation times, but with an opposite scaling behavior.

The estimation of the spin relaxation time (as well as spin diffusion coefficient) is generally achieved through spin valve measurements and Hanle precession effects (Fig. 8.6), which are nonlocal transport measurements in which the spin diffusion far from the source/drain contact is tuned with an external and perpendicular magnetic field, inducing spin precession. The measured nonlocal magnetoresistance is usually compared with a one-dimensional spin-Bloch diffusion equation, which assumes a diffusive

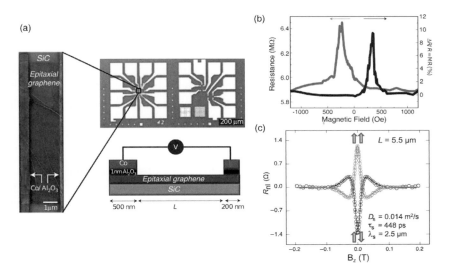

Figure 8.6 (a, left) Plan-view scanning electron microscope image of a two-terminal local spin valve. The width of the epitaxial graphene channel on SiC is 10 μm, and the distance between the two Al$_2$O$_3$/CO electrodes is $L = 2$ μm. (a, right) Optical image of the entire structure, including contact pads. (b) Large local ΔR spin signals measured at 4 K. (c) Oscillating and decaying device resistances as a function of the applied magnetic field (Hanle measurement). ((a) and (b) are adapted by permission of Macmillan Publishers Ltd: *Nature Physics* (Dlubak *et al.*, 2012), copyright (2012). (c) is adapted from Seneor *et al.* (2012) with permission.)

(random walk) propagation of spin and relates the resistance to microscopic parameters through

$$R_{nl} \sim \int_0^{+\infty} \frac{1}{\sqrt{4\pi Dt}} e^{-\frac{L^2}{4D_s t}} \cos(\omega_L t) e^{-\frac{t}{\tau_s}},$$

with $D_s = v_F \tau_s^2$ (τ_s the spin relaxation time) the spin diffusion coefficient, ω the Larmor frequency, and L the distance between electrodes. An important observation is that such an approach cannot tackle a situation of ballistic (or quasiballistic) spin motion, and needs further generalization particularly for describing clean graphene, for which mean free paths can be several hundreds of nanometers long (Du *et al.*, 2008), and so in the order of typical electrode spacing. Additionally, for more disordered graphene, the contributions of quantum interferences and localization phenomena (which in certain materials have shown to be robust up to 100 K) are neglected and could affect any estimation. Finally τ_s can also be estimated independently from two-terminal spin valve measurements (using a phenomenological approach), but this turns out to yield contradictory estimations which differ by orders of magnitudes from Hanle measurements (both types of measurement setups are illustrated in Fig. 8.6 and adapted from Dlubak *et al.*, 2012, Seneor *et al.*, 2012).

Even more puzzling, a recent experiment carried out on monolayer graphene on top of boron nitride substrates shows that neither Elliot–Yaffet nor Dyakonov–Perel alone allows for a consistent description of spin relaxation mechanisms in high quality graphene (Zomer *et al.*, 2012). A tentative crossover is established depending on charge density, and different mechanisms for spin relaxation are assumed to coexist, but without any clue concerning their respective origin. All these results clearly point out a lack of theoretical understanding of spin propagation and spin relaxation mechanisms in graphene, demanding further theoretical inspection and more quantitative and quantum simulation of spin transport, to explore the regimes out of reach of perturbative treatments and phenomenology.

Beyond such fundamental understanding of spin relaxation in graphene, another key issue that needs to be addressed in forthcoming years is the engineering of *spin gating* for progressing towards the manipulation of spin currents in graphene devices. The fundamental challenge in designing spin logic devices lies in developing external ways to control (gate) the propagation of spin-(polarized) currents at room temperature (Semenov, Kim & Zavada, 2007). Tuning spin transport signals could be achieved by magnetic proximity effects, including the deposition of magnetic insulators such as europium oxide (Swartz *et al.*, 2012, Yang *et al.*, 2013), or the creation of local magnetic ordering (Soriano *et al.*, 2011, McCreary *et al.*, 2012).

Recently, the deposition of certain types of heavy atoms in two-dimensional graphene has been predicted to considerably enhance local spin–orbit coupling and trigger the formation of topological insulating phases in the material (Kane & Mele, 2005, Jiang *et al.*, 2012, Weeks *et al.*, 2011). Additionally, several possibilities for generating photo-induced band gaps and the formation of states akin to those of topological insulators have been reported recently (Calvo *et al.*, 2011, Gu *et al.*, 2011, Kitagawa *et al.*, 2011). Although these results have established a possible foundation for groundbreaking spin manipulation, much work remains to be accomplished to make the long-standing expected spin-based devices emerge as a reality.

8.9   Further reading

- For a review on graphene applications we refer to Novoselov *et al.* (2012).

Appendix A Electronic structure calculations: the density functional theory (DFT)

A.1 Introduction

Over the last few decades, there has been a significant increase in the use of computational simulation within the scientific community. Through a combination of the phenomenal boost in computational processing power and continuing algorithm development, atomistic scale modeling has become a valuable asset, providing a useful insight into the properties of atoms, molecules, and solids on a scale "often inaccessible" to traditional experimental investigation.

Atomistic simulations can be divided into two main categories, quantum mechanical calculations and classical calculations based on empirical parameters. Quantum mechanical simulations (often referred to as *ab initio* or first-principles) aims at solving the many-body Schrödinger equation (Schrödinger, 1926). The original reformulation of the Schrödinger equation offered by the DFT provides valuable information on the electronic structure of the system studied.

The very essence of DFT is to deal with noncorrelated single-particle wavefunctions. Many of the chemical and electronic properties of molecules and solids are determined by electrons interacting with each other and with the atomic nuclei. In DFT, the knowledge of the average electron density of the electrons at all points in space is enough to determine the total energy from which other properties of the system can also be deduced. DFT is based on the one-electron theory and shares many similarities with the Hartree–Fock method. DFT is presently the most successful and promising (also the most widely used) approach to computing the electronic structure of matter. In this appendix, the basics of DFT modeling techniques are explained.

A.2 Overview of the approximations

A.2.1 The Schrödinger equation

In principle, an exact theory for a system of ions and interacting electrons is based on solving the many-body Schrödinger equation for the corresponding wavefunction (Schrödinger, 1926). The wavefunction of a many-body system consisting of interacting electrons and nuclei can be defined as $\Psi(\{\mathbf{r}_i\},\{\mathbf{R}_I\})$ where $\{\mathbf{r}_i\}$ and $\{\mathbf{R}_I\}$ correspond to the electronic and nuclear coordinates, respectively. In the framework of a nonrelativistic, time-independent approximation, the Schrödinger equation of a system is as follows:

$$\hat{H}\Psi(\{\mathbf{r}_i\}, \{\mathbf{R}_I\}) = E\Psi(\{\mathbf{r}_i\}, \{\mathbf{R}_I\}), \qquad (A.1)$$

where $\hat{H}$ is the time-independent Hamiltonian operator and E is the energy of the system. The Hamiltonian that describes the physics of this many-body system, neglecting the relativistic effects, is given by

$$\hat{H} = -\frac{\hbar^2}{2m_e}\sum_i^{N_e}\nabla_i^2 - \frac{\hbar^2}{2M_I}\sum_I^{N_n}\nabla_I^2 + \frac{1}{2}\sum_{I\neq J}^{N_n}\frac{Z_IZ_Je^2}{|\mathbf{R}_I - \mathbf{R}_J|}$$

$$+ \sum_i^{N_e}\frac{Z_Ie^2}{|\mathbf{r}_i - \mathbf{R}_I|} + \sum_I^{N_n}\frac{Z_Ie^2}{|\mathbf{r}_i - \mathbf{R}_I|} + \frac{1}{2}\sum_{i\neq j}^{N_e}\frac{e^2}{|\mathbf{r}_i - \mathbf{r}_j|}. \qquad (A.2)$$

Here, the atomic cores and the electrons are referred to by the capital and small indexes, respectively. The index i sums over the number of electrons N_e with mass m_e, and the index I sums over the number of nuclei N_n, with corresponding mass M_I. The first two terms of Eq. (A.2) represent the kinetic energies of all the electrons and nuclei, respectively. The remaining terms represent the electrostatic interactions that occur among the particles of the system: the repulsion between the nuclei, the electron–nucleus and the repulsive electron–electron Coulomb interactions, respectively. Equation (A.2) can be written in a compact way when considering T and V for kinetic and potential energies, respectively:

$$\hat{H} = T_e(\{\mathbf{r}_i\}) + T_n(\{\mathbf{R}_I\}) + V_{n-n}(\{\mathbf{R}_I\})$$

$$+ V_{e-n}(\{\mathbf{r}_i\}, \{\mathbf{R}_I\}) + V_{e-e}(\{\mathbf{r}_i\}). \qquad (A.3)$$

Ever since the Schrödinger equation was discovered, it has been a dream of researchers to find reasonable approximations to reduce its complexity. The first important approximation is obtained by decoupling the dynamics of the electrons and the nuclei, which is known as the Born–Oppenheimer approximation (Born & Oppenheimer, 1927).

A.2.2 The Born–Oppenheimer approximation

The Born–Oppenheimer approximation exploits the fact that the nuclei are much heavier than the electrons (Born & Oppenheimer, 1927). This is true even for the lightest nucleus, a proton whose mass is approximately 1800 times larger than the electron. Hence, in most cases the timescale of the electron response is a few orders of magnitude faster than that of nuclei, which allows the dynamics of the electrons and nuclei to be decoupled. As a consequence of this approximation, nuclei and electrons can be treated separately. The electrons are evolving in the field of fixed nuclei with Hamiltonian ($\hat{H}_e$) expressed as

$$\hat{H}_e = T_e(\{\mathbf{r}_i\}) + V_{e-n}(\{\mathbf{r}_i\}, \{\mathbf{R}_I\}) + V_{e-e}(\{\mathbf{r}_i\}). \qquad (A.4)$$

The solution to a Schrödinger equation involving the previous $\hat{H}_e$ is

$$\hat{H}_e \Psi_e(\{\mathbf{r}_i, \sigma_i\}, \{\mathbf{R}_I\}) = E_e \Psi_e(\{\mathbf{r}_i, \sigma_i\}, \{\mathbf{R}_I\}), \tag{A.5}$$

where Ψ_e represents the electronic wavefunction. The latter is a function only of the electronic coordinates $\{\mathbf{r}_i\}$, while it depends parametrically on the set of nuclear coordinates $\{\mathbf{R}_I\}$ for a fixed configuration of the nuclei. Furthermore, for simplicity the electronic spatial and spin coordinates $\{\mathbf{r}_i, \sigma_i\}$ are placed into one variable $\{\mathbf{x}_i\}$, so that Eq. (A.5) can be rewritten as

$$\hat{H}_e \Psi_e(\{\mathbf{x}_i\}) = E_e \Psi_e(\{\mathbf{x}_i\}). \tag{A.6}$$

The total energy E_{tot}, for given positions of the nuclei, corresponds to the sum of the E_e and the nuclear repulsion energy from the third term in Eq. (A.2), leading to

$$E_{tot} = E_e + V_{n-n}(\{\mathbf{R}_I\}). \tag{A.7}$$

In summary, the Born–Oppenheimer approximation allows one to treat separately the nuclear and electronic degrees of freedom in the many-body problem. The major difficulty in solving Eq. (A.6) is the interaction between electrons, where all the many-body quantum effects are hidden. Since the movements of electrons are correlated, the instantaneous coordinates of each electron should be known, which essentially requires the treatment of 3^{N_e} variables for an N_e-electron system. Even after applying this simplification, the many-body problem remains intractable. Hence, further approximations are needed to efficiently solve Eq. (A.6).

A.2.3   The Hartree approximation

The Hartree approximation (Hartree, 1957) provides one way to reduce Eq. (A.6), the many-electron wavefunction problem, to a product of N_e one-electron wavefunctions. Each electron moves independently within its own orbital and sees only the average potential generated by all the other electrons. This Hartree potential can be approximated by an average single-particle potential and is expressed by the Coulomb repulsion between the ith electron and the electron density produced by all other electrons $(n(\mathbf{x}_j))$:

$$V_H(\mathbf{x}_i) = \int \frac{n(\mathbf{x}_j)}{|\mathbf{r}_i - \mathbf{r}_j|} d\mathbf{x}_j, \quad n(\mathbf{x}_j) = \sum_{j=1}^{N_e} |\phi_j(\mathbf{x}_j)|^2. \tag{A.8}$$

The solution to the one-particle wave-equation $(\psi_i(\mathbf{x}_i))$ is

$$\left[-\frac{\hbar}{2m} \nabla^2 + V_{ext}(\mathbf{x}_i) + V_H(\mathbf{x}_i) \right] \psi_i(\mathbf{x}_i) = \epsilon_i \psi_i(\mathbf{x}_i), \tag{A.9}$$

where the first term corresponds to the one-electron kinetic energy. In this equation, the effective potential experienced by the single electron includes two terms: the Hartree potential and the external Coulomb potential, given by

$$V_{ext}(\mathbf{x}_i) = V_{e-n}(\{\mathbf{x}_i\}, \{\mathbf{R}_I\}) + V_{n-n}(\{\mathbf{R}_I\}). \tag{A.10}$$

However, the equations are still nonlinear and the fermionic character of the electrons is ignored. According to the Pauli exclusion principle, two electrons cannot occupy the same quantum state. However, the wavefunction in Hartree theory is

$$\Psi(\mathbf{x}_i) = \prod_i^{N_e} \psi_i(\mathbf{x}_i), \tag{A.11}$$

and is not antisymmetric under the interchange of two electrons, which is incompatible with the Pauli principle. This problem is rectified by the Hartree–Fock theory.

A.2.4 The Hartree–Fock approximation

The Hartree–Fock approach is considered as the fundamental first step in quantum chemistry. Indeed, the Hartree–Fock theory is derived by invoking the variational principle which states that the expected value of the electronic Hamiltonian ($\hat{H}_e$) for any guessed or trial wavefunction is always greater than or equal to the electronic ground state energy $E_0[\Psi_0]$. This remains true when the wavefunction is in the true ground state Ψ_0, i.e. ($E[\Psi] \geq E_0[\Psi_0]$). The advantage of the variational principle is that the ground state energy $E_0[\Psi_0]$ can be approached by starting with a trial function, and the quality of the wavefunction can be improved variationally in the restricted antisymmetrized space of *single-particle* wavefunctions. The wavefunctions of Eq. (A.11) can be approximately described with a single Slater determinant. A Slater determinant is a linear combination of the product of independent electron wavefunctions ($\psi_i(\mathbf{x}_i)$) with all possible combinations of the permutations of their coordinates. The Slater determinant satisfies the antisymmetric property of the wavefunction, hence obeys the exclusion principle of Pauli. The wavefunctions of Eq. (A.11) are replaced by

$$\Psi(\{\mathbf{x}_i\}) \approx \Psi_{HF}(\{\mathbf{x}_i\}) = \frac{1}{\sqrt{N_e!}} \begin{vmatrix} \psi_i(\mathbf{x}_i) & \psi_j(\mathbf{x}_i) & \cdots & \psi_{N_e}(\mathbf{x}_i) \\ \psi_i(\mathbf{x}_j) & \psi_j(\mathbf{x}_j) & \cdots & \psi_{N_e}(\mathbf{x}_j) \\ \vdots & \vdots & \ddots & \vdots \\ \psi_i(\mathbf{x}_{N_e}) & \psi_j(\mathbf{x}_{N_e}) & \cdots & \psi_{N_e}(\mathbf{x}_{N_e}) \end{vmatrix}, \tag{A.12}$$

constructed from a set of one-particle orbitals $\{\psi_i(\mathbf{x}_i)\}$ required to be mutually orthonormal $\langle \psi_i | \psi_j \rangle = \delta_{ij}$. Taking the expectation value of the electronic Hamiltonian $\hat{H}_e$ of Eq. (A.5) with the trial functions defined by the Slater determinant of Eq. (A.12), one gets the total electronic Hartree–Fock energy functional. The Hartree–Fock equation (Marx & Hutter, 2000) which comes from an energy-minimization of the Hartree–Fock energy functional, is given by

$$\left[-\frac{\hbar}{2m_e}\nabla^2 + V_{ext}(\mathbf{x}_i) + V_H(\mathbf{x}_i) + V_X(\mathbf{x}_i) \right] \psi_i(\mathbf{x}_i) = \epsilon_i \psi_i(\mathbf{x}_i), \tag{A.13}$$

where now the exchange term is properly taken into account as

$$V_X(\mathbf{x}_i)\psi_i(\mathbf{x}_i) = -e^2 \sum_j \psi_j(\mathbf{x}_i) \int d\mathbf{x}_j \frac{\psi_j^*(\mathbf{x}_j)\psi_i(\mathbf{x}_j)}{\mathbf{r}_i - \mathbf{r}_j}. \tag{A.14}$$

The exchange operator is defined via the action of the electrons on a particular orbital ψ_i. It is noticeable that upon action on orbital ψ_i, the exchange operator of the jth state "exchanges" $\psi_j(x_j) \rightarrow \psi_i(x_j)$ in the kernel as well as replaces $\psi_i(x_i) \rightarrow \psi_j(x_i)$ in its argument. Thus, the exchange term is a nonlocal operator, and in this sense the exchange operator does not possess a simple classical interpretation like the Hartree term (Marx & Hutter, 2000). A considerable amount of complexity is introduced in this nonlinear exchange operator due to the many-body interactions. This is true even for obtaining the first-order approximation of the total energy (Marx & Hutter, 2000).

A.3   Density functional theory

DFT differs from other wavefunction-based methods by using the electron density $n(\mathbf{r})$ as the central quantity. An important advantage of using the electron density over the wavefunction is the much reduced dimensionality. Regardless of how many electrons are present in the system, the density is always three-dimensional. This allows DFT to be readily applied to much larger systems; hundreds or even thousands of atoms become possible. This is one among the many reasons why DFT has become the most widely used electronic structure approach today. First, the electron density can be expressed as

$$n(\mathbf{r}) = N_e \int |\Psi(x_1, x_2, \ldots, x_{N_e})|^2 d\sigma_1 d\sigma_2 \ldots d\sigma_{N_e}, \tag{A.15}$$

where x_i represents both spatial and spin coordinates. $n(\mathbf{r})$ determines the probability of finding any of the N_e electrons within the volume $\mathbf{r}$. The electrons have arbitrary spin and the other $N_e - 1$ electrons have arbitrary positions and spin in the state represented by Ψ. This is a nonnegative simple function integrating the total number of electrons,

$$N_e = \int n(\mathbf{r}) d\mathbf{r}. \tag{A.16}$$

A.3.1   The Thomas–Fermi model

There have been many attempts to reformulate the problem based on the ground state charge density $n(\mathbf{r})$. Thomas and Fermi exploited first the fact that the electronic energy can be expressed in terms of electronic density (Thomas, 1927, Fermi, 1927). In their model, the kinetic energy of the electrons is derived from quantum statistical theory based on the uniform electron gas, whereas the electron–nucleus and electron–electron interactions are treated classically. Within this model, the total energy is a functional of $n(\mathbf{r})$ expressed as

$$E_{TF}[n(\mathbf{r})] = \frac{3}{10}(2\pi^2)^{2/3} \int n^{3/5}(\mathbf{r}) d\mathbf{r}$$
$$- Z \int \frac{n(\mathbf{r})}{\mathbf{r}} d\mathbf{r} + \frac{1}{2} \iint \frac{n_1(\mathbf{r})n_2(\mathbf{r})}{|\mathbf{r}_1 - \mathbf{r}_2|} d\mathbf{r}_1 d\mathbf{r}_2, \tag{A.17}$$

where the first term is the kinetic energy (i.e. the kinetic energy density for a system of noninteracting electrons with density n) while the second and third terms are the electron–nucleus and electron–electron interactions, respectively. Although it is not the most efficient, the Thomas–Fermi model illustrates that the ground state energy can be determined purely using the electron density.

A.3.2   The Hohenberg–Kohn theorem

The essential role played by the electron density in the search for the electronic ground state was pointed out in 1964 by Hohenberg and Kohn (Hohenberg & Kohn, 1964). They derived the fundamentals of DFT, which allows us to express the electronic Hamiltonian as a functional of $n(\mathbf{r})$. This formalism relies on two theorems: (i) there exists a one-to-one correspondence between external potential $v(\mathbf{r})$ and electron density $n(\mathbf{r})$, and (ii) the ground state electron density can be obtained by using a variational principle.

The electronic Hamiltonian depends explicitly on the configuration of the nuclei only through $v(\mathbf{r})$. Assuming that the first theorem is valid, then from $n(\mathbf{r})$ one can obtain $v(\mathbf{r})$ upto a trivial additive constant. The electronic Hamiltonian can be expressed as a functional of $n(r)$. Suppose there is a collection of electrons enclosed in a box influenced by two external potentials $v(\mathbf{r})$ and $v'(\mathbf{r})$, which differs from $v(\mathbf{r})$ by more than a constant in a nondegenerated system (local system). Assuming that these two potentials lead to the same electron density $n(\mathbf{r})$ for the ground state, two different Hamiltonians $\hat{H}$ and $\hat{H}'$, whose ground state electron density is the same, are present. However, their normalized wavefunctions Ψ and Ψ' would be different. As a consequence, the ground state energy E_0 would be

$$E_0 < \langle \Psi'|\hat{H}|\Psi'\rangle = \langle \Psi'|\hat{H}'|\Psi'\rangle + \langle \Psi'|\hat{H} - \hat{H}'|\Psi'\rangle$$

$$= E_0' + \int n(\mathbf{r})[v(\mathbf{r}) - v'(\mathbf{r})]d\mathbf{r}, \qquad (A.18)$$

where E_0 and E_0' are the ground state energies for $\hat{H}$ and $\hat{H}'$, respectively. Similarly to that, E_0' would be

$$E_0' < \langle \Psi|\hat{H}|\Psi\rangle = \langle \Psi|\hat{H}'|\Psi\rangle + \langle \Psi|\hat{H}' - \hat{H}|\Psi\rangle$$

$$= E_0 - \int n(r)[v(\mathbf{r}) - v'(\mathbf{r})]d\mathbf{r}. \qquad (A.19)$$

Summing Eq. (A.18) and Eq. (A.19) leads to $E_0 + E_0' < E_0' + E_0$, which is an obvious contradiction. This demonstrates that it is not possible to find two different external potentials that can give the same electron density $n(\mathbf{r})$. Consequently, $n(\mathbf{r})$ uniquely determines $v(\mathbf{r})$ (up to a constant) and all ground state properties.

The energy E_v can be explicitly written as a function of the electron density $n(\mathbf{r})$ for a given external potential $v(\mathbf{r})$:

$$E_v[n(\mathbf{r})] = T[n(\mathbf{r})] + V_{n-e}[n(\mathbf{r})] + V_{e-e}[n(\mathbf{r})]$$

$$= \int n(\mathbf{r})v(\mathbf{r})d\mathbf{r} + F_{HK}[n(\mathbf{r})], \qquad (A.20)$$

where $F_{HK}[n(\mathbf{r})]$ is dependent only on $n(\mathbf{r})$, independently of any external potential $v(\mathbf{r})$. Thus F_{HK} is a universal functional of $n(\mathbf{r})$ and is defined as

$$F_{HK}[n(\mathbf{r})] = T[n(\mathbf{r})] + V_{e-e}[n(\mathbf{r})]. \tag{A.21}$$

The second theorem of Hohenberg and Kohn demonstrates that the ground state energy can be obtained variationally from the density. The density that minimizes the total energy is the exact ground state density, thus rationalizing the original intuition of Thomas and Fermi (Thomas, 1927, Fermi, 1927). This is expressed as

$$E_0[n_0(\mathbf{r})] \leq E_v[n(\mathbf{r})]. \tag{A.22}$$

The total energy functional (following the first theorem) given by Eq. (A.20) is calculated for a trial density $v(\mathbf{r})$ and is consistent with charge conservation $\int n(\mathbf{r})d\mathbf{r} = N_e$. The total energy functional is always greater than or equal to the true ground state total energy E_0 of the system:

$$E_v[n(\mathbf{r})] = \int n(\mathbf{r})v(\mathbf{r})d\mathbf{r} + F_{HK}[n(\mathbf{r})] \geq E_0. \tag{A.23}$$

These two theorems demonstrate that the problem of solving the Schrödinger equation for the ground state can be exactly recast into the variational problem of minimizing the Hohenberg–Kohn functional, Eq. (A.21), with respect to the minimization of a functional of the three-dimensional density function. However, the major part of the complexities of the many-electron problem are associated with the determination of the universal Hohenberg–Kohn functional $F_{HK}[n(\mathbf{r})]$.

A.3.3   The Kohn–Sham equations

Kohn and Sham transformed the DFT into a practical electronic structure theory (Kohn & Sham, 1965). They recognized that the failure of the Thomas–Fermi theory mainly resulted from the bad description of the kinetic energy. In order to address this problem they came back to the picture of noninteracting electrons moving in an effective field.

$F_{HK}[n(\mathbf{r})]$ is written as a sum of the kinetic energy of noninteracting electrons (T_s), the classical electrostatic Hartree energy (E_H), and all the many-body quantum effects are put together into the exchange and correlation energy (E_{xc}). Thus, the energy functional of the previous section becomes

$$E[n(\mathbf{r})] = \int n(\mathbf{r})v(\mathbf{r})d\mathbf{r} + F_{HK}[n(\mathbf{r})]$$

$$= \int n(\mathbf{r})v(\mathbf{r})d\mathbf{r} + T_s[n(\mathbf{r})] + E_H[n(\mathbf{r})] + E_{xc}[n(\mathbf{r})]. \tag{A.24}$$

The constraint minimization of the $F_{HK}[n(\mathbf{r})]$ functional for N_e electrons can be rewritten, introducing the indeterminate multiplier μ as the variational problem:

$$\delta \left\{ F_{HK}[n(\mathbf{r})] + \int v(\mathbf{r})n(\mathbf{r})d\mathbf{r} - \mu \left(\int n(\mathbf{r})d\mathbf{r} - N_e \right) \right\} = 0. \tag{A.25}$$

Formally Eq. (A.25) leads to the Euler–Lagrange equation for the charge density:

$$\frac{\delta F_{\mathrm{HK}}[n(\mathbf{r})]}{\delta n(\mathbf{r})} + v(\mathbf{r}) = \mu. \tag{A.26}$$

This minimization under the constraint of orthonormality for the one-particle orbitals ψ_i leads to a set of N_e single-particle Schrödinger-like equations, the so-called *Kohn–Sham equations*, that are expressed as

$$\left[-\frac{\hbar^2}{2m_e}\nabla^2 + v_{\mathrm{KS}}[n(\mathbf{r})] \right]\psi_i(\mathbf{r}) = \hat{H}_{\mathrm{KS}}\psi_i(\mathbf{r}) = \epsilon_i\psi_i(\mathbf{r}). \tag{A.27}$$

Here, ψ_i are the Kohn–Sham one-electron orbitals and the electron density is defined as

$$n(\mathbf{r}) = \sum_{i=1}^{N_e}|\psi_i|^2, \tag{A.28}$$

where $v_{\mathrm{KS}}[n(\mathbf{r})]$ is the effective potential experienced by the electrons and is expressed as

$$v_{\mathrm{KS}}[n(\mathbf{r})] = \frac{\delta \int n(\mathbf{r})v(\mathbf{r})\mathrm{d}\mathbf{r} + T_s[n(\mathbf{r})] + E_{\mathrm{H}}[n(\mathbf{r})] + E_{\mathrm{xc}}[n(\mathbf{r})]}{\delta n(\mathbf{r})}$$

$$= v[n(\mathbf{r})] + \int \frac{n(\mathbf{r}')}{|\mathbf{r}-\mathbf{r}'|}\mathrm{d}\mathbf{r}' + v_{\mathrm{xc}}[n(\mathbf{r})]. \tag{A.29}$$

The exchange–correlation potential is given by the functional derivative of the exchange–correlation energy:

$$v_{\mathrm{xc}}[n(\mathbf{r})] = \frac{\delta E_{\mathrm{xc}}[n(\mathbf{r})]}{\delta n(\mathbf{r})}. \tag{A.30}$$

Finally, the total energy can be determined from the resulting equations of density and potentials through

$$E = \sum_{i=1}^{N_e}\epsilon_i - \frac{1}{2}\iint \frac{n(\mathbf{r})n(\mathbf{r}')}{|\mathbf{r}-\mathbf{r}'|} + E_{\mathrm{xc}}[n] - \int v_{\mathrm{xc}}[n(\mathbf{r})]n(\mathbf{r})\mathrm{d}\mathbf{r}. \tag{A.31}$$

The Kohn–Sham equations, Eq. (A.27), require to be solved self-consistently due to the density dependence on the one-electron Kohn–Sham effective potential v_{KS}. The general procedure consists in starting with an initial guess of the electron density, and constructing the effective potential $v(\mathbf{r})$ from Eq. (A.24) in order to extract the Kohn–Sham orbitals. Based on these orbitals, a new density is obtained from Eq. (A.28) and the process is repeated until convergence is achieved (Payne *et al.*, 1992). Finally, the total energy is calculated from Eq. (A.31) by using the obtained electron density, as illustrated in Fig. A.1.

The exact total energy can be extracted if each term in the Kohn–Sham energy functional is known. This is unfortunately not the case. Indeed, the exchange–correlation (xc) functional (E_{xc}) remains unknown. E_{xc} includes the non-classical aspects of the electron–electron interaction along with the component of the kinetic energy of the

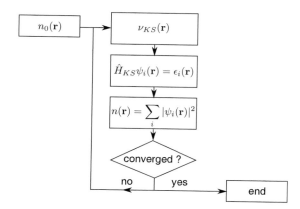

Figure A.1 The self-consistent resolution scheme for the Kohn–Sham equations. (Adapted from Nogueira, Castro & Marques, 2003).

real system, differently from the fictitious noninteracting system. Since E_{xc} cannot be determined exactly, this energy term has to be estimated using different possible approximations.

A.3.4   The exchange–correlation functionals

The simplest way to approximate the exchange–correlation (xc) energy of an electronic system is the local-density approximation (LDA) which was proposed by Kohn and Sham (Kohn & Sham, 1965). In this approximation, a real inhomogeneous system is divided into infinitesimal volumes in which the density is assumed to be constant. The exchange–correlation (xc) energy for the system is contructed by assuming that the exchange–correlation energy $\epsilon_{xc}[n(\mathbf{r})]$ per electron at a point $\mathbf{r}$ in the electron gas is equal to the exchange–correlation energy per electron in a homogeneous electron gas that has the same electron density as the electron gas at point $\mathbf{r}$. Thus, one can write

$$\epsilon_{xc}[n(\mathbf{r})] = \epsilon_{xc}^{hom}[n(\mathbf{r})], \tag{A.32}$$

and the total exchange–correlation energy is written as

$$E_{xc}^{LDA}[n(\mathbf{r})] = \int \epsilon_{xc}^{hom}[n(\mathbf{r})]n(\mathbf{r})d\mathbf{r}. \tag{A.33}$$

In practice, the exchange and correlation terms are calculated separately. The exchange part is given by the Dirac exchange-energy functional (Dirac, 1930):

$$E_{x}^{LDA}[n(\mathbf{r})] = -\frac{3}{4}\left(\frac{3n(\mathbf{r})}{\pi}\right)^{1/3}. \tag{A.34}$$

Here, the local Wigner–Seitz radius $r_s(\mathbf{r})$ is defined as $r_s(\mathbf{r}) = \left(\frac{3}{4\pi n(\mathbf{r})}\right)^{-1/3}$, which is the radius of the sphere that would contain exactly one electron in the homogeneous electron gas density $n(\mathbf{r})$.

As for the exchange term, the exact correlation term is unknown. However, an approximate expression can be determined by interpolating homogeneous electron gas data obtained by quantum Monte-Carlo calculations as reported by Ceperley and Alder (Ceperley & Alder, 1980). The parameterization proposed by Perdew and Zunger (Perdew & Zunger, 1981) is the most commonly used by the scientific community. The LDA approximation ignores corrections to the exchange–correlation energy at a point r due to the nearby inhomogeneities in the electron density. Strictly, the LDA is valid for slowly varying density. However, it is noticeable that calculations performed using the LDA have been remarkably successful (Payne *et al.*, 1992).

However, it was realized very early that only the local uniform density at each given point is not a reasonable approximation for the rapidly varying electron density of many materials. An attempt to improve the LDA approximation consists in taking into account not only the local uniform density $n(\mathbf{r})$, but also the gradient terms ($\nabla n(\mathbf{r})$) of the total charge density in the exchange–correlation energy term. Based on this idea, the exchange–correlation energy of the generalized gradient approximation (GGA) can be written as

$$E_{\text{xc}}^{\text{GGA}}[n(\mathbf{r}), \nabla n(\mathbf{r})] = \int F_{\text{xc}}[n(\mathbf{r}), \nabla n(\mathbf{r})] \mathrm{d}\mathbf{r}. \tag{A.35}$$

The GGA functionals are often called "semi-local" functionals due to their dependence on the gradient of the density $\nabla n(\mathbf{r})$. Typically for many properties, such as geometries and ground state energies of molecules and solids, GGA can yield better results than the LDA. The functional F_{xc} is taken as a correction to the LDA exchange and correlation relation, while ensuring again the consistency within exchange–correlation energy as in LDA. Within GGA, the exchange energy is then expressed as

$$E_{\text{x}}^{\text{GGA}}[n(\mathbf{r})] = \int \epsilon_{\text{x}}[n(\mathbf{r})] F_{\text{x}}^{\text{GGA}}(s) \mathrm{d}\mathbf{r}, \tag{A.36}$$

where $F_{\text{x}}^{\text{GGA}}(s)$ is the exchange enhanced factor which represents how much exchange energy is over the LDA exchange value for a given $n(\mathbf{r})$. One GGA functional differs from another according to the choice of this exchange enhanced factor F_{x}, which is a function of s, a dimensionless reduced gradient defined as

$$s = \frac{|\nabla n(\mathbf{r})|}{2(3\pi^2)^{1/3} n(\mathbf{r})^{4/3}}. \tag{A.37}$$

Various approximations have been proposed for the $F_{\text{x}}(s)$ functional, for instance, by Perdew and Wang (Perdew, 1991) and by Perdew, Burke and Ernzerhof (PBE) (Perdew, Burke & Ernzerhof, 1996). Generally, when materials properties have to be screened, the PBE functional is used, and the $F_{\text{x}}(s)$ is expressed as

$$F_{\text{x}}^{\text{PBE}}(s) = 1 + k - \frac{\kappa}{1 + \frac{\mu s^2}{\kappa}}. \tag{A.38}$$

In PBE, κ and μ are parameters obtained from (nonempirical) physical constraints (Perdew *et al.*, 1996). The functional form of the gradient-corrected correlation energy $E_{\text{c}}^{\text{GGA}}$ is also expressed as a complex function of s. For more details on PBE parameterizations see (Perdew *et al.*, 1996).

A.4   Practical calculations

For practical reasons, it is necessary to introduce crystal lattices (which exhibit a peri-odical symmetry) to solve numerically large periodic systems using DFT. The crystal lattice is typically used to reduce the amount of atoms (thus, of interacting particles) when using only the unit cell. Additionally to this, the equations derived in the previous section need to be projected onto a complete basis set. This basis set should be of finite size in order to allow us to perform computer calculations. Finally, band structures and **k**-point grids are introduced, followed by the pseudopotentials.

A.4.1   Crystal lattice and reciprocal space

A solid material is composed of many electrons and ionic cores per cm^3. In principle, all these positions should be taken into account to construct the Kohn–Sham Hamiltonian. Fortunately, the periodic symmetry of the crystal lattice allows us to reduce the interac-tion problem to only electrons and ionic cores that are present in the unit cell. A crystal is determined by its atoms' positions and follows the rules of symmetry (repeating them by performing translations). The set of translations that generates the complete crystal is called the Bravais lattice. This set of translations forms a group, as the sum of two translations is again a translation. Other symmetries (e.g. a rotation) leaving the crystal unchanged can also exist. These form a group called a point group which is a group of geometric symmetries (isometries) that keep at least one point fixed. The space group of a crystal is given by the sum of the translation group and the point group.

The set of all translations forms a lattice in space, in which each translation can be written as a linear combination of the primitive vectors $\mathbf{a}_1$, $\mathbf{a}_2$, $\mathbf{a}_3$:

$$\mathbf{t} = i_1\mathbf{a}_1 + i_2\mathbf{a}_2 + i_3\mathbf{a}_3, \tag{A.39}$$

where i_1, i_2, and i_3 are integers. The positions of the atoms in the unit cell can be described with respect to the primitive translation vectors. Due to the periodicity of the lattice, all periodic functions can be Fourier-transformed. The Fourier-transformed space is also called reciprocal space, where the set of reciprocal vectors $\mathbf{b}_i$ of the primi-tive translations $\mathbf{a}_j$ satisfy

$$\mathbf{b}_i \cdot \mathbf{a}_j = 2\pi\theta_{ij}, \tag{A.40}$$

which is the definition of the reciprocal lattice. A vector in the reciprocal space is usually denoted **G** and is given by

$$\mathbf{G} = i_1\mathbf{b}_1 + i_2\mathbf{b}_2 + i_3\mathbf{b}_3, \tag{A.41}$$

where i_1, i_2, and i_3 are integers. By using this reciprocal lattice, the first Brillouin zone can be defined as the Wigner–Seitz cell of the reciprocal space (Ashcroft & Mermin, 1976). This "minimum representation" of the system to be studied using the periodic boundary conditions of large systems is the basic idea of this approach, to be used for the DFT calculations throughout the various examples of carbon nanostructures inves-tigated in this book.

A.4.2 The plane wave representation

For such periodic systems, the external potential satisfies the relation $V(\mathbf{r} + \mathbf{t}) = V(\mathbf{r})$, imposed by the periodic boundary conditions, and therefore the corresponding effective one-electron Hamiltonian obeys the translation invariance. The Bloch theorem applies to the electronic wavefunctions of the system. The eigenfunctions can be written as a product of a plane wave ($e^{i\mathbf{k}\cdot\mathbf{r}}$) and a function ($u_{n,\mathbf{k}}(\mathbf{r})$) having the same periodicity as the potential $V(\mathbf{r})$:

$$\psi_{n,\mathbf{k}}(\mathbf{r}) = e^{i\mathbf{k}\cdot\mathbf{r}} u_{n,\mathbf{k}}(\mathbf{r}), \qquad \text{where} \quad u_{n,\mathbf{k}}(\mathbf{r} + \mathbf{t}) = u_{n,\mathbf{k}}(\mathbf{r}). \tag{A.42}$$

Here, $\mathbf{k}$ represents the wave vector, and n, the band index. The Bloch theorem allows us to expand the electronic wavefunction in terms of a discrete set of plane waves to the periodic function $u_{n,\mathbf{k}}(\mathbf{r})$, whose wave vectors are the reciprocal lattice vector ($\mathbf{G}$) of the periodic crystal:

$$u_{n,\mathbf{k}}(\mathbf{r}) = \frac{1}{\sqrt{\Omega}} \sum_{\mathbf{G}} C_{n,\mathbf{k}+\mathbf{G}}\, e^{i\mathbf{G}\cdot\mathbf{r}}, \tag{A.43}$$

where Ω is the volume of the unit cell. The electronic wavefunction can thus be rewritten as

$$\psi_{n,\mathbf{k}}(\mathbf{r}) = \frac{1}{\sqrt{\Omega}} \sum_{\mathbf{G}} C_{n,\mathbf{k}+\mathbf{G}+\mathbf{r}}\, e^{i(\mathbf{k}+\mathbf{G})\cdot\mathbf{r}}. \tag{A.44}$$

Using the above expressions to solve the one electron Schrödinger-like equation with an effective periodic potential, e.g. the Kohn–Sham potential defined in Eq. (A.29), the Kohn–Sham wavefunction can be expanded with the plane wave basis sets as described in Eq. (A.44). As a result, Eq. (A.27) can be rewritten as

$$\sum_{\mathbf{G}'} \left[\frac{\hbar^2}{2m} |\mathbf{k} + \mathbf{G}|^2 \delta_{\mathbf{G}\mathbf{G}'} + v_{\text{n-n}}(\mathbf{G} - \mathbf{G}') + v_{\text{H}}(\mathbf{G} - \mathbf{G}') \right.$$

$$\left. + v_{\text{xc}}(\mathbf{G} - \mathbf{G}') \right] C_{n,\mathbf{k}+\mathbf{G}} = \epsilon_{n,\mathbf{k}} C_{n,\mathbf{k}+\mathbf{G}}, \tag{A.45}$$

where $v(\mathbf{G} - \mathbf{G}')$ are the Fourier transforms of the potential in real space (which can exhibit different functional forms depending on the pseudopotential method): here, $v_{\text{n-n}}(\mathbf{G} - \mathbf{G}')$ is the nuclei–nuclei interaction Coulomb potential, $v_{\text{H}}(\mathbf{G} - \mathbf{G}')$ is the Hartree potential, $v_{\text{xc}}(\mathbf{G} - \mathbf{G}')$ is the exchange–correlation potential, $\delta_{\mathbf{G}\mathbf{G}'}$ is the Kronecker symbol δ, and reflects that the kinetic energy is diagonal, whereas ϵ_n are the electronic energies.

The solution of this secular equation is obtained by diagonalizing the Hamiltonian matrix of elements $H_{\mathbf{k}+\mathbf{G},\mathbf{k}+\mathbf{G}'}$, given by the terms in brackets of Eq. (A.45). The size of the matrix is determined by the choice of the cutoff energy. The plane wave expansion is truncated to include terms with a kinetic energy only up to a certain cutoff value:

$$\frac{\hbar^2}{2m_{\text{e}}} |\mathbf{k} + \mathbf{G}|^2 < E_{\text{cutoff}}. \tag{A.46}$$

Employing a finite basis set introduces a new source of inaccuracy, which can be reduced by increasing the number of plane waves or the kinetic energy cutoff (E_{cutoff}). Therefore, appropriate convergence tests have to be performed in order to find an E_{cutoff} that is sufficient to compute the property of interest with the required accuracy. Despite the E_{cutoff}, it is often computationally heavy to determine the size of the matrix for systems that contain both valence and core electrons. This problem can be overcome by using the pseudopotential approximation as discussed in the following sections.

A.4.3   k-point grids and band structures

In a periodic solid, the number of electrons is of the order of Avogadro's number. Only a set of **k** points is allowed and determined by the periodic boundary conditions or generalized Born–von Karman boundary conditions to the wavefunctions. The latter can be interpreted by saying that a particle which leaves one surface of the crystal simultaneously enters the crystal at the opposite surface. The density of allowed **k** points is proportional to the volume of the solid. The infinite number of electrons in the solid is accounted for by an infinite number of **k** points, and only a finite number of electronic states are occupied. The spacing of the **k** points goes to zero and **k** can be considered as a continuous variable. The occupied states at each **k** point contribute to the ground state properties of the solid such as the electronic potential, electron density, and the total energy. However, the electronic wavefunctions at **k** points that are very close together will be almost identical. Hence, it is possible to represent them by a single **k** point instead of over a region of **k** space. In order to obtain accurate electronic potential, electron density, and total energy, efficient methods have been used to choose the finite sets of **k** points. Generally, the method proposed by Monkhorst and Pack (Monkhorst & Pack, 1976) is used, in which a uniform mesh of **k** points is generated along the three lattice vectors in reciprocal space. The magnitude of any error in the total energy or the total energy difference due to inadequacy of the **k** points sampling can always be reduced to zero by using a denser set of **k** points. Therefore, it is crucial to test the convergence of the results with respect to the number of **k** points in general.

Due to the translational symmetry and to the continuous nature of **k** points, only the first Brillouin zone with its band eigenvalues and energy gaps is taken into account to plot these eigenvalues for the so-called band structures. In order to calculate the electronic density and other properties, it is necessary to integrate (all) over the **k** points in the Brillouin zone. The number of these **k** points necessarily depends on the material: for insulators only a few points are needed as all bands are filled, while for metals more points are needed for the bands that cross the Fermi level. The presence of additional symmetries, such as rotations or mirror reflections, allows one to consider only a part of the Brillouin zone. The smallest possible part that can be mapped to the complete Brillouin zone by applying all symmetries is called the irreducible Brillouin zone (IBZ).

A.4.4   The pseudopotential approximation

Most physically interesting properties of solids are largely determined by the valence electrons rather than the core. The valence electrons can be thought of as loosely bound orbitals, delocalized over the crystal, which strongly influence the formation of bands in a solid. In contrast, the core electrons, that are tightly bound around each atomic nucleus and largely unperturbed by the surrounding environment, are essentially not involved in chemical bondings. Moreover, the deeply bound core electrons within plane-wave basis sets require a huge amount of basis functions for their description, which implies a significant computational cost. This can be avoided by using the pseudopotential approximation (Phillips, 1958) which replaces the strong ionic potential with a weaker pseudopotential. In general, the pseudopotential formalism is used for two main reasons which are: (i) to reduce the number of plane waves to describe the core electrons as a weaker pseudopotential due to their deep potential, and (ii) to eliminate the fast oscillations of the wavefunctions of the valence electrons. These two issues are illustrated in Fig. A.2, where the pseudopotential is much weaker than the all-electron potential and where the pseudo wavefunction has no radial node inside the core region. It is essential within the pseudopotential scheme that outside the core region, the pseudopotential and pseudo wavefunction become the same with the corresponding all-electron functions.

The most common form of a pseudopotential is:

$$V^{\mathrm{PS}}(\mathbf{r}) = \sum |Y_{\mathrm{l,m}}\rangle V_{\mathrm{l}}(\mathbf{r}) \langle Y_{\mathrm{l,m}}|, \tag{A.47}$$

where $|Y_{\mathrm{l,m}}\rangle$ are the spherical harmonics and V_{l} are the angular momentum dependent components of the pseudopotential. A pseudopotential that uses the same potential in each angular momentum is called a local pseudopotential, i.e. it only depends on the distance from the nucleus. A fundamental requirement for the pseudopotential is the ability to accurately describe the valence electrons in different chemical environments. Such a property is called the transferability of the pseudopotential.

Based on this idea, a systematic and successful procedure for the development of accurate and transferable pseudopotentials has been developed, where the added constraint of norm-conservation is introduced (Hamann, Schlüter & Chiang, 1979,

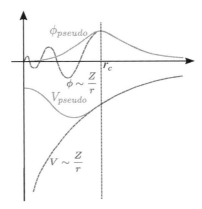

Figure A.2 All-electron potential (dashed line) and pseudopotential (solid line) and their corresponding wavefunctions. The radius at which the all-electron and pseudo-electron values match is designated r_{c}. (Adapted from Rignanese (1998).)

Bachelet, Hamann & Schlüter, 1982). Modern norm-conserving pseudopotentials are obtained by inverting the free atom Schrödinger equation for a given electronic configuration of reference and forcing the pseudo wavefunctions to match the true valence wavefunctions beyond a certain chosen distance from the nucleus. In summary, a norm-conserving pseudopotential is built on a given reference atomic configuration, to meet the following conditions:

(1) The pseudo energy-eigenvalues should match the true (all-electron) valence eigenvalues.

(2) The pseudo wavefunctions ϕ_l^{PS} should be nodeless, and the functions and their first derivatives must be differentiable.

(3) The pseudo wavefunctions should match the all-electron wavefunctions beyond a chosen core radius r_c^l:

$$\phi_l^{PS}(\mathbf{r}) = \phi_l^{AE}(\mathbf{r}), \qquad \text{for } r > r_c^l. \tag{A.48}$$

(4) The total integrated pseudo-charge density from a given $\phi_{l PS(\mathbf{r})}$ and the corresponding all-electron charge density are identical inside the core radius r_c^l:

$$\int_{r<r_c^l} |\phi_l^{PS}|^2 r^2 \mathrm{d}r = \int_{r<r_c^l} |\phi_l^{AE}|^2 r^2 \mathrm{d}r. \tag{A.49}$$

Various parameterization schemes exist to generate the norm-conserving pseudopotentials (e.g. Troullier and Martins (Troullier & Martins, 1991) parameterizations). Note that in the original formulations, the pseudopotentials are semi-local but they can easily be transformed into a separable form using the Kleinmann–Bylander scheme (Kleinman & Bylander, 1982). The non-local pseudopotential is used to accurately represent the combined effect of the nucleus and the core electrons, since different angular momenta can be scattered differently. The pseudopotential can be conventionally rewritten in a form that separates the long and short range components, where the long range component is local and corresponds to the Coulomb tail.

The norm-conserving requirement ensures that the logarithmic derivative of ϕ_l^{PS} (related to the phase shifts in the scattering) has the same behavior (up to the first order changes in energy) as in the all-electron case, which is often related to the transferability. For elements with strongly localized orbitals (i.e. $3d$ elements) the norm-conserving pseudopotentials require a large basis set of plane waves. To overcome this limitation one usually increases the cutoff radius, but this is in general not a good solution because the transferability is always considerably affected when r_c is increased. Many attempts to improve the norm-conserving pseudopotentials have been centered around the logarithmic derivative of the atomic and pseudo wavefunctions (Rignanese, 1998).

Vanderbilt proposed a radically new concept based on relaxing the norm-conserving constraint, introducing a so-called ultra soft pseudopotential (USPP) (Vanderbilt, 1990). As with norm-conserving pseudopotentials, the all-electron and pseudo wavefunctions are required to be equal outside r_c, but inside r_c they are allowed to be as smooth as possible. As a consequence, the pseudo wavefunctions are not normalized inside r_c, resulting in a charge deficit. This problem is solved by introducing a localized atom-centered augmentation charge, in which the correct pseudo-charge density accounts for the part of the charge (in the core region) that is not described by the pseudo wavefunctions ψ_i (Rignanese, 1998). The augmentation of the pseudo-density with appropriate functions (denoted $Q_{nm}^I(\mathbf{r})$) localized in the core region is defined as

$$n(\mathbf{r}) = \sum_i [\psi_i^*(\mathbf{r})\psi_i(\mathbf{r}) + \sum_{I,lm} Q_{nm}^I \langle \psi_i | \beta_n^I \rangle \langle \beta_m^I | \psi_i \rangle], \qquad (A.50)$$

where the functions β_n^I are strictly localized in the core region and are also used to define the nonlocal pseudopotential. The functions β_n^I and $Q_{nm}^I(\mathbf{r})$ are related to the atomic functions β_n and Q_{nm} by

$$\beta_n^I(\mathbf{r}) = \beta_n(\mathbf{r} - \mathbf{R}_I), \qquad (A.51)$$

$$Q_{nm}^I(\mathbf{r}) = Q_{nm}(\mathbf{r} - \mathbf{R}_I). \qquad (A.52)$$

The Q_{nm} are constructed in the atomic "pseudization" procedure in such way that, at the reference energies, the electron density of the pseudo wavefunctions as defined by Eq. A.50 is the same as the all-electron density. The functions β_n and Q_{nm} are obtained from first principles in the USPP scheme, and characterize the atomic species (Rignanese, 1998).

Although the pseudopotential approach has worked reliably for three decades, a new method called projector-augmented wavefunctions (PAW) has been introduced by Blöchl (Blöchl, 1994) to replace these pseudopotentials. In principle, this PAW approach is more accurate compared to others such as USPP for two main reasons. First, the cutoff distance or the extension in space of the pseudopotential is smaller in PAW than in USPP. Second, the PAW approach reconstructs the real wavefunction with all its nodes in the core region, while the USPP do not.

The central idea in the PAW method is to express the all-electron valence states ψ_i in terms of a smooth pseudo wavefunction $\tilde{\psi}_n$, which is augmented by a local basis set expansion and restricted to a small region, called the augmentation sphere, around each atom (Blöchl, 1994). The core states of the atoms are considered frozen. Given a smooth pseudo wavefunction, the corresponding all-electron wavefunction (that is orthogonal to the set of core orbitals) can be obtained through a linear transformation operator $\hat{\tau}$:

$$\psi_i(\mathbf{r}) = \hat{\tau}\tilde{\psi}_i(\mathbf{r}), \text{ with } \hat{\tau} = 1 + \sum_i |\phi_i^{\mathrm{AE}}\rangle \langle \tilde{p}_i | \tilde{\psi}_n - \sum_i |\tilde{\phi}_i^{\mathrm{PS}}\rangle \langle \tilde{p}_i | \tilde{\psi}_n, \qquad (A.53)$$

where $\tilde{p}_i$ are atom specific but system independent functions, which are only nonzero inside the augmentation sphere. The index i refers to the sum over the atomic sites, the angular momentum and reference energies. ϕ_i^{AE} and $\tilde{\phi}_i^{\mathrm{PS}}$ are the all-electron and pseudo-partial waves that match at the core radius. Finally, $\tilde{p}_i$ are the projector functions that have to be created in such a way as to be dual to the partial waves:

$$\langle \tilde{p}_i | \tilde{\psi}_j \rangle = \sigma_{ij}. \qquad (A.54)$$

A radial cutoff distance, r_c^{AE}, that defines the atomic augmentation sphere is selected, similarly to a cutoff radius for a pseudopotential. The larger is the augmentation sphere, the smoother are the pseudo wavefunctions. However, the overlapping with neighboring augmentation spheres should be avoided. For the all-electron valence states, smooth partial waves are constructed for $r > r_c^{AE}$ and one smooth projector is defined for each of the partial waves. In principle, an infinite number of projectors and partial waves are required for the PAW method to be exact (Blöchl, 1994). For practical calculations, an accurate dataset will need only one or two projection functions for each angular momentum. From the atomic frozen-core electron density $n_c^{AE}(r)$, a new smooth electron density $\tilde{n}_c^{AE}(\vec{r})$ is obtained. The latter must be identical to $n_c^{AE}(r)$ for radii larger than r_c^{AE}. The wavefunction and the atom-centered smooth-core electron density contribute to the whole pseudo-electron density. The true all-electron density is obtained from the pseudo-electron density. Finally, the PAW total energy is a function of the pseudo wavefunctions and of the occupation numbers (Blöchl, 1994).

A.4.5   Available DFT codes

Up to this point, the discussion has basically focused on the description of the DFT and its practical calculation to determine the electronic ground-state of an atomic system. Improvements in computer hardware and software allow simulations of materials with an increasing number of atoms. The computational load scales linearly (or with a higher power) with the number of atoms present in the simulation cells. It is thus of prime importance to choose the right software to be used in this context. Among the several existing DFT approaches, the choice of software is made according to the trade-off between numerical accuracy and system size. Several main DFT codes currently on the market are listed below as examples of software frequently used to predict the electronic and transport properties of carbon-based nanostructures:

(1) ABINIT (Gonze et al., 2005, Gonze et al., 2009) is a GPL licensed DFT package used worldwide. ABINIT is a package whose main program allows one to find the total energy, charge density and electronic structure of systems made of electrons and nuclei (molecules and periodic solids) within DFT, using pseudopotentials and a planewave or wavelet basis. Valence and core electrons are treated on different footings: the core electrons are frozen, and either replaced by norm-conserving pseudopotentials, or treated by the augmentation of plane waves by projectors (PAW method). The code also includes options to optimize the geometry according to the DFT forces and stresses, to perform molecular dynamics simulations using these forces, to generate dynamical matrices, Born effective charges, dielectric tensors based on density-functional perturbation theory (DFTP), and many more properties. Excited states can be computed within the many-body perturbation theory (the GW approximation and the Bethe–Salpeter equation), and time-dependent density functional theory (for molecules). More details are available on the website: **http://www.abinit.org**

(2) QUANTUM ESPRESSO (Giannozzi et al., 2009) is an integrated suite of open-source computer codes for electronic-structure calculations and materials modeling at the nanoscale. It is based on DFT, plane waves, and pseudopotentials. The

main program allows one to compute the ground-state properties within DFT using the pseudopotentials (norm-conserving and ultrasoft) and the PAW method. Several exchange–correlation functionals and some hybrid functionals are included in this code. More details and the code are available at: **http://www.quantum-espresso.org/**.

(3) SIESTA (Spanish initiative for electronic simulations with thousands of atoms) (Soler *et al.*, 2002) is both a method and its computer program implementation, to perform electronic structure calculations and *ab initio* molecular dynamics simulations of molecules and solids. This code uses the standard DFT with LDA and GGA approximations together with the norm-conserving pseudopotentials (in the form of Kleinman–Bylander). Atomic orbitals are used as a basis set, allowing unlimited multiple-zeta and angular momenta, polarization and off-site orbitals. SIESTA uses a finite 3D grid for the calculation of some integrals and the representation of charge densities and potentials. More details are available at: **http://www.icmab.es/dmmis/leem/siesta/**.

(4) CASTEP (Segall *et al.*, 2002, Clark *et al.*, 2005) is a leading code for calculating the properties of materials from first principles. Using density functional theory, it can simulate a wide range of materials properties including energetics, structure at the atomic level, vibrational properties, electronic response properties, etc. In particular, it has a wide range of spectroscopic features that link directly to experiment, such as infrared and Raman spectroscopies, NMR, and core level spectra. This software is a full-featured materials modeling code based on a first-principles quantum mechanical description of electrons and nuclei. It uses the robust methods of a plane-wave basis set and pseudopotentials. More details are available at: **http://www.castep.org**.

(5) VASP (The Vienna *ab initio* simulation package) (Kresse & Furthmüller 1996*a*, 1996*b*) is a computer program for atomic scale materials modeling, e.g. electronic structure calculations and quantum mechanical molecular dynamics, from first principles. It computes an approximate solution to the many-body Schrödinger equation, either within density functional theory (DFT), solving the Kohn–Sham equations, or within the Hartree–Fock (HF) approximation. Hybrid functionals that mix the Hartree–Fock approach with density functional theory are implemented as well. In VASP, central quantities, like the one-electron orbitals, the electronic charge density, and the local potential are expressed in plane-wave basis sets. The interactions between the electrons and ions are described using norm-conserving or ultrasoft pseudopotentials, or the projector-augmented-wave method. More details are available at: **http://www.vasp.at**.

Appendix B Electronic structure calculations: the many-body perturbation theory (MBPT)

B.1 Introduction

In Appendix A, a detailed description of the electronic structure calculation techniques based on the so-called density functional theory (DFT) was presented. As mentioned and illustrated in that section, DFT is widely used to investigate the electronic properties of materials, their defects, interfaces, etc. Unfortunately, the semi-local approximations of DFT, such as the local density approximation (LDA) and gradient generalized approximation (GGA), suffer from a well-known substantial underestimation of the band gap. This may be interpreted as a result of the fact that DFT does not properly describe excited states of a system. This failure of DFT may also induce a wrong estimation of the position of the electronic defect/dopant levels in the band gap.

Some empirical solutions exist to overcome the problem of DFT band gap underestimation. For example, the "scissor" technique consists in correcting the LDA/GGA gap error by shifting the conduction band up so as to match the gap relative to the experiment. However, such a method is not accurate enough for defining the accurate position of defect/dopant levels occurring in the band gap.

Another solution to the underestimation of the band gap in DFT consists in using the so-called hybrid functionals which have recently become very popular. Indeed, these functionals incorporate a fraction of Hartree–Fock (HF) exchange, which leads to improvement of the band gap compared to LDA/GGA (Curtiss et al., 1998, Muscat, Wander & Harrison, 2001, Paier et al., 2006). Yet, the fraction of HF exchange cannot be known in advance for all materials and its optimal value is material dependent (Ernzerhof, Perdew & Burke, 1997, Ernzerhof & Scuseria, 1999). Therefore the reliability of hybrid functionals cannot be assessed *a priori* (Kümmel & Kronik, 2008). Indeed, a recent theoretical work (Jain, Chelikowsky & Louie, 2011) demonstrates that the orbital energies from various existing hybrid functionals are not reliable in predicting the band gaps of all materials, either the optical or the quasiparticle gap. Even if a specific functional may give a good value for the bulk band gaps, the same functional in general does not yield accurate gap values for the same material in different configurations, such as at its surfaces or in nanostructures (Jain et al., 2011).

A more successful approximation for the determination of excited states is based on the many-body perturbation theory (MBPT) (Hedin & Lundqvist, 1970, Fetter & Walecka, 1971, Abrikosov, Gorkov & Dzyaloshinskii, 1975, Landau & Lifschitz, 1980, Onida, Reining & Rubio, 2002). Within the DFT scheme, the response of a system of interacting electrons to an external potential V_{ext} is that of independent particles responding to an "effective" potential. A similar idea is that the long-range, and relatively strong, Coulomb forces could screen the individual electrons, with a surrounding charge cloud of the other electrons. This leads to defining the quasiparticle as an electron plus its screening cloud. Thus, the response of strongly interacting particles can be described in terms of weakly interacting quasiparticles. Hence, MBPT offers an approach for obtaining quasiparticle (QP) energies in solids which is controlled and amenable to systematic improvements. The principal results of MBPT, as well as its practical use, are summarized in the following.

B.2   Many-body perturbation theory (MBPT)

B.2.1   Hedin's equations

In MBPT, the QP energies E_i^{QP} and wavefunctions ψ_i^{QP} are determined by solving the quasiparticle equation:

$$\left[-\frac{1}{2}\nabla^2 + V_{\text{ext}}(\mathbf{r}) + V_{\text{H}}(\mathbf{r}) \right] \psi_i^{\text{QP}}(\mathbf{r}) + \int \Sigma(\mathbf{r},\mathbf{r}';E_i^{\text{QP}})\psi_i^{\text{QP}}(\mathbf{r}')d\mathbf{r}'$$
$$= E_i^{\text{QP}} \psi_i^{\text{QP}}(\mathbf{r}), \tag{B.1}$$

where V_{ext} and V_{H} are the external and Hartree potentials, respectively. In this equation, the exchange and correlation effects are described by the electron self-energy operator $\Sigma(\mathbf{r},\mathbf{r}',E_i^{\text{QP}})$, which is nonlocal, energy dependent, and non-Hermitian. Hence, the eigenvalues E_i^{QP} are generally complex: their real part is the energy of the quasiparticle, while their imaginary part gives its lifetime.

The main difficulty resides in finding an adequate approximation for the self-energy operator Σ. Another key quantity is Green's function (Hedin & Lundqvist, 1970) $G(\mathbf{r},\mathbf{r}',t,t')$. It describes the probability of finding an electron with spin σ at time t and position $\mathbf{r}$, if another electron with spin σ' is added (or removed) at position $\mathbf{r}'$ at time t'. Considering the condition $\Sigma = 0$, the noninteracting (which still contains the Hartree potential) Green's function G_0 can be constructed from the one-particle wavefunctions ψ_i and energies E_i of the "zeroth-order" Hamiltonian as

$$G_0(\mathbf{r},\mathbf{r}',E) = \sum_i \frac{\psi_i(\mathbf{r})\psi_i^*(\mathbf{r}')}{E - E_i + i\eta\text{sgn}(E_i - \mu)} , \tag{B.2}$$

where μ is the chemical potential and η is a positive infinitesimal. Hedin (Hedin, 1965) proposed a systematic way to approximate the self-energy Σ by including a perturbation series expansion in the fully screened Coulomb interaction. The exact one-body Green's function G is thus written using the Dyson equation:

$$G(12) = G_0(12) + \int G_0(13)\Sigma(34)G(42)d(34). \tag{B.3}$$

Here, $1 \equiv (\mathbf{r}_1,\sigma_1,t_1)$ is used to denote space, spin, and time variables and the integral sign stands for summation or integration of all these where appropriate. 1^+ denotes

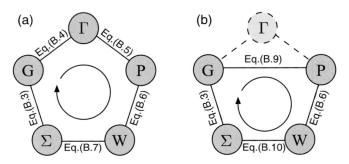

Figure B.1 (a) The self-consistent process for solving the complete Hedin's equations resulting from GW approximation. (b) Four coupled integro-differential equations resulting from the GW approximation. The so-called G_0W_0 approximation consists in performing the loop only once starting from $G = G_0$. (Adapted from Giantomassi *et al.* (2011)).

$t_1 + \eta$ where η is a positive infinitesimal. The self-energy Σ is obtained by solving self-consistently Hedin's closed set of coupled integro-differential equations:

$$\Gamma(12;3) = \delta(12)\delta(13) + \int \frac{\delta\Sigma(12)}{\delta G(45)} G(46)G(75)\Gamma(67;3)d(4567), \qquad \text{(B.4)}$$

$$P(12) = -i \int G(23)G(42^+)\Gamma(34;1)d(34), \qquad \text{(B.5)}$$

$$W(12) = v(12) + \int W(13)P(34)v(42)d(34), \qquad \text{(B.6)}$$

$$\Sigma(12) = i \int G(14)W(1^+3)\Gamma(42;3)d(34), \qquad \text{(B.7)}$$

where P is the polarizability, and W and v are the screened and the unscreened Coulomb interaction, respectively. Γ is the *vertex* function which describes higher-order corrections to the interaction between quasiholes and quasielectrons. The self-consistent iterative process is illustrated in Fig. B.1(a). The most complicated term in these equations is Γ, which contains a functional derivative and hence cannot in general be evaluated numerically.

B.2.2 GW approximation

In order to solve Hedin's equation, a possible strategy is to start with $\Sigma = 0$ and neglect the variation of the self-energy with respect to the Green's function $\delta\Sigma(12)/\delta G(45) = 0$ in Eq. (B.4). This leads to the Green's function G (at this step the Hartree independent-particle G_0) and the vertex function is set to a delta function as

$$\Gamma(12;3) = \delta(12)\delta(13). \qquad \text{(B.8)}$$

Thus, the polarizability in Eq. (B.5) writes

$$P(12) = -iG(12^+)G(21), \tag{B.9}$$

which corresponds to the random phase approximation (RPA) for the dielectric matrix. The screening also corresponds to RPA screening (W_0). The self-energy Σ in Eq. (B.7) is then a product of Green's function and of a screened Coulomb interaction:

$$\Sigma(12) = iG(12)W(1^+2), \tag{B.10}$$

where the Green's function is the one consistent with the Dyson's equation. In principle, this process should continue until self-consistency is reached (until the input Green's function equals the output one). However, in practice it has never been pursued. Instead, calculations usually stop once the self-energy $\Sigma = G_0 W_0$ (i.e. after one round), or search for the self-consistency of a reduced set of equations (see Fig. B.1(b)), short-cutting the vertex function. These approximations are called non-self-consistent (one-shot GW or $G_0 W_0$) and self-consistent GW approximation (GWA), respectively. More details can be found in Onida *et al.* (2002), and Giantomassi *et al.* (2011).

B.3 Practical implementation of $G_0 W_0$

B.3.1 Perturbative approach

In practical calculations, one needs a starting point for the independent-particle Green's function. The quasiparticle energies are more efficiently obtained from Eq. (B.1) than by solving the Dyson equation, Eq. (B.4). The approach consists in treating the difference of the self-energy and the Kohn–Sham potential (see Appendix A) as a perturbation. Despite some fundamental differences, the formal similarity is striking between the quasiparticle equation, Eq. (B.1), and the Kohn–Sham equation, Eq. (A.27), in Appendix A:

$$\left[-\frac{1}{2}\nabla^2 + V_{\text{ext}}(\mathbf{r}) + V_{\text{H}}(\mathbf{r}) \right] \psi_i^{\text{DFT}}(\mathbf{r}) + V_{\text{xc}}(\mathbf{r}) \psi_i^{\text{DFT}}(\mathbf{r}) = E_i^{\text{DFT}} \psi_i^{\text{DFT}}(\mathbf{r}), \tag{B.11}$$

where V_{xc} is the DFT exchange–correlation potential and E_i^{DFT} is the DFT energy. It turns out that the quasiparticle and the DFT wavefunctions are typically similar, at least for many simple bulk materials. For example, in silicon, the overlap between the quasiparticle and the DFT wavefunctions has been reported to be close to 99.9% (White *et al.*, 1998, Rohlfing *et al.*, 2003). Hence, E_i^{DFT} and ψ_i^{DFT} for the ith state are used as a zeroth-order approximation for their quasiparticle counterparts. The QP energy E_i^{QP} is then calculated by adding to E_i^{DFT} the first-order perturbation correction which comes from substituting the DFT exchange–correlation potential V_{xc} with the self-energy operator Σ:

$$E_i^{\text{QP}} = E_i^{\text{DFT}} + \langle \psi_i^{\text{DFT}} | \Sigma(E_i^{\text{QP}}) - V_{\text{xc}} | \psi_i^{\text{DFT}} \rangle. \tag{B.12}$$

To solve Eq. (B.12), the energy dependence of Σ must be known analytically, which is usually not the case. Under the assumption that the difference between QP and DFT

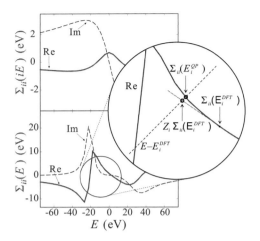

Figure B.2 The perturbative approach for finding the quasiparticle correction. In principle, the self-energy matrix element $\Sigma_{ii}(E) = \langle \psi_i^{\text{DFT}} | \Sigma(E) - V_{\text{xc}} | \psi_i^{\text{DFT}} \rangle$ and the true quasiparticle correction $\Sigma(E_i^{\text{QP}})$ are found from the solution of $E - E_i^{\text{DFT}} = \Sigma_{ii}(E)$, i.e. at the crossing of the dashed black line and $\Sigma_{ii}(E)$ in the circular zoom-in. In practice, the perturbative approach exploits the fact that it is computationally feasible to use the Taylor expansion around $\Sigma(E_i^{\text{DFT}})$ [Eqs. (B.14), B.15)] and to find an approximate value for the QP correction at the crossing of the grey and black dashed lines. (Extracted from Giantomassi *et al.* (2011).)

energies is relatively small, the matrix elements of the self-energy operator can be Taylor-expanded to the first-order around E_i^{DFT} to be evaluated at E_i^{QP}:

$$\Sigma(E_i^{\text{QP}}) \approx \Sigma(E_i^{\text{DFT}}) + (E_i^{\text{QP}} - E_i^{\text{DFT}}) \left. \frac{\partial \Sigma(E)}{\partial E} \right|_{E = E_i^{\text{DFT}}}. \tag{B.13}$$

In this expression, the QP energy E_i^{QP} can be solved for

$$E_i^{\text{QP}} = E_i^{\text{DFT}} + Z_i \langle \psi_i^{\text{DFT}} | \Sigma(E_i^{\text{DFT}}) - V_{\text{xc}} | \psi_i^{\text{DFT}} \rangle, \tag{B.14}$$

where Z_i is the *renormalization factor* defined by

$$Z_i^{-1} = 1 - \langle \psi_i^{\text{DFT}} | \left. \frac{\partial \Sigma(E)}{\partial E} \right|_{E = E_i^{\text{DFT}}} | \psi_i^{\text{DFT}} \rangle. \tag{B.15}$$

The principle is illustrated in Fig. B.2.

B.3.2   Plasmon pole

Another approximation that is often used concerns the screened Coulomb interaction W. In the calculation of $W = \epsilon^{-1} v$, the inverse dielectric function is a frequency-dependent matrix. The so-called plasmon-pole model of Godby–Needs (Godby & Needs, 1989) consists in substituting the frequency dependence of the imaginary part of every element of the matrix with just a narrow Lorentzian peak, which is related to the plasmon excitations of the system, since $\Im\left[\epsilon_{\mathbf{GG'}}^{-1}\right]$ is the loss function. This loss function is expressed as

$$\Im\left[\epsilon_{\mathbf{GG'}}^{-1}\right] = A_{\mathbf{GG'}}(\mathbf{q}) \times \left[\delta(\omega - \tilde{\omega}_{\mathbf{GG'}}(\mathbf{q})) - \delta(\omega + \tilde{\omega}_{\mathbf{GG'}}(\mathbf{q}))\right], \tag{B.16}$$

where $A_{\mathbf{GG'}}$ is the amplitude of a delta function centered at the plasmon frequency $\tilde{\omega}_{\mathbf{GG'}}(\mathbf{q})$. Using the Kramers–Kronig relations, the resulting dielectric function is then, in reciprocal space,

$$\Re\left[\epsilon_{\mathbf{GG'}}^{-1}\right] = \delta_{\mathbf{GG'}} + \frac{\Omega_{\mathbf{GG'}}^2(\mathbf{q})}{\omega^2 - \tilde{\omega}_{\mathbf{GG'}}^2(\mathbf{q})}, \tag{B.17}$$

where $\mathbf{G}$ is a reciprocal lattice vector and $\mathbf{q}$ a vector in the first Brillouin zone together with Ω and $\tilde{\omega}$ which are parameters giving the strength and the position of the poles, respectively. They can be obtained, for example, using the static screening and sum rules (Godby & Needs, 1989), or fitted to a full calculation along the imaginary energy axis. The exact formulation of the plasmon pole model is far beyond the scope of the present appendix. Note that other plasmon pole models have been proposed in the literature (Giantomassi *et al.*, 2011).

In summary, the theoretical bases of MBPT and Hedin's equations have been presented leading to the GW and G_0W_0 approximations. The perturbative approach, which is the most commonly used for obtaining the quasiparticle energies, is introduced together with the frequency dependence of the self-energy operator based on the plasmon pole model. This MBPT within the G_0W_0 approximation is frequently used in various chapters to model more accurately the electronic structures of various carbon-based nanostructures.

Appendix C Green's functions and *ab initio* quantum transport in the Landauer–Büttiker formalism

The Landauer–Büttiker (LB) formalism is widely used to simulate transport properties at equilibrium. The applications range from 1D conductors such as nanowires, nanotubes, nanoribbons, to 3D conductors such as molecular junctions with two or more contacts. At the *ab initio* level, this LB formalism is quite practical thanks to the Fisher–Lee relation, which connects the Landauer expression to the Green's function formalism. The transport properties of a given material can be simulated by finding the Green's function of the system within DFT (or even MBPT).

In this appendix the Green's function formalism is briefly reviewed. Section C.1 provides an introduction with a derivation of the trace formula starting from the Lippmann–Schwinger equations, then Section C.2 discusses recursive Green's function techniques, while Dyson's equation is introduced and applied to the case of a disordered system in Section C.3. Finally, Section C.4 is devoted to the implementation of LB formalism in conventional *ab initio* codes in order to investigate coherent electronic transport in nanoscale devices.

C.1 Phase-coherent quantum transport and the Green's function formalism

Green's functions are one of the most useful tools (Economou, 2006) for calculation of different physical quantities of interest such as the density of states or the quantum conductance and conductivity. In the context of phase-coherent quantum transport, they play a crucial role because their relation with the scattering matrix can be exploited to compute the quantum transmission probabilities as needed within the Landauer–Büttiker formalism presented in Section 3.3.

Although the real power of the Green's functions (GFs) appears when many-body effects such as electron–electron or electron–phonon interactions are taken into account, the subjects addressed in the present appendix are limited to a description of non-interacting systems.

Below, the particular transport setup depicted in Fig. C.1 is considered in a general scattering approach. This system is conveniently divided into three parts: (i) the left and (ii) right leads that are assumed to be made of perfect, defect-free crystalline materials, and (iii) a central region that can be any kind of set of atoms. This region, which potentially corresponds to the active part of the device, could thus be a single atom, a molecule, a section of carbon nanotubes, a finite-size graphene nanoribbon or even a

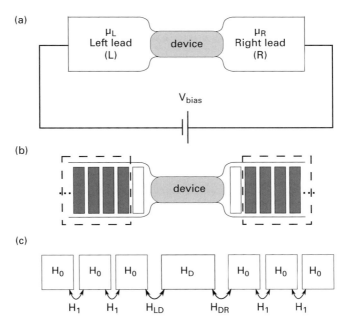

Figure C.1 A two-terminal transport setup: (a) connected to an external battery; (b) the device is contacted to two semi-infinite bulk electrodes assumed to be at thermal equilibrium; (c) block-decomposition of the Hamiltonian into sub-matrices associated with the central region and the principal layers of the semi-infinite leads. H_D contains all the interactions inside the central region. H_{LD} and H_{DR} describe the coupling between the central region and the leads. H_0 accounts for all interactions inside the principal layers (PLs) of the leads. H_1 describes the coupling between nearest-neighbors PLs.

bulk slab. In addition, the central part and the leads may either have the same cross-section dimensionality or not (e.g. a single molecule stretched between two massive gold electrodes). In summary, the system consists of two semi-infinite, defect-free, electrodes that are coupled to a central region, where all the scattering processes take place.

This system, composed of two semi-infinite, defect-free, electrodes that are coupled to a central region where all the scattering processes take place, can also be conveniently described by means of *principal layers*. A principal layer (PL) is the smallest set of atoms that represents a unit cell of the semi-infinite crystal forming the leads and which interacts only with the nearest-neighbor PLs. Therefore, owing to the decomposition into a set of localized basis functions, the Hamiltonian of the system under consideration can be written in matrix form as

$$
\mathcal{H} = \begin{pmatrix}
\cdots & \cdots & \cdots & \cdots & & & \\
H_{-1} & H_0 & H_1 & 0 & \cdots & & \\
0 & H_{-1} & H_0 & H_{LD} & 0 & \cdots & \\
\cdots & 0 & H_{DL} & H_D & H_{DR} & 0 & \cdots \\
& \cdots & 0 & H_{RD} & H_0 & H_1 & 0 \\
& & \cdots & 0 & H_{-1} & H_0 & H_1 \\
& & & \cdots & \cdots & \cdots & \cdots
\end{pmatrix}, \tag{C.1}
$$

where H_0 is the $[n \times n]$ matrix that describes all the interactions within the principal layers. Similarly (H_D) is the $[m \times m]$ matrix describing the interactions within the central region. Here, n and m are the dimensions of the basis set localized within a principal layer and the central region respectively. Finally, the interactions between nearest-neighboring PLs are contained in the $[n \times n]$ H_1 matrix, and the interactions between the central part and the left (right) lead is described in the $[n \times m]$ ($[m \times n]$) H_{LD} (H_{DR}) matrix. In the case of nonorthogonal localized orbitals, the overlap matrix (S) adopts the same block matrix form as the Hamiltonian $(\mathcal{H})$.

Such a transport setup presents an infinite hermitian problem whose solutions are not accessible by application of the Bloch theorem owing to the central region that breaks the translational symmetry. However, the problem can be solved by computing the retarded Green's function of the single-particle Schrödinger equation. Roughly speaking, this function gives the response at any point of the system due to a particular excitation at any other part. The retarded Green's function of the full system $(G^r(E))$ satisfies

$$[(E + i\eta) - \mathcal{H}] G^r(E) = \mathbb{I}, \tag{C.2}$$

where $\mathbb{I}$ is an infinite-dimensional identity matrix, $(E + i\eta)$ is the energy complemented with an infinitesimal positive imaginary part in order to ensure the causality (otherwise, if one takes $\eta \to 0^-$ one gets the advanced GF G^a). $\mathcal{H}$ and S are the infinite-dimensional matrix Hamiltonian and overlap matrix.

Another useful expression for the retarded GF can be given in terms of the complete, orthonormal, set of eigenfunctions $\{|\Psi_n\rangle\}$ of the defining operator,

$$\mathcal{H}|\Psi_n\rangle = E_n|\Psi_n\rangle, \tag{C.3}$$

$$G^r(E) = [(E + i\eta) - \mathcal{H}]^{-1} = \sum_n \frac{|\Psi_n\rangle\langle\Psi_n|}{E + i\eta - E_n}. \tag{C.4}$$

In addition to the retarded Green's function (G^r), it is convenient to introduce the system spectral function (A), defined as

$$A(E) = i\left(G(E) - G^\dagger(E)\right) = 2\pi \sum_n |\Psi_n(E)\rangle\langle\Psi_n(E)|. \tag{C.5}$$

The trace of the spectral function gives the density of states (times a constant factor 2π) and, within a spatial representation, its diagonal elements correspond to the local density of states (LDOS). Finally, it is interesting to mention some properties of the GF that allow us to establish a direct link between the retarded GF and the scattering theory. First, the retarded GF gives the response of the system to a source (f) added to the homogeneous Schrödinger equation:

$$(\mathcal{H} - (E + i\eta))|\psi(E)\rangle = |f\rangle. \tag{C.6}$$

Indeed,

$$|\psi(E)\rangle = -G^r(E)|f\rangle, \tag{C.7}$$

where $G^r = [\mathcal{H} - (E + i\eta)]^{-1}$.

A second interesting property arises when considering a perturbation V. Assuming that the solutions of the unperturbed Schrödinger equation are known, $(\mathcal{H}^0 |\Psi_n^0\rangle = E_n |\Psi_n^0\rangle)$, the solutions of the perturbed Schrödinger equation,

$$\left(\mathcal{H}^0 + V - (E + i\eta) \right) |\psi(\mathbf{r})\rangle = 0 \qquad (C.8)$$

are given by

$$|\psi(E)\rangle = |\psi^0(E)\rangle + G^0(E)\, V |\psi(E)\rangle, \qquad (C.9)$$

or equivalently

$$|\psi(E)\rangle = |\psi^0(E)\rangle + G(E)\, V |\psi^0(E)\rangle, \qquad (C.10)$$

where the $G^0(E)$ and $G(E)$ functions are respectively the retarded GFs of the unperturbed and perturbed system. These relations are called the Lippmann–Schwinger equations and can be applied to the scattering problem. Indeed, this procedure allows us to straightforwardly connect the bulk propagating states of the isolated leads ($|\phi_{k_j}^L\rangle$), to the stationary scattering states of the complete system ($|\Psi_{k_j}^L\rangle$). Assuming that the perturbation of the system V corresponds to the coupling between the leads and the central region, i.e. $V = H_{LD} + H_{DL} + H_{DR} + H_{RD}$, the Lippmann–Schwinger Eq. (C.10) can be written

$$|\Psi_{k_j}^L\rangle = |\phi_{k_j}^L\rangle + G(E_{k_j})\, V |\phi_{k_j}^L\rangle. \qquad (C.11)$$

Let us now apply the Green's function formalism to the transport problem illustrated in Fig. C.1. Though the matrix that has to be inverted is infinite-dimensional, the Green's function formalism allows us to account naturally for the open boundary conditions that rule the asymptotic behavior of the transport problem. In the following, the Green's function of the central part is shown to be easily calculated separately, without calculating the whole Green's function. Using the definition of the retarded Green's function, Eq. (C.2) reads

$$\begin{pmatrix} \epsilon^+ S_L - H_L & \epsilon^+ S_{LD} - H_{LD} & 0 \\ \epsilon^+ S_{DL} - H_{DL} & \epsilon^+ S_D - H_D & \epsilon^+ S_{DR} - H_{DR} \\ 0 & \epsilon^+ S_{RD} - H_{RD} & \epsilon^+ S_R - H_R \end{pmatrix}$$

$$\times \begin{pmatrix} G_L & G_{LD} & G_{LR} \\ G_{DL} & G_D & G_{DR} \\ G_{RL} & G_{RD} & G_R \end{pmatrix} = \mathbb{I}, \qquad (C.12)$$

where $\epsilon^+ = (E + i\eta)$, and the Hamiltonian, overlap, and Green's matrices have been divided into sub-matrices corresponding to the left/right leads and the central region. The $H_{L/R}$ sub-matrices account for all interaction inside the left and right leads. These blocks are thus infinite-dimensional $[\infty \times \infty]$, as well as $S_{L/R}$ and $G_{L/R}$. On the contrary, the sub-matrices of interest (i.e. S_D, H_D, and G_D) have the dimension $[m \times m]$, where m is the number of degrees of freedom (i.e. orbitals) in the central region. Finally, the sub-matrices H_{LD} and H_{DR} that represent the interactions between the leads and the central region have the dimension $[\infty \times m]$ and $[m \times \infty]$, respectively. However, these sub-matrices (H_{LD} and H_{DR}) are actually zero with the exception of the bottom/top n lines that represent the coupling between the last PL and the central region (i.e. the H_{LD} and H_{DR} matrices introduced in Eq. (C.1)). Here, n is the number of degrees of freedom (i.e. orbitals) in the PL.

Selecting the three block-equations that involve G_D, the following system of equations is readily derived:

$$\left(\epsilon^+ S_L - H_L\right) G_{LD} + \left(S_{LD} - H_{LD}\right) G_D = 0, \tag{C.13}$$

$$\left(S_{DL} - H_{DL}\right) G_{LD} + \left(\epsilon^+ S_D - H_D\right) G_D + \left(S_{DR} - H_{DR}\right) G_{RD} = 0, \tag{C.14}$$

$$\left(S_{RD} - H_{RD}\right) G_D + \left(S_R - H_R\right).G_{RD} = 0. \tag{C.15}$$

Substituting Eqs. (C.13) and (C.14) into Eq. (C.15), an explicit expression for the sub-matrix G_D is obtained, and has the following form:

$$G_D^r(E) = \left[\epsilon^+ S_D - H_D^{\text{eff}}(E)\right]^{-1}, \tag{C.16}$$

with

$$H_D^{\text{eff}}(E) = \left[H_D - \Sigma_L^r(E) - \Sigma_R^r(E)\right], \tag{C.17}$$

where the concept of the retarded self-energies associated with the left and right leads is introduced,

$$\begin{aligned} \Sigma_L^r(E) &= \left[\epsilon^+ S_{DL} - H_{DL}\right] G_L^{0r} \left[\epsilon^+ S_{LD} - H_{LD}\right], \\ \Sigma_R^r(E) &= \left[\epsilon^+ S_{DR} - H_{DR}\right] G_R^{0r} \left[\epsilon^+ S_{RD} - H_{RD}\right]. \end{aligned} \tag{C.18}$$

Here, G_α^{0r} are the retarded Green's function associated with the isolated leads $\alpha = L, R$,

$$G_\alpha^{0r} = \left[\epsilon^+ S_\alpha - H_\alpha\right]^{-1}. \tag{C.19}$$

Although this last equation still contains infinite matrices, the self-energy can be computed easily by exploiting the fact that the interaction between the central region and the leads only involves a finite number of atoms close to the interface. Besides, owing to the sparsity of $\{H_{DL}, S_{LD}, H_{DR}, S_{RD}\}$, only the n lines of $G_{L/R}^{0r}$ that correspond to the surface PL are actually needed in Eq. (C.18). This sub-matrix $G_{L/R}^{0r}$ has the finite dimension $[n \times n]$ and can be computed either semi-analytically or by recursion methods (see Section C.2).

In summary, the Green's function expressed in Eq. (C.16) is defined within the central region only while the effects of the contact are mapped into complex self-energies. The imaginary part of the self-energies derives from the finite lifetime of the electronic states inside the central region. Indeed, due to the interactions with the contact, eigenstates can leak out from the central region.

In order to set up a complete formalism, an explicit expression for the out-of-equilibrium density matrix in the central region is needed. In the absence of inelastic processes, the form of the out-of-equilibrium density matrix can be easily deduced from the physical meaning of the spectral function (Paulsson & Brandbyge, 2007). Indeed, the spectral function is by definition closely related to the density-of-states (see Eq. (C.5)). Using the projector operator P_D, that projects onto the central region, the spectral function of the central region can be written as $A_D = P_D A P_D$. At equilibrium ($\mu_L = \mu_R = \mu_D$) the complete system is characterized by a single Fermi–Dirac distribution function, and the electronic density can be expressed as

$$\rho = \frac{1}{2\pi} \int A_D f(E - \mu_D) \, dE. \tag{C.20}$$

In general, this expression is not valid once a bias is applied between the electrodes ($\mu_L - \mu_R \neq 0$, $V_{\text{bias}} \neq 0$ in Fig. C.1(a)) but remains valid in the noninteracting case. Each scattering eigenstate can be considered in equilibrium with one of the two reservoirs characterized by μ_L and μ_R. The spectral function of the device A_D can be decomposed in two components A_L and A_R, that are associated with the scattering states incoming from the left and right leads, respectively. Upon the assumption of coherent transport only, an expression for the out-of-equilibrium density matrix can thus be derived:

$$\begin{aligned}
A_D &= i\,(G_D - G_D^\dagger) = i\,G_D(G_D^{\dagger -1} - G_D^{-1})G_D^\dagger \\
&= G_D \Gamma_L G_D^\dagger + G_D \Gamma_R G_D^\dagger \\
&= A_L + A_R.
\end{aligned} \tag{C.21}$$

Here, the broadening $\Gamma_{L/R} = i\left[\Sigma_{L/R}^r - \Sigma_{L/R}^{r\dagger}\right]$ is introduced in order to account for the nonhermicity of the self-energy operators. At this point, the infinitesimal imaginary part introduced in Eq. (C.2) has been neglected, in order to ensure the causality of the retarded Green's function. As a consequence, the second equality in Eq. (C.21) is not correct when there are electronic states strictly localized in the device region that do not couple with the electrodes. In such a case, $\Gamma_L = \Gamma_R = 0$, and the information about the localized states is neither contained in A_L, nor in A_R.

Using the Lippmann–Schwinger equation, Eq. (C.10), the two parts of the device spectral function can be built up from states originating from the left and right electrodes respectively. Indeed, the part A_L can be expressed as

$$\begin{aligned}
A_L &= 2\pi \sum_j P_D |\Psi_j^L\rangle\langle\Psi_j^L| P_D \\
&= 2\pi \sum_j P_D \left(|\phi_j^L\rangle + G V_L |\phi_j^L\rangle\right) \left(\langle\phi_j^L| + \langle\phi_j^L| V_L^\dagger G^\dagger\right) P_D \\
&= G_D \Gamma_L G_D^\dagger.
\end{aligned} \tag{C.22}$$

Since the same procedure can be applied for the part A_R, in the absence of inelastic processes, the out-of-equilibrium density matrix of the central region can be expressed as

$$\rho = \frac{1}{2\pi} \int \left[A_L f(E - \mu_L) - A_R f(E - \mu_R) \right] dE, \qquad (C.23)$$

where the scattering states incoming from the left (right) are populated according to the equilibrium Fermi distribution of the left (right) electrode.

At this point, a complete scheme has been built in terms of the Green's functions, and the electric current I through the device can be computed. Indeed, once the density operator is defined, it is possible to evaluate the current operator. After some manipulations (Meir & Wingreen, 1992), the expression for the electric current through the junction can be easily derived:

$$I = \frac{2e}{h} \int Tr \left[\Gamma_L G_D^{r\dagger} \Gamma_R G_D^r \right] \left[f(E - \mu_L) - f(E - \mu_R) \right] dE, \qquad (C.24)$$

where the factor 2 on the right-hand side takes into account the spin degeneracy. Expression (C.24) is the Landauer formula (Landauer, 1957) for a two-terminal system where the transmission probability is written as a trace over the transmission matrix $[\Gamma_L G_D^{r\dagger} \Gamma_R G_D^r]$.

C.2   Self-energy corrections and recursive Green's functions techniques

Recursive Green's functions techniques must be in your toolbox, as they offer an efficient way to compute the self-energy corrections due to the semi-infinite leads and the Green's functions for the sample region (Guinea *et al.*, 1983, Pastawski, Weisz & Albornoz, 1983, Lopez Sancho, Sancho & Rubio, 1985). This is specially the case for tight-binding Hamiltonians, where their sparse nature can be fully exploited with these techniques. (For a detailed review of recursive Green's function methods see Lewenkopf & Mucciolo (2013).)

In the previous section we considered a tripartite system and expressed the GFs of the central region in terms of an effective [$m \times m$] Hamiltonian. The other two parts of the system are taken into account exactly through the self-energy corrections. If these two parts consisted, say, of only one layer, then Eq. (C.18) together with Eq. (C.19) would provide closed expressions for the self-energy corrections.

For semi-infinite leads further work is needed. One can take advantage of the fact that the layers are coupled following a nearest-neighbors structure. Indeed, the Hamiltonian has a tridiagonal block-matrix form that can be exploited to iteratively incorporate the effect of the leads layer by layer, following each time the procedure described before for the tripartite system. This is schematically represented in Fig. C.2. The upper line, step (0), represents the principal layers of the semi-infinite lead's Hamiltonian numerated from 0 on, H_0 is the intralayer block-matrix Hamiltonian while $H_{1(-1)}$ is the interlayer Hamiltonian connecting with the next layer to the left (right). Throughout this section an orthonormal basis will be considered, the case when the overlap matrix is different from the identity matrix can be worked out easily provided that it preserves the block tridiagonal structure.

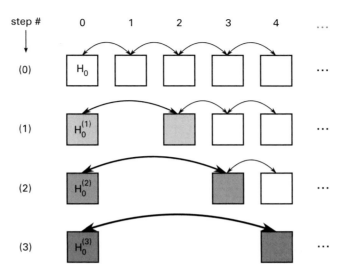

Figure C.2 Representing the recursive procedure described in the text for the calculation of the surface Green's functions. At each step one layer is eliminated and its effects included through a renormalization of the adjacent layer's Hamiltonian and the matrix elements between them.

At step (j), layer j is eliminated, thereby renormalizing the adjacent block Hamiltonian matrices. Their updated values are denoted by $H_0{}^{(j)}$ (marked with gray in Fig. C.2); the effective Hamiltonian is then

$$\mathcal{H}_{\text{lead}}^{\text{eff}(n)} = \begin{pmatrix} H_0^{(j)} & H_1^{(j)} & 0 & \dots & \\ H_{-1}^{(j)} & H_0^{(j)} & H_1 & 0 & \dots \\ 0 & H_{-1} & H_0 & H_1 & 0 & \dots \\ & \dots & \dots & \dots & \dots & \end{pmatrix}, \tag{C.25}$$

where

$$H_0^{(j)}(E) = H_0^{(j-1)} + H_{-1}^{(j-1)} G_{(j-1)}^{0r}(E) H_1^{(j-1)}, \tag{C.26}$$

$$H_{\pm 1}^{(j)}(E) = H_{\pm 1}^{(j-1)} G_{(j-1)}^{0r}(E) H_{\pm 1}^{(j-1)}, \tag{C.27}$$

$$G_{(j-1)}^{0r}(E) = ((E + i\eta)\mathbb{I} - H_0^{(j-1)}(E))^{-1}, \tag{C.28}$$

$$H_0^{(0)} = H_0, H_{\pm 1}^{(0)} = H_{\pm 1}. \tag{C.29}$$

By setting a small imaginary part η, after a large enough number of decimation steps $H_0^{(n)}(E)$ will converge and the associated surface GFs can be computed. This closes the loop for the calculation of the self-energy corrections Eq. (C.18). For N decimation steps, the computational cost will scale as $N \times n^3$ (since matrix inversions require n^3 operations at each step).

An alternative and much faster scheme (Lopez Sancho, Sancho & Rubio, 1985) that fully exploits the translational invariance within the leads is represented in Fig. C.3. Here, at each step, instead of one, half of the layers are eliminated. The effective block

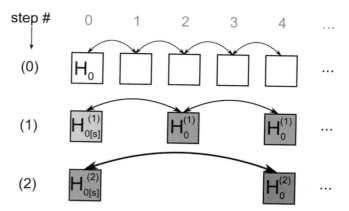

Figure C.3 An alternative and more efficient scheme for calculation of the surface Green's functions. In contrast to the procedure shown in Fig. C.2, at each step half of the layers are eliminated.

matrix Hamiltonian is not uniform, but distinguishes the surface layer from the bulk ones, and after j iteration steps it is given by

$$
\mathcal{H}_{\text{lead}}^{\text{eff}(n)} = \begin{pmatrix} H_{0[s]}^{(j)} & H_1^{(j)} & 0 & \cdots \\ H_{-1}^{(j)} & H_0^{(j)} & H_1^{(j)} & 0 & \cdots \\ 0 & H_{-1}^{(j)} & H_0^{(j)} & H_1 & 0 & \cdots \\ & \cdots & \cdots & \cdots & \cdots \end{pmatrix},
\tag{C.30}
$$

where

$$
H_{0[s]}^{(j)}(E) = H_{0[s]}^{(j-1)} + \Sigma_+(E),
\tag{C.31}
$$

$$
H_0^{(j)}(E) = H_0^{(j-1)} + \Sigma_+(E) + \Sigma_-(E),
\tag{C.32}
$$

$$
H_{\pm 1}^{(j)}(E) = H_{\pm 1}^{(j-1)} G_{(j-1)}^{0r}(E) H_{\pm 1}^{(j-1)},
\tag{C.33}
$$

$$
\Sigma_\pm(E) = H_{\mp 1}^{(j-1)} G_{(j-1)}^{0r}(E) H_\pm^{(j-1)},
\tag{C.34}
$$

$$
G_{(j-1)}^{0r}(E) = ((E + i\eta)\mathbb{I} - H_0^{(j-1)}(E))^{-1},
\tag{C.35}
$$

$$
H_0^{(0)} = H_{0[s]}^{(0)} = H_0, H_{\pm 1}^{(0)} = H_{\pm 1}.
\tag{C.36}
$$

After N steps, the surface layer Hamiltonian $H_{0[s]}^{(N)}$ incorporates the effect of $2^N - 1$ neighboring layers (to be compared with N in the previous scheme). The schemes presented here can be used as well to compute the Green's functions for the sample as needed for calculation of the conductance using the trace formula. In that case one must note that only GFs between the first and last layer in the device region are needed because of the sparsity of the self-energies, which only have nonvanishing elements on the sites of the device that are connected to the leads, as discussed in the paragraph after Eq. (C.19). Furthermore, when a mode decomposition such as the one explained in Section 4.2.2 is feasible, the reduction in the computational cost is enormous.

In cases where the layer dimensions are irregular across the sample or when additional leads are connected to it, the schemes above may fail or just be inefficient. A generalization for such cases, called a *knitting algorithm*, was presented in Kazymyrenko & Waintal (2008), and associated resources are also available on the web (http://inac.cea.fr/Pisp/xavier.waintal/KNIT.php). A new related project called Kwant is also under development (http://kwant-project.org/). Different codes related to this and other techniques are also available on the authors' website.

C.3   Dyson's equation and an application to treatment of disordered systems

Equation (C.4) provides a very elegant representation of the GFs in terms of the system eigenstates. The trouble is that most of the time those eigenstates are unknown. However, if the eigenstates of a given Hamiltonian $\mathcal{H}_0$ are known, then one expects that if an extra term $\mathbf{V}_{\mathrm{dis}}$, describing for example the disorder strength, remains weak enough to be treated as a perturbation, its effect can be captured (to a first approximation) through the induced elastic transition between states of the otherwise clean system. The exact GF for the system is

$$((E + \mathrm{i}\eta)\mathbb{I} - \hat{\mathcal{H}}_0 - \mathbf{V}_{\mathrm{dis}})G^r(E) = \mathbb{I}, \tag{C.37}$$

or

$$((E + \mathrm{i}\eta)\mathbb{I} - \hat{\mathcal{H}}_0)G^r(E) = (\mathbb{I} + \mathbf{V}_{\mathrm{dis}})G^r(E), \tag{C.38}$$

which can be recast in the so-called Dyson equation,

$$G^r(E) = G_0^r(E) + G_0^r(E)\mathbf{V}_{\mathrm{dis}}G^r(E), \tag{C.39}$$

which is a recursive equation valid whatever the strength of the potential $\mathbf{V}_{\mathrm{dis}}$. Given its form, the Dyson equation can be expanded to make explicit all multiple scattering events as

$$
\begin{aligned}
G^r(\varepsilon) &= G_0^r(E) + G_0^r(E)\mathbf{V}_{\mathrm{dis}}G_0^r(E) + G_0^r(E)\mathbf{V}_{\mathrm{dis}}G_0^r(E)\mathbf{V}_{\mathrm{dis}}G_0^r(E) + \cdots \\
&= G_0^r(E)\left(\mathbb{I} + \mathbf{V}_{\mathrm{dis}}G_0^r(E) + (\mathbf{V}_{\mathrm{dis}}G_0^r(E))^2 + (\mathbf{V}_{\mathrm{dis}}G_0^r(\varepsilon))^3 + \cdots\right). \quad (\mathrm{C}.40)
\end{aligned}
$$

In the presence of the disorder potential, the eigenstates of H_0 acquire a finite lifetime due to elastic scattering. To compute such a new timescale, one can consider the propagation probability amplitudes $\langle k|G^r(E)|k'\rangle$, given by

$$
\begin{aligned}
\langle k|G^r(E)|k'\rangle &= G_0^r(E, k)\delta_{k,k'} + G_0^r(E, k)\langle k|\mathbf{V}_{\mathrm{dis}}|k'\rangle G_0^r(E, k) \\
&+ \int \frac{d^d q}{(2\pi)^d}G_0^r(E, k)\langle k|\mathbf{V}_{\mathrm{dis}}|q\rangle G_0^r(E, q)\langle q|\mathbf{V}_{\mathrm{dis}}|k'\rangle G_0^r(E, k') + \cdots, \quad (\mathrm{C}.41)
\end{aligned}
$$

where the expansion should be made to all orders of the multiple scattering events (see Roche *et al.*, 2006). There are plenty of possible sources of disorder, such as a random distribution (with density n_i) of structural defects or charged impurities for instance, described by a screened Coulomb potential of the type $V(r - R_i) \sim -e^2/|r - R_i|^2 \times e^{-|r-R_i|/\lambda}$.

In practice, such a calculation for a particular disorder configuration is analytically and numerically out of reach without further approximation. Fortunately, when dealing with transport properties in disordered materials, one is first mainly interested by some disorder-average property, and if existing some universalities of transport features. Transport coefficients can be generally computed from disorder-averaged Green function elements such as $\langle k|G^r(E)|k'\rangle_{\text{dis}}$, where an average over a disorder statistics is performed. Disorder-averaging turns out to restore translational invariance of the computed quantity, an essential point to evaluate how averaged Green functions are renormalized in the presence of a weak disorder (treated perturbatively). The Dyson equation (also expressed using Feynman diagrams as illustrated in Fig. C.4) is essentially rewritten as (we take $k' = k$ for simplicity without loss of generality)

$$\langle k|G^r(E)|k\rangle_{\text{dis}} = G_0^r(E, k) + G_0^r(E, k) \Sigma_k(E) \langle k|G^r(E)|k\rangle_{\text{dis}}, \qquad (C.42)$$

where $\Sigma_k(E)$ encompasses the whole ensemble of so-called irreducible Feynman diagrams of $\langle k|G^r(E)|k\rangle_{\text{dis}}$, which cannot be decomposed into another subset of diagrams (see Fig. C.5). Equation (C.42) is solved easily and gives

$$\langle k|G^r(E)|k\rangle_{\text{dis}} = \frac{1}{E - \varepsilon_k - \Sigma_k(E)}. \qquad (C.43)$$

The calculation of $\Sigma_k(E)$ can be achieved to a given order of perturbation theory, which obviously depends on the disorder strength compared to the energy scale defining the unperturbed structure ($\pi - \pi$ hopping term in sp^2 carbon structures). At the lowest order of the so-called Born approximation, $\Sigma_k(E) = n_i V_0 = n_i \int d\mathbf{r} V(\mathbf{r})$ (the first diagram in Fig. C.5), the average Green function is just modified by a constant shift of all

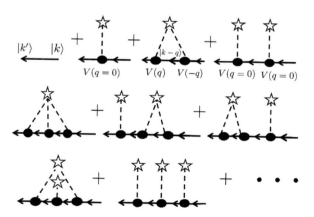

Figure C.4 Feynman diagrams representing probability amplitude $\langle k|G^r(E)|k'\rangle_{\text{dis}}$ up to the third order of perturbation. Star symbols denote the averaged scattering impurity potential, the number of stars giving the order of the perturbation. Filled circles pinpoint changes of the momentum direction upon scattering.

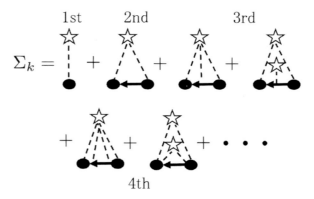

Figure C.5 Irreducible Feynman diagrams in the self-energy $\Sigma_k(\varepsilon)$ up to third order plus two fourth-order diagrams for illustration.

energy levels with no dynamical consequence. In contrast, at the first order of the Born approximation (second diagram of Fig. C.5), $\Sigma_k(E) = \sum_q |V(k-q)|^2 \frac{1}{E-\varepsilon_q+i\eta}$ (note that $V(k-q) = 1/\Omega \int d\mathbf{r} e^{(\mathbf{k}-\mathbf{q}).\mathbf{r}} V(\mathbf{r})$, which has real and imaginary parts, $\Im m\Sigma_k(E)$ moves the poles of the Green function away from the real axis, and is related to the finite lifetime of propagating states conveyed by the initial eigenstates of unperturbed Hamiltonian). In the Born approximation, to the second order of perturbation theory, $\Sigma_k(E)$ is derived as

$$\Re e\Sigma_k(E) = n_i V(q=0) + \sum_q \frac{|V(q)|^2}{\varepsilon_k - \varepsilon_q}, \tag{C.44}$$

$$\Im m\Sigma_k(E) = -\pi n_i \sum_q |V(k-q)|^2 \delta(\varepsilon_k - \varepsilon_q) = \frac{\hbar}{\tau_k}, \tag{C.45}$$

which is also known as the Fermi golden rule (τ_k is the elastic relaxation time). Using such an expression for the self-energy, the impurity average Green function in energy becomes

$$\langle k|G^r(E)|k\rangle_{\text{dis}} = \frac{1}{E - (\varepsilon_k + \Re e\Sigma_k(E)) + \dfrac{i\hbar}{\tau_k}}. \tag{C.46}$$

By Fourier-transforming such a Green function, one obtains the time- and space-dependent propagators as

$$\langle k|G^r(t)|k\rangle_{\text{dis}} = \int \frac{dE}{2\pi} \frac{e^{-i(E+i\eta)t}}{E - (\varepsilon_k + \Re e\Sigma_k(E)) + \dfrac{i\hbar}{\tau_k}} = -i\theta(t)e^{-i\varepsilon_k t}e^{-t/\tau_k},$$

$$\langle k|G^r(\mathbf{r})|k\rangle_{\text{dis}} = -\frac{\pi\rho(E)}{k_F.r} e^{ik|r|} e^{-|r|/\ell_{k_F}},$$

with $\ell_{k_F} = v_k \tau_{k_F}$. The impurity scattering transforms the free electrons into quasi-particles with a finite lifetime given by scattering time, and a finite elastic mean free path. When dealing with transport coefficients, one needs to evaluate the impurity-average of two-particle Green functions. Indeed, the Kubo–Greenwood formula (in d dimensions) for the quantum conductivity can be rewritten in terms of Green functions as (Kubo, 1966)

$$\sigma_{xx} = e^2 \int \frac{d^d k}{(2\pi)^d} \frac{dE}{2\pi} v_x^2(k) \frac{f(E) - f(E + \hbar\omega)}{\hbar\omega} G^r(k, E) G^a(k, E + \hbar\omega), \qquad \text{(C.47)}$$

developed on the basis of eigenstates of the unperturbed Hamiltonian. The perturbation introduced by a weak disorder potential yields a finite dissipation, which guarantees a finite conductivity. Similarly to the one-particle Green function case, the calculation of the propagator $G^r(k, E) G^a(k, E + \hbar\omega)$ is achieved using impurity-averaging and restored translational invariance. The Dyson equation generalized to the two-particles Green function is named the Bethe–Salpeter equation, which gives

$$\langle G^r(k, E) G^a(k', E + \hbar\omega)\rangle_{\text{dis}} = \delta(k, k') \langle G^r(k, E)\rangle_{\text{dis}} \langle G^a(k', E + \hbar\omega)\rangle_{\text{dis}}$$
$$+ \langle G^r(k, E)\rangle_{\text{dis}} \langle G^a(k, E + \hbar\omega)\rangle_{\text{dis}} \mathcal{C}(k, k', \omega) \langle G^r(k', E)\rangle_{\text{dis}} \langle G^a(k', E + \hbar\omega)\rangle_{\text{dis}}.$$
$$\text{(C.48)}$$

The first term is the classical (diffusion) term, which excludes all quantum interferences and obeys a classical diffusion equation. The second term contains all constructive quantum interferences that survive to the impurity-averaging process, and which are condensed in the so-called Cooperon term $\mathcal{C}(k, k', \omega)$ which contains all irreducible Feynman diagrams describing such interferences. This is the foundation of weak localization theory, and by developing the perturbation series for the two-particles Green function, the general form of $\mathcal{C}(k, k', \omega)$ is derived and shown to obey the equation

$$\mathcal{C}(k, k', \omega) = \langle G^r(k, E)\rangle_{\text{dis}} V(q = 0) \langle G^a(k', E + \hbar\omega)\rangle_{\text{dis}}$$
$$+ \sum_q |V(q)|^2 \langle G^r(k - q, E)\rangle_{\text{dis}} \langle G^a(k' + q, E + \hbar\omega)\rangle_{\text{dis}} |V(-q)|^2 + \cdots$$

Assuming isotropic scattering, $|V(q)|^2 = C_0 = \hbar/(2\pi\rho(E)\tau)$, and defining

$$\Pi = \sum_q \langle G^r(k - q, E)\rangle_{\text{dis}} \langle G^a(k' + q, E + \hbar\omega)\rangle_{\text{dis}}, \qquad \text{(C.49)}$$

the infinite series rewrites $\mathcal{C}(k, k', \omega) = C_0 + C_0 \Pi C_0 + C_0 \Pi C_0 \Pi C_0 + \cdots$, or $\mathcal{C}(k, k', \omega) = C_0(1 + \Pi C_0 + (\Pi C_0)^2 + \cdots) = C_0/(1 - \Pi C_0)$, which finally simplifies to

$$\mathcal{C}(k, k', \omega) = \frac{\hbar}{2\pi\rho(E)\tau^2} \frac{1}{D(k + k') - i\omega}. \qquad \text{(C.50)}$$

It is clear that such a Cooperon term presents a divergence when $k' = -k$, which pinpoints a mathematical pole of the interferences when considering the backscattering probability. The quantum correction of the conductivity is then given by

$$\delta\sigma(\omega) = -\frac{2e^2}{\pi\hbar} D\tau \int \frac{d^d q}{(2\pi)^d} \frac{1}{Dq^2\tau - i\omega\tau}. \qquad \text{(C.51)}$$

The integral remains finite owing to the physical cutoff (elastic τ_{el} and coherence times τ_φ) that need to be introduced. In two dimensions a straightforward integral calculation yields

$$\delta\sigma(\omega) = -\frac{e^2}{2\pi^2\hbar}\ln\frac{\tau_\varphi}{\tau_{el}}, \tag{C.52}$$

which is the basis of the scaling analysis of the localization theory. Equation (C.52) is used when applying the Kubo method in disordered graphene-based materials.

C.4   Computing transport properties within *ab initio* simulations

In Section C.1, the computation of the Green's functions for the central region is shown to rely on a proper evaluation of the self-energies associated with the left and right electrodes. In Section C.2 we had an overview of some of the most common recursive Green functions methods as typically used for tight-binding Hamiltonians. Here we revisit this issue and then present the main steps involved in *ab initio* simulations of quantum transport within a self-consistent Landauer–Büttiker scheme.

Since these self-energies have to be computed for each transverse k-vector, and at several energies, a stable and efficient computational algorithm is crucial for the code performance. The main computational cost in calculating the self-energies of the contacts is related to evaluation of the retarded Green's function of the isolated leads. Within several packages (SMEAGOL, TRANSIESTA, etc.; see below), these Green's functions are constructed following a semi-analytical scheme (Rocha, 2007) globally composed of three steps. First, the Bloch states of the infinite system are derived. Second, these Bloch states are used to build the retarded Green's function corresponding to the infinite leads. Then, appropriate boundary conditions are applied in order for the Green's function to vanish at the free surface of the semi-infinite leads. The three steps are described in more detail below.

(1) Owing to the division of the leads into "principal layers" (PLs – see Fig. C.1), the Hamiltonian and the overlap matrices are arranged in the trigonal form described in Eq. (C.1). In this formulation, H_0 and S_0 account for all the interactions and overlap integrals inside one PL. H_1 and S_1 are the off-diagonal blocks that correspond to the interactions and overlap integrals between the nearest-neighbor PLs. In order to simplify the notation, the matrices $K_\alpha = H_\alpha - ES_\alpha$ with ($\alpha = 0, 1, -1$) are introduced. In the case of an infinite periodic lead, the Bloch theorem applies along the transport direction (z axis) and the bulk electronic states can be mapped into Bloch states $\Psi_z = e^{ikz}\phi_k$, where $z = n*d$ is an integer multiple of the PL length (d) and k is the wave-vector along the transport axis. Within this PL notation, the time-independent Schrödinger equation of the isolated contact [$H|\psi\rangle = ES|\psi\rangle$] assumes the following form:

$$K_{-1}\,C_{\alpha-1} + K_0\,C_\alpha + K_1\,C_{\alpha+1} = 0, \tag{C.53}$$

where C_m is a vector of length n, and α labels the successive PLs. The different K matrices are thus defined as

$$K_0 = H_0 - ES_0,$$
$$K_1 = H_1 - ES_1,$$
$$K_{-1} = H_{-1} - ES_{-1} .$$
(C.54)

Note that, for some applications, it is also common to write Eq. (C.53) in the so-called transfer-matrix form (Sanvito *et al.*, 1999),

$$\begin{pmatrix} -(K_1)^{-1} K_0 & -(K_1)^{-1} K_{-1} \\ 1 & 0 \end{pmatrix} \cdot \begin{pmatrix} C_\alpha \\ C_{\alpha+1} \end{pmatrix} = \mathcal{T} \cdot \begin{pmatrix} C_\alpha \\ C_{\alpha+1} \end{pmatrix},$$
(C.55)

$$= \begin{pmatrix} C_{\alpha-1} \\ C_\alpha \end{pmatrix},$$
(C.56)

where $\mathcal{T}$ is the transfer matrix.

Since deep inside the contacts, the leads are periodic along the transport direction, the electronic scattering states of the system have to obey the asymptotic behavior of propagating Bloch states. The Schrödinger equation (Eq. C.53) inside the contact can therefore be re-expressed into the following eigenvalue problem:

$$\begin{pmatrix} -(K_1)^{-1} K_0 & -(K_1)^{-1} K_{-1} \\ 1 & 0 \end{pmatrix} \cdot \begin{pmatrix} C_\alpha \\ C_{\alpha+1} \end{pmatrix} = e^{ikd} \begin{pmatrix} C_\alpha \\ C_{\alpha+1} \end{pmatrix},$$
(C.57)

where d is the length of the PLs along the direction of transport. Solving this equation for both leads yields two sets of $2n$ complex wave-vectors ($k_j^{L/R} : j = 1, ..., 2m$) and their associated bulk complex state-vectors ($\phi_{k_j}^{L/R}$). These propagating (and decaying) Bloch states are the basis functions on which the transport problem is developed. For convenience, the wave-vectors are ordered such that the first m states are incoming to the scattering region, and the other m outgoing from the scattering region. Finally, the open boundary conditions of the original transport problem can be described in terms of the bulk states $\phi_{k_j}^{L/R}$ by imposing the proper asymptotic form on the stationary scattering states of $\mathcal{H}$:

$$\Psi_{k_j}^L(\mathbf{r}) = \begin{cases} \phi_{k_j}^L + \sum_{i=m+1}^{2m} r_{ij} \, \phi_{k_i}^L, & r_z \in L, \\ \sum_{i=m+1}^{2m} t_{ij} \, \phi_{k_i}^R, & r_z \in R, \end{cases}$$
(C.58)

where t_{ij} and r_{ij} are the transmission and the reflection amplitudes, respectively. Here, $\Psi_{k_j}^L$ are the scattering states that are incident from the left contact and are characterized by k_j in an asymptotic sense, i.e. these originate from the bulk states ϕ_{k_j}. A similar expression is easily derived for the scattering states $\Psi_{k_j}^R$ that are incident from the right contact. The transmission and reflection amplitudes found here are the usual quantities of the scattering theory and can be computed using transfer matrix techniques (Sanvito *et al.*, 1999).

As mentioned earlier, the Schrödinger equation (C.53) can be conveniently mapped into an eigenvalue calculation with the aid of the transfer matrix (Sanvito *et al.*, 1999):

$$\mathcal{T} = \begin{pmatrix} -(K_1)^{-1}K_0 & -(K_1)^{-1}K_{-1} \\ 1 & 0 \end{pmatrix}. \tag{C.59}$$

The eigenvalues of $\mathcal{T}$ are the $2n$ roots $e^{ik_l d}$ that define the complex wave-vectors of the Bloch states at energy E. The upper part (i.e. the n first elements) of the $2n$ eigenvectors are the expansion components of the Bloch electronic functions over the localized basis functions. In order to simplify the discussion, the eigenvectors are ordered such that the electronic states with indices ranging from 1 to n correspond to Bloch waves propagating/decaying in the right direction. On the contrary, the indices ranging from $n + 1$ to $2n$ are associated with Bloch waves propagating/decaying in the left direction. Finally, it is worth mentioning that the solution of the eigenvalue problem (Eq. C.59) assumes that K_1 is invertible. Moreover, the stability of the algorithm requires that K_1 is not ill-defined. Therefore, a regularization procedure for K_1 is highly desirable, though this point is crucial for the accuracy and stability of the code (Rocha, 2007, Sanvito *et al.*, 1999).

(2) In order to construct the retarded Green's function of the infinite leads, one notes that the Green's function ($G_{zz'}$) is a simple wave-function for all $z \leq z'$. Since it has to be retarded, and continuous at $z = z'$, one may assume the following form:

$$G_{zz'} = \begin{cases} \sum_{l=1}^{n} \phi_{k_l} e^{ik_l(z-z')} \alpha_{k_l}, & z \geq z', \\ \sum_{l=n+1}^{2n} \phi_{k_l} e^{ik_l(z-z')} \alpha_{k_l}, & z \leq z', \end{cases} \tag{C.60}$$

where the α_{k_l} coefficients have to be determined. Folding these expressions into the Schrödinger equation (C.53), the coefficients present the following form:

$$\alpha_{k_l} = \tilde{\phi}_{k_l}^{\dagger} V^{-1}, \tag{C.61}$$

with

$$V = (H_1^{\dagger} - ES_1^{\dagger}) \left[\sum_{l=1}^{n} \phi_{k_l} e^{ik_l(z-z')} \tilde{\phi}_{k_l}^{\dagger} - \sum_{l=n+1}^{2n} \phi_{k_l} e^{ik_l(z-z')} \tilde{\phi}_{k_l}^{\dagger} \right], \tag{C.62}$$

where the set of $\tilde{\phi}_{k_l}^{\dagger}$ is made from the duals of ϕ_{k_l} (i.e. $\tilde{\phi}_{k_l}^{\dagger} \phi_{k_m} = \delta_{lm}$).

(3) Finally, the Green's function of the semi-infinite leads can be obtained from the Green's function of the infinite leads by subtracting from $G_{zz'}$ a linear combination of eigenvectors that ensure the annihilation of the Green's function at the free surface. Considering, for example, the left lead which extends from $z = -\infty$ to $z = z_0 - 1$, subtraction of the term

$$\Delta_z(z' - z_0) = \sum_{l=1}^{n} \sum_{h=n+1}^{2n} \left(\phi_{k_h} e^{ik_l(z-z_0)} \tilde{\phi}_{k_l}^{\dagger} \right) \cdot \left(\phi_{k_l} e^{ik_l(z_0-z')} \tilde{\phi}_{k_l}^{\dagger} \right) V^{-1} \tag{C.63}$$

from $G_{zz'}$ gives a new retarded Green's function which vanishes at the principal layer $z = z_0$, taken as the surface. In this way, the retarded Green's functions of the semi-infinite leads have been computed and all the ingredients are gathered for calculation of the self-energies.

Finally, note that the derivation of the self-energies proposed above assumes that the entire leads are undisturbed by the central region. This means that the interfaces at which the surface Green's functions are computed have to be sufficiently far away from the scattering region in order for the scattering potential to be zero into the leads. Upon this assumption, the lead self-energies have to be computed only once and remain valid throughout the self-consistent determination of the scattering potential.

Before summarizing the self-consistent procedure used for calculation of the out-of-equilibrium scattering potential, calculation of the out-of-equilibrium density matrix should be briefly mentioned:

$$\rho = \frac{1}{2\pi} \int_{-\infty}^{\infty} \left[f_L(E - \mu_L) A_L(E) + f_R(E - \mu_R) A_R(E) \right] dE, \qquad (C.64)$$

where $f(E)$ is the equilibrium Fermi electronic distribution, and $\mu_{L/R}$ are the chemical potentials inside the left and right leads, respectively. It is worth noting that the integration in Eq. (C.64) is not a trivial computational task since the integral is unbound and the spectral functions $A_{L/R}$ are not analytical. However, the computational cost of the integration can be significantly reduced by rewriting the integral in Eq. (C.64) as the sum of two contributions, ρ_{equ} and ρ_Δ:

$$\begin{aligned}
\rho &= \frac{1}{2\pi} \int_{-\infty}^{\infty} \left[f(E - \mu_L) A_L(E) + f(E - \mu_R) A_R(E) \right] \\
&\quad + \left[f(E - \mu_L) A_R(E) - f(E - \mu_L) A_R(E) \right] dE \\
&= \frac{1}{2\pi} \int_{-\infty}^{\infty} A_D f(E - \mu_L)\, dE + \frac{1}{2\pi} \int_{-\infty}^{\infty} A_R(E) \left[f(E - \mu_R) - f(E - \mu_L) \right] dE \\
&= \rho_{\mathrm{equ}} + \rho_\Delta \, .
\end{aligned} \qquad (C.65)$$

The first term (ρ_{equ}) accounts for the "equilibrium" component of the electronic density. Indeed, it corresponds to what should be obtained if both leads have the same chemical potential ($\mu_L = \mu_R$). All poles of the spectral function A_D are lying on the real axis and the function is analytical elsewhere. Therefore, though unbound, the integration of A_D can be performed in the complex plane using contour integral techniques. The contour used conventionally is depicted in Fig. C.6, where E_{bottom} is chosen below the deepest valence electronic state. For finite temperatures, the contour encloses some poles of the Fermi distribution function located at $\{ z_\alpha = i(2\alpha + 1) \pi k_B T,\ \text{with}\ \alpha = 1, \ldots, n_\alpha \}$. Finally, the integral is computed as

$$\frac{1}{2\pi} \int_{-\infty}^{\infty} A_D(E) f_L(E - \mu_L)\, dE =$$

$$- \int_{C_{\mathrm{Im}}} A_D(z) f_L(z - \mu_L)\, dz - 2\pi k_B T \sum_{\alpha=1}^{n_\alpha} A_D(z_\alpha). \qquad (C.66)$$

The second term in Eq. (C.66) accounts for the corrections induced by the out-of-equilibrium conditions. The spectral functions $A_{L/R}$ are not analytical and the

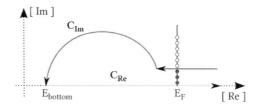

Figure C.6 The closed contour ($C_{\text{Im}} + C_{\text{Re}}$) used to compute the equilibrium component of the electronic density matrix. Filled dots account for the poles of the Fermi distribution function enclosed in the complex integration part.

integration cannot be obtained using complex contours techniques. However, the integration is bounded by the two Fermi distribution functions and can be evaluated on a dense energy grid. Note that the formula (C.66) does not account for the contribution to the density of the electronic states that are localized within the scattering device and do not couple with the leads. In the presence of such states, (C.66) is not valid anymore and additional information has to be supplied.

Finally, the self-consistent procedure related to calculation of the out-of-equilibrium electronic density and electric current is described, as implemented in several *ab initio* transport packages (SMEAGOL, TRANSIESTA, etc.; see below). The main feature of the Green's function DFT-based schemes is to extend the scope of the standard DFT codes based on localized basis sets, by calculating the out-of-equilibrium density matrix. As such, though the out-of-equilibrium electronic density does not minimize the density functional, these schemes rely on the Kohn–Sham Hamiltonian as the single particle Hamiltonian. Therefore, the self-consistency is very similar to the one encountered in standard DFT codes. The self-consistent procedure can be summarized in the diagram in Fig. C.7.

In many situations, the nonequilibrium condition brings only small corrections to the ground-state electronic density. Therefore, it is good practice to use a DFT computed ground-state electronic density as the guess input density for the nonequilibrium calculations. Then, the latter proceed in two steps. First, the electronic structure corresponding to the infinite leads is computed, in order for the contact self-energies to be built over the proper range of energies (i.e. the full bandwidth of the material). This step has to be performed only once, since the leads are assumed to remain in their equilibrium states. Second, the out-of-equilibrium density matrix is computed in a self-consistent way. The self-consistent cycle is accomplished by determining the Hartree (electrostatic) potential as a solution of Poisson's equation with appropriate boundary conditions. Though Poisson's equation can be solved in real space, it is common to use fast-Fourier transform algorithms that are computationally efficient. Actually, the electrostatic potential is calculated for a virtual periodic system obtained by repeating the effective Hamiltonian $H_D^{\text{eff}}(E)$ along the transport direction. Besides, a saw-like term, whose drop is identical to the bias applied, is added to the Hartree potential in order to recover the correct voltage drop across the device. Finally, when convergence is reached, the electric current is computed as a by product of the retarded Green's function.

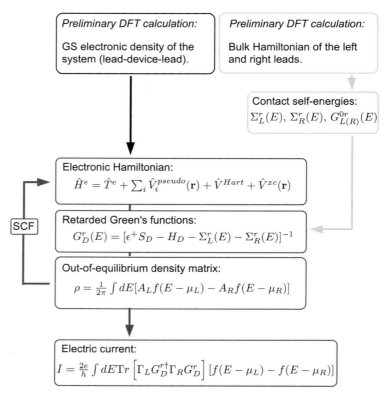

Figure C.7 Self-consistent cycle for computation of the out-of-equilibrium density matrix and electric current. Dark grey frame (top left) corresponds to the preliminary DFT calculation aiming at producing a good initial guess for the out-of-equilibrium electronic density. Light grey frames (top right) describe the preliminary calculation of the self-energies of the leads. Medium grey frames (bottom) correspond to the main step of the actual self-consistent cycle.

The Green's function formalism outlined in the present appendix has been implemented in several simulation codes. In the following we list some of these codes and their main features as specified by their developers:

1. SMEAGOL (Rocha *et al.*, 2006) has been designed to calculate transport properties of atomic scale devices. SMEAGOL is an *ab initio* electronic transport code based on a combination of density functional theory (DFT) and nonequilibrium Green's function transport methods (NEGF). The Kohn–Sham equations for an open non-periodic system are solved in the NEGF scheme, and the current is then extracted from the Landauer formula. The code has been designed to describe two terminal nanoscale devices, for which the potential drop must be calculated accurately. It has been specifically created to deal also with magnetic systems, since the code is fully spin-polarized and includes the possibility of performing noncolinear spin calculations (Rocha *et al.*, 2005). More details are available at: **http://www.smeagol.tcd.ie**.

2. TRANSIESTA (Brandbyge *et al.*, 2002): the present SIESTA release (Soler *et al.*, 2002) includes the possibility of performing calculations of electronic transport

properties using the Transiesta method, which is a procedure to solve the electronic structure of an open system formed by a finite structure sandwiched between two semi-infinite metallic leads. A finite bias can be applied between both leads, to drive a finite current. In practical terms, calculations using Transiesta involve the solution of the electronic density from the DFT Hamiltonian using Green's functions techniques, instead of the usual diagonalization procedure. Therefore, Transiesta calculations involve a Siesta run, in which a set of routines is invoked to solve the Green's functions and the charge density for the open system. These routines are packed in a set of modules, referred to as the "Transiesta module." More details are available at: **http://www.icmab.es/dmmis/leem/siesta/**.

3. Want (Calzolari *et al.*, 2004) is an open-source, GNU General Public License suite of codes that provides an integrated approach for the study of coherent electronic transport in nanostructures. The core methodology combines state-of-the-art DFT, plane-wave, norm-conserving pseudopotential calculations with a Green's function method based on the Landauer formalism to describe quantum conductance. The essential connection between the two, and a crucial step in the calculation, is use of the maximally localized Wannier function representation to introduce naturally the ground-state electronic structure into the lattice Green's function approach at the basis of the evaluation of the quantum conductance. Moreover, knowledge of the Wannier functions of the system allows direct linking between the electronic transport properties of the device and the nature of the chemical bonds, providing insight into the mechanisms that govern electron flow at the nanoscale. More details are available at: **http://www.wannier-transport.org**.

4. ONETEP (Skylaris *et al.*, 2005) (order-N electronic total energy package) is a linear-scaling code for quantum-mechanical calculations based on DFT. ONETEP uses a reformulation of the plane wave pseudopotential method, which exploits the electronic localization that is inherent in systems with a nonvanishing band gap. Direct optimization of strictly localized quantities expressed in terms of a delocalized plane wave basis allows division of the computational effort among many processors to allow calculations to be performed efficiently on parallel supercomputers. More details are available at: **http://www2.tcm.phy.cam.ac.uk/onetep/**.

Appendix D Recursion methods for computing the DOS and wavepacket dynamics

The Lanczos tridiagonalization method orthogonally transforms a real symmetric matrix A to symmetric tridiagonal form. Traditionally, this very simple algorithm is suitable when one needs only a few of the lower eigenvalues and the corresponding eigenvectors of very large Hermitian matrices, whose full diagonalization is technically impossible. We introduce here the basic ingredients of the recursion method based on the Lanczos tridiagonalization, and explain how calculation of the DOS as well as the dynamics of wavepackets (and related conductivity) can be performed efficiently.

D.1   Lanczos method for the density of states

The Lanczos method is a highly efficient recursive approach for calculation of the electronic structure (Lanczos, 1950). This method, first developed by Haydock, Heine, and Kelly (Haydock, Heine & Kelly, 1972, 1975), is based on an eigenvalue approach due to Lanczos. It relies on computation of Green functions matrix elements by continued fraction expansion, which can be implemented either in real or reciprocal space. These techniques are particularly well suited for treating disorder and defect-related problems, and were successfully implemented to tackle impurity-level calculations in semiconductors using a tight-binding approximation (Lohrmann, 1989), and for electronic structure investigations for amorphous semiconductors, transition metals, and metallic glasses based on linear-muffin-tin orbitals (Bose, Winer & Andersen, 1988). Recent developments include the exploration of a degenerated orbital extended Hubbard Hamiltonian of system size up to ten millions atoms, with the Krylov subspace method (Takayama, Hoshi & Fujiwara, 2004, Hoshi et al., 2012).

The recursion method is said to be of order N since the computational cost scales linearly with the total number of atoms defining the (disordered) system (Grosso & Parravicini, 2006). The key idea of the recursion method is to construct iteratively a Lanczos (or Krylov) basis which tridiagonalizes the Hamiltonian (initially defined in a localized basis set), and then to compute diagonal matrix elements of the Green function (to access the density of states) by using the continued fraction expansion method.

The recursion method is thus a basis transformation which turns out to be very suitable for dealing with tight-binding Hamiltonians for which strong disorder limits the use of diagonalization methods and perturbative treatments. The Lanczos method allows simulation of electronic (and transport as discussed in Section D.2) behavior of disordered systems up to the scale of 100 millions of orbitals using high performance computing resources. The limit is actually not really the total number of atoms, but rather the necessary computational time to access the transport regimes of interest. As shown below, a very efficient computational trick is to compute the trace of any operator related to electronic and transport properties on a reduced number of random phase states (N_{RP}) instead of a fully complete and orthogonal basis:

$$\text{Tr}\left[\delta(E - \hat{\mathcal{H}})\right] = \sum_{J=1}^{M} \langle \varphi_J | \delta(E - \hat{\mathcal{H}}) | \varphi_J \rangle = \frac{M}{N_{RP}} \times \sum_{i=1}^{N_{RP}} \langle \varphi_{RP}^i | \delta(E - \hat{\mathcal{H}}) | \varphi_{RP}^i \rangle, \quad \text{(D.1)}$$

where M is the dimension of $\hat{\mathcal{H}}$ in the TB basis and $|\varphi_{RP}\rangle$ is defined by

$$|\varphi_{RP}\rangle = \frac{1}{\sqrt{M}} \sum_{J=1}^{M} e^{i2\pi\theta_J} |\varphi_J\rangle, \quad \text{(D.2)}$$

with θ_J a random number between 0 and 1. This state $|\varphi_{RP}\rangle$ has a random phase on each orbital of the TB basis. Numerically, with only about ten of such states, a high-energy resolution can be obtained, even though the starting dimension M can be as high as several tens or hundreds of millions of orbitals. Concerning the Lanczos method, the basic algorithm is described as follows:

- The first step starts with $|\psi_1\rangle = |\varphi_{RP}\rangle$:

$$a_1 = \langle \psi_1 | \hat{\mathcal{H}} | \psi_1 \rangle, \quad \text{(D.3)}$$
$$|\tilde{\psi}_2\rangle = \hat{\mathcal{H}} | \psi_1 \rangle - a_1 | \psi_1 \rangle, \quad \text{(D.4)}$$
$$b_1 = \| |\tilde{\psi}_2\rangle \| = \sqrt{\langle \tilde{\psi}_2 | \tilde{\psi}_2 \rangle}, \quad \text{(D.5)}$$
$$|\psi_2\rangle = \frac{1}{b_1} |\tilde{\psi}_2\rangle. \quad \text{(D.6)}$$

- All other recursion steps ($\forall\, n \geq 2$) are identical and given through

$$a_n = \langle \psi_n | \hat{\mathcal{H}} | \psi_n \rangle, \quad \text{(D.7)}$$
$$|\tilde{\psi}_{n+1}\rangle = \hat{\mathcal{H}} | \psi_n \rangle - a_n | \psi_n \rangle - b_{n-1} | \psi_{n-1} \rangle, \quad \text{(D.8)}$$
$$b_n = \sqrt{\langle \tilde{\psi}_{n+1} | \tilde{\psi}_{n+1} \rangle}, \quad \text{(D.9)}$$
$$|\psi_{n+1}\rangle = \frac{1}{b_n} |\tilde{\psi}_{n+1}\rangle. \quad \text{(D.10)}$$

The coefficients a_n and b_n are named recursion coefficients, and are respectively the diagonal and off-diagonal of the matrix representation of $\hat{\mathcal{H}}$ in the Lanczos basis (that we write $\hat{\tilde{\mathcal{H}}}$):

$$
\hat{\tilde{\mathcal{H}}} = \begin{pmatrix}
a_1 & b_1 & & & \\
b_1 & a_2 & b_2 & & \\
& b_2 & \ddots & \ddots & \\
& & \ddots & \ddots & b_N \\
& & & b_N & a_N
\end{pmatrix}.
$$ (D.11)

Simple linear algebra shows that

$$
\langle \varphi_{RP} | \delta(E - \hat{\mathcal{H}}) | \varphi_{RP} \rangle = \langle \psi_1 | \delta(E - \hat{\mathcal{H}}) | \psi_1 \rangle
$$
$$
= \lim_{\eta \to 0} -\frac{1}{\pi} \Im m \left(\langle \psi_1 | \frac{1}{E + i\eta - \hat{\mathcal{H}}} | \psi_1 \rangle \right),
$$

while

$$
\langle \psi_1 | \frac{1}{E + i\eta - \hat{\tilde{\mathcal{H}}}} | \psi_1 \rangle = \cfrac{1}{E + i\eta - a_1 - \cfrac{b_1^2}{E + i\eta - a_2 - \cfrac{b_2^2}{E + i\eta - a_3 - \cfrac{b_3^2}{\ddots}}}}
$$ (D.12)

which is termed a continued fraction. To compute Eq. (D.12), however, in practice we must introduce a cutoff, or termination (named TERM), typically after $N \sim 1000$ recursion steps. Then $\langle \psi_1 | (E + i\eta - \hat{\tilde{\mathcal{H}}})^{-1} | \psi_1 \rangle$ becomes

$$
\cfrac{1}{E + i\eta - a_1 - \cfrac{b_1^2}{\cdots\cdots \cfrac{\ddots}{\cdots\cdots \cfrac{\ddots}{E + i\eta - a_{N-1} - \cfrac{b_{N-1}^2}{E + i\eta - a_N - b_N^2 \times \text{TERM}}}}}}
$$

Several types of terminations can be employed depending on the spectrum of the system under study (and mainly depending on the number of gaps and energy resolution that is needed). The energy resolution on the computed density of states depends on the total number of recursion steps and the stability of the orthogonality of the constructed Lanczos basis set. Additionally, to evaluate $\delta(E - \hat{\mathcal{H}})$, a Lorentzian function $(E + i\eta - \hat{\mathcal{H}})^{-1}$ is usually employed, although it is not suitable for a one-dimensional system because of the large number of van Hove singularities (for instance for studying carbon nanotubes or semiconducting nanowires (Persson *et al.*, 2008)). Another method, the kernel polynomials method (KPM) can cure these problems, and we refer to Weisse *et al.* (2006) and to the Ph.D. thesis of Aurélien Lherbier (Lherbier, 2008) for further details, and applications to complex materials (such as semiconducting silicon nanowires).

D.1.1 Termination of the continued fraction

Let us now show how one can concretely compute the diagonal matrix element from the continued fraction. The general form of the tridiagonalized Hamiltonian after the Lanczos algorithm reads

$$
\tilde{\mathcal{H}} =
\begin{pmatrix}
a_1 & b_1 & & & \\
b_1 & a_2 & b_2 & & \\
& b_2 & \ddots & \ddots & \\
& & \ddots & \ddots & b_N \\
& & & b_N & a_N
\end{pmatrix}.
\tag{D.13}
$$

The total density of states can then be computed as a continued fraction. In the tight-binding basis, one has

$$
\langle \psi_{RP} | \delta(E - \hat{\mathcal{H}}) | \psi_{RP} \rangle = \lim_{\eta \to 0} -\frac{1}{\pi} \Im_m \left(\langle \psi_{RP} | \frac{1}{E + i\eta - \hat{\mathcal{H}}} | \psi_{RP} \rangle \right),
\tag{D.14}
$$

while in the Lanczos basis one gets (with $|\psi_{RP}\rangle$ becoming $|\psi_1\rangle$, the seed vector of the Lanczos procedure),

$$
\langle \psi_1 | \delta(E - \tilde{\mathcal{H}}) | \psi_1 \rangle = \lim_{\eta \to 0} -\frac{1}{\pi} Im \cfrac{1}{E + i\eta - a_1 - \cfrac{b_1^2}{E + i\eta - a_2 - \cfrac{b_2^2}{E + i\eta - a_3 - \cfrac{b_3^2}{\ddots}}}}
\tag{D.15}
$$

We name G_1 the continued fraction and define G_n as

$$
G_1 = \cfrac{1}{E + i\eta - a_1 - \cfrac{b_1^2}{E + i\eta - a_2 - \cfrac{b_2^2}{E + i\eta - a_3 - \cfrac{b_3^2}{\ddots}}}},
\tag{D.16}
$$

$$
G_1 = \frac{1}{E + i\eta - a_1 - b_1^2 G_2},
\tag{D.17}
$$

$$
G_n = \frac{1}{E + i\eta - a_n - b_n^2 G_{n+1}}.
\tag{D.18}
$$

Since we compute a finite number of recursion coefficients, the subspace of Lanczos is of finite dimension (N), so it is crucial to terminate the continued fraction by an appropriate choice of the last $\{a_{n=N}, b_{n=N}\}$ elements. Let us rewrite the continued fraction as

$$G_1 = \cfrac{1}{E + i\eta - a_1 - \cfrac{b_1^2}{E + i\eta - a_2 - \cfrac{b_2^2}{E + i\eta - a_3 - \cfrac{b_3^2}{\ddots \cfrac{}{E + i\eta - a_N - b_N^2 G_{N+1}}}}}}, \tag{D.19}$$

where G_{N+1} denotes such a termination. The simplest case is when all the spectrum is contained in a finite bandwidth $[a - 2b; a + 2b]$, a is the spectrum center and $4b$ its bandwidth. Recursion coefficients a_n and b_n oscillate around their average value a and b, and the damping is usually fast after a few hundreds of recursion steps. The termination then satisfies

$$G_{N+1} = \frac{1}{E + i\eta - a - b^2 G_{N+2}} = \frac{1}{E + i\eta - a - b^2 G_{N+1}}, \tag{D.20}$$

from which a polynomial of second degree is found,

$$-(b^2) G_{N+1}^2 + (E + i\eta - a) G_{N+1} - 1 = 0, \tag{D.21}$$

and straightforwardly solved

$$\Delta = (E + i\eta - a)^2 - (2b)^2, \tag{D.22}$$

$$G_{N+1} = \frac{(E + i\eta - a) \mp i\sqrt{-\Delta}}{2b^2}, \tag{D.23}$$

$$G_{N+1} = \frac{(E + i\eta - a) - i\sqrt{(2b)^2 - (E + i\eta - a)^2}}{2b^2}. \tag{D.24}$$

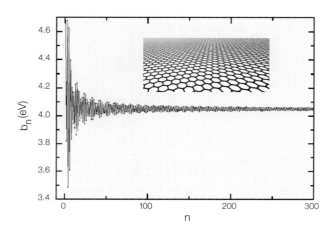

Figure D.1 Recursion coefficient b_n for pristine graphene with bandwith $\sim [-8; 8]$ eV.

Figure D.1 shows $b(n)$ for the case of pristine graphene. The value of b_n is seen to quickly tend towards $b \sim 4$ eV. As seen in Fig. D.1, a closed form of the termination term can be typically introduced after a few hundreds of recursion steps ($N \sim 100-1000$) depending on the spectrum complexity.

D.2   Wavepacket propagation method

We shown here how to apply the Lanczos method to compute the wavepacket spreading in an arbitary complex disordered material. This is the central technical ingredient of the (real space and order N) computational implementation of the Kubo–Greenwood method used throughout this book. It has been pioneered by Roche and Mayou for studying quantum transport in quasicrystals (Roche, 1996, Roche & Mayou, 1997, Roche, 1999), and then further improved for achieving higher energy resolution and optimized computational cost (Triozon, 2002, Lherbier, 2008). The most recent review of its applicability to graphene-related systems has been provided in Roche *et al.* (2012).

The main quantity to compute is the mean square spread ($\Delta X^2(E, t)$) introduced in Section 3.4.4. First, starting from Eq. (3.58), we have

$$\Delta X^2(E, t) = \langle |\hat{X}(t) - \hat{X}(0)|^2 \rangle_E, \tag{D.25}$$

with $\hat{X}(t)$ the position operator in the Heisenberg representation. In Section 3.4, the average of any operator (Eq. 3.46) was introduced, which we can rewrite as

$$\Delta X^2(E, t) = \frac{\mathrm{Tr}\left[\delta(E - \hat{\mathcal{H}})|\hat{X}(t) - \hat{X}(0)|^2\right]}{\mathrm{Tr}\left[\delta(E - \hat{\mathcal{H}})\right]}, \tag{D.26}$$

$$\Delta X^2(E, t) = \frac{\mathrm{Tr}\left[(\hat{X}(t) - \hat{X}(0))^\dagger \delta(E - \hat{\mathcal{H}})(\hat{X}(t) - \hat{X}(0))\right]}{\mathrm{Tr}\left[\delta(E - \hat{\mathcal{H}})\right]}. \tag{D.27}$$

We then use several identities and definitions to rewrite ($\hat{X}(t) - \hat{X}(0)$):

$$\hat{X}(t) = e^{\frac{i\hat{\mathcal{H}}t}{\hbar}} \hat{X}(0) e^{\frac{-i\hat{\mathcal{H}}t}{\hbar}}, \tag{D.28}$$

$$\hat{U}(t) = e^{\frac{-i\hat{\mathcal{H}}t}{\hbar}}, \tag{D.29}$$

where $\hat{U}(t)$ is the evolution operator,

$$\hat{X}(t) - \hat{X}(0) = \hat{U}^\dagger(t)\hat{X}\hat{U}(t) - \hat{X}, \tag{D.30}$$

$$\hat{X}(t) - \hat{X}(0) = \hat{U}^\dagger(t)\hat{X}\hat{U}(t) - \hat{U}^\dagger(t)\hat{U}(t)\hat{X}, \tag{D.31}$$

$$\hat{X}(t) - \hat{X}(0) = \hat{U}^\dagger(t)[\hat{X}, \hat{U}(t)], \tag{D.32}$$

using $\hat{U}^\dagger(t)\hat{U}(t) = \mathbb{I}$, and $[\cdots,\cdots]$ the commutator. Then by replacing these quantities in Eq. (D.27), one gets

$$\Delta X^2(E,t) = \frac{\text{Tr}\left[[\hat{X},\hat{U}(t)]^\dagger \hat{U}(t)\delta(E-\hat{\mathcal{H}})\hat{U}^\dagger(t)[\hat{X},\hat{U}(t)]\right]}{\text{Tr}\left[\delta(E-\hat{\mathcal{H}})\right]}, \tag{D.33}$$

$$\Delta X^2(E,t) = \frac{\text{Tr}\left[[\hat{X},\hat{U}(t)]^\dagger \delta(E-\hat{\mathcal{H}})[\hat{X},\hat{U}(t)]\right]}{\text{Tr}\left[\delta(E-\hat{\mathcal{H}})\right]}. \tag{D.34}$$

Using the random phase states as initial states, we find

$$\Delta X^2(E,t) = \frac{\langle\varphi_{RP}|[\hat{X},\hat{U}(t)]^\dagger \delta(E-\hat{\mathcal{H}})[\hat{X},\hat{U}(t)]|\varphi_{RP}\rangle}{\langle\varphi_{RP}|\delta(E-\hat{\mathcal{H}})|\varphi_{RP}\rangle}, \tag{D.35}$$

$$\Delta X^2(E,t) = \frac{\langle\varphi'_{RP}(t)|\delta(E-\hat{\mathcal{H}})|\varphi'_{RP}(t)\rangle}{\langle\varphi_{RP}|\delta(E-\hat{\mathcal{H}})|\varphi_{RP}\rangle}. \tag{D.36}$$

The techniques used for the computation of the density of states can thus also be employed for the computation of $\Delta X^2(E,t)$, provided that one first evaluates $|\varphi'_{RP}(t)\rangle$. The evaluation of $|\varphi'_{RP}(t)\rangle$ needs $\hat{U}(t)|\varphi_{RP}\rangle$ together with $[\hat{X},\hat{\mathcal{H}}]$. Let us start by explaining the calculation of $[\hat{X},\hat{\mathcal{H}}]$. By definition $[\hat{X},\hat{\mathcal{H}}] = \hat{X}\hat{\mathcal{H}} - \hat{\mathcal{H}}\hat{X}$, given that $\hat{X}$ is diagonal:

$$[\hat{X},\hat{\mathcal{H}}] = \begin{pmatrix} 0 & & & & \\ & \ddots & & \mathcal{H}_{ij}(X_i - X_j) & \\ & & \ddots & & \\ & \mathcal{H}_{ij}(X_i - X_j) & & \ddots & \\ & & & & 0 \end{pmatrix}, \tag{D.37}$$

where $(X_i - X_j)$ is the distance between orbitals $|\varphi_i\rangle$ and $|\varphi_j\rangle$. We now focus on the calculation of $\hat{U}(t)|\varphi_{RP}\rangle = |\varphi_{RP}(t)\rangle$. The time evolution of the random phase wavepacket is followed through use of the evolution operator $\hat{U}(t)$, which can be efficiently approximated using a basis of orthogonal polynomials, with the Chebyshev polynomials as the most computationally efficient choice. For a given time step (T) we can write such a decomposition as

$$\hat{U}(T) = e^{\frac{-i\hat{\mathcal{H}}T}{\hbar}} = \sum_{n=0}^{\infty} c_n(T)Q_n(\hat{\mathcal{H}}), \tag{D.38}$$

where Q_n is a Chebyshev polynomial of order n. The Chebyshev polynomials (T_n) usually act on the interval $[-1:1]$, whereas the Hamiltonians considered here have larger bandwiths $[-1 : 1]$ so some rescaling of the polynomials needs to be performed to use their recurrent properties. Some useful equations are

$$T_n(\cos(\theta)) = \cos(n\theta), \tag{D.39}$$

$$T_n(\tilde{E}) = \frac{n}{2} \sum_{k=0}^{Pi(\frac{n}{2})} (-1)^k \frac{(n-k-1)!}{k!(n-2k)!} (2\tilde{E})^{n-2k} \; (\forall \, n \neq 0), \tag{D.40}$$

where Pi denotes the integer part and $\tilde{E} \in [-1 : 1]$. Then

$$T_0(\tilde{E}) = 1, \tag{D.41}$$

$$T_1(\tilde{E}) = \tilde{E}, \tag{D.42}$$

$$T_2(\tilde{E}) = 2\tilde{E}^2 - 1, \tag{D.43}$$

$$\vdots$$

additionally for $n \geq 1$,

$$T_{n+1}(\tilde{E}) = 2\tilde{E} \, T_n(\tilde{E}) - T_{n-1}(\tilde{E}), \tag{D.44}$$

while for the rescaled Chebyshev polynomials ($\forall \, E \in [a - 2b : a + 2b]$) we get

$$Q_n(E) = \sqrt{2} T_n \left(\frac{E-a}{2b} \right) \; (\forall \, n \geq 1), \tag{D.45}$$

$$Q_0(E) = 1, \tag{D.46}$$

$$Q_1(E) = \sqrt{2} \frac{E-a}{2b}, \tag{D.47}$$

$$Q_2(E) = 2\sqrt{2} \left(\frac{E-a}{2b} \right)^2 - \sqrt{2}, \tag{D.48}$$

$$\vdots$$

with the recurrence relation $n \geq 2$,

$$Q_{n+1}(E) = 2 \left(\frac{E-a}{2b} \right) Q_n(E) - Q_{n-1}(E). \tag{D.49}$$

Once the Q_n polynomials are well defined, one can compute the related $c_n(T)$ coefficients:

$$c_n(T) = \int dE \, p_Q(E) Q_n(E) e^{\frac{-iET}{\hbar}}, \tag{D.50}$$

$$\text{or } p_Q(E) = \frac{1}{2b\sqrt{1 - \left(\frac{E-a}{2b}\right)^2}}, \tag{D.51}$$

introducing the weight $p_Q(E)$ to get an orthonormalized basis. Practically, $c_n(T)$ are computed starting from a fictitious tridiagonal Hamiltonian ($\hat{\mathcal{H}}_f$), where all diagonal elements are identical and taken as a, while all identical off-diagonal elements are b except the first one, being $\sqrt{2}b$, so that

$$\hat{\mathcal{H}}_f = \begin{pmatrix} a & \sqrt{2}b & & & \\ \sqrt{2}b & a & b & & \\ & b & \ddots & & \\ & & & \ddots & b \\ & & & b & a \end{pmatrix}. \tag{D.52}$$

Using this fictitious Hamiltonian in Eq. (D.50),

$$c_n(T) = \int dE \, p_Q(E) Q_n(\hat{\mathcal{H}}_f) e^{\frac{-iET}{\hbar}}, \tag{D.53}$$

$$c_n(T) = \int dE \, \langle 0|\delta(E - \hat{\mathcal{H}}_f)|0\rangle Q_n(\hat{\mathcal{H}}_f) e^{\frac{-iET}{\hbar}}, \tag{D.54}$$

$$c_n(T) = \int dE \, \langle n|\delta(E - \hat{\mathcal{H}}_f)|0\rangle e^{\frac{-iET}{\hbar}}, \tag{D.55}$$

$$c_n(T) = \langle n|e^{\frac{-i\hat{\mathcal{H}}_f T}{\hbar}}|0\rangle, \tag{D.56}$$

$$c_n(T) = \sum_{i=0}^{N} \langle n|E_i\rangle e^{\frac{-iE_i T}{\hbar}} \langle E_i|0\rangle, \tag{D.57}$$

where we have used $Q_n(\hat{\mathcal{H}}_f)|0\rangle = |n\rangle$ and where E_i and $|E_i\rangle$ are the eigenvalues and eigenvectors of $\hat{\mathcal{H}}_f$. We can now calculate $|\varphi_{RP}(T)\rangle$:

$$|\varphi_{RP}(T)\rangle = \hat{U}(T)|\varphi_{RP}\rangle, \tag{D.58}$$

$$|\varphi_{RP}(T)\rangle \simeq \sum_{n=0}^{N} c_n(T) Q_n(\hat{\mathcal{H}})|\varphi_{RP}\rangle = \sum_{n=0}^{N} c_n(T)|\alpha_n\rangle, \tag{D.59}$$

where $|\alpha_n\rangle = Q_n(\hat{\mathcal{H}})|\varphi_{RP}\rangle$. With the definitions introduced in Eqs. (D.46–48) and the recurrence relation Eq. (D.49), we obtain

$$|\alpha_0\rangle = |\varphi_{RP}\rangle, \tag{D.60}$$

$$|\alpha_1\rangle = \left(\frac{\hat{\mathcal{H}} - a}{\sqrt{2}b}\right)|\alpha_0\rangle, \tag{D.61}$$

$$|\alpha_2\rangle = \left(\frac{\hat{\mathcal{H}} - a}{b}\right)|\alpha_1\rangle - \sqrt{2}|\alpha_0\rangle, \tag{D.62}$$

$$|\alpha_{n+1}\rangle = \left(\frac{\hat{\mathcal{H}} - a}{b}\right)|\alpha_n\rangle - |\alpha_{n-1}\rangle \ (\forall \, n \geq 2). \tag{D.63}$$

Following the same reasoning as for $|\varphi_{RP}(T)\rangle$, $|\varphi'_{RP}(T)\rangle$ can be evaluated first, writing

$$|\varphi'_{RP}(T)\rangle = [\hat{X}, \hat{U}(T)]|\varphi_{RP}\rangle, \tag{D.64}$$

$$|\varphi'_{RP}(T)\rangle \simeq \sum_{n=0}^{N} c_n(T)[\hat{X}, Q_n(\hat{\mathcal{H}})]|\varphi_{RP}\rangle = \sum_{n=0}^{N} c_n(T)|\beta_n\rangle, \tag{D.65}$$

with $|\beta_n\rangle = [\hat{X}, Q_n(\hat{\mathcal{H}})]|\varphi_{RP}\rangle$. Using Eq. (D.49), we deduce

$$[\hat{X}, Q_{n+1}(\hat{\mathcal{H}})] = [\hat{X}, \left(\frac{\hat{\mathcal{H}} - a}{b}\right) Q_n(\hat{\mathcal{H}})] - [\hat{X}, Q_{n-1}(\hat{\mathcal{H}})], \qquad (D.66)$$

which is rewritten using $[A, BC] = B[A, C] + [A, B]C$,

$$[\hat{X}, Q_{n+1}(\hat{\mathcal{H}})] = \left(\frac{\hat{\mathcal{H}} - a}{b}\right)[\hat{X}, Q_n(\hat{\mathcal{H}})] - [\hat{X}, Q_{n-1}(\hat{\mathcal{H}})] + [\hat{X}, \left(\frac{\hat{\mathcal{H}} - a}{b}\right)]Q_n(\hat{\mathcal{H}}).$$
$$\qquad (D.67)$$

Multiplying the right term by $|\varphi_{RP}\rangle$ using $|\beta_n\rangle$ and $|\alpha_n\rangle$ we get

$$|\beta_{n+1}\rangle = \left(\frac{\hat{\mathcal{H}} - a}{b}\right)|\beta_n\rangle - |\beta_{n-1}\rangle + [\hat{X}, \left(\frac{\hat{\mathcal{H}} - a}{b}\right)]|\alpha_n\rangle. \qquad (D.68)$$

The commutator is rewritten as $[\hat{X}, \left(\frac{\hat{\mathcal{H}} - a}{b}\right)]$,

$$[\hat{X}, \left(\frac{\hat{\mathcal{H}} - a}{b}\right)] = \frac{1}{b}[\hat{X}, \left(\hat{\mathcal{H}} - a\right)] = \frac{1}{b}\left(\hat{X}\left(\hat{\mathcal{H}} - a\right) - \left(\hat{\mathcal{H}} - a\right)\hat{X}\right)$$

$$= \frac{1}{b}\left(\hat{X}\hat{\mathcal{H}} - \hat{\mathcal{H}}\hat{X} + a(\hat{X} - \hat{X})\right) = \frac{1}{b}[\hat{X}, \hat{\mathcal{H}}]. \qquad (D.69)$$

So we finally obtain the recurrence relation for $|\beta_n\rangle$:

$$|\beta_{n+1}\rangle = \left(\frac{\hat{\mathcal{H}} - a}{b}\right)|\beta_n\rangle - |\beta_{n-1}\rangle + \frac{1}{b}[\hat{X}, \hat{\mathcal{H}}]|\alpha_n\rangle. \qquad (D.70)$$

One notes that computation of $|\beta_n\rangle$ requires evaluation of $|\alpha_n\rangle$ and the commutator $[\hat{X}, \hat{\mathcal{H}}]$. Using such a real space approach, simulations of charge mobility in disordered graphene samples can be achieved for systems with several tens of millions of atomic orbitals (typical graphene area of $1\mu m^2$).

Note that a similar methodology has also been developed for following phonon propagation in the harmonic approximation and computing the associated thermal conductivity of material of any complexity (Li *et al.*, 2010, Li *et al.*, 2011, Sevincli *et al.*, 2011).

D.3   Lanczos method for computing off-diagonal Green's functions

One inconvenient feature of the Lanczos approach presented so far is its restriction to calculation of diagonal Green function matrix elements. Although some generalization is possible, as recently proposed by Ortmann and Roche (Ortmann & Roche, 2013), for calculation of the Hall Kubo conductivity, the situation of transport through open systems (heterojunctions) becomes more problematic.

A similar order-N method for calculation of Landauer–Büttiker conductance has been achieved, however, by Triozon and Roche (Triozon & Roche, 2005). This formula has the advantages of being general, and independent of the dimensionality of the system and its eventual geometrical complications.

To implement the recursion approach in the Landauer framework (Imry & Landauer, 1999), a generalization of the Lanczos approach to *nonsymmetric matrices* is necessary. The Green's function is obtained from an effective Hamiltonian which is nonsymmetric, $\mathcal{H} = \mathcal{H}_0 + \Sigma_L + \Sigma_R$ (with $\mathcal{H}_0$ the Hamiltonian of the system connected to electrodes left (L) and right (R)) because of the presence of complex self-energy matrix elements ($\Sigma_{L,R}$) describing the finite system coupled to the electrodes. This requires us to implement a so-called *bi-orthogonalization process* as summarized below. The basic principle of the algorithm is to start from the normalized vector $|\psi\rangle$ and from the non-hermitian matrix $\mathcal{H}$, from which a bi-orthogonal basis $\{|\psi_n\rangle, \langle\phi_n|\}$ is constructed following

$$|\psi_{n+1}\rangle = \mathcal{H}|\psi_n\rangle - a_{n+1}|\psi_n\rangle - b_n|\psi_{n-1}\rangle, \tag{D.71}$$

$$\langle\phi_{n+1}| = \langle\phi_n|\mathcal{H} - \langle\phi_n|a_{n+1} - \langle\phi_{n-1}|b_n, \tag{D.72}$$

with the initial conditions $|\psi_{-1}\rangle = |\phi_{-1}\rangle = 0$, $|\psi_0\rangle = |\phi_0\rangle = |\psi\rangle$, and the bi-orthogonality condition $\langle\phi_n|\psi_m\rangle = 0$ if $n \neq m$. This last condition is equivalent to the following relations for a_n and b_n:

$$a_n = \frac{\langle\phi_n|\mathcal{H}|\psi_n\rangle}{\langle\phi_n|\psi_n\rangle}, \tag{D.73}$$

$$b_n = \frac{\langle\phi_{n-1}|\mathcal{H}|\psi_n\rangle}{\langle\phi_{n-1}|\psi_{n-1}\rangle} = \frac{\langle\phi_n|\psi_n\rangle}{\langle\phi_{n-1}|\psi_{n-1}\rangle}. \tag{D.74}$$

The four Eqs. (D.71–74) allow a recursive determination of the bi-orthogonal basis and of the coefficients a_n, b_n. Note that in "ket" notation, Eq. (D.72) must be understood as $|\phi_{n+1}\rangle = \mathcal{H}^\dagger|\phi_n\rangle - a_{n+1}^*|\phi_n\rangle - b_n^*|\phi_{n-1}\rangle$. One starts from $|\phi_0\rangle = |\psi_0\rangle = |\psi\rangle$. At step 0, one computes $\mathcal{H}|\psi_0\rangle$ and $a_1 = \langle\phi_0|\mathcal{H}|\psi_0\rangle/\langle\phi_0|\psi_0\rangle$ by expanding all the amplitudes within the tight-binding localized basis.

Values of $|\psi_1\rangle$ and $|\phi_1\rangle$ are then obtained by computing $\mathcal{H}|\psi_0\rangle - a_1|\psi_0\rangle$ and $\mathcal{H}^\dagger|\phi_0\rangle - a_1^*|\phi_0\rangle$, while the first coefficient b_1 is subsequently deduced from Eq. (D.74). At step 1, $\mathcal{H}|\psi_1\rangle$ is computed together with $a_2 = \langle\phi_1|\hat{\mathcal{H}}|\psi_1\rangle/\langle\phi_1|\psi_1\rangle$.

Then $|\psi_2\rangle$ and $|\phi_2\rangle$ result from the computation of vectors $\mathcal{H}|\psi_1\rangle - a_2|\psi_1\rangle - b_1|\psi_0\rangle$ and $\mathcal{H}^\dagger|\phi_1\rangle - a_2^*|\phi_1\rangle - b_1^*|\phi_0\rangle$. Finally, the coefficient b_2 is deduced from Eq. (D.74). Steps $n \geq 2$ are fully similar to step 1. In the basis $\{|\psi_n\rangle\}$, $\mathcal{H}$ thus has a tridiagonal form:

$$\mathcal{H} = \begin{pmatrix} a_1 & b_1 & & & \\ 1 & a_2 & b_2 & & \\ & 1 & a_3 & b_3 & \\ & & 1 & \cdot & \cdot \\ & & & \cdot & \cdot \end{pmatrix}. \tag{D.75}$$

Hence the recurrence relations (D.71) and (D.72) lead to a nonsymmetric matrix and to a nonnormalized bi-orthogonal basis. By choosing a different convention, a symmetric tridiagonal matrix and/or a normalized basis could be obtained. The quantity $\langle\psi|G^r(z = E \pm 0^+)|\psi\rangle = \langle\phi_0|\frac{1}{z-\mathcal{H}}|\psi_0\rangle$ can then be computed by the continued fraction method.

This quantity is actually equal to the first diagonal element of $(z - \mathcal{H})^{-1}$ where $\mathcal{H}$ is the tridiagonal matrix (D.75). Let us call $G_0(z)$ this matrix element and define $G_n(z)$, the first diagonal element of the matrix $(z - \mathcal{H}_n)^{-1}$, with $\mathcal{H}_n$ the matrix $\mathcal{H}$ without its n first lines and columns:

$$\mathcal{H}_n = \begin{pmatrix} a_n & b_n & & & \\ 1 & a_{n+1} & b_{n+1} & & \\ & 1 & a_{n+2} & b_{n+2} & \\ & & 1 & \cdot & \cdot \\ & & & \cdot & \cdot \end{pmatrix}. \tag{D.76}$$

From standard linear algebra, it can be shown that

$$G_0(z) = \frac{1}{z - a_1 - b_1 G_1(z)}, \tag{D.77}$$

and replicating such an algorithm, one gets a continued fraction of $G_0(z)$:

$$G_0(z) = \cfrac{1}{z - a_1 - \cfrac{b_1}{z - a_2 - \cfrac{b_2}{\dots}}}. \tag{D.78}$$

In contrast with the standard recursion method, the recursion coefficients a_n and b_n do not show any simple behavior for large n. Simple truncation of the continued fraction at sufficiently large n has been shown to yield reasonably good convergence. This method was tested on carbon nanotube based heterojunctions (Triozon & Roche, 2005), with perfect agreement with the decimation techniques presented in Appendix C.

References

Abanin, D. A., Lee, P. A. & Levitov, L. S. (2006), 'Spin-filtered edge states and quantum Hall effect in graphene', *Phys. Rev. Lett.* **96**, 176803.

Abanin, D. A., Novoselov, K. S., Zeitler, U., *et al.* (2007), 'Dissipative quantum Hall effect in graphene near the Dirac point', *Phys. Rev. Lett.* **98**, 196806.

Abergel, D. S. L. & Chakraborty, T. (2009), 'Generation of valley polarized current in bilayer graphene', *Appl. Phys. Lett.* **95**, 062107.

Abrahams, E., Anderson, P. W., Licciardello, D. C. & Ramakrishnan, T. V. (1979), 'Scaling theory of localization: Absence of quantum diffusion in two dimensions', *Phys. Rev. Lett.* **42**, 673–676.

Abrikosov, A., Gorkov, L. & Dzyaloshinskii, E. (1975), *Methods of Quantum Field Theory in Statistical Physics*, Dover, New York.

Adam, S., Hwang, E. H., Galitski, V. M. & Sarma, S. D. (2007), 'A self-consistent theory for graphene transport', *PNAS* **104**, 18392.

Adessi, C., Roche, S. & Blase, X. (2006), 'Reduced backscattering in potassium-doped nanotubes: *ab initio* and semi-empirical simulations', *Phys. Rev. B* **73**, 125414.

Aharonov, Y. & Bohm, D. (1959), 'Significance of electromagnetic potentials in the quantum theory', *Phys. Rev.* **115**, 485–491.

Ajiki, H. & Ando, T. (1993), 'Electronic states of carbon nanotubes', *J. Phys. Soc. Jpn.* **62**, 1255–1266.

Ajiki, H. & Ando, T. (1996), 'Energy bands of carbon nanotubes in magnetic fields', *J. Phys. Soc. Jpn.* **65**, 505–514.

Akhmerov, A. (2011), Dirac and Majorana edge states in graphene and topological superconductors, Ph.D. thesis, Leiden University.

Akhmerov, A. R. & Beenakker, C. W. J. (2008), 'Boundary conditions for Dirac fermions on a terminated honeycomb lattice', *Phys. Rev. B* **77**, 085423.

Akkermans, E. & Montambaux, G. (2007), *Mesoscopic Physics of Electrons and Photons*, Cambridge University Press, Cambridge.

Alam, A. & Mookerjee, A. (2005), 'Lattice thermal conductivity of disordered binary alloys', *Phys. Rev. B* **72**, 214207.

Aleiner, I. L. & Efetov, K. B. (2006), 'Effect of disorder on transport in graphene', *Phys. Rev. Lett.* **97**, 236801.

Alhassid, Y. (2000), 'The statistical theory of quantum dots', *Rev. Mod. Phys.* **72**, 895–968.

Allain, P. & Fuchs, J. (2011), 'Klein tunneling in graphene: Optics with massless electrons', **83**, 301–317.

Allen, M. T., Martin, J. & Yacoby, A. (2012), 'Gate-defined quantum confinement in suspended bilayer graphene', *Nature Communications* **3**, 934–936.

Allen, P. B. & Feldman, J. L. (1989), 'Thermal conductivity of glasses: Theory and application to amorphous silicon', *Phys. Rev. Lett.* **62**, 645–648.

Alos-Palop, M. & Blaauboer, M. (2011), 'Adiabatic quantum pumping in normal-metal-insulator-superconductor junctions in a monolayer of graphene', *Phys. Rev. B* **84**, 073402.

Altland, A. (2006), 'Low-energy theory of disordered graphene', *Phys. Rev. Lett.* **97**, 236802.

Altshuler, B. & Aronov, A. G. (1985), 'Electron–electron interaction in disordered conductors', in *Electron–Electron Interactions in Disordered Systems*, Efros, A.L. & Pollak, M. eds, Elsevier, Amsterdam, pp. 1–75.

Altshuler, B. L., Aronov, A. G. & Spivak, B. Z. (1981), 'The Aharonov–Bohm effect in disordered conductors', *JETP Letters* **33**, 94.

Altshuler, B. L. & Glazman, L. I. (1999), 'Pumping electrons', *Science* **283**, 1864–1865.

Amara, H., Latil, S., Meunier, V., Lambin, P. & Charlier, J.-C. (2007), 'Scanning tunneling microscopy fingerprints of point defects in graphene: A theoretical prediction', *Phys. Rev. B* **76**, 115423.

Amorim, R. G., Fazzio, A., Antonelli, A., Novaes, F. D. & da Silva, A. J. R. (2007), 'Divacancies in graphene and carbon nanotubes', *Nano Lett.* **7**, 2459–2462.

An, J., Voelkl, E., Suk, J. W., *et al.* (2011), 'Domain (grain) boundaries and evidence of "twinlike" structures in chemically vapor deposited grown graphene', *ACS Nano* **5**, 2433.

Anantram, M. P. (2000), 'Current-carrying capacity of carbon nanotubes', *Phys. Rev. B* **62**, R4837–R4840.

Anantram, M. P. & Léonard, F. (2006), 'Physics of carbon nanotube electronic devices', *Reports on Progress in Physics* **69**, 507.

Anda, E. V., Makler, S., Pastawski, H. M., & Barrera, R. G. (1994), 'Electron–phonon effects on transport in mesoscopic heterostructures', *Braz. J. Phys.* **24**, 330.

Anderson, P. W. (1958), 'Absence of diffusion in certain random lattices', *Phys. Rev.* **109**, 1492–1505.

Anderson, P. W., Thouless, D. J., Abrahams, E. & Fisher, D. S. (1980), 'New method for a scaling theory of localization', *Phys. Rev. B* **22**, 3519–3526.

Ando, T. (1991), 'Quantum point contacts in magnetic fields', *Phys. Rev. B* **44**, 8017.

Ando, T., Nakanishi, T. & Saito, R. (1998), 'Berry's phase and absence of back scattering in carbon nanotubes', *J. Phys. Soc. Jpn.* **67**, 2857–2862.

Ando, T. & Seri, T. (1997), 'Quantum transport in a carbon nanotube in magnetic fields', *J. Phys. Soc. Jpn.* **66**, 3558–3565.

Appenzeller, J., Radosavljević, M., Knoch, J. & Avouris, P. (2004), 'Tunneling versus thermionic emission in one-dimensional semiconductors', *Phys. Rev. Lett.* **92**, 048301.

Apsel, S. E., Emmert, J. W., Deng, J. & Bloomfield, L. A. (1996), 'Surface-enhanced magnetism in nickel clusters', *Phys. Rev. Lett.* **76**, 1441–1444.

Areshkin, D. A., Gunlycke, D. & White, C. T. (2007), 'Ballistic transport in graphene nanostrips in the presence of disorder: importance of edge effects', *Nano Letters* **7**, 204–210. PMID: 17212465.

Areshkin, D. A. & White, C. T. (2007), 'Building blocks for integrated graphene circuits', *Nano Lett.* **7**, 3253–3259.

Arrachea, L. & Moskalets, M. (2006), 'Relation between scattering-matrix and Keldysh formalisms for quantum transport driven by time-periodic fields', *Phys. Rev. B* **74**, 245322.

Ashcroft, N. W. & Mermin, N. D. (1976), *Solid State Physics*, Holt Saunders, Philadelphia.

Avouris, P. (2010), 'Graphene: Electronic and photonic properties and devices', *Nano Lett.* **10**, 4285–4294.

Avouris, P., Chen, Z. & Perebeinos, V. (2007), 'Carbon-based electronics', *Nature Nanotechnology* **2**, 605–615.

Avriller, R. (2008), Contribution à la modélisation théorique et à l'étude du transport quantique dans les dispositifs à base de nanotubes de carbone, Ph.D. thesis, Université Joseph-Fourier.

Avriller, R., Latil, S., Triozon, F., Blase, X. & Roche, S. (2006), 'Chemical disorder strength in carbon nanotubes: Magnetic tuning of quantum transport regimes', *Phys. Rev. B* **74**, 121406.

Avriller, R., Roche, S., Triozon, F., Blase, X. & Latil, S. (2007), 'Low-dimensional quantum transport properties of chemically-disordered carbon nanotubes: From weak to strong localization regimes', *Modern Physics Letters B* **21**, 1955.

Avsar, A., Yang, T.-Y., Bae, S., *et al.* (2011), 'Toward wafer scale fabrication of graphene based spin valve devices', *Nano Lett.* **11**, 2363–2368.

Babic, B. & Schönenberger, C. (2004), 'Observation of Fano resonances in single-wall carbon nanotubes', *Phys. Rev. B* **70**, 195408.

Bachelet, G. B., Hamann, D. R. & Schlüter, M. (1982), 'Pseudopotentials that work: From H to Pu', *Phys. Rev. B* **26**, 4199–4228.

Bachilo, S. M., Strano, M. S., Kittrell, C., *et al.* (2002), 'Structure-assigned optical spectra of single-walled carbon nanotubes', *Science* **298**, 2361–2366.

Bachtold, A., Strunk, C., Salvetat, J.-P., *et al.* (1999), 'Aharonov–Bohm oscillations in carbon nanotubes', *Nature* **397**, 673–675.

Bae, S., Kim, H., Lee, Y., *et al.* (2010), 'Roll-to-roll production of 30-inch graphene films for transparent electrodes', *Nature Nanotechnology* **5**, 574–578.

Bahreyni, B. (2008), *Fabrication and Design of Resonant Microdevices*, Elsevier, New York, Chapter 10: 'Survey of applications'.

Balandin, A. A., Ghosh, S., Bao, W., *et al.* (2008), 'Superior thermal conductivity of single-layer graphene', *Nano Lett.* **8**, 902–907.

Balasubramanian, K., Lee, E. J. H., Weitz, R. T., Burghard, M. & Kern, K. (2008), 'Carbon nanotube transistors – chemical functionalization and device characterization', *Phys. Stat. Sol. (A)* **205**, 633–646.

Baldoni, M., Sgamellotti, A. & Mercuri, F. (2008), 'Electronic properties and stability of graphene nanoribbons: An interpretation based on Clar sextet theory', *Chemical Physics Letters* **464**, 202–207.

Baletto, F. & Ferrando, R. (2005), 'Structural properties of nanoclusters: Energetic, thermodynamic, and kinetic effects', *Rev. Mod. Phys.* **77**, 371–423.

Banhart, F., Kotakoski, J. & Krasheninnikov, A. V. (2011), 'Structural defects in graphene', *ACS Nano* **5**, 26–41.

Bardarson, J. H., Tworzydło, J., Brouwer, P. W. & Beenakker, C. W. J. (2007), 'One-parameter scaling at the Dirac point in graphene', *Phys. Rev. Lett.* **99**, 106801.

Barone, V., Hod, O. & Scuseria, G. E. (2006), 'Electronic structure and stability of semiconducting graphene nanoribbons', *Nano Lett.* **6**, 2748–2754.

Beenakker, C. W. J. (1991), 'Theory of Coulomb-blockade oscillations in the conductance of a quantum dot', *Phys. Rev. B* **44**, 1646–1656.

Beenakker, C. W. J. (1997), 'Random-matrix theory of quantum transport', *Rev. Mod. Phys.* **69**, 731–808.

Beenakker, C. W. J. (2008), 'Colloquium: Andreev reflection and Klein tunneling in graphene', *Rev. Mod. Phys.* **80**, 1337–1354.

Begliarbekov, M., Sasaki, K.-I., Sul, O., Yang, E.-H. & Strauf, S. (2011), 'Optical control of edge chirality in graphene', *Nano Lett.* **11**, 4874–4878.

Berger, C., Song, Z., Li, X., *et al.* (2006), 'Electronic confinement and coherence in patterned epitaxial graphene', *Science* **312**, 1191–1196.

Bergman, G. (1984), 'Weak localization in thin films: A time-of-flight experiment with conduction electrons', *Physics Reports* **107**, 1–58.

Bernal, J. D. (1924), 'The structure of graphite', *Proceedings of the Royal Society of London, Series A* **106**, 749–773.

Berry, M. V. & Mondragon, R. J. (1987), 'Neutrino billiards: Time-reversal symmetry-breaking without magnetic fields', *Proceedings of the Royal Society of London, A. Mathematical and Physical Sciences* **412**, 53–74.

Bethune, D. S., Klang, C. H., de Vries, M. S., *et al.* (1993), 'Cobalt-catalysed growth of carbon nanotubes with single-atomic-layer walls', *Nature* **363**, 605–607.

Biel, B., Triozon, F., Blase, X. & Roche, S. (2009*a*), 'Chemically induced mobility gaps in graphene nanoribbons: A route for upscaling device performances', *Nano Lett.* **9**, 2725–2729.

Biel, B., Triozon, F., Niquet, Y. & Roche, S. (2009*b*), 'Anomalous doping effects on charge transport in graphene nanoribbons', *Phys. Rev. Lett.* **102**, 096803.

Biró, L. P., Márk, G. I., Koós, A. A., Nagy, B. J. & Lambin, P. (2002), 'Coiled carbon nanotube structures with supraunitary nonhexagonal to hexagonal ring ratio', *Phys. Rev. B* **66**, 165405.

Blanter, Y. & Büttiker, M. (2000), 'Shot noise in mesoscopic conductors', *Physics Reports* **336**, 1–166.

Blase, X., Benedict, L. X., Shirley, E. L. & Louie, S. G. (1994), 'Hybridization effects and metallicity in small radius carbon nanotubes', *Phys. Rev. Lett.* **72**, 1878–1881.

Blöchl, P. E. (1994), 'Projector augmented-wave method', *Phys. Rev. B* **50**, 17953–17979.

Bockrath, M., Cobden, D. H., Lu, J., *et al.* (1999), 'Luttinger-liquid behaviour in carbon nanotubes', *Nature* **397**, 598–601.

Boehm, H. P., Clauss, A., Fischer, G. O. & Hofmann, U. (1962), 'Das adsorptionsverhalten sehr diinner kohlenstoff-folien', *Z. Anorg. Allg. Chem.* **316**, 119–127.

Bolotin, K. I., Sikes, K. J., Hone, J., Stormer, H. L. & Kim, P. (2008), 'Temperature-dependent transport in suspended graphene', *Phys. Rev. Lett.* **101**, 096802.

Bonaccorso, F., Sun, Z., Hasan, T. & Ferrari, A. C. (2010), 'Graphene photonics and optoelectronics', *Nature Photonics* **4**, 611–622.

Bonča, J. & Trugman, S. A. (1995), 'Effect of inelastic processes on tunneling', *Phys. Rev. Lett.* **75**, 2566–2569.

Born, M. & Oppenheimer, M. (1927), 'Zur quantentheorie der molekeln', *Ann. Physik* **84**, 457.

Bose, S. K., Winer, K. & Andersen, O. K. (1988), 'Electronic properties of a realistic model of amorphous silicon', *Phys. Rev. B* **37**, 6262.

Bostwick, A., McChesney, J. L., Emtsev, K. V., *et al.* (2009), 'Quasiparticle transformation during a metal–insulator transition in graphene', *Phys. Rev. Lett.* **103**, 056404.

Botello-Méndez, A. R., Cruz-Silva, E., Romo-Herrera, J., *et al.* (2011*a*), 'Quantum transport in graphene nanonetworks', *Nano Lett.* **11**, 3058–3064.

Botello-Mendez, A. R., Declerck, X., Terrones, M., Terrones, H. & Charlier, J.-C. (2011*b*), 'One-dimensional extended lines of divacancy defects in graphene', *Nanoscale* **3**, 2868–2872.

Bouilly, D., Cabana, J. & Martel, R. (2012), 'Unaltered electrical conductance in single-walled carbon nanotubes functionalized with divalent adducts', *Appl. Phys. Lett.* **101**, 053116.

Brandbyge, M., Mozos, J.-L., Ordejn, P., Taylor, J. & Stokbro, K. (2002), 'Density-functional method for nonequilibrium electron transport', *Phys. Rev. B* **65**, 165401.

Brenner, D. W. (1990), 'Empirical potential for hydrocarbons for use in simulating the chemical vapor deposition of diamond films', *Phys. Rev. B* **42**, 9458–9471.

Brey, L. & Fertig, H. A. (2006), 'Electronic states of graphene nanoribbons studied with the Dirac equation', *Phys. Rev. B* **73**, 235411.

Brouwer, P. W. (1998), 'Scattering approach to parametric pumping', *Phys. Rev. B* **58**, R10135–R10138.

Bunch, J. S., Yaish, Y., Brink, M., Bolotin, K. & McEuen, P. L. (2005), 'Coulomb oscillations and Hall effect in quasi-2D graphite quantum dots', *Nano Lett.* **5**, 287–290.

Busl, M., Platero, G. & Jauho, A.-P. (2012), 'Dynamical polarizability of graphene irradiated by circularly polarized AC electric fields', *Phys. Rev. B* **85**, 155449.

Büttiker, M. (1988*a*), 'Absence of backscattering in the quantum Hall effect in multiprobe conductors', *Phys. Rev. B* **38**, 9375–9389.

Büttiker, M. (1988*b*), 'Symmetry of electrical conduction', *IBM Journal of Research and Development* **32**, 317–334.

Büttiker, M., Imry, Y., Landauer, R. & Pinhas, S. (1985), 'Generalized many-channel conductance formula with application to small rings', *Phys. Rev. B* **31**, 6207–6215.

Büttiker, M. & Moskalets, M. (2006), 'Scattering theory of dynamic electrical transport', in Asch J. & Joye A., eds., *Mathematical Physics of Quantum Mechanics*, Vol. 690 of *Lecture Notes in Physics*, Springer, Berlin / Heidelberg, pp. 33–44. 10.1007/3-540-34273-7_5.

Büttiker, M., Thomas, H. & Pretre, A. (1994), 'Current partition in multiprobe conductors in the presence of slowly oscillating external potentials', *Zeitschrift für Physik B Condensed Matter* **94**, 133–137.

Cabana, J. & Martel, R. (2007), 'Probing the reversibility of sidewall functionalization using carbon nanotube transistors', *J. Am. Chem. Soc.* **129**, 2244–2245.

Cai, J., Ruffieux, P., Jaafar, R., *et al.* (2010), 'Atomically precise bottom-up fabrication of graphene nanoribbons', *Nature* **466**, 470–473.

Calandra, M. & Mauri, F. (2007), 'Electron–phonon coupling and electron self-energy in electron-doped graphene: Calculation of angular-resolved photoemission spectra', *Phys. Rev. B* **76**, 205411.

Calleja, M., Rey, C., Alemany, M. M. G., *et al.* (1999), 'Self-consistent density-functional calculations of the geometries, electronic structures, and magnetic moments of Ni-Al clusters', *Phys. Rev. B* **60**, 2020–2024.

Calvo, H. L., Pastawski, H. M., Roche, S. & Foa Torres, L. E. F. (2011), 'Tuning laser-induced band gaps in graphene', *Appl. Phys. Lett.* **98**, 2321033.

Calvo, H. L., Perez-Piskunow, P. M., Pastawski, H. M., Roche, S. & Foa Torres, L. E. F. (2013), 'Non-perturbative effects of laser illumination on the electrical properties of graphene nanoribbons', *Journal of Physics: Condensed Matter* **25**, 144202.

Calvo, H. L., Perez-Piskunow, P. M., Roche, S. & Foa Torres, L. E. F. (2012), 'Laser-induced effects on the electronic features of graphene nanoribbons', *Appl. Phys. Lett.* **101**, 253506.

Calzolari, A., Marzari, N., Souza, I. & Buongiorno Nardelli, M. (2004), '*Ab initio* transport properties of nanostructures from maximally localized Wannier functions', *Phys. Rev. B* **69**, 035108.

Campidelli, S., Ballesteros, B., Filoramo, A., *et al.* (2008), 'Facile decoration of functionalized single-wall carbon nanotubes with phthalocyanines via click chemistry', *J. Am. Chem. Soc.* **130**, 11503–11509.

Campos-Delgado, J., Romo-Herrera, J. M., Jia, X., *et al.* (2008), 'Bulk production of a new form of sp^2 carbon: Crystalline graphene nanoribbons', *Nano Letters* **8**, 2773–2778. PMID: 18700805.

Cançado, L. G., Pimenta, M. A., Neves, B. R. A., Dantas, M. S. S. & Jorio, A. (2004), 'Influence of the atomic structure on the Raman spectra of graphite edges', *Phys. Rev. Lett.* **93**, 247401.

Castro, E. V., Novoselov, K. S., Morozov, S. V., *et al.* (2007), 'Biased bilayer graphene: Semiconductor with a gap tunable by the electric field effect', *Phys. Rev. Lett.* **99**, 216802.

Castro Neto, A. H., Guinea, F., Peres, N. M. R., Novoselov, K. S. & Geim, A. K. (2009), 'The electronic properties of graphene', *Rev. Mod. Phys.* **81**, 109–162.

Cataldo, F., ed. (2005), *Polyynes: Synthesis, Properties, and Applications*, Taylor & Francis, London.

Cayssol, J., Dóra, B., Simon, F. & Moessner, R. (2013), 'Floquet topological insulators', *Physica Status Solidi (RRL) – Rapid Research Letters* **7**, 101.

Cazalilla, M. A., Iucci, A., Guinea, F. & Neto, A. H. C. (2012), 'Local moment formation and Kondo effect in defective graphene', (unpublished) arXiv:1207.3135 [cond-mat.str-el].

Ceperley, D. M. & Alder, B. J. (1980), 'Ground state of the electron gas by a stochastic method', *Phys. Rev. Lett.* **45**, 566–569.

Chakravarty, S. & Schmid, A. (1986), 'Weak localization: The quasiclassical theory of electrons in a random potential', *Physics Reports* **140**, 193–236.

Chang, A.M., Baranger, H.U., Pfeiffer, L.N., West, K.W. & Chang, T.Y. (1996), 'Non-Gaussian distribution of Coulomb blockade peak heights in quantum dots', *Phys. Rev. Lett.* **76**, 1695.

Charlier, J.-C., Arnaud, L., Avilov, I. V., *et al.* (2009), 'Carbon nanotubes randomly decorated with gold clusters: From nano 2 hybrid atomic structures to gas sensing prototypes', *Nanotechnology* **20**, 375501.

Charlier, J.-C., Blase, X. & Roche, S. (2007), 'Electronic and transport properties of nanotubes', *Rev. Mod. Phys.* **79**, 677–732.

Charlier, J.-C., Ebbesen, T. W. & Lambin, P. (1996), 'Structural and electronic properties of pentagon–heptagon pair defects in carbon nanotubes', *Phys. Rev. B* **53**, 11108–11113.

Charlier, J.-C., Gonze, X. & Michenaud, J.-P. (1994a), 'First-principles study of the stacking effect on the electronic properties of graphite(s)', *Carbon* **32**, 289–299.

Charlier, J.-C., Gonze, X. & Michenaud, J.-P. (1994b), 'Graphite interplanar bonding: Electronic delocalization and van der Waals interaction', *EPL (Europhysics Letters)* **28**, 403.

Charlier, J.-C., Gonze, X. & Michenaud, J.-P. (1995), 'First-principles study of carbon nanotube solid-state packings', *EPL (Europhysics Letters)* **29**, 43.

Charlier, J.-C. & Lambin, P. (1998), 'Electronic structure of carbon nanotubes with chiral symmetry', *Phys. Rev. B* **57**, R15037–R15039.

Charlier, J.-C., Michenaud, J.-P. & Gonze, X. (1992), 'First-principles study of the electronic properties of simple hexagonal graphite', *Phys. Rev. B* **46**, 4531–4539.

Charlier, J.-C., Michenaud, J.-P., Gonze, X. & Vigneron, J.-P. (1991), 'Tight-binding model for the electronic properties of simple hexagonal graphite', *Phys. Rev. B* **44**, 13237–13249.

Chaste, J., Eichler, A., Moser, J., *et al.* (2012), 'A nanomechanical mass sensor with yoctogram resolution', *Nature Nanotechnology* **7**, 301–304.

Che, J., Çagin, T. & Goddard, W.A., III (2000), 'Thermal conductivity of carbon nanotubes', *Nanotechnology* **11**, 65.

Checkelsky, J. G., Li, L. & Ong, N. P. (2008), 'Zero-energy state in graphene in a high magnetic field', *Phys. Rev. Lett.* **100**, 206801.

Cheianov, V. V. & Fal'ko, V. I. (2006), 'Selective transmission of Dirac electrons and ballistic magnetoresistance of n-p junctions in graphene', *Phys. Rev. B* **74**, 041403.

Chen, C., Rosenblatt, S., Bolotin, K. I., *et al.* (2009), 'Performance of monolayer graphene nanomechanical resonators with electrical readout', *Nature Nanotechnology* **4**, 861–867.

Chen, J., Badioli, M., Alonso-Gonzalez, P., *et al.* (2012), 'Optical nano-imaging of gate-tunable graphene plasmons', *Nature.* **487**, (July), 77–81.

Chen, J.-H., Cullen, W. G., Jang, C., Fuhrer, M. S. & Williams, E. D. (2009), 'Defect scattering in graphene', *Phys. Rev. Lett.* **102**, 236805.

Chen, J.-H., Jang, C., Adam, S., *et al.* (2008), 'Charged-impurity scattering in graphene', *Nature Physics* **4**, 377–381.

Chen, J.-H., Li, L., Cullen, W. G., Williams, E. D. & Fuhrer, M. S. (2011), 'Tunable Kondo effect in graphene with defects', *Nat Phys* **7**, 535–538.

Chen, Z., Lin, Y.-M., Rooks, M. & Avouris, P. (2007), 'Graphene nanoribbon electronics', *Physica E: Low-dimensional Systems and Nanostructures* **40**, 228–232.

Chenaiov, V., Fal'ko, V., Altshuler, B. I. & Aleiner, I. (2007), 'Random resistor network model of minimal conductivity in graphene', *Phys. Rev. Lett.* **99**, 176801.

Chico, L., Crespi, V. H., Benedict, L. X., Louie, S. G. & Cohen, M. L. (1996), 'Pure carbon nanoscale devices: Nanotube heterojunctions', *Phys. Rev. Lett.* **76**, 971–974.

Chiu, H.-Y., Hung, P., Postma, H. W. C. & Bockrath, M. (2008), 'Atomic-scale mass sensing using carbon nanotube resonators', *Nano Lett.* **8**, 4342–4346.

Choi, H., Ihm, J., Louie, S. & Cohen, M. (2000), 'Defects quasibound states, and quantum conductance in metallic carbon nanotubes', *Phys. Rev. Lett.* **84**, 9172920.

Chuvilin, A., Kaiser, U., Bichoutskaia, E., Besley, N. A. & Khlobystov, A. N. (2010), 'Direct transformation of graphene to fullerene', *Nature Chemistry* **2**, 450–453.

Chuvilin, A., Meyer, J. C., Algara-Siller, G. & Kaiser, U. (2009), 'From graphene constrictions to single carbon chains', *New Journal of Physics* **11**, 083019.

Ci, L., Song, L., Jariwala, D., *et al.* (2009), 'Graphene shape control by multistage cutting and transfer', *Adv. Mater.* **21**, 4487–4491.

Ci, L., Xu, Z., Wang, L., *et al.* (2008), 'Controlled nanocutting of graphene', *Nano Research* **1**, 116–122.

Clar, E. (1964), *Polycyclic Hydrocarbons*, Academic Press, London.

Clar, E. (1972), *The Aromatic Sextet*, Wiley, New York.

Clark, S. J., Segall, M. D., Pickard, C. J., *et al.* (2005), 'First principles methods using castep', *Zeitschrift für Kristallographie – Crystalline Materials* **220**, 567–570.

Cockayne, E., Rutter, G. M., Guisinger, N. P., *et al.* (2011), 'Grain boundary loops in graphene', *Phys. Rev. B* **83**, 195425.

Coiffic, J. C., Fayolle, M., Maitrejean, S., Foa Torres, L. E. F. & Le Poche, H. (2007), 'Conduction regime in innovative carbon nanotube via interconnect architectures', *Appl. Phys. Lett.* **91**, 252107–3.

Collins, A., Kanda, H., Isoya, J. & van Wyk, C. A. J. (1998), 'Correlation between optical absorption and EPR in high-pressure diamond grown from a nickel solvent catelyst', *Diam. Relat. Mater.* **7**, 333–338.

Connétable, D., Rignanese, G.-M., Charlier, J.-C. & Blase, X. (2005), 'Room temperature Peierls distortion in small diameter nanotubes', *Phys. Rev. Lett.* **94**, 015503.

Connolly, M.R., Chiu, K.L., Giblin, S.P., *et al.* (2013), 'Gigahertz quantised charge pumping in graphene quantum dots', *Nature Nanotechnology* **8**, 417–420.

Cornaglia, P. S., Usaj, G. & Balseiro, C. A. (2009), 'Localized spins on graphene', *Phys. Rev. Lett.* **102**, 046801.

Crespi, A., Corrielli, G., Valle, G. D., Osellame, R. & Longhi, S. (2013), 'Dynamic band collapse in photonic graphene', *New Journal of Physics* **15**, 013012.

Crespi, V. H., Benedict, L. X., Cohen, M. L. & Louie, S. G. (1996), 'Prediction of a pure-carbon planar covalent metal', *Phys. Rev. B* **53**, R13303–R13305.

Cresti, A., Grosso, G. & Parravicini, G. (2007), 'Numerical study of electronic transport in gated graphene ribbons', *Phys. Rev. B* **76**, 205433.

Cresti, A., Lopez-Bezanilla, A., Ordejon, P. & Roche, S. (2011), 'Oxygen surface functionalization of graphene nanoribbons for transport gap engineering', *ACS Nano* **5**, 9271–9277.

Cresti, A., Ortmann, F., Louvet, Th., Van Tuan, D. & Roche, S. (2013), 'Broken symmetries, zero-energy modes, and quantum transport in disordered graphene', *Phys. Rev. Lett.* **110**, 196601.

Cresti, A., Nemec, N., Biel, B., *et al.* (2008), 'Charge transport in disordered graphene-based low dimensional materials', *Nano Research* **1**, 361–394.

Cresti, A. & Roche, S. (2009), 'Edge-disorder-dependent transport length scales in graphene nanoribbons: From Klein defects to the superlattice limit', *Phys. Rev. B* **79**, 233404.

Cruz-Silva, E., Cullen, D. A., Gu, L., *et al.* (2008), 'Heterodoped nanotubes: Theory, synthesis, and characterization of phosphorus-nitrogen-doped multiwalled carbon nanotubes', *ACS Nano* **2**, 441–448.

Cruz-Silva, E., Lopez-Urias, F., Munoz-Sandoval, E., *et al.* (2011), 'Phosphorus and phosphorus-nitrogen doped carbon nanotubes for ultrasensitive and selective molecular detection', *Nanoscale* **3**, 1008–1013.

Cruz-Silva, E., Lopez-Urias, F., Munoz-Sandoval, E., *et al.* (2009), 'Electronic transport and mechanical properties of phosphorus- and phosphorus-nitrogen-doped carbon nanotubes', *ACS Nano* **3**, 1913–1921.

Curtiss, L. A., Raghavachari, K., Redfern, P. C., Rassolov, V. & Pople, J. A. (1998), 'Gaussian-3 (g3) theory for molecules containing first and second-row atoms', *The Journal of Chemical Physics* **109**, 7764–7776.

D'Amato, J. L. & Pastawski, H. M. (1990), 'Conductance of a disordered linear chain including inelastic scattering events', *Phys. Rev. B* **41**, 7411–7420.

D'Amato, J. L., Pastawski, H. M. & Weisz, J. F. (1989), 'Half-integer and integer quantum-flux periods in the magnetoresistance of one-dimensional rings', *Phys. Rev. B* **39**, 3554–3562.

Das Sarma, S., Adam, S., Hwang, E. H. & Rossi, E. (2011), 'Electronic transport in two-dimensional graphene', *Rev. Mod. Phys.* **83**, 407–470.

Das Sarma, S., Hwang, E. H. & Li, Q. (2012), 'Disorder by order in graphene', *Phys. Rev. B* **85**, 195451.

Datta, S. (1995), *Electronic Transport in Mesoscopic Systems*, Cambridge University Press, Cambridge.

Datta, S. S., Strachan, D. R., Khamis, S. M. & Johnson, A. T. C. (2008), 'Crystallographic etching of few-layer graphene', *Nano Letters* **8**, 1912–1915. PMID: 18570483.

Dean, C. R., Young, A. F., Meric, I., *et al.* (2010), 'Boron nitride substrates for high-quality graphene electronics', *Nature Nanotechnology* **5**, 722–726.

Dean, C. R., Wang, L., Maher, C., *et al.* (2013), 'Hofstadter's butterfly and the fractal quantum Hall effect in Moiré superlattices', *Nature* **497**, 598–602.

Delaney, P., Choi, H. J., Ihm, J., Louie, S. G. & Cohen, M. L. (1998), 'Broken symmetry and pseudogaps in ropes of carbon nanotubes', *Nature* **391**, 466–468.

Dery, H., Wu, H., Ciftcioglu, B., *et al.* (2012), 'Nanospintronics based on magnetologic gates', *IEEE Transactions on Electron Devices* **59**, 259–262.

Derycke, V., Martel, R., Appenzeller, J. & Avouris, P. (2001), 'Carbon nanotube inter- and intramolecular logic gates', *Nano Lett.* **1**, 453–456.

Di Ventra, M. (2008), *Electrical Transport in Nanoscale Systems*, Cambridge University Press, Cambridge.

Dinh, V. T., Kotakoski, J. & Louvet, T., *et al.* (2013), 'Scaling properties of charge transport in polycrystalline graphene', *Nano Letters* **13**, 1730–1735.

Dion, M., Rydberg, H., Schröder, E., Langreth, D. C. & Lundqvist, B. I. (2004), 'Van der Waals density functional for general geometries', *Phys. Rev. Lett.* **92**, 246401.

Dirac, P. A. M. (1930), 'Note on exchange phenomena in the Thomas atom', *Mathematical Proceedings of the Cambridge Philosophical Society* **26**, 376–385.

DiVicenzo, D. P. & Mele, E. J. (1984), 'Self-consistent effective mass theory for intralayer screening in graphite intercalation compounds', *Phys. Rev. B* **29**, 1685.

Dlubak, B., Martin, M.-B., Deranlot, C., *et al.* (2012), 'Highly efficient spin transport in epitaxial graphene on SiC', *Nature Physics* **8**, 557–561.

Dresselhaus, G., Pimenta, M., Saito, R., *et al.* (2000), *Science and Application of Nanotubes*, Kluwer Academic/Plenum Publishers, New York, chapter 'On the $\pi - \pi$ overlap energy in carbon nanotubes', pp. 275–295.

Dresselhaus, M., Dresselhaus, G. & Eklund, P. (1996), *Science of Fullerenes and Carbon Nanotubes: Their Properties and Applications*, Academic Press, New York.

Dresselhaus, M. S. (2011), 'On the past and present of carbon nanostructures', *Phys. Status Solidi B* **248**, 1566–1574.

Dresselhaus, M. S. & Dresselhaus, G. (2002), 'Intercalation compounds of graphite', *Advances in Physics* **51**, 1–186.

Dresselhaus, M. S., Dresselhaus, G. & Avouris, P., eds (2001), *Carbon Nanotubes: Synthesis, Structure, Properties, and Applications*, Topics in Applied Physics Vol. 80, Springer, Berlin.

Drexler, C., Tarasenko, S. A., Olbrich, P., *et al.* (2013), 'Magnetic quantum ratchet effect in graphene', *Nature Nanotechnology* **8**, 104–107.

Du, X., Skachko, I., Barker, A. & Andrei, E. Y. (2008), 'Approaching ballistic transport in suspended graphene', *Nature Nanotechnology* **3**, 491–495.

Dubois, S. M.-M. (2009), Quantum transport in graphene-based nanostructures, Ph.D. thesis, Université Catholique de Louvain.

Dubois, S. M.-M., Lopez-Bezanilla, A., Cresti, A., *et al.* (2010), 'Quantum transport in graphene nanoribbons: Effects of edge reconstruction and chemical reactivity', *ACS Nano* **4**, 1971–1976.

Dubois, S. M.-M., Zanolli, Z., Declerck, X. & Charlier, J.-C. (2009), 'Electronic properties and quantum transport in graphene-based nanostructures', *The European Physical Journal B* **72**, 1–24.

Dunlap, B. I. (1994), 'Relating carbon tubules', *Phys. Rev. B* **49**, 5643–5651.

Economou, E. N. (2006), *Green's Functions in Quantum Physics*, Springer Berlin / Heidelberg.

Egger, R. (1999), 'Luttinger liquid behavior in multiwall carbon nanotubes', *Phys. Rev. Lett.* **83**, 5547–5550.

Egger, R. & Gogolin, A. O. (1997), 'Effective low-energy theory for correlated carbon nanotubes', *Phys. Rev. Lett.* **79**, 5082–5085.

Elias, D. C., Gorbachev, R. V., Mayorov, A. S., *et al.* (2011), 'Dirac cones reshaped by interaction effects in suspended graphene', *Nature Physics* **7**, 701–704.

Elias, D. C., Nair, R. R., Mohiuddin, T. M. G., *et al.* (2009), 'Control of graphene's properties by reversible hydrogenation: Evidence for graphane', *Science* **323**, 610–613.

Enoki, T., Kobayashi, Y. & Fukui, K.-I. (2007), 'Electronic structures of graphene edges and nanographene', *International Reviews in Physical Chemistry* **26**, 609–645.

Entin-Wohlman, O., Aharony, A. & Levinson, Y. (2002), 'Adiabatic transport in nanostructures', *Phys. Rev. B* **65**, 195411.

Ernzerhof, M., Perdew, J. P. & Burke, K. (1997), 'Coupling-constant dependence of atomization energies', *International Journal of Quantum Chemistry* **64**, 285–295.

Ernzerhof, M. & Scuseria, G. E. (1999), 'Assessment of the Perdew–Burke–Ernzerhof exchange–correlation functional', *The Journal of Chemical Physics* **110**, 5029–5036.

Esconjauregui, S., Fouquet, M., Bayer, B. C., *et al.* (2010), 'Growth of ultrahigh density vertically aligned carbon nanotube forests for interconnects', *ACS Nano* **4**, 7431–7436.

Evaldsson, M., Zozoulenko, I. V., Xu, H. & Heinzel, T. (2008), 'Edge-disorder-induced Anderson localization and conduction gap in graphene nanoribbons', *Phys. Rev. B* **78**, 161407.

Evers, F. & Mirlin, A. D. (2008), 'Anderson transitions', *Rev. Mod. Phys.* **80**, 1355–1417.

Ezawa, M. (2006), 'Peculiar width dependence of the electronic properties of carbon nanoribbons', *Phys. Rev. B* **73**, 045432.

Fabian, J., Matos-Abiague, A., Ertler, C., Stano, P. & Zutic, I. (2007), 'Semiconductor spintronics', *Acta Physica Slovaca* **57**, 565.

Fal'ko, V. I., Kechedzhi, K., McCann, E., *et al.* (2007), 'Weak localization in graphene', *Solid State Communications* **143**, 3338.

Fano, U. (1935), 'Sullo spettro di assorbimento dei gas nobili presso il limite dello spettro d'arco', *Il Nuovo Cimento* **12**, 154–161.

Farhat, H., Son, H., Samsonidze, G. G., *et al.* (2007), 'Phonon softening in individual metallic carbon nanotubes due to the Kohn anomaly', *Phys. Rev. Lett.* **99**, 145506.

Farmer, D. B., Golizadeh-Mojarad, R., Perebeinos, V., *et al.* (2008), 'Chemical doping and electronhole conduction asymmetry in graphene devices', *Nano Lett.* **9**, 388–392.

Fedorov, G., Tselev, A., Jimnez, D., *et al.* (2007), 'Magnetically induced field effect in carbon nanotube devices', *Nano Lett.* **7**, 960–964.

Fermi, E. (1927), 'Un metodo statistico per la determinazione di alcune priorietà dell'atomo', *Rend. Accad. Naz. Lincei* **6**, 602–607.

Ferreira, A., Xu, X., Tan, C.-L., *et al.* (2011), 'Transport properties of graphene with one-dimensional charge defects', *EPL (Europhysics Letters)* **94**, 28003.

Fetter, A. & Walecka, J. (1971), *Quantum Theory of Many-Particle Systems*, McGraw-Hill, New York.

Fisher, D. S. & Lee, P. A. (1981), 'Relation between conductivity and transmission matrix', *Phys. Rev. B* **23**, 6851–6854.

Floquet, G. (1883), 'Sur les équations différentielles linéaires à coefficients périodiques', *Annales Scientifiques de l'École Normale Supérieure*, Sér. 2 **12**, 47–88.

Foa Torres, L. E. F. (2005), 'Mono-parametric quantum charge pumping: Interplay between spatial interference and photon-assisted tunneling', *Phys. Rev. B* **72**, 245339.

Foa Torres, L. E. F., Avriller, R. & Roche, S. (2008), 'Nonequilibrium energy gaps in carbon nanotubes: Role of phonon symmetries', *Phys. Rev. B* **78**, 035412.

Foa Torres, L. E. F., Calvo, H. L., Rocha, C. G. & Cuniberti, G. (2011), 'Enhancing single-parameter quantum charge pumping in carbon-based devices', *Appl. Phys. Lett.* **99**, 092102.

Foa Torres, L. E. F. & Cuniberti, G. (2009), 'Controlling the conductance and noise of driven carbon-based Fabry–Pérot devices', *Applied Physics Letters* **94**, 222103.

Foa Torres, L. E. F., Lewenkopf, C. H. & Pastawski, H. M. (2003), 'Coherent versus sequential electron tunneling in quantum dots', *Phys. Rev. Lett.* **91**, 116801.

Foa Torres, L. E. F. & Roche, S. (2006), 'Inelastic quantum transport and Peierls-like mechanism in carbon nanotubes', *Phys. Rev. Lett.* **97**, 076804.

Franklin, A. D. & Chen, Z. (2010), 'Length scaling of carbon nanotube transistors', *Nature Nanotechnology* **5**, 858–862.

Gabor, N. M., Song, J. C. W., Ma, Q., *et al.* (2011), 'Hot carrier-assisted intrinsic photoresponse in graphene', *Science* **334**, 648–652.

Gao, B., Komnik, A., Egger, R., Glattli, D. C. & Bachtold, A. (2004), 'Evidence for Luttinger-liquid behavior in crossed metallic single-wall nanotubes', *Phys. Rev. Lett.* **92**, 216804.

Geim, A. K. (2011), 'Nobel lecture: Random walk to graphene', *Rev. Mod. Phys.* **83**, 851–862.

Geim, A. K. & Novoselov, K. S. (2007), 'The rise of graphene', *Nature Materials* **6**, 183–191.

Gheorghe, M., Gutiérrez, R., Ranjan, N., *et al.* (2005), 'Vibrational effects in the linear conductance of carbon nanotubes', *EPL (Europhysics Letters)* **71**, 438.

Giaever, I. & Zeller, H. R. (1968), 'Superconductivity of small tin particles measured by tunneling', *Phys. Rev. Lett.* **20**, 1504–1507.

Giannozzi, P., Baroni, S., Bonini, N., *et al.* (2009), 'Quantum Espresso: A modular and open-source software project for quantum simulations of materials', *Journal of Physics: Condensed Matter* **21**, 395502.

Giantomassi, M., Stankovski, M., Shaltaf, R., *et al.* (2011), 'Electronic properties of interfaces and defects from many-body perturbation theory: Recent developments and applications', *Physica Status Solidi (B)* **248**, 275–289.

Giesbers, A. J. M., Ponomarenko, L. A., Novoselov, K. S., *et al.* (2009), 'Gap opening in the zeroth Landau level of graphene', *Phys. Rev. B* **80**, 201403.

Girit, C., Meyer, J. C., Erni, R., *et al.* (2009), 'Graphene at the edge: Stability and dynamics', *Science* **323**, 1705–1708.

Glazov, M. M. & Ganichev, S. D., 'High frequency electric field-induced nonlinear effects in graphene', arXiv:1306.2049[cond-mat.mes-hall].

Gmitra, M., Konschuh, S., Ertler, C., Ambrosch-Draxl, C. & Fabian, J. (2009), 'Band-structure topologies of graphene: Spin–orbit coupling effects from first principles', *Phys. Rev. B* **80**, 235431.

Godby, R. W. & Needs, R. J. (1989), 'Metal–insulator transition in Kohn–Sham theory and quasiparticle theory', *Phys. Rev. Lett.* **62**, 1169–1172.

Goerbig, M. (2011), 'Electronic properties of graphene in a strong magnetic field', *Rev. Mod. Phys.* **83**, 1193.

Goldoni, A., Petaccia, L., Lizzit, S. & Larciprete, R. (2010), 'Sensing gases with carbon nanotubes: A review of the actual situation', *Journal of Physics: Condensed Matter* **22**, 013001.

Gómez-Navarro, C., de Pablo, P., Biel, B., *et al.* (2005), 'Tuning the conductance of single-walled carbon nanotubes by ion irradiation in the Anderson localization regime', *Nature Materials* **4**, 534.

Gonze, X., Rignanese, G. M., Verstraete, M. J., *et al.* (2005), 'A brief introduction to the Abinit software package', *Zeitschrift für Kristallographie – Crystalline Materials* **220**, 558–562.

Gonze, X., Amadon, B., Anglade, P.-M., *et al.* (2009), 'Abinit: First-principles approach to material and nanosystem properties', *Computer Physics Communications* **180**, 2582–2615.

Grichuk, E. & Manykin, E. (2010), 'Quantum pumping in graphene nanoribbons at resonant transmission', *EPL (Europhysics Letters)* **92**, 47010.

Grossmann, F., Dittrich, T., Jung, P. & Hänggi, P. (1991), 'Coherent destruction of tunneling', *Phys. Rev. Lett.* **67**, 516–519.

Grosso, G. & Parravicini, G. P. (2006), *Solid State Physics*, Elsevier, London.

Gruber, M., Heimel, G., Romaner, L., Brédas, J.-L. & Zojer, E. (2008), 'First-principles study of the geometric and electronic structure of Au_{13} clusters: Importance of the prism motif', *Phys. Rev. B* **77**, 165411.

Grüneis, A., Attaccalite, C., Wirtz, L., *et al.* (2008), 'Tight-binding description of the quasiparticle dispersion of graphite and few-layer graphene', *Phys. Rev. B* **78**, 205425.

Grushina, A. L. & Morpurgo, A. F. (2013) 'A ballistic *pn* junction in suspended graphene with split bottom gates', *Appl. Phys. Lett.* **102**, 223102.

Gu, Z., Fertig, H. A., Arovas, D. P. & Auerbach, A. (2011), 'Floquet spectrum and transport through an irradiated graphene ribbon', *Phys. Rev. Lett.* **107**, 216601.

Guinea, F., Tejedor, C., Flores, F. & Louis, E. (1983), 'Effective two-dimensional Hamiltonian at surfaces', *Phys. Rev. B* **28**, 4397–4402.

Guinea, F. & Vergés, J. A. (1987), 'Localization and topological disorder', *Phys. Rev. B* **35**, 979–986.

Gunlycke, D. & White, C. T. (2008), 'Tight-binding energy dispersions of armchair-edge graphene nanostrips', *Phys. Rev. B* **77**, 115116.

Gunlycke, D. & White, C. T. (2011), 'Graphene valley filter using a line defect', *Phys. Rev. Lett.* **106**, 136806.

Güttinger, J., Molitor, F., Stampfer, C., *et al.* (2012), 'Transport through graphene quantum dots', *Reports on Progress in Physics* **75**, 126502.

Haeckel, E. (1862), *Die Radiolarien*, Georg Reimer, Berlin.

Haering, R. R. (1958), 'Band structure of rhombohedral graphite', *Canadian Journal of Physics* **36**, 352–362.

Haldane, F. D. M. (1988), 'Model for a quantum Hall effect without Landau levels: Condensed-matter realization of the *parity anomaly*', *Phys. Rev. Lett.* **61**, 2015–2018.

Hamada, N., Sawada, S.-I. & Oshiyama, A. (1992), 'New one-dimensional conductors: Graphitic microtubules', *Phys. Rev. Lett.* **68**, 1579–1581.

Hamann, D. R., Schlüter, M. & Chiang, C. (1979), 'Norm-conserving pseudopotentials', *Phys. Rev. Lett.* **43**, 1494–1497.

Han, M. Y., Özyilmaz, B., Zhang, Y. & Kim, P. (2007), 'Energy band-gap engineering of graphene nanoribbons', *Phys. Rev. Lett.* **98**, 206805.

Han, W. & Kawakami, R. K. (2011), 'Spin relaxation in single-layer and bilayer graphene', *Phys. Rev. Lett.* **107**, 047207.

Harris, P. (1999), *Carbon Nanotubes and Related Structures: New Materials for the XXI Century*, Cambridge University Press, Cambridge.

Harrison, W. (1989), *Electronic Structure and the Properties of Solids: The Physics of the Chemical Bond*, Dover Publications, New York.

Hartree, D. R. (1957), *The Calculation of Atomic Structures*, John Wiley and Sons, New York.

Hasan, M. Z. & Kane, C. L. (2010), 'Colloquium: Topological insulators', *Rev. Mod. Phys.* **82**, 3045–3067.

Haydock, R., Heine, V. & Kelly, M. J. (1972), 'Electronic structure based on the local atomic environment for tight-binding bands', *Journal of Physics C: Solid State Physics* **5**, 2845.

Haydock, R., Heine, V. & Kelly, M. J. (1975), 'Electronic structure based on the local atomic environment for tight-binding bands. II', *Journal of Physics C: Solid State Physics* **8**, 2591.

Hedin, L. (1965), 'New method for calculating the one-particle Green's function with application to the electron-gas problem', *Phys. Rev.* **139**, A796–A823.

Hedin, L. & Lundqvist, S. (1970), *Effects of Electron–Electron and Electron–Phonon Interactions on the One-Electron States of Solids*, Vol. 23 of *Solid State Physics*, Academic Press, New York, pp. 1–181.

Heimann, R., Evsyukov, S. & Kavan, L. (1999), *Carbyne and Carbynoid Structures*, Kluwer Academic, Dordrecht, The Netherlands.

Heinze, S., Tersoff, J., Martel, R., *et al.* (2002), 'Carbon nanotubes as Schottky barrier transistors', *Phys. Rev. Lett.* **89**, 106801.

Hemstreet, Louis A., J., Fong, C. Y. & Cohen, M. L. (1970), 'Calculation of the band structure and optical constants of diamond using the nonlocal-pseudopotential method', *Phys. Rev. B* **2**, 2054–2063.

Herrmann, L. G., Delattre, T., Morfin, P., *et al.* (2007), 'Shot noise in Fabry-Perot interferometers based on carbon nanotubes', *Phys. Rev. Lett.* **99**, 156804.

Hikami, S., Larkin, A. I. & Nagaoka, Y. (1980), 'Spin–orbit interaction and magnetoresistance in the two dimensional random system', *Progress of Theoretical Physics* **63**, 707–710.

Hjort, M. & Stafstrom, S. (2001), 'Disorder-induced electron localization in metallic carbon nanotubes', *Phys. Rev. B* **63**, 113406.

Hofstadter, D. R. (1976), 'Energy levels and wave functions of Bloch electrons in rational and irrational magnetic fields', *Phys. Rev. B* **14**, 2239–2249.

Hohenberg, P. & Kohn, W. (1964), 'Inhomogeneous electron gas', *Phys. Rev.* **136**, B864–B871.

Holmström, E., Fransson, J., Eriksson, O., *et al.* (2011), 'Disorder-induced metallicity in amorphous graphene', *Phys. Rev. B* **84**, 205414.

Hong, A. J., Song, E. B., Yu, H. S., *et al.* (2011), 'Graphene flash memory', *ACS Nano* **5**, 7812–7817.

Hong, S. K., Kim, J. E., Kim, S. O., Choi, S.-Y. & Cho, B. J. (2010), 'Flexible resistive switching memory device based on graphene oxide', *Electron Device Letters, IEEE* **31**, 1005–1007.

Horsell, D. W., Tikhonenko, F. V., Gorbachev, R. V. & Savchenko, A. K. (2008), 'Weak localization in monolayer and bilayer graphene', *Phil. Trans. Roy. Soc. A* **366**, 245.

Hoshi, T., Yamamoto, S., Fujiwara, T., Sogabe, T. & Zhang, S.-L. (2012), 'An order-N electronic structure theory with generalized eigenvalue equations and its application to a ten-million-atom system', *J. Phys.: Condens. Matter* **24**, 165502.

Hossain, M. Z., Johns, J. E., Bevan, K. H., *et al.* (2012), 'Chemically homogeneous and thermally reversible oxidation of epitaxial graphene', *Nature Chemistry* **4**, 305–309.

Huang, B., Liu, M., Su, N., *et al.* (2009), 'Quantum manifestations of graphene edge stress and edge instability: A first-principles study', *Phys. Rev. Lett.* **102**, 166404.

Huang, L., Lai, Y.-C. & Grebogi, C. (2010), 'Relativistic quantum level-spacing statistics in chaotic graphene billiards', *Phys. Rev. E* **81**, 055203.

Huang, P., Ruiz-Vargas, C. S., van der Zande, A. M., *et al.* (2011), 'Grains and grain boundaries in single-layer graphene atomic patchwork quilts', *Nature* **469**, 389–392.

Huertas-Hernando, D., Guinea, F. & Brataas, A. (2006), 'Spin–orbit coupling in curved graphene, fullerenes, nanotubes, and nanotube caps', *Phys. Rev. B* **74**, 155426.

Hueso, L. E., Pruneda, J. M., Ferrari, V., *et al.* (2007), 'Transformation of spin information into large electrical signals using carbon nanotubes', *Nature* **445**, 410–413.

Hwang, C., Park, C.-H., Siegel, D. A., *et al.* (2011), 'Direct measurement of quantum phases in graphene via photoemission spectroscopy', *Phys. Rev. B* **84**, 125422.

Hwang, E. H. & Sarma, S. D. (2008), 'Acoustic phonon scattering limited carrier mobility in two-dimensional extrinsic graphene', *Phys. Rev. B* **77**, 115449.

Hwang, W. S., Tahy, K., Li, X., *et al.* (2012), 'Transport properties of graphene nanoribbon transistors on chemical-vapor-deposition grown wafer-scale graphene', *Appl. Phys. Lett.* **100**, 203107.

Iijima, S. (1991), 'Helical microtubules of graphitic carbon', *Nature* **354**, 56–58.

Iijima, S. & Ichihashi, T. (1993), 'Single-shell carbon nanotubes of 1 nm diameter', *Nature* **363**, 603–605.

Iijima, S., Yudasaka, M., Yamada, R., *et al.* (1999), 'Nano-aggregates of single-walled graphitic carbon nano-horns', *Chemical Physics Letters* **309**, 165–170.

Imry, Y. & Landauer, R. (1999), 'Conductance viewed as transmission', *Rev. Mod. Phys.* **71**, S306–S312.

Ingaramo, L. H. & Foa Torres, L. E. F. (2013), 'Quantum charge pumping in graphene-based devices: When lattice defects do help', *Appl. Phys. Lett.* **103**, 123508 (2013).

Ishii, H., Roche, S., Kobayashi, N. & Hirose, K. (2010), 'Inelastic transport in vibrating disordered carbon nanotubes: Scattering times and temperature-dependent decoherence effects', *Phys. Rev. Lett.* **104**, 116801.

Ishii, H., Triozon, F., Kobayashi, N., Hirose, K. & Roche, S. (2009), 'Charge transport in carbon nanotube based materials, a Kubo Greenwood computational approach', *Comptes Rendus Physique* **10**, 283–296.

Jain, M., Chelikowsky, J. R. & Louie, S. G. (2011), 'Reliability of hybrid functionals in predicting band gaps', *Phys. Rev. Lett.* **107**, 216806.

Jalabert, R. A., Stone, A. D. & Alhassid, Y. (1992), 'Statistical theory of Coulomb blockade oscillations: Quantum chaos in quantum dots', *Phys. Rev. Lett.* **68**, 3468–3471.

Javey, A., Guo, J., Paulsson, M., *et al.* (2004), 'High-field quasiballistic transport in short carbon nanotubes', *Phys. Rev. Lett.* **92**, 106804.

Javey, A., Guo, J., Wang, Q., Lundstrom, M. & Dai, H. (2003), 'Ballistic carbon nanotube field-effect transistors', *Nature* **424**, 654–657.

Jayasekera, T. & Mintmire, J. W. (2007), 'Transport in multiterminal graphene nanodevices', *Nanotechnology* **18**, 424033.

Jensen, K., Kim, K. & Zettl, A. (2008), 'An atomic-resolution nanomechanical mass sensor', *Nature Nanotechnology* **3**, 533–537.

Jia, X., Campos-Delgado, J., Terrones, M., Meunier, V. & Dresselhaus, M. S. (2011), 'Graphene edges: A review of their fabrication and characterization', *Nanoscale* **3**, 86–95.

Jia, X., Goswami, P. & Chakravarty, S. (2008), 'Dissipation and criticality in the lowest Landau level of graphene', *Phys. Rev. Lett.* **101**, 036805.

Jia, X., Hofmann, M., Meunier, V., *et al.* (2009), 'Controlled formation of sharp zigzag and armchair edges in graphitic nanoribbons', *Science* **323**, 1701–1705.

Jiang, H., Qiao, Z., Liu, H., Shi, J. & Niu, Q. (2012), 'Stabilizing topological phases in graphene via random adsorption', *Phys. Rev. Lett.* **109**, 116803.

Jiang, J., Dong, J. & Xing, D. Y. (2003), 'Quantum interference in carbon-nanotube electron resonators', *Phys. Rev. Lett.* **91**, 056802.

Jiang, Z., Zhang, Y., Stormer, H. L. & Kim, P. (2007), 'Quantum Hall states near the charge-neutral Dirac point in graphene', *Phys. Rev. Lett.* **99**, 106802.

Jiao, L., Wang, X., Diankov, G., Wang, H. & Dai, H. (2010), 'Facile synthesis of high-quality graphene nanoribbons', *Nature Nanotechnology* **5**, 321–325.

Jiao, L., Zhang, L., Ding, L., Liu, J. & Dai, H. (2010), 'Aligned graphene nanoribbons and cross-bars from unzipped carbon nanotubes', *Nano Research* **3**, 387–394.

Jiao, L., Zhang, L., Wang, X., Diankov, G. & Dai, H. (2009), 'Narrow graphene nanoribbons from carbon nanotubes', *Nature* **458**, 877–880.

Jin, C., Lan, H., Peng, L., Suenaga, K. & Iijima, S. (2009), 'Deriving carbon atomic chains from graphene', *Phys. Rev. Lett.* **102**, 205501.

Jonson, M. & Grincwajg, A. (1987), 'Effect of inelastic scattering on resonant and sequential tunneling in double barrier heterostructures', *Applied Physics Letters* **51**, 1729–1731.

Jorio, A., Dresselhaus, M. S., Saito, R. & Dresselhaus, G. (2011), *Raman Spectroscopy in Graphene Related Systems*, Wiley-VCH, Singapore.

Jorio, A., Souza Filho, A. G., Dresselhaus, G., *et al.* (2001), 'Joint density of electronic states for one isolated single-wall carbon nanotube studied by resonant Raman scattering', *Phys. Rev. B* **63**, 245416.

Kaestner, B., Kashcheyevs, V., Amakawa, S., *et al.* (2008), 'Single-parameter nonadiabatic quantized charge pumping', *Phys. Rev. B* **77**, 153301.

Kane, C. L. & Mele, E. J. (1997), 'Size, shape, and low energy electronic structure of carbon nanotubes', *Phys. Rev. Lett.* **78**, 1932–1935.

Kane, C. L. & Mele, E. J. (2005), 'Z_2 topological order and the quantum spin Hall effect', *Phys. Rev. Lett.* **95**, 146802.

Kapko, V., Drabold, D. & Thorpe, M. (2010), 'Electronic structure of a realistic model of amorphous graphene', *Phys. Stat. Sol. B* **247**, 1197.

Karch, J., Drexler, C., Olbrich, P., *et al.* (2011), 'Terahertz radiation driven chiral edge currents in graphene', *Phys. Rev. Lett.* **107**, 276601.

Kashcheyevs, V., Aharony, A. & Entin-Wohlman, O. (2004), 'Resonance approximation and charge loading and unloading in adiabatic quantum pumping', *Phys. Rev. B* **69**, 195301.

Kastner, M. A. (1992), 'The single-electron transistor', *Rev. Mod. Phys.* **64**, 849–858.

Kato, T. & Hatakeyama, R. (2012), 'Site- and alignment-controlled growth of graphene nanoribbons from nickel nanobars', *Nature Nanotechnology* **7**, 651–656.

Katsnelson, M. I. (2012), *Graphene: Carbon in Two Dimensions*, Cambridge University Press, Cambridge.

Katsnelson, M. I., Novoselov, K. S. & Geim, A. K. (2006), 'Chiral tunnelling and the Klein paradox in graphene', *Nature Physics* **2**, 620–625.

Kauffman, D. R., Sorescu, D. C., Schofield, D. P., *et al.* (2010), 'Understanding the sensor response of metal-decorated carbon nanotubes', *Nano Lett.* **10**, 958–963.

Kavan, L. & Kastner, J. (1994), 'Carbyne forms of carbon: Continuation of the story', *Carbon* **32**, 1533–1536.

Kawai, T., Miyamoto, Y., Sugino, O. & Koga, Y. (2000), 'Graphitic ribbons without hydrogen-termination: Electronic structures and stabilities', *Phys. Rev. B* **62**, R16349–R16352.

Kazymyrenko, K. & Waintal, X. (2008), 'Knitting algorithm for calculating Green functions in quantum systems', *Phys. Rev. B* **77**, 115119.

Keating, P. N. (1966), 'Effect of invariance requirements on the elastic strain energy of crystals with application to the diamond structure', *Phys. Rev.* **145**, 637.

Kechedzhi, K., McCann, E. I. F. V., Suzuura, H., Ando, T. & Altshuler, B. (2007), 'Weak localization in monolayer and bilayer graphene', *Eur. Phys. J. Special Topics* **148**, 39.

Kibis, O. V. (2010), 'Metal–insulator transition in graphene induced by circularly polarized photons', *Phys. Rev. B* **81**, 165433.

Kim, K., Lee, Z., Regan, W., *et al.* (2011), 'Grain boundary mapping in polycrystalline graphene', *ACS Nano* **5**, 2142.

Kim, K. Su, Walter, A. L., Moreschini, L., *et al.* (2013), 'Co-existing massive and massless Dirac fermions in symmetry-broken bi-layer graphene', *Nature Materials*, advance online publication, DOI: 10.1038/NMAT3717.

Kim, N. Y., Recher, P., Oliver, W. D., *et al.* (2007), 'Tomonaga–Luttinger liquid features in ballistic single-walled carbon nanotubes: Conductance and shot noise', *Phys. Rev. Lett.* **99**, 036802.

Kim, S. Y. & Park, H. S. (2010), 'On the utility of vacancies and tensile strain-induced quality factor enhancement for mass sensing using graphene monolayers', *Nanotechnology* **21**, 105710.

Kim, W., Javey, A., Tu, R., *et al.* (2005), 'Electrical contacts to carbon nanotubes down to 1 nm in diameter', *Applied Physics Letters* **87**, 173101.

Kirwan, D. F., Rocha, C. G., Costa, A. T. & Ferreira, M. S. (2008), 'Sudden decay of indirect exchange coupling between magnetic atoms on carbon nanotubes', *Phys. Rev. B* **77**, 085432.

Kitagawa, T., Oka, T., Brataas, A., Fu, L. & Demler, E. (2011), 'Transport properties of nonequilibrium systems under the application of light: Photoinduced quantum Hall insulators without Landau levels', *Phys. Rev. B* **84**, 235108.

Klein, O. (1929), 'Die reflexion von elektronen an einem potentialsprung nach der relativistischen dynamik von Dirac', *Zeitschrift fr Physik A Hadrons and Nuclei* **53**, 157–165.

Kleinman, L. & Bylander, D. M. (1982), 'Efficacious form for model pseudopotentials', *Phys. Rev. Lett.* **48**, 1425–1428.

Klitzing, K. V., Dorda, G. & Pepper, M. (1980), 'New method for high-accuracy determination of the fine-structure constant based on quantized Hall resistance', *Phys. Rev. Lett.* **45**, 494–497.

Klos, J. W. & Zozoulenko, I. V. (2010), 'Effect of short- and long-range scattering in the conductivity of graphene: Boltzmann approach vs tight-binding calculations', *Phys. Rev. B* **82**, 081414.

Kobayashi, Y., Fukui, K.-I., Enoki, T., Kusakabe, K. & Kaburagi, Y. (2005), 'Observation of zigzag and armchair edges of graphite using scanning tunneling microscopy and spectroscopy', *Phys. Rev. B* **71**, 193406.

Kohler, S., Lehmann, J. & Hänggi, P. (2005), 'Driven quantum transport on the nanoscale', *Physics Reports* **406**, 379–443.

Kohn, W. (1959), 'Image of the Fermi surface in the vibration spectrum of a metal', *Phys. Rev. Lett.* **2**, 393–394.

Kohn, W. & Sham, L. J. (1965), 'Self-consistent equations including exchange and correlation effects', *Phys. Rev.* **140**, A1133–A1138.

Konschuh, S., Gmitra, M. & Fabian, J. (2010), 'Tight-binding theory of the spin–orbit coupling in graphene', *Phys. Rev. B* **82**, 245412.

Konstantatos, G., Badioli, M., Gaudreau, L., *et al.* (2012), 'Hybrid graphene-quantum dot phototransistors with ultrahigh gain', *Nature Nanotechnology* **7**, 363–368.

Koppens, F. H. L., Chang, D. E. & Garcia de Abajo, F. J. (2011), 'Graphene plasmonics: A platform for strong light-matter interactions', *Nano Lett.* **11**, 3370–3377.

Koskinen, P., Malola, S. & Häkkinen, H. (2008), 'Self-passivating edge reconstructions of graphene', *Phys. Rev. Lett.* **101**, 115502.

Kostyrko, T., Bartkowiak, M. & Mahan, G. D. (1999), 'Localization in carbon nanotubes within a tight-binding model', *Phys. Rev. B* **60**, 10735–10738.

Kosynkin, D. V., Higginbotham, A. L., Sinitskii, A., *et al.* (2009), 'Longitudinal unzipping of carbon nanotubes to form graphene nanoribbons', *Nature* **458**, 872–876.

Kotakoski, J., Krasheninnikov, A. V., Kaiser, U. & Meyer, J. C. (2011), 'From point defects in graphene to two-dimensional amorphous carbon', *Phys. Rev. Lett.* **106**, 105505.

Kotakoski, J. & Meyer, J. C. (2012), 'Mechanical properties of polycrystalline graphene based on a realistic atomistic model,', *Phys. Rev. B* **85**, 195447.

Kouwenhoven, L. & Glazman, L. (2001), 'Revival of the Kondo effect', *Physics World* (Jan.), 33–38.

Kouwenhoven, L. P., Marcus, C. M., McEuen, P. L., *et al.* (1997), *Nato ASI conference proceedings*, Kluwer Academic, Dordrecht, The Netherlands, chapter 'Electron transport in quantum dots', pp. 105–214.

Kowalczyk, P., Holyst, R., Terrones, M. & Terrones, H. (2007), 'Hydrogen storage in nanoporous carbon materials: Myth and facts', *Phys. Chem. Chem. Phys.* **9**, 1786–1792.

Krasheninnikov, A. V. & Banhart, F. (2007), 'Engineering of nanostructured carbon materials with electron or ion beams', *Nature Materials* **6**, 723–733.

Kresse, G. & Furthmüller, J. (1996*a*), 'Efficiency of *ab initio* total energy calculations for metals and semiconductors using a plane-wave basis set', *Computational Materials Science* **6**, 15–50.

Kresse, G. & Furthmüller, J. (1996*b*), 'Efficient iterative schemes for *ab initio* total energy calculations using a plane-wave basis set', *Phys. Rev. B* **54**, 11169–11186.

Krishnan, A., Dujardin, E., Treacy, M. M. J., *et al.* (1997), 'Graphitic cones and the nucleation of curved carbon surfaces', *Nature* **388**, 451–454.

Kroto, H. W., Heath, J. R., O'Brien, S. C., Curl, R. F. & Smalley, R. E. (1985), 'C60: Buckminsterfullerene', *Nature* **318**, 162–163.

Kroto, H. W. & McKay, K. (1988), 'The formation of quasi-icosahedral spiral shell carbon particles', *Nature* **331**, 328–331.

Kubo, R. (1966), 'The fluctuation–dissipation theorem', *Reports on Progress in Physics* **29**, 255.

Kümmel, S. & Kronik, L. (2008), 'Orbital-dependent density functionals: Theory and applications', *Rev. Mod. Phys.* **80**, 3–60.

Kurasch, S., Kotakoski, J., Lehtinen, O., *et al.* (2012), 'Atom-by-atom observation of grain boundary migration in graphene', *Nano Lett.* **12**, 3168–3173.

Kurganova, E. V., van Elferen, H. J., McCollam, A., *et al.* (2011), 'Spin splitting in graphene studied by means of tilted magnetic-field experiments', *Phys. Rev. B* **84**, 121407.

Kwon, Y.-K. & Tománek, D. (1998), 'Electronic and structural properties of multiwall carbon nanotubes', *Phys. Rev. B* **58**, R16001–R16004.

Lagendijk, A., van Tiggelen, B. & Wiersma, D. S. (2009), 'Fifty years of Anderson localization', *Phys. Today* **62**, 24–29.

Lagow, R. J., Kampa, J. J., Wei, H.-C., *et al.* (1995), 'Synthesis of linear acetylenic carbon: The "sp" carbon allotrope', *Science* **267**, 362–367.

Lahiri, J., Lin, Y., Bozkurt, P., Oleynik, I. I. & Batzill, M. (2010), 'An extended defect in graphene as a metallic wire', *Nature Nanotechnology* **5**, 326–329.

Lambin, P., Fonseca, A., Vigneron, J., Nagy, J. & Lucas, A. (1995), 'Structural and electronic properties of bent carbon nanotubes', *Chemical Physics Letters* **245**, 85–89.

Lambin, P., Philippe, L., Charlier, J. & Michenaud, J. (1994), 'Electronic band structure of multilayered carbon tubules', *Computational Materials Science* **2**, 350–356.

Lanczos, C. (1950), 'Solution of systems of linear equations by minimized iterations', *J. Res. Natl. Bur. Stand.* **45**, 255.

Landau, L. & Lifschitz, E. (1980), *Statistical Physics Part II*, Pergamon, Oxford.

Landauer, R. (1957), 'Spatial variation of currents and fields due to localized scatterers in metallic conduction', *IBM J. Res. Dev.* **1**, 223.

Landauer, R. (1970), 'Electrical resistance of disordered one-dimensional lattices', *Philosophical Magazine* **21**, 863–867.

Landman, U. (2005), 'Materials by numbers: Computations as tools of discovery', *Proceedings of the National Academy of Sciences of the United States of America* **102**, 6671–6678.

Lassagne, B., Garcia-Sanchez, D., Aguasca, A. & Bachtold, A. (2008), 'Ultrasensitive mass sensing with a nanotube electromechanical resonator', *Nano Lett.* **8**, 3735–3738.

Latil, S. & Henrard, L. (2006), 'Charge carriers in few-layer graphene films', *Phys. Rev. Lett.* **97**, 036803.

Latil, S., Meunier, V. & Henrard, L. (2007), 'Massless fermions in multilayer graphitic systems with misoriented layers: *Ab initio* calculations and experimental fingerprints', *Phys. Rev. B* **76**, 201402.

Latil, S., Roche, S. & Charlier, J.-C. (2005), 'Electronic transport in carbon nanotubes with random coverage of physisorbed molecules', *Nano Lett.* **5**, 2216–2219.

Latil, S., Roche, S., Mayou, D. & Charlier, J.-C. (2004), 'Mesoscopic transport in chemically doped carbon nanotubes', *Phys. Rev. Lett.* **92**, 256805.

Laughlin, R. B. (1983), 'Anomalous quantum Hall effect: An incompressible quantum fluid with fractionally charged excitations', *Phys. Rev. Lett.* **50**, 1395–1398.

Lazzeri, M. & Mauri, F. (2006), 'Coupled dynamics of electrons and phonons in metallic nanotubes: Current saturation from hot-phonon generation', *Phys. Rev. B* **73**, 165419.

Lazzeri, M., Piscanec, S., Mauri, F., Ferrari, A. C. & Robertson, J. (2006), 'Phonon linewidths and electron–phonon coupling in graphite and nanotubes', *Phys. Rev. B* **73**, 155426.

Leconte, N., Lherbier, A., Varchon, F., *et al.* (2011), 'Quantum transport in chemically modified two-dimensional graphene: From minimal conductivity to Anderson localization', *Phys. Rev. B* **84**, 235420.

Leconte, N., Moser, J., Ordejon, P., *et al.* (2010), 'Damaging graphene with ozone treatment: A chemically tunable metal–insulator transition', *ACS Nano* **4**, 4033–4038.

Lee, G.-D., Wang, C. Z., Yoon, E., *et al.* (2005), 'Diffusion, coalescence, and reconstruction of vacancy defects in graphene layers', *Phys. Rev. Lett.* **95**, 205501.

Lee, H., Son, Y.-W., Park, N., Han, S. & Yu, J. (2005), 'Magnetic ordering at the edges of graphitic fragments: Magnetic tail interactions between the edge-localized states', *Phys. Rev. B* **72**, 174431.

Lee, P. A. & Ramakrishnan, T. V. (1985), 'Disordered electronic systems', *Rev. Mod. Phys.* **57**, 287–337.

Lee, P. A. & Stone, A. D. (1985), 'Universal conductance fluctuations in metals', *Phys. Rev. Lett.* **55**, 1622–1625.

Lee, S., Jagannathan, B., Narasimha, S., *et al.* (2007), Record RF performance of 45 nm SOI CMOS technology, in *Electron Devices Meeting, 2007.* IEDM 2007. IEEE International, pp. 255–258.

Lee, Y.-S. & Marzari, N. (2006), 'Cycloaddition functionalizations to preserve or control the conductance of carbon nanotubes', *Phys. Rev. Lett.* **97**, 116801.

Leek, P. J., Buitelaar, M. R., Talyanskii, V. I., *et al.* (2005), 'Charge pumping in carbon nanotubes', *Phys. Rev. Lett.* **95**, 256802.

Lefebvre, J., Homma, Y. & Finnie, P. (2003), 'Bright band gap photoluminescence from unprocessed single-walled carbon nanotubes', *Phys. Rev. Lett.* **90**, 217401.

Leghrib, R., Felten, A., Demoisson, F., *et al.* (2010), 'Room-temperature, selective detection of benzene at trace levels using plasma-treated metal-decorated multiwalled carbon nanotubes', *Carbon* **48**, 3477–3484.

Lehtinen, P. O., Foster, A. S., Ayuela, A., Vehvilinen, T. T. & Nieminen, R. M. (2004*a*), 'Structure and magnetic properties of adatoms on carbon nanotubes', *Phys. Rev. B* **69**, 155422.

Lehtinen, P. O., Foster, A. S., Ma, Y., Krasheninnikov, A. V. & Nieminen, R. M. (2004*b*), 'Irradiation-induced magnetism in graphite: A density functional study', *Phys. Rev. Lett.* **93**, 187202.

Léonard, F. & Talin, A. A. (2011), 'Electrical contacts to one- and two-dimensional nanomaterials', *Nature Nanotechnology* **6**, 773–783.

Léonard, F. & Tersoff, J. (1999), 'Novel length scales in nanotube devices', *Phys. Rev. Lett.* **83**, 5174–5177.

Léonard, F. & Tersoff, J. (2000*a*), 'Negative differential resistance in nanotube devices', *Phys. Rev. Lett.* **85**, 4767–4770.

Léonard, F. & Tersoff, J. (2000*b*), 'Role of Fermi-level pinning in nanotube Schottky diodes', *Phys. Rev. Lett.* **84**, 4693–4696.

Léonard, F. & Tersoff, J. (2002), 'Multiple functionality in nanotube transistors', *Phys. Rev. Lett.* **88**, 258302.

Lepro, X., Vega-Cantu, Y., Rodriguez-Macias, F., *et al.* (2007), 'Production and characterization of coaxial nanotube junctions and networks of cnx/cnt', *Nano Lett.* **7**, 2220–2226.

Lewenkopf, C. H. & Mucciolo, E. R. (2013), 'The recursive Green's function method for graphene', *J. Comput. Electron.*, **12**, 203.

Lherbier, A. (2008), Étude des propriétés électroniques et des propriétés de transport de nanofils semiconducteurs et de plans de graphène, Ph.D. thesis, Université Joseph-Fourier.

Lherbier, A., Biel, B., Niquet, Y.-M. & Roche, S. (2008), 'Transport length scales in disordered graphene-based materials: Strong localization regimes and dimensionality effects', *Phys. Rev. Lett.* **100**, 036803.

Lherbier, A., Blase, X., Niquet, Y. M., Triozon, F. & Roche, S. (2008), 'Charge transport in chemically doped 2D graphene', *Phys. Rev. Lett.* **101**, 036808.

Lherbier, A., Dubois, S. M.-M., Declerck, X., *et al.* (2012), 'Transport properties of graphene containing structural defects', *Phys. Rev. B* **86**, 075402.

Lherbier, A., Dubois, S. M.-M., Declerck, X., *et al.* (2011), 'Two-dimensional graphene with structural defects: Elastic mean free path, minimum conductivity, and Anderson transition', *Phys. Rev. Lett.* **106**, 046803.

Lherbier, A., Roche, S., Restrepo, O. A., *et al.* (2013), *Nano Research*, in press.

Li, D., Muller, M. B., Gilje, S., Kaner, R. B. & Wallace, G. G. (2008), 'Processable aqueous dispersions of graphene nanosheets', *Nature Nanotechnology* **3**, 101–105.

Li, G., Luican, A. & Andrei, E. Y. (2009), 'Scanning tunneling spectroscopy of graphene on graphite', *Phys. Rev. Lett.* **102**, 176804.

Li, G., Luican, A., Lopes dos Santos, J. M. B., *et al.* (2010), 'Observation of Van Hove singularities in twisted graphene layers', *Nat Phys* **6**, 109–113.

Li, S. & Jiang, Y. (1995), 'Bond lengths, reactivities, and aromaticities of benzenoid hydrocarbons based on valence bond calculations', *J. Am. Chem. Soc.* **117**, 8401–8406.

Li, W., Sevinçli, H., Cuniberti, G. & Roche, S. (2010), 'Phonon transport in large scale carbon-based disordered materials: Implementation of an efficient order-N and real-space Kubo methodology', *Phys. Rev. B* **82**, 041410.

Li, W., Sevinçli, H., Roche, S. & Cuniberti, G. (2011), 'Efficient linear scaling method for computing the thermal conductivity of disordered materials', *Phys. Rev. B* **83**, 155416.

Li, X., Magnuson, C. W., Venugopal, A., *et al.* (2010), 'Graphene films with large domain size by a two-step chemical vapor deposition process', *Nano Lett.* **10**, 4328–4334.

Li, X., Wang, X., Zhang, L., Lee, S. & Dai, H. (2008), 'Chemically derived, ultrasmooth graphene nanoribbon semiconductors', *Science* **319**, 1229–1232.

Liang, W., Bockrath, M., Bozovic, D., *et al.* (2001), 'Fabry–Perot interference in a nanotube electron waveguide', *Nature* **411**, 665–669.

Liao, L., Lin, Y.-C., Bao, M., *et al.* (2010), 'High-speed graphene transistors with a self-aligned nanowire gate', *Nature* **467**, 305–308.

Libisch, F., Stampfer, C. & Burgdörfer, J. (2009), 'Graphene quantum dots: Beyond a Dirac billiard', *Phys. Rev. B* **79**, 115423.

Lieb, E. H. (1989), 'Two theorems on the Hubbard model', *Phys. Rev. Lett.* **62**, 1201–1204.

Lin, Y.-M., Dimitrakopoulos, C., Jenkins, K. A., *et al.* (2010), '100 GHz transistors from wafer-scale epitaxial graphene', *Science* **327**, 662.

Lin, Y.-M., Jenkins, K. A., Valdes-Garcia, A., *et al.* (2008), 'Operation of graphene transistors at gigahertz frequencies', *Nano Lett.* **9**, 422–426.

Lindner, N. H., Refael, G. & Galitski, V. (2011), 'Floquet topological insulator in semiconductor quantum wells', *Nature Physics* **7**, 490–495.

Lipson, H. & Stokes, A. R. (1942), 'The structure of graphite', *Proc. Roy. Soc. of London. Series A. Mathematical and Physical Sciences* **181**, 101–105.

Liu, J., Dai, H., Hafner, J. H., *et al.* (1997), 'Fullerene "crop circles"', *Nature* **385**, 780–781.

Liu, J. & Hersam, M. C. (2010), 'Recent developments in carbon nanotube sorting and selective growth', *MRS Bulletin* **35**, 315–321.

Liu, K., Avouris, P., Martel, R. & Hsu, W. K. (2001), 'Electrical transport in doped multiwalled carbon nanotubes', *Phys. Rev. B* **63**, 161404.

Liu, X., Oostinga, J. B., Morpurgo, A. F. & Vandersypen, L. M. K. (2009), 'Electrostatic confinement of electrons in graphene nanoribbons', *Phys. Rev. B* **80**, 121407.

Liu, Y., Bian, G., Miller, T. & Chiang, T.-C. (2011), 'Visualizing electronic chirality and Berry phases in graphene systems using photoemission with circularly polarized light', *Phys. Rev. Lett.* **107**, 166803.

Liu, Z., Suenaga, K., Harris, P. J. F. & Iijima, S. (2009), 'Open and closed edges of graphene layers', *Phys. Rev. Lett.* **102**, 015501.

Lohmann, T., von Klitzing, K. & Smet, J. H. (2009), 'Four-terminal magneto-transport in graphene p-n junctions created by spatially selective doping', *Nano Lett.* **9**, 1973–1979.

Lohrmann, D. (1989), 'Shallow and deep impurity levels in multivalley semiconductors: A Green function study of a cubic model by the recursion method', *Phys. Rev. B* **40**, 8404.

Lopez-Bezanilla, A. (2009), Etude à partir des premiers principes de l'effet de la fonctionnalisation sur le transport de charge dans les systemes à base de carbone à l'echelle mesoscopique, Ph.D. thesis, Université Joseph Fourier.

López-Bezanilla, A., Blase, X. & Roche, S. (2010), 'Quantum transport properties of chemically functionalized long semiconducting carbon nanotubes', *Nano Research* **3**, 288–295.

López-Bezanilla, A., Triozon, F., & Roche, S. (2009*a*), 'Chemical functionalization effects on armchair graphene nanoribbons transport', *Nano Lett.* **9**, 2527.

López-Bezanilla, A., Triozon, F., Latil, S., Blase, X. & Roche, S. (2009*b*), 'Effect of the chemical functionalization on charge transport in carbon nanotubes at the mesoscopic scale', *Nano Lett.* **9**, 940–944.

Lopez Sancho, M. P., Sancho, J. M. L., & Rubio, J. (1985), 'Highly convergent schemes for the calculation of bulk and surface Green functions', *Journal of Physics F: Metal Physics* **15**, 851.

Low, T., Jiang, Y., Katsnelson, M. & Guinea, F. (2012), 'Electron pumping in graphene mechanical resonators', *Nano Lett.* **12**, 850–854.

Luryi, S. (1989), 'Coherent versus incoherent resonant tunneling and implications for fast devices', *Superlattices and Microstructures* **5**, 375–382.

Luttinger, J. M. (1951), 'The effect of a magnetic field on electrons in a periodic potential', *Phys. Rev.* **84**, 814–817.

Luttinger, J. M. (1963), 'An exactly soluble model of a many-fermion system', *J. Math. Phys.* **4**, 1154–1162.

Maassen, J., Zahid, F. & Guo, H. (2009), 'Effects of dephasing in molecular transport junctions using atomistic first principles', *Phys. Rev. B* **80**, 125423.

Mak, K. F., Lui, C. H., Shan, J. & Heinz, T. F. (2009), 'Observation of an electric-field-induced band gap in bilayer graphene by infrared spectroscopy', *Phys. Rev. Lett.* **102**, 256405.

Marconcini, P. & Macucci, M. (2011), 'The k.p method and its application to graphene, carbon nanotubes and graphene nanoribbons: The Dirac equation', *La Rivista del Nuovo Cimento* **34**, 489–584.

Margine, E. R., Bocquet, M.-L. & Blase, X. (2008), 'Thermal stability of graphene and nanotube covalent functionalization', *Nano Lett.* **8**, 3315–3319.

Martel, R., Derycke, V., Lavoie, C., *et al.* (2001), 'Ambipolar electrical transport in semiconducting single-wall carbon nanotubes', *Phys. Rev. Lett.* **87**, 256805.

Martinez, D. F. (2003), 'Floquet–Green function formalism for harmonically driven Hamiltonians', *Journal of Physics A: Mathematical and General* **36**, 9827.

Marx, D. & Hutter, J. (2000), *Modern Methods and Algorithms of Quantum Chemistry*, chapter '*Ab initio* molecular dynamics: Theory and implementation', pp. 329–477.

Marzari, N. & Vanderbilt, D. (1997), 'Maximally localized generalized Wannier functions for composite energy bands', *Phys. Rev. B* **56**, 12847–12865.

Matsumura, H. & Ando, T. (2001), 'Conductance of carbon nanotubes with a Stone–Wales defect', *J. Phys. Soc. Jpn.* **70**, 2657–2665.

Matsuo, Y., Tahara, K. & Nakamura, E. (2003), 'Theoretical studies on structures and aromaticity of finite-length armchair carbon nanotubes', *Org. Lett.* **5**, 3181–3184.

Mayorov, A. S., Gorbachev, R. V., Morozov, S. V., *et al.* (2011), 'Micrometer-scale ballistic transport in encapsulated graphene at room temperature', *Nano Lett.* **11**, 2396–2399.

McCann, E., Abergel, D. S. & Fal'ko, V. I. (2007), 'The low energy electronic band structure of bilayer graphene', *The European Physical Journal Special Topics* **148**, 91–103.

McCann, E. & Fal'ko, V. I. (2006), 'Landau-level degeneracy and quantum Hall effect in a graphite bilayer', *Phys. Rev. Lett.* **96**, 086805.

McCann, E., Kechedzhi, K., Fal'ko, V. I., *et al.* (2006), 'Weak-localization magnetoresistance and valley symmetry in graphene', *Phys. Rev. Lett.* **97**, 146805.

McClure, J. (1969), 'Electron energy band structure and electronic properties of rhombohedral graphite', *Carbon* **7**, 425–432.

McClure, J. W. (1956), 'Diamagnetism of graphite', *Phys. Rev.* **104**, 666–671.

McClure, J. W. (1957), 'Band structure of graphite and de Haas-van Alphen effect', *Phys. Rev.* **108**, 612–618.

McCreary, K. M., Swartz, A. G., Han, W., Fabian, J. & Kawakami, R. K. (2012), 'Magnetic moment formation in graphene detected by scattering of pure spin currents', *Phys. Rev. Lett.* **109**, 186604.

Meir, Y. & Wingreen, N. S. (1992), 'Landauer formula for the current through an interacting electron region', *Phys. Rev. Lett.* **68**, 2512–2515.

Mello, P. A., Pereyra, P. & Kumar, N. (1988), 'Macroscopic approach to multichannel disordered conductors', *Ann. Phys.* **181**, 290–317.

Meric, I., Baklitskaya, N., Kim, P. & Shepard, K. (2008), 'RF performance of top-gated, zero-bandgap graphene field-effect transistors', in 'Electron devices' meeting, 2008. IEEE International, pp. 1–4.

Meyer, J. C., Girit, C. O., Crommie, M. F. & Zettl, A. (2008), 'Imaging and dynamics of light atoms and molecules on graphene', *Nature* **454**, 319–322.

Meyer, J. C., Kisielowski, C., Erni, R., *et al.* (2008), 'Direct imaging of lattice atoms and topological defects in graphene membranes', *Nano Lett.* **8**, 3582–3586.

Miao, F., Wijeratne, S., Zhang, Y., *et al.* (2007), 'Phase-coherent transport in graphene quantum billiards', *Science* **317**, 1530–1533.

Mingo, N. (2006), 'Anharmonic phonon flow through molecular-sized junctions', *Phys. Rev. B* **74**, 125402.

Mingo, N., Esfarjani, K., Broido, D. A. & Stewart, D. A. (2010), 'Cluster scattering effects on phonon conduction in graphene', *Phys. Rev. B* **81**, 045408.

Mingo, N. & Han, J. (2001), 'Conductance of metallic carbon nanotubes dipped into metal', *Phys. Rev. B* **64**, 201401.

Mingo, N., Yang, L., Han, J. & Anantram, M. (2001), 'Resonant versus anti-resonant tunneling at carbon nanotube ABA heterostructures', *Physica Status Solidi (B)* **226**, 79–85.

Mintmire, J. W., Dunlap, B. I. & White, C. T. (1992), 'Are fullerene tubules metallic?', *Phys. Rev. Lett.* **68**, 631–634.

Mintmire, J. W. & White, C. T. (1998), 'Universal density of states for carbon nanotubes', *Phys. Rev. Lett.* **81**, 2506–2509.

Miroshnichenko, A. E., Flach, S. & Kivshar, Y. S. (2010), 'Fano resonances in nanoscale structures', *Rev. Mod. Phys.* **82**, 2257–2298.

Miyake, T. & Saito, S. (2003), 'Quasiparticle band structure of carbon nanotubes', *Phys. Rev. B* **68**, 155424.

Miyamoto, Y., Nakada, K. & Fujita, M. (1999), 'First-principles study of edge states of h-terminated graphitic ribbons', *Phys. Rev. B* **59**, 9858–9861.

Miyamoto, Y., Saito, S. & Tománek, D. (2001), 'Electronic interwall interactions and charge redistribution in multiwall nanotubes', *Phys. Rev. B* **65**, 041402.

Monkhorst, H. J. & Pack, J. D. (1976), 'Special points for Brillouin-zone integrations', *Phys. Rev. B* **13**, 5188–5192.

Monteverde, M., Ojeda-Aristizabal, C., Weil, R., *et al.* (2010), 'Transport and elastic scattering times as probes of the nature of impurity scattering in single-layer and bilayer graphene', *Phys. Rev. Lett.* **104**, 126801.

Moon, J., Curtis, D., Hu, M., *et al.* (2009), 'Epitaxial-graphene RF field-effect transistors on Si-face 6h-SiC substrates', *Electron Device Letters, IEEE* **30**, 650–652.

Moore, A. (1974), *Chemistry and Physics of Carbon, Vol. 11*, P. L. Walker and P. A. Thrower (eds.), Marcel Dekker Inc., New York, chapter 'Highly oriented pyrolytic graphite'.

Moser, J., Tao, H., Roche, S., *et al.* (2010), 'Magnetotransport in disordered graphene exposed to ozone: From weak to strong localization', *Phys. Rev. B* **81**, 205445.

Moskalets, M. & Buttiker, M. (2002), 'Floquet scattering theory of quantum pumps', *Phys. Rev. B* **66**, 205320.

Mott, N. F. (1990), *Metal–Insulator Transitions*, 2nd edition. Taylor and Francis, UK.

Mpourmpakis, G., Andriotis, A. N. & Vlachos, D. G. (2010), 'Identification of descriptors for the co-interaction with metal nanoparticles', *Nano Lett.* **10**, 1041–1045.

Mucciolo, E. R., Castro Neto, A. H. & Lewenkopf, C. H. (2009), 'Conductance quantization and transport gaps in disordered graphene nanoribbons', *Phys. Rev. B* **79**, 075407.

Mueller, T., Xia, F. & Avouris, P. (2010), 'Graphene photodetectors for high-speed optical communications', *Nature Photonics* **4**, 297–301.

Mukhopadhyay, P. & Gupta, R. (2011), 'Trends and frontiers in graphene-based polymer nanocomposites', *Plastics Engineeering* (Jan.), 32–42.

Muñoz, E. (2012), 'Phonon-limited transport coefficients in extrinsic graphene', *J. Phys.: Condens. Matter* **24**, 195302.

Muscat, J., Wander, A. & Harrison, N. (2001), 'On the prediction of band gaps from hybrid functional theory', *Chemical Physics Letters* **342**, 397–401.

Nair, R. R., Blake, P., Grigorenko, A. N., *et al.* (2008), 'Fine structure constant defines visual transparency of graphene', *Science* **320**, 1308.

Nakada, K., Fujita, M., Dresselhaus, G. & Dresselhaus, M. S. (1996), 'Edge state in graphene ribbons: Nanometer size effect and edge shape dependence', *Phys. Rev. B* **54**, 17954–17961.

Nakaharaim, S., Iijima, T., Ogawa, S., *et al.* (2013), 'Conduction tuning of graphene based on defect-induced localization', *ACS Nano* **7**, 5694–5700.

Nathan, A., Ahnood, A., Cole, M., *et al.* (2012), 'Flexible electronics: The next ubiquitous platform', *Proceedings of the IEEE* **100** (Special Centennial Issue), 1486–1517.

Nemec, N., Tománek, D. & Cuniberti, G. (2006), 'Contact dependence of carrier injection in carbon nanotubes: An *ab initio* study', *Phys. Rev. Lett.* **96**, 076802.

Nobunaga, D., Abedifard, E., Roohparvar, F., *et al.* (2008), 'A 50 nm 8GB nand flash memory with 100 MB/s program throughput and 200 MB/s ddr interface', in 'Solid-state circuits' Conference, 2008. Digest of Technical Papers. IEEE International, pp. 426–625.

Nogueira, F., Castro, A. & Marques, M. A. L. (2003), 'A tutorial on density functional theory', *A Primer in Density Functional Theory*, Fiolhais, C., Nogueira, F. & Marques, M. A. L., eds., *Lecture Notes in Physics* **620**, Springer, Berlin, pp. 218–256.

Nomura, K. & MacDonald, A. H. (2006), 'Quantum Hall ferromagnetism in graphene', *Phys. Rev. Lett.* **96**, 256602.

Nomura, K. & MacDonald, A. H. (2007), 'Quantum transport of massless Dirac fermions', *Phys. Rev. Lett.* **98**, 076602.

Novoselov, K. S., Fal'ko, V. I., Colombo, L., *et al.* (2012), 'A roadmap for graphene', *Nature* **490**, 192–200.

Novoselov, K. S., Geim, A. K., Morozov, S. V., *et al.* (2005*b*), 'Two-dimensional gas of massless Dirac fermions in graphene', *Nature* **438**, 197–200.

Novoselov, K. S., Geim, A. K., Morozov, S. V., *et al.* (2004), 'Electric field effect in atomically thin carbon films', *Science* **306**, 666–669.

Novoselov, K. S., Jiang, D., Schedin, F., *et al.* (2005*a*), 'Two-dimensional atomic crystals', *Proceedings of the National Academy of Sciences of the United States of America* **102**, 10451–10453.

Novoselov, K. S., Jiang, Z., Zhang, Y., *et al.* (2007), 'Room-temperature quantum Hall effect in graphene', *Science* **315**, 1379.

Nozaki, D., Girard, Y. & Yoshizawa, K. (2008), 'Theoretical study of long-range electron transport in molecular junctions', *The Journal of Physical Chemistry C* **112**, 17408–17415.

Nyakiti, L., Wheeler, V., Garces, N., *et al.* (2012), 'Enabling graphene-based technologies: Toward wafer-scale production of epitaxial graphene', *MRS Bulletin* **37**, 1149–1157.

Oberlin, A., Endo, M. & Koyama, T. (1976), 'Filamentous growth of carbon through benzene decomposition', *Journal of Crystal Growth* **32**, 335–349.

Ochoa, H., Castro Neto, A. H. & Guinea, F. (2012), 'Elliot–Yafet mechanism in graphene', *Phys. Rev. Lett.* **108**, 206808.

O'Connell, M. J., Bachilo, S. M., Huffman, C. B., *et al.* (2002), 'Band gap fluorescence from individual single-walled carbon nanotubes', *Science* **297**, 593–596.

Odom, T. W., Huang, J.-L., Kim, P. & Lieber, C. M. (1998), 'Atomic structure and electronic properties of single-walled carbon nanotubes', *Nature* **391**, 62–64.

Odom, T. W., Huang, J.-L. & Lieber, C. M. (2002), 'STM studies of single-walled carbon nanotubes', *Journal of Physics: Condensed Matter* **14**, R145.

Oezyilmaz, B., Jarillo-Herrero, P., Efetov, D., *et al.* (2007), 'Electronic transport and quantum Hall effect in bipolar graphene *p-n-p* junctions', *Phys. Rev. Lett.* **99**, 166804.

Oka, T. & Aoki, H. (2009), 'Photovoltaic Hall effect in graphene', *Phys. Rev. B* **79**, 081406.

Okada, S. & Oshiyama, A. (2001), 'Magnetic ordering in hexagonally bonded sheets with first-row elements', *Phys. Rev. Lett.* **87**, 146803.

Oksanen, M., Uppstu, A., & Laitinen, A., *et al.* (undated), 'Single-and multi-mode Fabry–pérot interference in suspended graphene', arXiv: 1306.1212 [cond-mat.mes-hall].

Onida, G., Reining, L. & Rubio, A. (2002), 'Electronic excitations: Density-functional versus many-body Green's-function approaches', *Rev. Mod. Phys.* **74**, 601–659.

Orellana, P. A. & Pacheco, M. (2007), 'Photon-assisted transport in a carbon nanotube calculated using Green's function techniques', *Phys. Rev. B* **75**, 115427.

Ormsby, J. L. & King, B. T. (2004), 'Clar valence bond representation of bonding in carbon nanotubes', *J. Org. Chem.* **69**, 4287–4291.

Ortmann, F., Cresti, A., Montambaux, G. & Roche, S. (2011), 'Magnetoresistance in disordered graphene: The role of pseudospin and dimensionality effects unraveled', *EPL (Europhysics Letters)* **94**, 47006.

Ortmann, F. & Roche, S. (2011), 'Polaron transport in organic crystals: Temperature tuning of disorder effects', *Phys. Rev. B* **84**, 180302.

Ortmann, F. & Roche, S. (2013), 'Splitting of the zero-energy Landau level and universal dissipative conductivity at critical points in disordered graphene', *Phys. Rev. Lett.* **110**, 086602.

Ostrovsky, P. M., Gornyi, I. V. & Mirlin, A. D. (2006), 'Electron transport in disordered graphene', *Phys. Rev. B* **74**, 235443.

Ostrovsky, P. M., Gornyi, I. V. & Mirlin, A. D. (2008), 'Theory of anomalous quantum Hall effects in graphene', *Phys. Rev. B* **77**, 195430.

Ostrovsky, P. M., Titov, M., Bera, S., Gornyi, I. V. & Mirlin, A. D. (2010), 'Diffusion and criticality in undoped graphene with resonant scatterers', *Phys. Rev. Lett.* **105**, 266803.

Ouyang, M., Huang, J.-L., Cheung, C. L. & Lieber, C. M. (2001*a*), 'Atomically resolved single-walled carbon nanotube intramolecular junctions', *Science* **291**, 97–100.

Ouyang, M., Huang, J.-L., Cheung, C. L. & Lieber, C. M. (2001*b*), 'Energy gaps in "metallic" single-walled carbon nanotubes', *Science* **292**, 702–705.

Ouyang, Y., Dai, H. & Guo, J. (2010), 'Projected performance advantage of multilayer graphene nanoribbons as a transistor channel material', **3**, 8–15.

Paier, J., Marsman, M., Hummer, K., *et al.* (2006), 'Screened hybrid density functionals applied to solids', *The Journal of Chemical Physics* **124**, 154709.

Palacios, J. J., Pérez-Jiménez, A. J., Louis, E., SanFabián, E. & Vergés, J. A. (2003), 'First-principles phase-coherent transport in metallic nanotubes with realistic contacts', *Phys. Rev. Lett.* **90**, 106801.

Park, H., Zhao, J. & Lu, J. P. (2006), 'Effects of sidewall functionalization on conducting properties of single wall carbon nanotubes', *Nano Lett.* **6**, 916–919.

Park, J. K., Song, S. M., Mun, J. H. & Cho, B. J. (2011), 'Graphene gate electrode for MOS structure-based electronic devices', *Nano Lett.* **11**, 5383–5386.

Park, J.-Y., Rosenblatt, S., Yaish, Y., *et al.* (2004), 'Electron–phonon scattering in metallic single-walled carbon nanotubes', *Nano Letters* **4**, 517–520.

Park, K.-T., Kang, M., Kim, D., *et al.* (2008), 'A zeroing cell-to-cell interference page architecture with temporary lsb storing and parallel msb program scheme for mlc nand flash memories', *IEEE Journal of Solid-State Circuits* **43**, 919–928.

Park, N., Sung, D., Lim, S., Moon, S. & Hong, S. (2009), 'Realistic adsorption geometries and binding affinities of metal nanoparticles onto the surface of carbon nanotubes', *Appl. Phys. Lett.* **94**, 073105.

Parks, E. K., Zhu, L., Ho, J. & Riley, S. J. (1994), 'The structure of small nickel clusters. I. Ni_3–Ni_15', *J. Chem. Phys.* **100**, 7206–7222.

Pastawski, H. M. (1991), 'Classical and quantum transport from generalized Landauer Buttiker equations', *Phys. Rev. B* **44**, 6329–6339.

Pastawski, H. M. & Medina, E. (2001), 'Tight binding methods in quantum transport through molecules and small devices: From the coherent to the decoherent description', *Revista Mexicana de Fisica* **47 S1**, 1–23.

Pastawski, H. M., Weisz, J. F. & Albornoz, S. (1983), 'Matrix continued-fraction calculation of localization length', *Phys. Rev. B* **28**, 6896–6903.

Patel, S. R., Stewart, D. R., Marcus, C. M., *et al.* (1998), 'Non-Gaussian distribution of Coulomb blockade peak heights in quantum dots', *Phys. Rev. Lett.* **81**, 5900.

Paulsson, M. & Brandbyge, M. (2007), 'Transmission eigenchannels from nonequilibrium Green's functions', *Phys. Rev. B* **76**, 115117.

Payne, M. C., Teter, M. P., Allan, D. C., Arias, T. A. & Joannopoulos, J. D. (1992), 'Iterative minimization techniques for *ab initio* total-energy calculations: Molecular dynamics and conjugate gradients', *Rev. Mod. Phys.* **64**, 1045–1097.

Peierls, R. (1933), 'On the theory of the diamagnetism of conduction electrons', *Z. Phys* **80**, 763.

Peng, S. & Cho, K. (2003), '*Ab initio* study of doped carbon nanotube sensors', *Nano Lett.* **3**, 513–517.

Perdew, J. P. (1991), *Electronic Structure of Solids '91*, Akademie Verlag, Berlin, p. 11.

Perdew, J. P., Burke, K. & Ernzerhof, M. (1996), 'Generalized gradient approximation made simple', *Phys. Rev. Lett.* **77**, 3865–3868.

Perdew, J. P. & Zunger, A. (1981), 'Self-interaction correction to density-functional approximations for many-electron systems', *Phys. Rev. B* **23**, 5048–5079.

Pereira, V. M., dos Santos, L. & Neto, A. H. C. (2008), 'Modeling disorder in graphene', *Phys. Rev. B* **77**, 115109.

Perez-Piskunow, P. M., Usaj, G., Balseiro, C. A. & Foa Torres, L. E. F. (2013), 'Unveiling laser-induced chiral edge states in graphene', arxiv: 1308.4362 [cond-mat.mes-hall].

Perfetto, E., Stefanucci, G. & Cini, M. (2010), 'Time-dependent transport in graphene nanoribbons', *Phys. Rev. B* **82**, 035446.

Persson, M. P., Lherbier, A., Niquet, Y.-M., Triozon, F. & Roche, S. (2008), 'Orientational dependence of charge transport in disordered silicon nanowires', *Nano Lett.* **8**, 4146–4150.

Phillips, J. C. (1958), 'Energy-band interpolation scheme based on a pseudopotential', *Phys. Rev.* **112**, 685–695.

Pisana, S., Lazzeri, M., Casiraghi, C., *et al.* (2007), 'Breakdown of the adiabatic Born–Oppenheimer approximation in graphene', *Nature Materials* **6**, 198–201.

Platero, G. & Aguado, R. (2004), 'Photon-assisted transport in semiconductor nanostructures', *Physics Reports* **395**, 1–157.

Poncharal, P., Berger, C., Yi, Y., Wang, Z. L. & de Heer, W. A. (2002), 'Room temperature ballistic conduction in carbon nanotubes', *The Journal of Physical Chemistry B* **106**, 12104–12118.

Ponomarenko, L. A., Geim, A. K., Zhukov, A. A., *et al.* (2011), 'Tunable metal-insulator transition in double-layer graphene heterostructures', *Nature Physics* **7**, 958–961.

Ponomarenko, L. A., Gorbachev, R. V., Yu, G. L., *et al.* (2013), 'Cloning of Dirac fermions in graphene superlattices', *Nature* **497**, 594–597.

Ponomarenko, L. A., Schedin, F., Katsnelson, M. I., *et al.* (2008), 'Chaotic Dirac billiard in graphene quantum dots', *Science* **320**, 356–358.

Poumirol, J.-M., Cresti, A., Roche, S., *et al.* (2010), 'Edge magnetotransport fingerprints in disordered graphene nanoribbons', *Phys. Rev. B* **82**, 041413.

Prada, E., San-Jose, P. & Schomerus, H. (2009), 'Quantum pumping in graphene', *Phys. Rev. B* **80**, 245414.

Prasher, R. S., Hu, X. J., Chalopin, Y., *et al.* (2009), 'Turning carbon nanotubes from exceptional heat conductors into insulators', *Phys. Rev. Lett.* **102**, 105901.

Purewal, M. S., Hong, B. H., Ravi, A., *et al.* (2007), 'Scaling of resistance and electron mean free path of single-walled carbon nanotubes', *Phys. Rev. Lett.* **98**, 186808.

Querlioz, D., Apertet, Y., Valentin, A., *et al.* (2008), 'Suppression of the orientation effects on bandgap in graphene nanoribbons in the presence of edge disorder', *Applied Physics Letters* **92**, 042108.

Radchenko, T. M., Shylau, A. A. & Zozoulenko, I. V. (2012), 'Influence of correlated impurities on conductivity of graphene sheets: Time-dependent real-space Kubo approach', *Phys. Rev. B* **86**, 035418.

Raquet, B., Avriller, R., Lassagne, B., *et al.* (2008), 'Onset of Landau-level formation in carbon-nanotube-based electronic Fabry–Perot resonators', *Phys. Rev. Lett.* **101**, 046803.

Ravagnan, L., Piseri, P., Bruzzi, M., *et al.* (2007), 'Influence of cumulenic chains on the vibrational and electronic properties of sp-sp^2 amorphous carbon', *Phys. Rev. Lett.* **98**, 216103.

Ravagnan, L., Siviero, F., Lenardi, C., *et al.* (2002), 'Cluster-beam deposition and *in situ* characterization of carbyne-rich carbon films', *Phys. Rev. Lett.* **89**, 285506.

Rechtsman, M. C., Zeuner, J. M., Plotnik, Y., *et al.* (2013), 'Photonic Floquet topological insulators', *Nature* **496**, 196–200.

Reich, S., Maultzsch, J., Thomsen, C. & Ordejón, P. (2002), 'Tight-binding description of graphene', *Phys. Rev. B* **66**, 035412.

Ribeiro, R., Poumirol, J.-M., Cresti, A., *et al.* (2011), 'Unveiling the magnetic structure of graphene nanoribbons', *Phys. Rev. Lett.* **107**, 086601.

Rickhaus, P., Maurand, R., Liu, M.-H., *et al.* (2013), 'Ballistic interferences in suspended graphene', *Nature Communications* **4**, 2342, DOI 10.1038/ncomms 2342.

Rignanese, G.-M. (1998), First-principles molecular dynamics study of SiO_2: Surface and interface with Si, Ph.D. thesis, Université Catholique de Louvain.

Ritter, K. A. & Lyding, J. W. (2009), 'The influence of edge structure on the electronic properties of graphene quantum dots and nanoribbons', *Nature Materials* **8**, 235–242.

Rocha, A. R. (2007), Theoretical and computational aspects of electronic transport at the nanoscale, Ph.D. thesis, University of Dublin, Trinity College.

Rocha, A. R., Garcia-Suarez, V. M., Bailey, S. W., *et al.* (2005), 'Towards molecular spintronics', *Nature Materials* **4**, 335–339.

Rocha, A. R., Garca-Surez, V. M., Bailey, S., *et al.* (2006), 'Spin and molecular electronics in atomically generated orbital landscapes', *Phys. Rev. B* **73**, 085414.

Rocha, A. R., Rossi, M., Fazzio, A. & da Silva, A. J. R. (2008), 'Designing real nanotube-based gas sensors', *Phys. Rev. Lett.* **100**, 176803.

Rocha, C. G., Foa Torres, L. E. F. & Cuniberti, G. (2010), 'AC transport in graphene-based Fabry–Pérot devices', *Phys. Rev. B* **81**, 115435.

Roche, S. (1996), Contribution à l'étude théorique du transport électronique dan les quasicristaux, Ph.D. thesis, Université Joseph-Fourier.

Roche, S. (1999), 'Quantum transport by means of o(n) real-space methods', *Phys. Rev. B* **59**, 2284–2291.

Roche, S. (2011), 'Nanoelectronics: Graphene gets a better gap', *Nature Nanotechnology* **6**, 8–9.

Roche, S., Akkermans, E., Chauvet O., *et al.* (2006), *Understanding Carbon Nanotubes, from Basics to Application*, Lect. Notes Phys., chapter 'Transport properties', pp. 335–437.

Roche, S., Dresselhaus, G., Dresselhaus, M. S. & Saito, R. (2000), 'Aharonov–Bohm spectral features and coherence lengths in carbon nanotubes', *Phys. Rev. B* **62**, 16092–16099.

Roche, S., Jiang, J., Triozon, F. & Saito, R. (2005), 'Quantum dephasing in carbon nanotubes due to electron–phonon coupling', *Phys. Rev. Lett.* **95**, 076803.

Roche, S., Leconte, N., Ortmann, F., *et al.* (2012), 'Quantum transport in disordered graphene: A theoretical perspective', *Solid State Commun.* **152**, 1404–1410.

Roche, S. & Mayou, D. (1997), 'Conductivity of quasiperiodic systems: A numerical study', *Phys. Rev. Lett.* **79**, 2518–2521.

Roche, S. & Saito, R. (2001), 'Magnetoresistance of carbon nanotubes: From molecular to mesoscopic fingerprints', *Phys. Rev. Lett.* **87**, 246803.

Roche, S., Triozon, F., Rubio, A. & Mayou, D. (2001), 'Conduction mechanisms and magneto-transport in multiwalled carbon nanotubes', *Phys. Rev. B* **64**, 121401.

Rohlfing, M., Wang, N.-P., Krüger, P. & Pollmann, J. (2003), 'Image states and excitons at insulator surfaces with negative electron affinity', *Phys. Rev. Lett.* **91**, 256802.

Romo-Herrera, J. M., Terrones, M., Terrones, H., Dag, S. & Meunier, V. (2006), 'Covalent 2D and 3D networks from 1D nanostructures: Designing new materials', *Nano Lett.* **7**, 570–576.

Romo-Herrera, J. M., Terrones, M., Terrones, H. & Meunier, V. (2008), 'Guiding electrical current in nanotube circuits using structural defects: A step forward in nanoelectronics', *ACS Nano* **2**, 2585–2591.

Rudner, M. S., Lindner, N. H., Berg, E. & Levin, M. (2013), 'Anomalous edge states and the bulk-edge correspondence for periodically-driven two dimensional systems', *Phys. Rev. X* **3**, 031005.

Rycerz, A., Tworzydo, J. & Beenakker, C. W. J. (2007), 'Anomalously large conductance fluctuations in weakly disordered graphene', *Europhys. Lett.* **79**, 57003.

Saito, R., Dresselhaus, G. & Dresselhaus, M. (1998), *Physical Properties of Carbon Nanotubes*, Imperial College Press, London.

Saito, R., Dresselhaus, G. & Dresselhaus, M. S. (1994), 'Magnetic energy bands of carbon nanotubes', *Phys. Rev. B* **50**, 14698–14701.

Saito, R., Dresselhaus, G. & Dresselhaus, M. S. (1996*a*), 'Erratum: Magnetic energy bands of carbon nanotubes', *Phys. Rev. B* **53**, 10408.

Saito, R., Dresselhaus, G. & Dresselhaus, M. S. (1996*b*), 'Tunneling conductance of connected carbon nanotubes', *Phys. Rev. B* **53**, 2044–2050.

Saito, R., Dresselhaus, G. & Dresselhaus, M. S. (2000), 'Trigonal warping effect of carbon nanotubes', *Phys. Rev. B* **61**, 2981–2990.

Saito, R., Fujita, M., Dresselhaus, G. & Dresselhaus, M. S. (1992*a*), 'Electronic structure of chiral graphene tubules', *Applied Physics Letters* **60**, 2204–2206.

Saito, R., Fujita, M., Dresselhaus, G. & Dresselhaus, M. S. (1992*b*), 'Electronic structure of graphene tubules based on C_{60}', *Phys. Rev. B* **46**, 1804–1811.

Sakhaee-Pour, A., Ahmadian, M. & Vafai, A. (2008), 'Applications of single-layered graphene sheets as mass sensors and atomistic dust detectors', *Solid State Communications* **145**, 168–172.

Sambe, H. (1973), 'Steady states and quasienergies of a quantum-mechanical system in an oscillating field', *Phys. Rev. A* **7**, 2203–2213.

San-Jose, P., Prada, E., Kohler, S. & Schomerus, H. (2011), 'Single-parameter pumping in graphene', *Phys. Rev. B* **84**, 155408.

San-Jose, P., Prada, E., Schomerus, H. & Kohler, S. (2012), 'Laser-induced quantum pumping in graphene', *Appl. Phys. Lett.* **101**, 153506.

Sanvito, S., Lambert, C. J., Jefferson, J. H. & Bratkovsky, A. M. (1999), 'General Greens-function formalism for transport calculations with spd Hamiltonians and giant magnetoresistance in Co- and Ni-based magnetic multilayers', *Phys. Rev. B* **59**, 11936–11948.

Sasaki, K., Murakami, S. & Saito, R. (2006), 'Stabilization mechanism of edge states in graphene', *Applied Physics Letters* **88**, 113110.

Savelev, S. E. & Alexandrov, A. S. (2011), 'Massless Dirac fermions in a laser field as a counterpart of graphene superlattices', *Phys. Rev. B* **84**, 035428.

Savelev, S. E., Häusler, W. & Hänggi, P. (2012), 'Current resonances in graphene with time-dependent potential barriers', *Phys. Rev. Lett.* **109**, 226602.

Savić, I., Mingo, N. & Stewart, D. A. (2008), 'Phonon transport in isotope-disordered carbon and boron-nitride nanotubes: Is localization observable?', *Phys. Rev. Lett.* **101**, 165502.

Scholz, A., Lopez, A. & Schliemann, J. (2013), 'Interplay between spin–orbit interactions and a time-dependent electromagnetic field in monolayer graphene', *Phys. Rev. B* **88**, 045118.

Schrödinger, E. (1926), 'An undulatory theory of the mechanics of atoms and molecules', *Phys. Rev.* **28**, 1049–1070.

Schwierz, F. (2010), 'Graphene transistors', *Nature Nanotechnology* **5**, 487–496.

Segall, M. D., Lindan, P. J. D., Probert, M. J., *et al.* (2002), 'First-principles simulation: Ideas, illustrations and the Castep code', *Journal of Physics: Condensed Matter* **14**, 2717.

Sela, I. & Cohen, D. (2008), 'Quantum stirring in low-dimensional devices', *Phys. Rev. B* **77**, 245440.

Semenov, Y. G., Kim, K. W. & Zavada, J. M. (2007), 'Spin field effect transistor with a graphene channel', *Appl. Phys. Lett.* **91**, 153105.

Seneor, P., Dlubak, B., Martin, M.-B., *et al.* (2012), 'Spintronics with graphene', *MRS Bulletin* **37**, 1245–1254.

Seol, J. H., Jo, I., Moore, A. L., *et al.* (2010), 'Two-dimensional phonon transport in supported graphene', *Science* **328**, 213–216.

Sevincli, H., Li, W., Mingo, N., Cuniberti, G. & Roche, S. (2011), 'Effects of domains in phonon conduction through hybrid boron nitride and graphene sheets', *Phys. Rev. B* **84**, 205444.

Sheng, D. N., Sheng, L. & Weng, Z. Y. (2006), 'Quantum Hall effect in graphene: Disorder effect and phase diagram', *Phys. Rev. B* **73**, 233406.

Shibayama, Y., Sato, H., Enoki, T. & Endo, M. (2000), 'Disordered magnetism at the metal–insulator threshold in nano-graphite-based carbon materials', *Phys. Rev. Lett.* **84**, 1744–1747.

Shimizu, T., Haruyama, J., Marcano, D., *et al.* (2011), 'Large intrinsic energy bandgaps in annealed nanotube-derived graphene nanoribbons', *Nature Nanotechnology* **6**, 45–50.

Shin, Y. (2005), Non-volatile memory technologies for beyond 2010, in 'VLSI circuits, 2005'. Digest of Technical Papers, pp. 156–159.

Shirley, J. H. (1965), 'Solution of the Schrödinger equation with a Hamiltonian periodic in time', *Phys. Rev.* **138**, B979–B987.

Shon, N. H. & Ando, T. (1998), 'Quantum transport in two-dimensional graphite system', *J. Phys. Soc. Jpn.* **67**, 2421–2429.

Shytov, A. V., Rudner, M. S. & Levitov, L. S. (2008), 'Klein backscattering and Fabry–Perot interference in graphene heterojunctions', *Phys. Rev. Lett.* **101**, 156804.

Siegel, D. A., Park, C.-H., Hwang, C., *et al.* (2011), 'Many-body interactions in quasi-freestanding graphene', *Proceedings of the National Academy of Sciences* **108**, 11365–11369.

Simmons, J. M., In, I., Campbell, V. E., *et al.* (2007), 'Optically modulated conduction in chromophore-functionalized single-wall carbon nanotubes', *Phys. Rev. Lett.* **98**, 086802.

Sire, C., Ardiaca, F., Lepilliet, S., *et al.* (2012), 'Flexible gigahertz transistors derived from solution-based single-layer graphene', *Nano Lett.* **12**, 1184–1188.

Skylaris, C.-K., Haynes, P. D., Mostofi, A. A. & Payne, M. C. (2005), 'Introducing [ONETEP]: Linear-scaling density functional simulations on parallel computers', *J. Chem. Phys.* **122**, 084119.

Slonczewski, J. C. & Weiss, P. R. (1958), 'Band structure of graphite', *Phys. Rev.* **109**, 272–279.

Sluiter, M. H. F. & Kawazoe, Y. (2003), 'Cluster expansion method for adsorption: Application to hydrogen chemisorption on graphene', *Phys. Rev. B* **68**, 085410.

Smith, B. W., Monthioux, M. & Luzzi, D. E. (1998), 'Encapsulated C60 in carbon nanotubes', *Nature* **396**, 323–324.

Sofo, J. O., Chaudhari, A. S. & Barber, G. D. (2007), 'Graphane: A two-dimensional hydrocarbon', *Phys. Rev. B* **75**, 153401.

Soler, J. M., Artacho, E., Gale, J. D., *et al.* (2002), 'The siesta method for *ab initio* order-N materials simulation', *Journal of Physics: Condensed Matter* **14**, 2745.

Son, Y.-W., Cohen, M. L. & Louie, S. G. (2006a), 'Energy gaps in graphene nanoribbons', *Phys. Rev. Lett.* **97**, 216803.

Son, Y.-W., Cohen, M. L. & Louie, S. G. (2006b), 'Half-metallic graphene nanoribbons', *Nature* **444**, 347–349.

Soriano, D., Leconte, N., Ordejón, P., *et al.* (2011), 'Magnetoresistance and magnetic ordering fingerprints in hydrogenated graphene', *Phys. Rev. Lett.* **107**, 016602.

Spataru, C. D., Ismail-Beigi, S., Benedict, L. X. & Louie, S. G. (2004), 'Excitonic effects and optical spectra of single-walled carbon nanotubes', *Phys. Rev. Lett.* **92**, 077402.

Sprinkle, M. M. R., Hu, Y., Hankinson, J., *et al.* (2010), 'Scalable templated growth of graphene nanoribbons on SiC', *Nature Nanotechnology* **5**, 727–731.

Stampfer, C., Güttinger, J., Hellmüller, S., *et al.* (2009), 'Energy gaps in etched graphene nanoribbons', *Phys. Rev. Lett.* **102**, 056403.

Stampfer, C., Guttinger, J., Molitor, F., *et al.* (2008), 'Tunable Coulomb blockade in nanostructured graphene', *Appl. Phys. Lett.* **92**, 012102.

Stander, N., Huard, B. & Goldhaber-Gordon, D. (2009), 'Evidence for Klein tunneling in graphene *p-n* junctions', *Phys. Rev. Lett.* **102**, 026807.

Star, A., Gabriel, J.-C. P., Bradley, K. & Grüner, G. (2003), 'Electronic detection of specific protein binding using nanotube FET devices', *Nano Lett.* **3**, 459–463.

Stauber, T., Peres, N. M. R. & Guinea, F. (2007), 'Electronic transport in graphene: A semiclassical approach including midgap states', *Phys. Rev. B* **76**, 205423.

Stefanucci, G., Kurth, S., Rubio, A. & Gross, E. K. U. (2008), 'Time-dependent approach to electron pumping in open quantum systems', *Phys. Rev. B* **77**, 075339.

Stern, A., Aharonov, Y. & Imry, Y. (1990), 'Phase uncertainty and loss of interference: A general picture', *Phys. Rev. A* **41**, 3436–3448.

Stojetz, B., Miko, C., Forró, L. & Strunk, C. (2005), 'Effect of band structure on quantum interference in multiwall carbon nanotubes', *Phys. Rev. Lett.* **94**, 186802.

Stone, A. & Wales, D. (1986), 'Theoretical studies of icosahedral C60 and some related species', *Chemical Physics Letters* **128**, 501–503.

Strano, M. S., Dyke, C. A., Usrey, M. L., *et al.* (2003), 'Electronic structure control of single-walled carbon nanotube functionalization', *Science* **301**, 1519–1522.

Strunk, C., Stojetz, B. & Roche, S. (2006), 'Quantum interference in multiwall carbon nanotubes', *Semiconductor Science and Technology* **21**, S38.

Stützel, E. U., Burghard, M., Kern, K., *et al.* (2010), 'A graphene nanoribbon memory cell', *Small* **6**, 2822–2825.

Suárez Morell, E. & Foa Torres, L. E. F. (2012), 'Radiation effects on the electronic properties of bilayer graphene', *Phys. Rev. B* **86**, 125449.

Suárez Morell, E. E., Correa, J. D., Vargas, P., Pacheco, M. & Barticevic, Z. (2010), 'Flat bands in slightly twisted bilayer graphene: Tight-binding calculations', *Phys. Rev. B* **82**, 121407.

Suenaga, K., Wakabayashi, H., Koshino, M., *et al.* (2007), 'Imaging active topological defects in carbon nanotubes', *Nature Nanotechnology* **2**, 358–360.

Suzuura, H. & Ando, T. (2002), 'Crossover from symplectic to orthogonal class in a two-dimensional honeycomb lattice', *Phys. Rev. Lett.* **89**, 266603.

Svensson, J. & Campbell, E. E. B. (2011), 'Schottky barriers in carbon nanotube–metal contacts', *Journal of Applied Physics* **110**, 111101.

Swartz, A. G., Odenthal, P. M., Hao, Y., Ruoff, R. S. & Kawakami, R. K. (2012), 'Integration of the ferromagnetic insulator EuO onto graphene', *ACS Nano* **6**, 10063–10069.

Switkes, M., Marcus, C. M., Campman, K. & Gossard, A. C. (1999), 'An adiabatic quantum electron pump', *Science* **283**, 1905–1908.

Syzranov, S. V., Fistul, M. V. & Efetov, K. B. (2008), 'Effect of radiation on transport in graphene', *Phys. Rev. B* **78**, 045407.

Takayama, R., Hoshi, T. & Fujiwara, T. (2004), 'Krylov subspace method for molecular dynamics simulation based on large-scale electronic structure theory', *J. Phys. Soc. Jpn* **73**, 1519.

Tamura, R. & Tsukada, M. (1994), 'Disclinations of monolayer graphite and their electronic states', *Phys. Rev. B* **49**, 7697–7708.

Tan, Y.-W., Zhang, Y., Bolotin, K., *et al.* (2007), 'Measurement of scattering rate and minimum conductivity in graphene', *Phys. Rev. Lett.* **99**, 246803.

Tang, Z. K., Zhang, L., Wang, N., *et al.* (2001), 'Superconductivity in 4 angstrom single-walled carbon nanotubes', *Science* **292**, 2462–2465.

Tapaszto, L., Dobrik, G., Lambin, P. & Biro, L. P. (2008), 'Tailoring the atomic structure of graphene nanoribbons by scanning tunnelling microscope lithography', *Nature Nanotechnology* **3**, 397–401.

Terrones, H. & Terrones, M. (2003), 'Curved nanostructured materials', *New Journal of Physics* **5**, 126.

Terrones, H., Terrones, M., Hernández, E., *et al.* (2000), 'New metallic allotropes of planar and tubular carbon', *Phys. Rev. Lett.* **84**, 1716–1719.

Terrones, M. (2009), 'Materials science: Nanotubes unzipped', *Nature* **458**, 845–846.

Terrones, M., Banhart, F., Grobert, N., *et al.* (2002), 'Molecular junctions by joining single-walled carbon nanotubes', *Phys. Rev. Lett.* **89**, 075505.

Terrones, M., Botello-Mendez, A. R., Campos-Delgado, J., *et al.* (2010), 'Graphene and graphite nanoribbons: Morphology, properties, synthesis, defects and applications', *Nano Today* **5**, 351–372.

Terrones, M., Terrones, H., Banhart, F., Charlier, J.-C. & Ajayan, P. M. (2000), 'Coalescence of single-walled carbon nanotubes', *Science* **288**, 1226–1229.

Tersoff, J. (2003), 'Nanotechnology: A barrier falls', *Nature* **424**, 622–623.

Thess, A., Lee, R., Nikolaev, P., Dai, H., *et al.* (1996), 'Crystalline ropes of metallic carbon nanotubes', *Science* **273**, 483–487.

Thomas, L. (1927), 'On the capture of electrons by swiftly moving electrified particles', *Proc. Roy. Soc. London. Series A* **114**, 561–576.

Thouless, D. (1998), *Topological Quantum Numbers in Nonrelativistic Physics*, World Scientific, Singapore.

Thouless, D. J. (1973), 'Localization distance and mean free path in one-dimensional disordered systems', *Journal of Physics C: Solid State Physics* **6**, L49.

Thouless, D. J. (1977), 'Maximum metallic resistance in thin wires', *Phys. Rev. Lett.* **39**, 1167–1169.

Thouless, D. J. (1983), 'Quantization of particle transport', *Phys. Rev. B* **27**, 6083–6087.

Tielrooij, K. J., Song, J. C. W., Jensen, S. A., *et al.* (2013), 'Photoexcitation cascade and multiple hot-carrier generation in graphene', *Nature Physics* advance online publication.

Tien, P. K. & Gordon, J. P. (1963), 'Multiphoton process observed in the interaction of microwave fields with the tunneling between superconductor films', *Phys. Rev.* **129**, 647–651.

Tikhonenko, F. V., Horsell, D. W., Gorbachev, R. V. & Savchenko, A. K. (2008), 'Weak localization in graphene flakes', *Phys. Rev. Lett.* **100**, 056802.

Tikhonenko, F. V., Kozikov, A. A., Savchenko, A. K. & Gorbachev, R. V. (2009), 'Transition between electron localization and antilocalization in graphene', *Phys. Rev. Lett.* **103**, 226801.

Todd, K., Chou, H.-T., Amasha, S. & Goldhaber-Gordon, D. (2009), 'Quantum dot behavior in graphene nanoconstrictions', *Nano Lett.* **9**, 416–421.

Tombros, N., Jozsa, C., Popinciuc, M., Jonkman, H. T. & van Wees, B. J. (2007), 'Electronic spin transport and spin precession in single graphene layers at room temperature', *Nature* **448**, 571–574.

Torrisi, F., Hasan, T., Wu, W., *et al.* (2012), 'Inkjet-printed graphene electronics', *ACS Nano* **6**, 2992–3006.

Tournus, F., Latil, S., Heggie, M. I. & Charlier, J.-C. (2005), 'Stacking interaction between carbon nanotubes and organic molecules', *Phys. Rev. B* **72**, 075431.

Trevisanutto, P. E., Giorgetti, C., Reining, L., Ladisa, M. & Olevano, V. (2008), '*Ab initio* gw many-body effects in graphene', *Phys. Rev. Lett.* **101**, 226405.

Triozon, F. (2002), Diffusion quantique et conductivité dans les systèmes aperiodiques, Ph.D. thesis, Universit Joseph-Fourier.

Triozon, F., Lambin, P. & Roche, S. (2005), 'Electronic transport properties of carbon nanotube based metal/semiconductor/metal intramolecular junctions', *Nanotechnology* **16**, 230.

Triozon, F., Roche, S., Rubio, A. & Mayou, D. (2004), 'Electrical transport in carbon nanotubes: Role of disorder and helical symmetries', *Phys. Rev. B* **69**, 121410.

Triozon, F. & Roche, S. (2005), 'Efficient linear scaling method for computing the Landauer–Buttiker conductance', *The European Physical Journal B – Condensed Matter and Complex Systems* **46**, 427–431.

Troullier, N. & Martins, J. L. (1991), 'Efficient pseudopotentials for plane-wave calculations', *Phys. Rev. B* **43**, 1993–2006.

Tsen, A. W., Brown, L., Levendorf, M. P., *et al.* (2012), 'Tailoring electrical transport across grain boundaries in polycrystalline graphene', *Science* **336**, 1143.

Tsui, D. C., Stormer, H. L. & Gossard, A. C. (1982), 'Two-dimensional magnetotransport in the extreme quantum limit', *Phys. Rev. Lett.* **48**, 1559–1562.

Tsukagoshi, K., Alphenaar, B. W. & Ago, H. (1999), 'Coherent transport of electron spin in a ferromagnetically contacted carbon nanotube', *Nature* **401**, 572–574.

Tuan, D. V., Kotakoski, J., Louvet, T., *et al.* (2013), 'Scaling properties of charge transport in polycrystalline graphene: Role of grain boundary morphologies and atomic scale electron-hole puddles', *Nano Letters* **13**, 1730–1735.

Tworzydło, J., Trauzettel, B., Titov, M., Rycerz, A. & Beenakker, C. W. J. (2006), 'Sub-Poissonian shot noise in graphene', *Phys. Rev. Lett.* **96**, 246802.

Ugarte, D. (1992), 'Curling and closure of graphitic networks under electron-beam irradiation', *Nature* **359**, 707–709.

Ugeda, M. M., Brihuega, I., Guinea, F. & Rodríguez, J. M. G. (2010), 'Missing atom as a source of carbon magnetism', *Phys. Rev. Lett.* **104**, 096804.

Ugeda, M. M., Brihuega, I., Hiebel, F., *et al.* (2012), 'Electronic and structural characterization of divacancies in irradiated graphene', *Phys. Rev. B* **85**, 121402.

Usaj, G. (2009), 'Edge states interferometry and spin rotations in zigzag graphene nanoribbons', *Phys. Rev. B* **80**, 081414.

van der Zande, A. M., Barton, R. A., Alden, J. S., *et al.* (2010), 'Large-scale arrays of single-layer graphene resonators', *Nano Lett.* **10**, 4869–4873.

Vanderbilt, D. (1990), 'Soft self-consistent pseudopotentials in a generalized eigenvalue formalism', *Phys. Rev. B* **41**, 7892–7895.

Varchon, F., Feng, R., Hass, J., *et al.* (2007), 'Electronic structure of epitaxial graphene layers on SiC: Effect of the substrate', *Phys. Rev. Lett.* **99**, 126805.

Venema, L. C., Wilder, J. W. G., Janssen, J. W., *et al.* (1999), 'Imaging electron wave functions of quantized energy levels in carbon nanotubes', *Science* **283**, 52–55.

Venezuela, P., Muniz, R. B., Costa, A. T., *et al.* (2009), 'Emergence of local magnetic moments in doped graphene-related materials', *Phys. Rev. B* **80**, 241413.

Vogt, P., De Padova, P., Quaresima, C., *et al.* (2012), 'Silicene: Compelling experimental evidence for graphene-like two-dimensional silicon', *Phys. Rev. Lett.* **108**, 155501.

Wakabayashi, K., Fujita, M., Ajiki, H. & Sigrist, M. (1999), 'Electronic and magnetic properties of nanographite ribbons', *Phys. Rev. B* **59**, 8271–8282.

Wallace, P. R. (1947), 'The band theory of graphite', *Phys. Rev.* **71**, 622–634.

Wang, H., Nezich, D., Kong, J. & Palacios, T. (2009), 'Graphene frequency multipliers', *Electron Device Letters, IEEE* **30**, 547–549.

Wang, H., Yu, L., Lee, Y.-H., *et al.* (2012), 'Integrated circuits based on bilayer MOS2 transistors', *Nano Lett.* **12**, 4674–4680.

Wang, N., Tang, Z. K., Li, G. D. & Chen, J. S. (2000), 'Materials science: Single-walled 4 Å carbon nanotube arrays', *Nature* **408**, 50–51.

Wang, X. & Dai, H. (2010), 'Etching and narrowing of graphene from the edges', *Nature Chemistry* **2**, 661–665.

Wang, X., Ouyang, Y., Jiao, L., *et al.* (2011), 'Graphene nanoribbons with smooth edges behave as quantum wires', *Nature Nanotechnology* **6**, 563–567.

Wang, X., Ouyang, Y., Li, X., *et al.* (2008), 'Room-temperature all-semiconducting sub-10 nm graphene nanoribbon field-effect transistors', *Phys. Rev. Lett.* **100**, 206803.

Wassmann, T., Seitsonen, A. P., Saitta, A. M., Lazzeri, M. & Mauri, F. (2008), 'Structure, stability, edge states, and aromaticity of graphene ribbons', *Phys. Rev. Lett.* **101**, 096402.

Weeks, C., Hu, J., Alicea, J., Franz, M. & Wu, R. (2011), 'Engineering a robust quantum spin Hall state in graphene via adatom deposition', *Phys. Rev. X* **1**, 021001.

Weil, T. & Vinter, B. (1987), 'Equivalence between resonant tunneling and sequential tunneling in double-barrier diodes', *Applied Physics Letters* **50**, 1281–1283.

Weisse, A., Wellein, G., Alvermann, A. & Fehske, H. (2006), 'The kernel polynomial method', *Rev. Mod. Phys.* **78**, 275.

White, C. T., Li, J., Gunlycke, D. & Mintmire, J. W. (2007), 'Hidden one-electron interactions in carbon nanotubes revealed in graphene nanostrips', *Nano Lett.* **7**, 825–830.

White, C. T. & Mintmire, J. W. (1998), 'Density of states reflects diameter in nanotubes', *Nature* **394**, 29–30.

White, C. T. & Todorov, T. N. (1998), 'Carbon nanotubes as long ballistic conductors', *Nature* **393**, 240–242.

White, I. D., Godby, R. W., Rieger, M. M. & Needs, R. J. (1998), 'Dynamic image potential at an Al(111) surface', *Phys. Rev. Lett.* **80**, 4265–4268.

Wilder, J. W. G., Venema, L. C., Rinzler, A. G., Smalley, R. E. & Dekker, C. (1998), 'Electronic structure of atomically resolved carbon nanotubes', *Nature* **391**, 59–62.

Wimmer, M. (2009), Quantum transport in nanostructures: From computational concepts to spintronics in graphene and magnetic tunnel junctions, Ph.D. thesis, Universität Regensburg.

Wimmer, M., Adagideli, I. Berber, Tománek, D. & Richter, K. (2008), 'Spin currents in rough graphene nanoribbons: Universal fluctuations and spin injection', *Phys. Rev. Lett.* **100**, 177207.

Wong, C.-L., Annamalai, M., Wang, Z.-Q. & Palaniapan, M. (2010), 'Characterization of nanomechanical graphene drum structures', *Journal of Micromechanics and Microengineering* **20**, 115029.

Wooten, F., Winer, K. & Weaire, D. (1985), 'Computer generation of structural models of amorphous Si and Ge', *Phys. Rev. Lett.* **54**, 1392.

Wu, C., Li, F., Zhang, Y. & Guo, T. (2012), 'Recoverable electrical transition in a single graphene sheet for application in nonvolatile memories', *Appl. Phys. Lett.* **100**, 042105.

Wu, F., Queipo, P., Nasibulin, A., *et al.* (2007), 'Shot noise with interaction effects in single-walled carbon nanotubes', *Phys. Rev. Lett.* **99**, 156803.

Wu, Y., Lin, Y.-M., Bol, A. A., *et al.* (2011), 'High-frequency, scaled graphene transistors on diamond-like carbon', *Nature* **472**, 74–78.

Wu, Y., Perebeinos, V., Lin, Y.-M., *et al.* (2012), 'Quantum behavior of graphene transistors near the scaling limit', *Nano Lett.* **12**, 1417–1423.

Xia, F., Farmer, D. B., Lin, Y.-M. & Avouris, P. (2010), 'Graphene field-effect transistors with high on/off current ratio and large transport band gap at room temperature', *Nano Lett.* **10**, 715–718.

Xia, F., Mueller, T., Lin, Y.-M., Valdes-Garcia, A. & Avouris, P. (2009), 'Ultrafast graphene photodetector', *Nature Nanotechnology* **4**, 839–843.

Xia, F., Perebeinos, V., Lin, Y.-M., Wu, Y. & Avouris, P. (2011), 'The origins and limits of metal–graphene junction resistance', *Nature Nanotechnology* **6**, 179–184.

Xiao, D., Chang, M.-C. & Niu, Q. (2010), 'Berry phase effects on electronic properties', *Rev. Mod. Phys.* **82**, 1959–2007.

Yacoby, A. (2011), 'Graphene: Tri and tri again', *Nature Physics* **7**, 925–926.

Yamamoto, T. & Watanabe, K. (2006), 'Nonequilibrium Greens function approach to phonon transport in defective carbon nanotubes', *Phys. Rev. Lett.* **96**, 255503.

Yan, J. & Fuhrer, M. S. (2011), 'Correlated charged impurity scattering in graphene', *Phys. Rev. Lett.* **107**, 206601.

Yang, H. X., Hallal, A., Terrade, D., *et al.* (2013), 'Magnetic insulator-induced proximity effects in graphene: Spin filtering and exchange splitting gaps', *Phys. Rev. Lett.* **110**, 046603.

Yang, L., Park, C.-H., Son, Y.-W., Cohen, M. L. & Louie, S. G. (2007), 'Quasiparticle energies and band gaps in graphene nanoribbons', *Phys. Rev. Lett.* **99**, 186801.

Yao, Z., Kane, C. L. & Dekker, C. (2000), 'High-field electrical transport in single-wall carbon nanotubes', *Phys. Rev. Lett.* **84**, 2941–2944.

Yao, Z., Postma, H. W. C., Balents, L. & Dekker, C. (1999), 'Carbon nanotube intramolecular junctions', *Nature* **402**, 273–276.

Yazyev, O. V. (2008), 'Magnetism in disordered graphene and irradiated graphite', *Phys. Rev. Lett.* **101**, 037203.

Yazyev, O. V. (2010), 'Emergence of magnetism in graphene materials and nanostructures', *Reports on Progress in Physics* **73**, 056501.

Yazyev, O. V. & Helm, L. (2007), 'Defect-induced magnetism in graphene', *Phys. Rev. B* **75**, 125408.

Yazyev, O. V. & Louie, S. (2010*a*), 'Electronic transport in polycrystalline graphene', *Nature Materials* **9**, 806.

Yazyev, O. V. & Louie, S. G. (2010*b*), 'Topological defects in graphene: Dislocations and grain boundaries', *Phys. Rev. B* **81**, 195420.

Young, A. F. & Kim, P. (2009), 'Quantum interference and Klein tunnelling in graphene hetero-junctions', *Nature Physics* **5**, 222–226.

Young, A. F. & Kim, P. (2011), 'Electronic transport in graphene heterostructures', *Annual Review of Condensed Matter Physics* **2**, 101–120.

Yu, Q., Jauregui, L. A., Wu, W., *et al.* (2011), 'Control and characterization of individual grains and grain boundaries in graphene grown by chemical vapour deposition', *Nature Materials* **10**, 443–449.

Yuan, S., De Raedt, H. & Katsnelson, M. I. (2010), 'Modeling electronic structure and transport properties of graphene with resonant scattering centers', *Phys. Rev. B* **82**, 115448.

Zanolli, Z. & Charlier, J.-C. (2009), 'Defective carbon nanotubes for single-molecule sensing', *Phys. Rev. B* **80**, 155447.

Zanolli, Z. & Charlier, J.-C. (2010), 'Spin transport in carbon nanotubes with magnetic vacancy-defects', *Phys. Rev. B* **81**, 165406.

Zanolli, Z. & Charlier, J.-C. (2012), 'Single-molecule sensing using carbon nanotubes decorated with magnetic clusters', *ACS Nano* **6**, 10786–10791.

Zanolli, Z., Leghrib, R., Felten, A., *et al.* (2011), 'Gas sensing with Au-decorated carbon nano-tubes', *ACS Nano* **5**, 4592–4599.

Zaric, S., Ostojic, G. N., Kono, J., *et al.* (2004), 'Optical signatures of the Aharonov–Bohm phase in single-walled carbon nanotubes', *Science* **304**, 1129–1131.

Zhan, N., Olmedo, M., Wang, G. & Liu, J. (2011), 'Graphene based nickel nanocrystal flash memory', *Appl. Phys. Lett.* **99**, 113112.

Zhang, L., Zhang, Y., Khodas, M., Valla, T. & Zaliznyak, I. A. (2010), 'Metal to insulator transition on the $n = 0$ Landau level in graphene', *Phys. Rev. Lett.* **105**, 046804.

Zhang, Y., Jiang, Z., Small, J. P., *et al.* (2006), 'Landau-level splitting in graphene in high magnetic fields', *Phys. Rev. Lett.* **96**, 136806.

Zhang, Y., Tan, Y.-W., Stormer, H. L. & Kim, P. (2005), 'Experimental observation of the quantum Hall effect and Berry's phase in graphene', *Nature* **438**, 201–204.

Zhang, Y., Tang, T.-T., Girit, C., *et al.* (2009), 'Direct observation of a widely tunable bandgap in bilayer graphene', *Nature* **459**, 820–823.

Zhang, Y.-Y., Hu, J., Bernevig, B. A., *et al.* (2009), 'Localization and the Kosterlitz–Thouless transition in disordered graphene', *Phys. Rev. Lett.* **102**, 106401.

Zhao, J., Park, H., Han, J. & Lu, J. P. (2004), 'Electronic properties of carbon nanotubes with covalent sidewall functionalization', *J. Phys. Chem. B* **108**, 4227–4230.

Zhao, P. & Guo, J. (2009), 'Modeling edge effects in graphene nanoribbon field-effect transistors with real and mode space methods', *J. Appl. Phys.* **105**, 034503.

Zhao, Y., Cadden-Zimansky, P., Ghahari, F. & Kim, P. (2012), 'Magnetoresistance measurements of graphene at the charge neutrality point', *Phys. Rev. Lett.* **108**, 106804.

Zheng, L. X., O'Connell, M. J., Doorn, S. K., *et al.* (2004), 'Ultralong single-wall carbon nanotubes', *Nature Materials* **3**, 673–676.

Zhou, S. Y., Gweon, G.-H., Fedorov, A. V., *et al.* (2007), 'Substrate-induced bandgap opening in epitaxial graphene', *Nature Materials* **6**, 770–775.

Zhou, Y. & Wu, M. W. (2011), 'Optical response of graphene under intense terahertz fields', *Phys. Rev. B* **83**, 245436.

Zhou, Y. & Wu, M. W. (2012), 'Single-parameter quantum charge and spin pumping in armchair graphene nanoribbons', *Phys. Rev. B* **86**, 085406.

Zhu, R. & Chen, H. (2009), 'Quantum pumping with adiabatically modulated barriers in graphene', *Applied Physics Letters* **95**, 122111.

Zhu, W., Li, W., Shi, Q. W., *et al.* (2012), 'Vacancy-induced splitting of the Dirac nodal point in graphene', *Phys. Rev. B* **85**, 073407.

Zimmermann, J., Pavone, P. & Cuniberti, G. (2008), 'Vibrational modes and low-temperature thermal properties of graphene and carbon nanotubes: Minimal force-constant model', *Phys. Rev. B* **78**, 045410.

Zomer, P. J., Guimares, M. H. D., Tombros, N. & van Wees, B. J. (2012), 'Long-distance spin transport in high-mobility graphene on hexagonal boron nitride', *Phys. Rev. B* **86**, 161416.

Zurek, W. H. (2003), 'Decoherence and the transition from quantum to classical revisited', arXiv:quant-ph/0306072.

Index